OXFORD'S OWN

ALSO BY THE AUTHOR

Desert Flyer: The Log and Journal of Flying Officer William Marsh

OXFORD'S OWN

Men and Machines
of No.15/XV Squadron
Royal Flying Corps/Royal Air Force

Martyn R. Ford-Jones
& Valerie A. Ford-Jones

For/ David P. Williams,

A friend, fellow author and enthusiast,

With Best Wishes,

Martyn R. Ford-Jones

Valerie Ford-Jones.

Schiffer Military History
Atglen, PA

Dust jacket artwork by Keith Aspinall.

OXFORD'S OWN

Flying Officer Alan Fleming and his crew had a lucky escape from enemy nightfighters when Stettin was again selected as the target by Bomber Command for the night of 29th/30th August 1944. The Australian pilot, flying Lancaster, NF953, LS-A, took-off from Mildenhall at 20.54 hours, and followed the fourteen other bombers detailed by the Squadron for the raid.

During the outward leg of the mission, NF953 was attacked by an Me 410 twin-engine nightfighter. However, the crew of the bomber were vigilant and gave warning of its approach. Flying Officer Fleming threw the Lancaster into a violent evasive maneuver, which succeeded in losing the aggressor. Unfortunately, in taking this action, they had flown into the path of a prowling Bf 109. Instinctively the German pilot opened-up with a burst of cannon fire which, without taking proper aim, struck the bomber in the starboard aileron. As the single-engine fighter made a passing turn aft of the Lancaster, Sergeant Rarp, the rear gunner, fired a four second burst which is thought to have struck the nightfighter. Discretion being the better part of valour, the enemy pilot broke off the attack, and unaware of the extent of the damage to the aircraft, F/O Fleming elected to abort his mission and return to base. However, his troubles were not completely over as on the return trip, NF953 was attacked by a third enemy nightfighter. Against the odds, the crew of LS-A fought off another attack and returned to Mildenhall without injury to themselves.

Book design by Robert Biondi.

Printed in China
ISBN: 0-7643-0954-4

We are interested in hearing from authors with book ideas on related topics.

Published by Schiffer Publishing Ltd.
4880 Lower Valley Road
Atglen, PA 19310
Phone: (610) 593-1777
FAX: (610) 593-2002
E-mail: Schifferbk@aol.com.
Visit our web site at: www.schifferbooks.com
Please write for a free catalog.
This book may be purchased from the publisher.
Please include $3.95 postage.
Try your bookstore first.

In Europe, Schiffer books are distributed by:
Bushwood Books
6 Marksbury Road
Kew Gardens
Surrey TW9 4JF
England
Phone: 44 (0)181 392-8585
FAX: 44 (0)181 392-9876
E-mail: Bushwd@aol.com.

Try your bookstore first.

Foreword

by Air Marshal Sir Michael Simmons KCB, AFC

In its relatively short existence, the Royal Air Force (and before it, the Royal Flying Corps) has had a critical influence over the destiny of the free world. In two world wars, various regional conflicts and the cold war No.15/XV Squadron has made a very substantial contribution to the national effort. This book documents a proud record in great detail. As a history it presents a compendium of information and anecdotal record that will be of value to the serious student and of interest to the casual reader.

Unusually for a work of this sort, the narrative includes a very human touch. The inspiring story behind *MacRobert's Reply* and the many examples of great gallantry among those who repeatedly carried the battle to the enemy, particularly during the Second World War, provide a heart-warming reminder of the indomitable spirit of those who believe in their cause. At the same time we are also reminded that, despite the pressures over the years, members of No.15/XV Squadron have always enjoyed life. Long may this continue.

Finally, this work is a particular tribute to those who flew with the Squadron and did not return. The vast majority of those served during the first thiry years of the Squadron's existence. Over fifty years later this book provides a timely reminder to those of us who enjoy the benefits of their sacrifice, of the huge price that was paid by so many young men who died while serving with No.15/XV Squadron.

Contents

Acknowledgments

Being an Honorary Member of the No.XV Squadron Association, when Robert 'Bob' Biondi, Senior Editor at Schiffer Publishing Ltd., offered me the opportunity to produce a new book on the Squadron, I eagerly accepted the offer.

The writing and compiling of a book of this nature cannot be undertaken without the help, assistance and guidance of an enormous number of people. The majority of those who helped are proud to affirm that they served with the Squadron during the Second World War or the post war period. Likewise, some of the relatives of those who did not return from either the Great War or World War Two have also offered help, proud in the knowledge that their husband, father, brother or uncle served with No.15/XV.

In some cases, I have drawn on information and photographs supplied to me for inclusion in the book *'Bomber Squadron – Men who flew with XV'*, which was published some years ago. The reason for this is that the said material is also pertinent to this work, or is the only available first hand account of an event in the Squadron's history. Unfortunately, some of the people who supplied such material and assisted me at that time, I have either lost touch with, or they are no longer with us. In either case, I eagerly acknowledge and record their names.

To all of those who have allowed me to invade their privacy, assisted me with stories and anecdotes or loaned me photographs, I express my sincere thanks and record your names below. I only hope I have not forgotten anybody. If I have, I can assure you it was not intentional and further hope you will accept my most sincere apology. Likewise, if any photograph has been incorrectly credited efforts will be made to rectify the error in any future editions the publisher may deem to issue.

Special thanks are extended to the following: Victoria and Richard Hansen, of Bushwood Books, Kew, London, whose continued advice and help is appreciated. I am greatly indebted to Air Marshal Sir Michael Simmons, K.C.B, A.F.C., a former commanding officer of No.XV, for agreeing to write the foreword to this book. To Wing Commander Graham Bowerman, O.B.E., ex-O.C. No.XV Squadron, who welcomed Valerie and myself to RAF Lossiemouth and allowed us access to the Squadron albums, some unrestricted files and ephemera. To Flight Lieutenant Torben Harris (later Squadron Leader), No.XV Squadron Historian, who gave up his time to ensure those visits were both rewarding and enjoyable. To Flight Lieutenant Chris Stradling who put up with, and answered many of my questions during the preparation of this book. To Lt.Col Robert "Bob" Lyles, USAFE, who gave freely of his time and acted as host, guide, chauffeur and friend to Valerie and myself during our visit to RAF Mildenhall. To Lt.Col David A Pfeifer, Commander, 100 Operations Support Squadron, USAFE, and the staff at RAF Mildenhall for their help and support. To Don Clarke, M.B.E, of the Mildenhall Register, and his wife Win, both of whom responded kindly to my numerous phone calls. Their endeavors in arranging the Mildenhall reunion every year brings together many people, some of whom have contributed to this book. To my good friend Keith Aspinall, who labored with my thoughts to produce the dustcover artwork for this book. His work as an aviation artist I greatly admire and strongly recommend. I also eagerly record a debt of gratitude to another good friend Steve Smith (No.3 Group Historian), who gave up much of his valuable time to carry out research on my behalf.

No.XV Squadron Personnel, past and present: To George Allen, D.F.C.; Leslie Bentley, D.F.M,; To S/L Peter Boggis, D.F.C (Scotland), whose special friendship over many years has been highly valued; G.H. Broome; Les Butcher (Groundcrew); S/L Philip Camp, D.F.M,; Fred Coney, Hon Chairman, The Mildenhall Register; 'Gary' Cooper; W/C Eddie Cox; Graham 'Mick' Cullen (ex.RNZAF); Ken Dougan (Canada); W/C Barry Dove, A.F.C, ex-O.C. No.XV Squadron and ex-Secretary of the No.XV Squadron Association; Bernard Dye; Arthur Edgley; Doug Fry; William 'Bill' Garrioch (Eire); W/C Hugh George, D.F.C; Jim Glasspool; Vincent Gowland; William Gundry; Geoff Hill; Arthur Horton (No.622 Squadron); G/C Alan Hudson, O.B.E.; Ken James; Ken Lewis, D.F.C (Australia); Don 'Ike' McFadden (America); Gil Marsh; Len Miller, D.F.C.; John Pratt (Groundrew); S/L John Peters; Norman Roberson (former XV Sqdn Historian); 'Robbie' Roberts (Groundcrew); Reg Rose; Tom Rudall, C.de.G.; Air/Cdre Mike Rudd, A.F.C.; Robert 'Bob' Smith (Australia); W/C Gordon Swanson; Theo Thorkilsen (Canada); Jack Trend; R.J. 'Bob' Weeks; George Wright; G/C Michael Wyatt, D.F.C.

Relatives and Friends of those who served with the Squadron: Patricia Banks (Australia), daughter of AVM J Cox, who has entrusted me with her father's papers, letters and photograph albums etc; Kay Boggis, a lovely and charming lady, for allowing me to 'raid' her photograph albums for pictures relating to Lady MacRobert; Ian Bone; Renee Bysouth; Jean Campden; Ros Dengate; Lyn Hicks; Trevor Lloyd; Mary Evelyn McCarten; Moira Michie (Scotland); Irene Murphy; Brian Parker; Ralph Skilbeck (Australia); June and Terry Watkins; Peter G Willis.

Those who previosly supplied information and with whom I have lost contact: the late Oliver Brooks, D.F.C.; the late Harry Bysouth, D.F.M; Glyn Thomas, D.F.M.; the late AVM Stewart Menaul, C.B., C.B.E., D.F.C., A.F.C.; the late Ken Rogers (Groundcrew); the late Pat Russell; the late John Sparrow; Stanley Smith (Groundrew); the late Frederick 'Steve' Stevens; G/C T H 'Brian' Tayler.

Authors and Historians: A special thank you is extended to my friend Chas Bowyer for allowing me full and unrestricted access to both his *'History of No.15 Squadron, RFC'* and relevant photographic collection. To Norman Franks, Peter Hinchliffe, Don Neate and Ray Vann all of whom have also allowed use of photographs from their respective collections. To Squadron Leader Charles Lofthouse, O.B.E, D.F.C, ex-Secretary, No.7 Squadron Association. To another friend Chris Ward, who gave freely of his specialist knowledge of Bomber Command. To Rod Priddle, for sharing his expert knowledge of Wiltshire airfields. To Bob Collis of the Norfolk and Suffolk Aviation Museum. To Mme Josselyne Lejeune-Piche (France), for her help regarding photographs and information relating to cemeteries in France.

Luftwaffe Information: to Hans-Joachim Jabs (Germany); Andreas Wachtel (Germany); David Williams, another friend and budding author, whose knowledge and guidance regarding Luftwaffe pilots is appreciated.

Public Sources: The Staff of the Royal Air Force Museum, London; The Staff of the Public Record Office, London; The Staff of the Commonwealth War Graves Commission, Berkshire; The Staff of the Imperial War Museum, London; The Staff of the New Zealand High Commission, London.

Others who have helped: To Rose and Tony Excell, who located the land owners of the Friston airfield site; To Sam Stanistreet and Wycliff Stutchbury both of whom allowed Valerie and myself access to photograph the former Friston airfield (now part of their respective farms); To G/C David Luck, R.A.F (Retired). To Ad van Zantvoort (Netherlands), who provided a wealth of information relating to RAF bomber losses in Holland.

I must extend sincere thanks to Robert 'Bob' Biondi, Senior Editor, Schiffer Publishing, for his professional guidance, constructive comments and input during the preparation of this book. I should also like to thank the staff of Schiffer Publishing, who have always shown great courtesy when accepting my trans-Atlantic phone calls, and messages.

On the personal side I would like to thank my two daughters, Emma whose continued guidance with regard to modern technology is greatly appreciated and Alexandra for her continued help and support in many other ways. Finally to Valerie (my wife and co-author), who has shared my life for over thirty-three years, nearly half of which she has endured my passion for No.XV Squadron. Throughout that period she has 'worn many hats' including those of researcher, secretary, agent, sales manager, museum curator, hotel proprietor, bookkeeper, copier of documents, photographic assistant, traveling companion and navigator to various locations and all things No.XV Squadron related. She has spent many hours proof reading, amending paragraphs and formulating the manuscript. Her input has been vital to the completion of this tome. Together, we have produced "Oxford's 'Own'".

Martyn R. Ford-Jones
Swindon, Wiltshire, England. 1999.

In Memory of

FLIGHT LIEUTENANT STEPHEN HICKS, RAF
Killed in Action on 14th February 1991
And All Those Who Went Before

Introduction

On learning of my interest of No.XV Squadron, people invariably ask me "What did they do?".

This book sets out to answer that question by relating the human aspect of the Squadron's history, rather than the technicalities of the aircraft and equipment used.

It is the story of a Squadron formed during the war commonly labeled 'The War to end all Wars'. Through the courage and devotion to duty of the men who undertook the tasks imposed upon it, No.XV emerged from that conflict having earned many accolades. But, a generation later it was embroiled in another global conflict. Like their fathers before them, young men from the Dominion and Commonwealth countries responded to the call.

It is the story of those young men who, like all who served in Bomber Command, took the war to the enemy. Flying night after night over enemy occupied Europe, with darkness as their only shield. As with the earlier conflict they emerged with honor.

It is the story of a Squadron who, during the cold war period, helped keep peace in Europe, albeit a fragile one.

It is the story of 80 years of service, loyalty, courage and sacrifice; qualities which each new generation of Squadron personnel holds in high esteem.

It is the story of a Squadron which reunites and welcomes those who formed part of its history, whilst not forgetting to pay homage to its war dead.

The story related in this book is incomplete and, whilst No.XV Squadron continues to exist, it will remain so ...

1

The Fledgling Squadron

Three bullets, fired from a browning automatic pistol, by a young student named Gavrilo Prinzep, in the streets of Sarajevo, were to cause the greatest loss of life in the known military history of mankind.

The assassinations of the Grand Archduke Ferdinand, heir to the Austro-Hungarian thrones, and his wife, the Duchess of Hohenberg, with those three bullets, on Sunday, 28th June 1914, was the excuse the German Government was waiting for. To use the political situation which followed to its own advantage and ultimately, declare war on Serbia.

The Countries of Europe responded according to their respective allegiances until finally, without a declaration of war, Germany invaded France by attacking through the neutral country of Belgium.

Belgian neutrality having been violated, Britain, bounded by a treaty to protect that neutrality, declared war on Germany, on Tuesday, 4th August 1914.

Just over five weeks after those three shots were fired, virtually every country in Europe was at, or ready for, war.

Britain mobilised her Forces, sent an expeditionary army to France, and began to expand her military might.

The cogs of the machinery of war had been turning for over six months when No.15 Squadron was established as part of the expansion of the Royal Flying Corps, at South Farnborough, Hampshire, on 1st March 1915.

The Royal Flying Corps itself had been established, by Royal Warrant, on 13th April 1912, from the Air Battalions of the Corps of Royal Engineers.

As an independent service, not controlled by either the Army or the Royal Navy, it was decided in 1913 that the new service should have its own uniform and flying badges; double 'wings' to signify pilots, whilst observers would wear single, or half 'wings'. In both cases, the 'wings' were to be worn over the left breast. Sergeant pilots, who would wear the same 'wings' as officers, were to be identified by a four-bladed propeller badge worn on the upper sleeve of the tunic, immediately above the rank stripes. Initially, any serviceman transferring to the new air arm continued to wear the uniform of the unit from which he was transferring.

The fledgling squadron was manned by personnel drawn from No.1 Reserve Squadron and the Recruits Depot, both already based at Farnborough. The men were billeted in a timber built airplane shed, with their beds drawn-up in ranks along the entire length of the building.

In order to allow the new unit to work up to operational standards it was initially equipped, for training and instructional purposes, with an assortment of aircraft, including Henri and Maurice Farmans, Bleriots, Avros, a few Moranes and B.E.2cs aeroplanes. It also acquired some motor transport in the form of Leyland Lorries and a Crossley tender.

The commanding officer of No.1 Reserve Squadron, Captain P. B. Joubert de la Ferte (later to become Air Chief Marshal Sir Philip Joubert de la Ferte, K.C.B, C.M.G, D.S.O), who had carried out the first aerial reconnaissance patrol undertaken by the British Expeditionary Force on 19th August 1914, was appointed as No.15's first C.O. during April. He was also, at the time of his appointment, promoted to the rank of Major.

Major Joubert de la Ferte's first main task as the new C.O. was to oversee, on 13th April, the Squadron's move to Hounslow, approximately twelve miles west of Central London.

Two days short of a month later, on 11th May, another move was made. On this occasion the Squadron headed south to Swingate Down.

It was from this airfield, on top of the cliffs to the east of Dover, in Kent, that Nos.2, 3, 4, and 5 Squadrons, which together formed the RFC British Expeditionary Force, took-off between 13th and 15th August 1914 and flew to Amiens, France, on the outbreak of the war.

Three and a half months after its arrival at Dover, the Squadron was absorbed into No.6 Wing, when the latter unit was formed there on the penultimate day of August, under the command of Lieutenant-Colonel CAH Longcroft.

Although No.15's primary duty, whilst based at Swingate Down, was to train its pilots to operational standards, it also found itself tasked with home defense duties.

However, during the late summer a detachment from the Squadron was posted to an airfield at Lydd, approximately twenty nautical miles south-west of Dover.

The facilities at both these airfields were very basic, and not much to write home about, but the airmen were not there on vacation.

Although they were not aware of it at the time, the personnel who made up the fledgling squadron were laying the foundations for what was to become in future years a much respected unit, with a tradition of being to the fore.

The war in France was taking a high toll on trained aircrews, and the supply of replacements being sent to the front was not comparable to the loss ratio.

New squadrons were being formed to meet the demand, but again the early fliers who were to man these squadrons needed to be trained quickly. It was due to these circumstances that No.15 Squadron was detailed to undertake the training of pilots for other units, whilst training its own personnel.

One such airman who undertook his Service Flying Training with No.15 was Captain John Aiden Liddell, who had gained flying experience prior to the outbreak of war by undertaking private tuition at the Vickers Flying School, in Surrey.

Furthermore Liddell was no stranger to the war in France, having served there from August 1914 with the 3rd Argyll and Sutherland Highlanders. It was due to his health that he transferred to the Royal Flying Corps, and found himself undertaking pilot training at Swingate Down.

On completion of his course Captain Liddell was posted back to France to join No.7 Squadron, arriving there during the third week of July 1915. However, his time with the RFC was to be short. On his second mission, a reconnaissance patrol over Belgium on 31st July, Liddell's aircraft was attacked from above by a enemy machine. In the ensuing fight Liddell received severe wounds from which he was to die a month later. For his courage in flying his crippled machine back to Allied lines thus saving it, himself and his observer from capture, Captain Liddell was awarded the Victoria Cross, Britain's highest award for bravery.

During August 1915 No.15 Squadron came under the temporary command of Captain Dawes, when Major Joubert de la Ferte was posted to an overseas unit. Captain Dawes remained in office until 13th September, when Major Edgar Ludlow-Hewitt was officially appointed as the new C.O.

Major Ludlow-Hewitt's tenure as C.O. was also a short one, lasting only until the end of October when he was succeeded by Major H de L Brock, D.S.O. Major Brock officially took command of the Squadron on 7th November.

During the late summer period the Squadron had began to re-equip with additional B.E.2c aircraft, on which a variety of training tasks were undertaken, including aerial photography and artillery observation. Both of which would be the Squadron's main duties when sent to the Front.

Orders to prepare for the move to France were received on 13th December. Ten days later ten B.E.2c's climbed away from the English coast and headed for St Omer, south-east of Calais. The groundcrews, support equipment and transport, who had been split into two sections, joined the aircrew three days later, having made the crossing by sea and landing in France at Le Havre and Rouen.

Having re-established itself as a single unit, and become accustomed to its new environment, a further move was made to Droglandt, north-east of St Omer, on 5th January 1916. Here the Squadron came under the jurisdiction of the 2nd Wing, which was part of the Second Army. Immediately No.15 was ordered to detail one aircraft, on a daily basis, for escort duty to an Army reconnaissance mission. Five days later, on 10th January, the Squadron undertook the first of its own reconnaissance sorties. Many more similar operations were to follow, and it was to become a role for which No.15 would later receive high praise. It was also later to prove the foundation on which the Squadron's traditions were built.

In later years, when it adopted a Squadron Crest, No.15 took the motto, 'Aim Sure'. A motto which could be equally applied to the aiming of a reconnaissance camera or the discharging of a bomb load. The latter of which it was to become all too familiar with in later years.

The first operational loss to the Squadron came on 17th January 1916 during a reconnaissance mission, and illustrated the mettle of the men who served within the Unit.

The officer Commanding "A" Flight, Captain Vivian Wadham, was flying as one of three B.E.2c's escorting a similar reconnaissance aircraft, also from the Squadron, when an enemy machine attacked. In order to protect their charge, the three B.E.2c's dived onto the German Fokker. During the aerial combat which took place over the forest of Houthulst, Captain Wadham received bullet wounds to the neck and died instantly.

As the B.E.2c, serial 2705, began to fall, Sergeant Piper, the Observer, climbed out of his own cockpit and, with the determination of self preservation, edged his way towards the fatally injured pilot's position. Achieving his aim, Sergeant Piper climbed in, placed himself on the dead pilot's lap, and attempted to fly the aircraft home. However, this proved impossible and the Observer crash-landed the aircraft behind enemy lines. Having jumped clear of the machine as it struck the ground, Sergeant Piper was immediately captured and taken prisoner of war.

Two days later on 19th January, another combat with enemy aircraft took place, the outcome of which was to see the award of the Squadron's first decoration for bravery.

A B.E.2c, serial 4123, piloted by Captain G Henderson, was leading a flight of three aircraft who were acting as escort to a fourth B.E.2c engaged on a reconnaissance patrol. During the mission two enemy aircraft swooped down on the British formation and opened fire. During the ensuing dog-fight, B.E.2c, 4123, was attacked from the rear. Henderson's observer/gunner, Corporal C Nott, was struck in the eye and knocked insensible by a fragment of flak just as he was about to return fire.

The aircraft also sustained damage during the attack, causing Captain Henderson to break-away and make for the Allied lines. However, a short while later the B.E.2c was attacked again, when a German Albatross joined one of the original attacking machines in an effort to complete the task of shooting down the British aircraft. The German pilots positioned themselves for the kill, but overcoming the pain of his injury, Corporal Nott employed his Lewis machine gun to good effect and drove the attackers off. For his courage and devotion to duty, Corporal Nott was later awarded a Distinguished Conduct Medal.

The annals of No.15 Squadron had been opened and many names, some of whom would become well-known in aviation history, would be recorded within their pages.

No.XV Squadron's Birthplace

The timber constructed airplane sheds at Royal Flying Corps, Farnborough, where No.15 Squadron was formed on 1st March 1915. *Courtesy of and via Flt/Lt T. N. Harris, XV (R) Squadron.*

The interior of No.1 Sheds at Farnborough, used as billets by the men of the newly formed Squadron, with their camp beds arranged in ranks the length of the building. Personal possessions, including at least three rifles, have been left on beds to the right of the picture. *Courtesy of and via Flt/Lt T. N. Harris, XV (R) Squadron.*

The Squadron's Transport

One of the Leyland Lorries used by the Squadron, photographed at Farnborough, in March 1915. *Courtesy of and via Flt/Lt T. N. Harris, XV (R) Squadron.*

The Squadron's Crossley Tender, registration number 2905C, photographed at Hounslow, west of London, during April 1915. The driver is unidentified. *Courtesy of and via Flt/Lt T. N. Harris, XV (R) Squadron.*

The First Commanding Officer

Right: Air Chief Marshal Philip Joubert de le Ferte, KCB, CMG, DSO, who as Captain P B Joubert de le Ferte, was posted to No.15 Squadron, RFC, on 13th April 1915. Twelve days after his arrival, he was officially appointed Squadron Commander and promoted to the rank of Major. *Courtesy of and via Flt/Lt T. N. Harris, XV (R) Squadron.*

On the Move

An airman at the top of a ladder adjusts the canvas roof covering, whilst erecting a tent hangar, possibly at Hounslow, during April 1915. Some of the Squadron's vehicles are parked in the background. *Courtesy of and via Flt/Lt T. N. Harris, XV (R) Squadron.*

The site of the former Royal Flying Corps landing ground at Swingate Down, near Dover, Kent. It was from here the first British Squadrons flew to France, a fact which is commemorated on the memorial to the right of the entrance gate. *Author's Collection.*

First Aircraft

A B.E.2c, identified only by the code 169, photographed at an unknown location. *Courtesy of and via Flt/Lt T. N. Harris, XV (R) Squadron.*

An air mechanic prepares to swing the propeller of a Martinsyde aircraft, which was on charge for a short while to No.15 Squadron. Both the mechanic and pilot are unidentified. *Courtesy of and via Flt/Lt T. N. Harris, XV (R) Squadron.*

The Facilities at Lydd

Right: The cookhouse facilities on the airfield at Lydd, Kent, where 15 Squadron had a detachment of men and machines during the late summer of 1915. *Courtesy of and via Flt/Lt T. N. Harris, XV (R) Squadron.*

Below: Two air mechanics make use of the washing facilities, prior to taking their meal break, at Lydd, Kent. It was here that special instruction in artillery observation was undertaken towards the end of summer 1915. *Courtesy of and via Flt/Lt T. N. Harris, XV (R) Squadron.*

Below right: Meal time for 15 Squadron, RFC, personnel at Lydd, Kent. The majority of the men appear to be wearing the 'maternity' style tunic, with Royal Flying Corps shoulder flashes. *Courtesy of and via Flt/Lt T. N. Harris, XV (R) Squadron.*

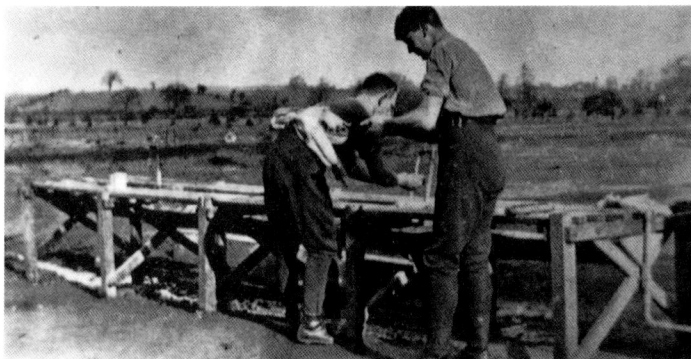

Early Flyers   **A New C.O.**

Left: One of Fifteen's early flyers. 2780 Sergeant Billy Gibson, suitably attired in fur lined flying helmet and leather coat. *Courtesy of and via Flt/Lt T. N. Harris, XV (R) Squadron.* Center: John Aiden Liddell, undertook his Service Flying Training with No.15 Squadron during June/July 1915. On 23rd July he reported for duty, in France, with No.7 Squadron. He was to be awarded the Victoria Cross following an action during his second mission, in which he was seriously wounded. He subsequently died as a result of those wounds. *Courtesy of and via S/L C. Lofthouse, OBE, DFC.* Right: Air Chief Marshal Sir Edgar Ludlow-Hewitt, GCB, GBE, DSO, MC, who as Major Ludlow-Hewitt, commanded No.15 Squadron, Royal Flying Corps, from 13th September to 30th October 1915. He saw service during the Second World War and retired from the Royal Air Force on 19th November 1945. *Courtesy of Flt/Lt T. N. Harris, XV (R) Squadron.*

Preparing for War

A B.E.2c of No.15 Squadron shares a hangar with two unidentified Bristol M1 aircraft. The photograph was possibly taken at Lydd, Kent, toward the end of summer 1915. *Courtesy of and via Flt/Lt T. N. Harris, XV (R) Squadron.*

An armored B.E.2c, of No.15 Squadron, possibly photographed at Swingate Down, near Dover, prior to the Squadron's move to France. *Courtesy of and via Flt/Lt T. N. Harris, XV (R) Squadron.*

A line-up of approximately nine B.E.2c aircraft, photographed during October 1915, prior to their departure for St Omer, France. *Courtesy of Ray Vann.*

A B.E.2c, of No.15 Squadron, photographed from a similar machine, flies out over the English Channel, on 23rd October 1915, heading for France and the war. *Courtesy of Ray Vann.*

The B.E.2c at War

Close-up photograph of the flare chute on the port side, and adjacent to the rear cockpit, of an unidentified B.E.2c, of No.15 Squadron. *Courtesy of and via Flt/Lt T. N. Harris, XV (R) Squadron.*

B.E.2c, serial 2705, is inspected by members of the German army, following the aircraft's crash landing behind enemy lines, on 17th January 1916. Sergeant Piper, the observer, had attempted to land the machine after his pilot had been killed in aerial combat. *Courtesy of Don Neate.*

2

No.15's Battle of the Somme

At the end of February the Squadron received orders to prepare for a move to the Somme front, for which purpose it would be based at Vert Galand.

The groundcrew initiated the move on 4th March, when they left Droglandt with the Squadron's transport. They stayed overnight at St. Omer and continued their journey the following morning. The aircraft and aircrew followed four days later, arriving at Vert Galand on 8th March.

Apart from moving to a new base the Squadron also, around this period, took delivery of two Bristol Scout C, single seat aircraft, which it operated during the spring of 1916.

Four days after the commencement of operations from Vert Galand, just before mid-day on the 14th March, another crew were lost.

Second Lieutenant James Campbell Cunningham, and Air Mechanic First Class John William Newton, were engaged in an escort duty on a reconnaissance patrol when their B.E.2c, serial 4153, was attacked over the Achiet-le-Grand area of the Somme. Flying at an altitude of approximately 6,500', the B.E.2c was attacked by a German Fokker Scout aircraft. The enemy pilot, Leutnant G Leffers, dived down from his patrolling altitude of 8,000', opened fire and despatched the R.F.C aircraft from the sky. Both 2nd/Lt Cunningham and A.M.I. Newton were killed in the ensuing crash. John William Newton was the first airman, of such lowly rank, whose name was to be recorded in the Squadron's Roll of Honor. Others would follow.

Upon its arrival at Vert Galand, the Squadron was immediately incorporated into the 3rd Wing, but came under the command of the Fourth Army. The former, which was commanded by Lieutenant Colonel E Ludlow-Hewitt, M.C. who, as a Major, had been C.O. of No.15 during September 1915, comprised of Nos 3, 4, 9 and 15 Squadrons. The arrangement however was to be a short term one, as nearly three weeks later No.15 moved again. On Monday, 27th March 1916, the Squadron vacated Vert Galand and headed seven miles north-east to Marieux where, although it remained in the 3rd Wing, it relinquished its Army role and became a Corps squadron attached to VII Corps.

The Squadron's duties included reconnaissance patrols, escort duty patrols, artillery observation and aerial photography, all of which were not without elements of danger.

The sixth member of the Squadron to be lost whilst undertaking one of these missions, was Lieutenant Geoffrey Welsford.

Geoffrey Joseph Lightbourn Welsford had been educated at Marlborough College, twelve miles south of Swindon, Wiltshire, and was studying Law at Cambridge when war was declared. He enlisted in the army and was immediately nominated for officer training at Sandhurst, where he spent four months. On completion of his training, Geoffrey Welsford joined the 3rd Battalion, Middlesex Regiment, and was posted to France where he was wounded on 9th May 1915, during the second battle of Ypres.

It was during his convalescence in England that Welsford undertook flying lessons at Hendon, north of London, and was granted Royal Aero Club Certificate No.1702 on 6th September 1915.

Promoted to Lieutenant, Geoffrey Welsford had only been back in France three weeks when he was recalled to England for transfer to the Royal Flying Corps.

Further pilot training was undertaken at Farnborough and the Central Flying School at Upavon, where Welsford received his 'wings'. After a short period at Bristol, Lt Welsford returned to the front on 24th February 1916, and reported for duty with No.15 Squadron.

Exactly five weeks later, on 30th March, Geoffrey Welsford was piloting a B.E.2c on a reconnaissance patrol mission. His observer was his old school friend Lieutenant Wayland Joyce, with whom Welsford had been at Marlborough College between 1909 and 1912. During the mission Welsford's machine was set upon by an enemy Fokker aircraft piloted by the German ace Max Immelmann, who dived into the attack with machine guns blazing. As the B.E.2c fell earthwards, Immelmann turned his attention to another victim, but later claimed to have killed 20 year old Welsford with a bullet through the heart. Although Immelmann also recorded that Joyce had been killed in the ensuing crash, it is known that Lieutenant Joyce survived and was taken prisoner of war.

With plans being made for an offensive on the Somme, the need for reconnaissance and photographic missions increased, and became a priority. In continuing their specified role, the aircrew of No.15 Squadron undertook a number of these missions.

Raymond John Ward, from Chippenham, Wiltshire, was a member of the Corps of Royal Engineers who, during 1916, transferred to the Royal Flying Corps and became an observer; he flew on some of these missions, and a number of others, with No.15.

Raymond Ward's work with the Squadron during the First World War was never recognised, at least not with an award. However, on return to civilian life, he qualified as an architect and was later to become a technical adviser to the War Damage Commission between 1940 and 1964. For this work he was awarded the Order of the British Empire, in the Queen's Birthday Honors, in June 1954.

The plans and preparations completed, the first phase of the battle of the Somme commenced on 1st July, but the inclement weather during the early part of that morning precluded many missions from being completed satisfactorily. No.15 Squadron, who on the 3rd Wing, Order of Battle, were recorded as having an establishment of sixteen B.E.2c aircraft, remained on line and observed the fighting north of the River Ancre, near Beaumont Hamel and Serre. Later in the day, however, they were able to get airborne and carry out constant low-level patrols over the battlefield. The fact that many of the Squadron's aircraft returned with battle damage, bore witness to the low altitude at which the pilots were flying.

The next day the Squadron was transferred to the 15th Wing, (Reserve Army) which was formed that day, under the command of Lieutenant-Colonel J G Hearson. Their work, along with No's.4, 7 and 32 Squadrons, was to provide air protection along the Reserve Army's front.

During the middle of the month a new observer reported for duty with the Squadron. The airman, a Canadian from Manitoba, was one of those whose name would be recorded, not only in the annals of No.15 Squadron in general, but also those of aerial warfare in particular.

By the time William G Barker joined No.15 at Marieux on 7th July 1916, he had already served with No's.9 and 4 Squadrons and arrived at his new unit with the victory of a German Fokker aircraft to his credit.

During his time with the Squadron, Barker claimed to have driven down two enemy aircraft although the first, a Roland two-seater on 21st July, was not included in his subsequent score tally.

On 16th November, after almost exactly four months with No.15, William Barker was posted back to England, to Narborough, Norfolk, for pilot training.

His ability as an observer was recognised on 10th January 1917, when he was awarded the Military Cross.

On completion of his training, Lieutenant Barker returned to No.15 Squadron, arriving back on his old unit on 24th February.

The following day he undertook a reconnaissance and trench patrol, which he carried out at an altitude of between 50' and 500'. It was a standard which Barker and his observer set for the majority of their patrols, and between them achieved some valuable results for Corps Headquarters.

The courage and devotion to duty of Lieutenant William Barker, M.C., was recognised again on 18th July 1917, with the award of a Bar to the Military Cross. On completion of his tour of duty with No.15 Squadron, William Barker was posted to England as an instructor. However, he returned to the front with No.28 Squadron during October 1917 and continued the fight. By the end of the war Barker had been awarded the V.C., D.S.O and Bar, M.C. and two Bars and Croix de Guerre.

Such was the quality of the work carried out by the pilots and observers that, by the end of the first phase of the Battle of the Somme on 17th July, army commanders in some areas were reporting the destruction of German artillery and acknowledging the role played by the aerial artillery spotters.

The reconnaissance, artillery observation and aerial photographic duties continued on a daily basis throughout the rest of July and into August, when the Squadron was transferred to the control of V Corps.

September saw an almost continuous workload for the Squadron, with the third phase of the battle starting on the 15th of the month, with artillery ranging, bomber escort duties, trench strafing and the inevitable photo-reconnaissance missions.

As only to be expected these missions were not without casualties, which continued to mount, both for the R.F.C in general and No.15 Squadron in particular. On 16th September, a B.E.2c, serial 2617, piloted by 2nd/Lt Albert Vinson, was attacked by three enemy aircraft during an artillery ranging mission. The observer, Captain Kenneth Brooke-Murray, opened fire with his Lewis machine gun, which unfortunately jammed. Before he could clear the stoppage, the observer was hit twice in the left leg by return fire. Breaking away from the fight, Vinson made an immediate landing near Albert in an effort to secure medical attention for his wounded observer. Unfortunately, such was the nature of his wounds, that Captain Kenneth Algernon Brooke-Murray died as a result of them on 23rd September.

On 2nd October, after nearly six month at Marieux, the Squadron received orders to re-locate to Lealvillers between three and four miles to the south. The move also required V Corps to move their Headquarters from Marieux to Acheux.

Further casualties to the Squadron were recorded on the day of the move, when Lieutenant Walter Miller, pilot, and 2nd/Lieutenant W R Carmichael, observer, failed to return from an artillery observation mission. B.E.2c, serial 4190, was reportedly shot down behind enemy lines and crashed between St Pierre Divion and Grandcourt. It later transpired that Lt Miller was killed, whilst 2nd/Lt Carmichael, who had been wounded, had survived the crash and had been taken prisoner of war.

The losses continued to rise during October, with Sergeant Frederick Barton, pilot, and Lieutenant Edward Carre, observer, being killed on the 16th and 2nd/Lt Fawkner on the 28th of the month. Barton and Carre, who made up the crew of B.E.2d, serial 6745, had the misfortune to fly into the same patch of sky as the German ace Hauptmann Oswald Boelcke, of Jasta 2. The German pilot shot down the No.15 Squadron aircraft, which crashed into No.XIII Corps barbed wire near Hebuterne. 2nd/Lieutenant Leslie Fawkner, an Australian whose real surname name was Focken, was killed in aerial combat whilst piloting B.E.2c, serial 4205, over the Somme. Although Leslie Fawkner's parents resided in Melbourne, Australia, he was born in Hong Kong. Unfortunately he was never to see either country again, and was laid to rest in the French soil over which he had flown. On this same day a new pilot, who had qualified for his 'wings' nine days earlier, joined the Squadron. He was Lieutenant F F Wessel, and he boasted nearly 21 hours solo flying experience. Four days later he increased that total by flying a familiarization flight over No.15's operational area.

The loss of some of their colleagues, and depressing weather conditions during October/November, did not deter the flyers of No.15 Squadron from carrying out their allotted duties to the best of their ability. On 15th November 1916, the Squadron undertook a number of low-level reconnaissance missions, during which valuable information and progress reports were obtained and passed to Corps H.Q.

No.5385, Air Mechanic First Class Alexander Allardice, who was attached to No.91 Siege Battery, Royal Garrison Artillery, was killed, during an artillery observation mission, on 13th November.

The names of 2nd/Lieutenants' Philip Haarer, pilot, and Andrew Laird, observer, were added to the casualty list as accidental deaths, when their B.E.2d, serial 7172, crashed to earth on 22nd November. Philip Haarer, who joined the Squadron less than a month previously, had had less than five and a half hours flying instruction, out of a total of just over 23 hours flying training.

Andrew Laird, who was 22 years of age when he died, was born in Glasgow. He had enlisted in the Glasgow Highlanders on the outbreak of war, and received a commission in the Black Watch in June 1915. Just over a year later, in August 1916, he transferred to the Royal Flying Corps and was posted to No.15 Squadron. Alas, as with so many of those who took to the skies, his time on the Squadron was to be short.

However the deaths of Haarer and Laird were not in vain, as during 1916 an instruction was issued by the War Office setting out certain criteria, which prevented 'rookie' pilots from being sent to squadrons without sufficient training; thus hopefully saving them from death or injury in unnecessary accidents.

On 29th of November, V Corps relinguished its position at the front line and handed over to XIII Corps. As a result, No.15 Squadron then become attached to the latter.

Further changes were made on 19th December, when Major George Carmichael, D.S.O, took over command of the Squadron from Major Brock. The latter having been posted home to England two days earlier.

The year was to end on a tragic note with the death of 22 year old Lieutenant Herbert Brereton, who was killed in an accident on 21st December. Brereton, who was piloting B.E.2d, serial 7094, had been on active service for nineteen months, and had previously held a commission in the 11th Battalion, Kings Liverpool Regiment.

Scout on a Bomber Squadron

A Bristol Scout, serial 5313, of No.15 Squadron. The aircraft, which was flown by 2nd/Lt Elliot, was photographed during May 1916. *Courtesy of and via Ray Vann.*

2nd Lieutenant Geoffrey Welsford

The photograph and entry of, and relating to, Lt Geoffrey Welsford (top left and right respectively), in the Marlborough College, Roll of Honor. *Author's Collection via Marlborough College.*

Lieutenant Geoffrey Joseph Lightbourne Welsford, pilot, Royal Flying Corps, who previously saw service with the 3rd Battalion, Middlesex Regiment. He was killed at 11.00 hours (British Time), on 30th March 1916, following aerial combat with Max Immelmann. Lieutenant Welsford was the German ace's 13th victory. *Author's Collection via Marlborough College.*

Raymond Ward, Observer

Raymond John Ward (standing centre), photographed with three unidentified colleagues, whilst serving with the Corps of Royal Engineers. Raymond Ward transferred to the Royal Flying Corps, where he flew as an observer with No.15 Squadron between 1916 and 1917. He was responsible for a number of aerial photographs now included in the Squadron's photographic collection. *Courtesy of Trevor Lloyd.*

A series of three aerial photographs, taken by Raymond Ward, of the village of Beaucourt-Sur-Ancre. On 4th May 1916 the village was reasonable intact ...*Courtesy of and via Flt/Lt T. N. Harris, XV (R) Squadron.*

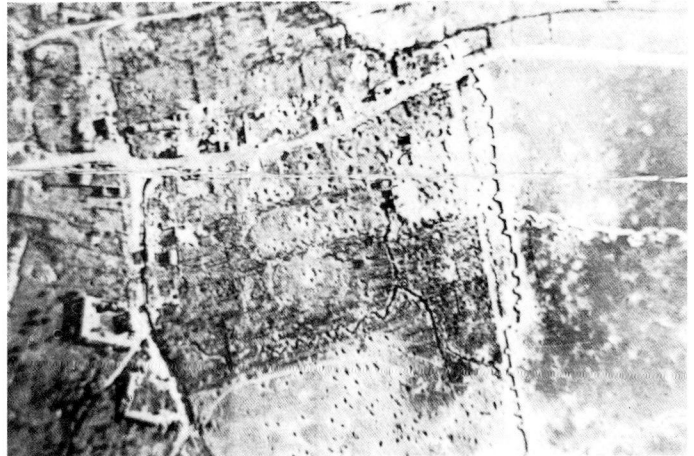

... On 30th September 1916 the village was surrounded by a number of shell and bomb craters ... *Courtesy of and via Flt/Lt T. N. Harris, XV (R) Squadron.*

... By 17th November 1916 the area was devastated. The crossroads being the only identifiable landmark. *Courtesy of and via Flt/Lt T. N. Harris, XV (R) Squadron.*

The Known and the Unknown

Lieutenant William Barker, originally an observer with No.15 Squadron, undertook pilot training in Norfork, England, and returned to the Squadron as a pilot on 24th February 1917. By the end of the war Barker was credited with the destruction of 57 German aircraft, and had been awarded the Victoria Cross, the Distinguished Service Order and the Military Cross. The V.C. was awarded whilst Barker was serving with No.201 Squadron. *Courtesy of and via Flt/Lt T. N. Harris, XV (R) Squadron.*

A fine portrait of an unidentified officer, of No.15 Squadron, Royal Flying Corps. Note the RFC badges on the tunic collar and cap, together with the rank 'pip' on the shoulder epaulette. This officer must have been a non-flying type, as there is an absence of aircrew badges on the tunic. *Author's Collection via Don Neate*

The 'Workhorse' and the Workload

A B.E.2c, coded 2637, of No.15 Squadron, bearing the name 'Punjab 15 Kashmir'. The aircraft photographed during August 1916, crashed on 15th of the following month and was struck off charge at No.2 Aircraft Depot. *Via Chaz Bowyer.*

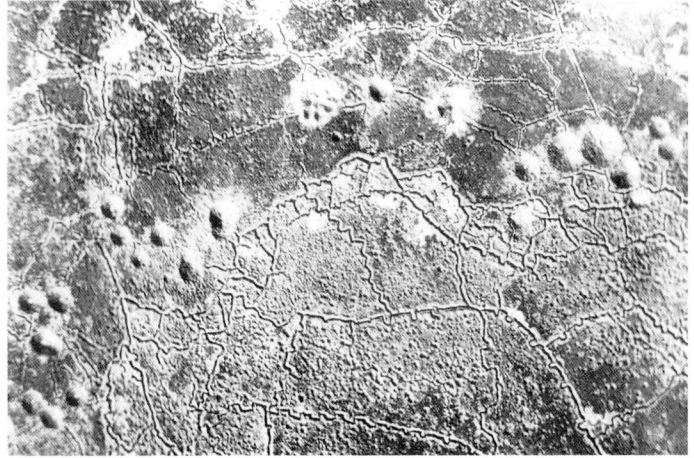

Another example of Raymond Ward's photographic work. An aerial photograph, taken on 26th September 1916, of the Front Line to the south of Nord St Vaart, France. The shell and bomb craters are clearly visible. *Courtesy of and via Flt/Lt T. N. Harris, XV (R) Squadron.*

The Fifth C.O.

Major George Carmichael, DSO, ex-Commanding Officer, No.16 Squadron, RFC, assumed command of No.15 Squadron on 19th December 1916. He remained as C.O. until he was posted back to the United Kingdom, in February 1917. The original photograph bears the signature 'George' in the bottom right-hand corner, and is dated 1st April 1918; the day the Royal Air Force was created. *Courtesy of and via Flt/Lt T. N. Harris, XV (R) Squadron.*

3

Arrival of the "Harry Tates"

The successes promised in the early stages of the Somme campaign failed to materialise by the time the battle came to an end in October 1916. The air supremacy the Allies had gained during the battle was being challenged by the Imperial German Air Service, who were forming new special pursuit squadrons and developing new tactics.

No.15 Squadron, which had discharged its duty honorably, was taken out of the line and rested for a week. The time was used profitably, repairing and overhauling the aircraft, engines and equipment, and well as giving the personnel a chance to recuperate from the effects of their recent labors.

The future was looking as bleak as the weather, with the latter being dismal and very wet. The inclement conditions, to some degree, impeded No.15 from carrying out its allotted tasks on its return to duty, but it did manage to get aircraft away during breaks in the weather.

Seven days into 1917, 2nd Lieutenants' Pateman and Goodfellow, pilot and observer respectively, took advantage of one of those breaks and took-off for a trench reconnaissance mission. They carried out their mission with such vigour that at one point, 2nd/Lt Pateman brought the aircraft down to 400' and managed to disperse a group of approximately sixty German infantry.

No.XIII Corps, who had taken over from V Corps the previous November when the latter were relieved from the line, were themselves relieved, when V Corps returned on 12th January to re-assumed command of the forward area.

The rain, which turned the ground into a sea of mud over the new year period, turned to snow during the latter part of January. Although the conditions restricted operational duties the snow had at least one benefit, that of being an aid to aerial photography. When the photographs taken by Raymond Ward over the front line, south of Arras on 23rd January were developed, the trench systems, courses and shell craters were all clearly defined on the prints.

The Squadron undertook a change of operation on the last day of January, when it was ordered to carry out a bombing mission. It was the first time that No.15 had been considered for such a role, but history was to prove it would not be the last, and was the foundation of the Squadron's role in later years.

Four B.E.2's were detailed for an attack against a dump at Fremicourt, at the end of the day on 31st January. The four aircraft, which on this occasion were being flown as single-seaters, comprised two B.E.2c's, piloted by Captain R J Hudson and 2/Lt C L Milburn, and two B.E.2e's, flown by Lt F M Carter and Lt J H Sayer. Although the four aircraft started out on the mission, Lt Sayer's machine developed ignition trouble and he was forced to abort; taking his bombs home with him.

The three other aircraft continued on to their allotted target, where one of Captain Hudson's two 112lb bombs exploded thirty yards from the northern edge of the aiming point, whilst the other failed to explode. Lt Milburn fared slightly better with two near-misses, whilst Lt Carter saw his two bombs fall on Fremicourt itself. Again, one of the latter's bombs failed to explode.

February got off to a bad start for the Squadron, with a number of casualties being recorded. On the 1st of the month Captain G M Moore was wounded by small arms fire whilst trench strafing. Three days later, on 4th February, the incident was of a more serious nature when Sergeant Shaw, pilot, and 2nd/Lt Bradford, observer, were lost. The pair were flying on an artillery patrol, in B.E.2e, serial 7105, when they were attacked and shot down by Leutnant Erwin Bohme, of Jasta Bolcke. The B.E.2e, which fell in flames, was Bohme's 11th victory.

Second Lieutenant Henry Lewis Pateman, who on 7th January had dispersed the group of German infantry, was lost on 6th February. He had been participating in an Artillery observation patrol with his observer, 2nd/Lt Horace Davis, from which the pair failed to return. It later transpired that their aircraft, B.E.2e, serial 7144, fell in flames and crashed inside British lines, having been shot down by Halberstadt scouts. The German scouts also attacked a B.E.2 in which Lieutenant William Henry Ritter was flying as observer, but Ritter was ready for the Halberstadts and opened fire with his machine gun. It is not known whether Ritter's aim was good, but he certainly drove off the attacking aircraft which did not bother to return to the fray.

Finally, on 11th February, 2nd/Lt L R Brown was wounded in the forearm whilst carrying out an artillery patrol. He returned to base without further incident.

Another change of command occurred on 19th February, when Major H S Walker took over from Major George Carmichael, who had been transferred to Home Establishment two days earlier. The former had previously been attached to II Brigade, where he had been Officer Commanding, No.46 Squadron.

The last couple of weeks of February saw some outstanding work undertaken by the Squadron aircrews. Lt George Barker who, as previously mentioned had returned from pilot training on 24th February, carried out some exceedingly low-level trench reconnaissance patrols with his observers. Likewise the information gleaned on patrols flown by Captain J R Hudson, 2nd/Lt James Herbert Sayer and 2nd/Lt W H Ritter respectively, was recorded as being, "Exceedingly good work".

The arrival of March did not induce any change in the weather, which still consisted of low cloud, mist, rain and the occasional snowstorm.

Due to the fact the Germans had pulled back to the Hindenburg Line in order to prepare defenses from Arras to St. Quentin, the Royal Flying Corps was able to fly numerous reconnaissance patrols over the enemy lines, despite the weather conditions. Needless to say some of these patrols were flown by No.15 Squadron.

The information obtained during the aforementioned missions assisted in finalising plans for a general advance, which General Sir Douglas Haig had ordered to commence on the morning of 16th March. That same morning V Corps issued orders which directed No.15 Squadron to work in association with the 7th and 62nd Division. Furthermore, with No.5 Squadron moving north to La Gorgue, where it came under the command of I Brigade, No.15 was left with sole responsibility for all V Corps air co-operation work.

Fifteen's flyers responded to the challenge and carried out their task with the usual efficiency and determination for which they were becoming known. Captain A H Wynne, pilot, and his observer, Lt A S Mackenzie, obtained information relating to the location of vacated enemy battery positions, as well as strafing retiring German infantry with machine-gun fire.

Having failed in their task on 6th March, three German Halberstadt scout aircraft again attacked a B.E.2 in which Henry Ritter was flying as observer, but again Lieutenant Ritter was ready for them. A short aerial battle ensued whereby Ritter defended his aircraft but, on return to base, it was later found to have sustained damage to the rudder and propeller.

The name of 2nd/Lt Edward John Hare was added to the Squadron Roll of Honor, after he was killed during a photographic reconnaissance mission near Heinel, on 24th March. John Hare was flying as observer on B.E.2e, serial 7254, piloted by Sergeant J F Ridgway, when the aircraft was attacked by an enemy machine. Whilst Ridgway continued to take photographs, Hare opened fire and succeeded in driving off the attacker. However, anti-aircraft fire immediately opened-up and, with its first salvo, hit the B.E.2e between the pilot and observer's seats. With severe damage incurred to the center section of the aircraft, a broken rudder-bar, a ruptured fuel tank, and dead observer on board, the wounded pilot managed to land his stricken machine at Beugny.

William Henry Ritter was again in trouble this same day, but on this occasion his problem had nothing to do with the German Air Service. Ritter was engaged, as usual, on a photographic patrol, and had just started taking pictures when the string to the camera shutter broke!!

The inclement weather which had greeted March stayed around to herald the arrival of April, and the start of the Allied air offensive for the battle of Arras; and with it came the loss of another Squadron member.

Captain A M Wynne and his observer, Lt Adrian Mackenzie, were engaged on a ground support operation when the latter was killed by a bullet through the heart. The slug came from the guns of the German fighter ace, Werner Voss, who had attacked the B.E.2e, serial 2561, over St. Leger. The British aircraft, which was recorded as going down at 11.42 hours, was the German ace's 23rd victory. Captain Wynne, who survived, was fortunate to escape with only a leg wound.

A similar occurrence happened two days later, on 3rd April, when 2nd/Lt James Sayer was attacked whilst also carrying out a ground support mission. Enemy fighters from Jasta 12 bounced the British B.E.2e, serial 7236, and shot it down near Croisilles; Hauptmann T. von Osteroht claimed the victory. Second Lieutenant Sayer was killed, whilst 2nd/Lt V C Morris survived to be taken prisoner of war. The latter's incarceration lasting until his repatriation on 2nd January 1919.

A photographic reconnaissance mission on 6th April could have ended with disastrous results but, fortunately for the crew, the outcome was successful. At approximately 09.00 hours that morning, 2nd/Lieutenant Albert Vinson and 2nd/Lt E C Gwilt were photographing Bullescourt from an altitude of 7,000'. During the course of their task, they were attacked by six German Albatros DIIIs. During the ensuing fight both machine guns from the British observation aircraft fell overboard, but not before five of the German machines had been driven off. The remaining German fighter was determined not to give up the fight so easily, and forced the unarmed B.E.2e, serial A3517, to land near Lagnicourt. He then proceeded to strafe the R.F.C aircraft, as its crew hastily sought protection in a shell crater. British Army machine gunners entered the fight and drove off the Albatros, by opening fire at it. Determined not to let the results of his mornings work go to waste, 2nd/Lt Vinson broke cover from his shell crater and retrieved the photographic plates from the stricken aircraft, as German artillery decided to join the fray. He had just succeeded in his task, when a direct hit from a German artillery shell destroyed the B.E.2e. Vinson's efforts were not in vain, as the exposed pictures revealed some interesting details.

Another incident which occurred on 6th April 1917, which was to have far more reaching consequences, was the declaration of war by the United States of America, against Germany.

The Declaration document signed by President Woodrow Wilson at 13.18 hours that afternoon, put the US regular army force of 5,000 officers and 123,000 men, supported by 8,500 part-time soldiers of the National Guard, at readiness for war supporting the Allies.

During April No.15 Squadron took advantage of the use of an advanced landing ground at Courcelles-le-Comte. Each morning the aircrew would fly their aircraft from Lealvillers the few short miles to the ALG, from where they would operate and then return to Lealviller at the end of the day. Little did they realise that soon the ALG would be 'home' to them.

Another bombing mission was undertaken by the Squadron during the early morning of 11th April, when Lt F F Wessel and 2nd/Lt Vachell attacked Cambrai railway station. Each aircraft was armed with one 112lb and two 20lb bombs, all of which were recorded as having struck the target.

Preparations for the ground offensive continued despite the on-going adverse weather conditions. Occasionally the weather broke, allowing No.15 Squadron to obtain good quality reconnaissance photographs, thus allowing the plans and preparations to be up-dated on an almost daily basis.

The 15th April was to prove exceedingly busy for No.15 Squadron, with crews continually over the battle area.

As was becoming expected, the weather conditions on this day left a lot to be desired, with cloud at 500' accompanied by heavy, driving rain. These conditions however did not deter 2nd/Lt W Buckingham and 2nd/Lt W R Cox from directing artillery fire at a German breakthrough near Lagnicourt, where enemy troops had by-passed British field gun positions. Following strafing attacks by the B.E.2 crew, and a counter-attack by British infantry, the Germans incurred heavy losses as they were pushed back into the line of the artillery fire.

Although the R.F.C had been engaged in conflict over and around Arras for a couple of weeks, the spring offensive against the Hindenburg Line, known as the battle of Arras, did not commence for the ground forces until 16th April. That day, Easter Monday, broke with driving rain and mist, but the British, Australian and Canadian forces secured their respective objectives and took over 200,000 prisoners of war.

No.15 Squadron was never far from the front, and its crews were continually over the battlefield. Many returned from missions with battle damage to their aircraft, whilst others endured personal injuries. Unfortunately, Lieutenant F F Wessel, the only known Danish pilot in the Royal Flying Corps, went one better; he returned from an operation minus his aircraft and with a wounded foot. He had just taken-off during the early morning of 20th April, for an artillery patrol, when he was attacked by three German Albatros scout aircraft. Being at an altitude of only 500', his immediate instinct was to gain height and seek protection in the low cloud above him. As Lt Wessel pulled back on the 'stick', putting his B.E.2 into a climb, machine gun fire raked the aircraft, rupturing the fuel tank and wounding the pilot in the right foot. The Danish airman's 94th mission came to an unceremonious end, when the aircraft force-landed in front of British trenches in No Man's Land. Both the wounded pilot and his observer managed to clamber out of the stricken machine and gain the safety of their own lines. Due to the nature of his wound, Lt Wessel was repatriated to England the following day.

Other personnel who sustained injuries included Sergeant W G Bennett, who was hit in the leg by machine gun fire, whilst Air Mechanic L A Blunden and Lt F E Elliot, received injuries of an unspecified nature.

April concluded, for No.15 squadron, with the loss of two more members of aircrew, when B.E.2e, serial 7660 was shot down. Lieutenant D K Paris and 2nd/Lt A E Fereman, survived the ordeal, and were taken prisoners of war.

Although the R.F.C had entered the battle of Arras with numerical superiority over the German Air Service, such was the nature of losses in aircraft and aircrew, to the Royal Flying Corps in general, that the last month became known as "Bloody April".

The advent of May brought forth little change in either the weather or general conditions. In fact, for No.15 Squadron things only got worse, by virtue of the fact that their aircraft were now feeling the strain of constant operations, and engine failures were becoming the norm. Lieutenants' H W Girdleston and Hoskins were made aware of this fact when their B.E.2 crashed on a road on 7th May, due to this very reason. However, even with these problems the mechanics managed to keep some of the aircraft flying, enabling the aircrews to continue with their respective photographic and artillery patrols etc.

It soon however became obvious that the mechanics were 'fighting a losing battle', for although they could get the aircraft into the air one of the problems was keeping them there. A defective engine with a burned out valve forced one B.E.2 to land near Fremicourt on 12th May, whilst seven days later a similar situation forced another B.E.2 to abandon an artillery observation. On 21st May, the mechanics 'fought' with one particular engine which completely refused to start!

With the current problems the Squadron was experiencing with its aircraft, No.15 was pleased to receive a notification from Headquarters, Royal Flying Corps, implying that the Squadron was to be re-equipped with R.E.8 aircraft.

Six days after receipt of the notification, on 28th May, the first five R.E.8's were delivered. Many more were to follow, and the aircraft stayed in service with the Squadron until the end of the war in November 1918.

Lieutenant William Henry Ritter, the veteran observer who had accumulated over 100 operational flying hours, was killed in action on 2nd June, depriving the Squadron of a very experienced flyer.

Having been re-equipped with new aircraft at the end of May, a change of base, and a change of command followed for the Squadron on 6th June, when No.15 moved to Courcelles-le-Comte, the airfield it used as an advanced landing ground. This move placed the Squadron under the command of 12th Wing, III Brigade.

In comparison to the previous months, June was a fairly quiet period for No.15, but nevertheless losses were still recorded. On the 15th of the month Lieutenant John de Conway, pilot, and 2nd/Lt Cecil Powell, observer, failed to return from a reconnaissance patrol. Although they have no known graves, their R.E.8, serial A4310, was reported as having been shot down by Vizefeldwebel Riessinger, of Jasta 12.

A month after its move to Courcelles-le-Comte, the Squadron was temporarily withdrawn from the command of III Brigade and transferred to I Brigade, as a component of the General Headquarters Reserve in the First Army area. This transfer, made on 7th July, necessitated a move north to La Gorgue, although "C" Flight were moved again five days later. Whilst "A" and "B" Flights remained at La Gorgue, ostensibly for a rest period, "C" Flight were sent to Clairmarais airfield for, of all things, practice in contact patrol!

During this period a new pilot, 2nd/Lt James William Schofield, arrived on the Squadron and was allotted to "B" Flight where, like the experienced pilots, he got to know the handling characteristics of the R.E.8. Whilst flying the new machine, Schofield also had to learn the layout of the locality to get his bearings.

Leaving the Reserve on 18th August "A" and "B" Flights, now completely re-equipped with R.E.8 aircraft, moved to a new base at Savy, northwest of Arras.

Second Lieutenant Schofield flew his first operational mission on 19th August when, accompanied by Lt G Gibb as his observer, he undertook an artillery patrol of the Lens – Arras region. His reaction and feelings to his first encounter with anti-aircraft fire were not recorded.

The arrival of the *'Harry Tates'* of "C" Flight the following day, brought the Squadron back up to full compliment. *'Harry Tate'* was the name given by the aircrews, and others, to the R.E.8 aircraft. It was also the name of a famous British Music Hall star, who sported a large 'handlebar' mustache resembling the wings of an aircraft. Also of course, there was the fact that the name fitted rather nicely in East London's cockney rhyming slang; R.E.8. – Harry Tate.

Having had his first encounter with enemy anti-aircraft fire, or 'Archie' as it was more familiarly known, Schofield had the dubious pleasure of meeting the German Air Service itself on 26th August. With him on this occasion, flying as his observer, was Lt Cambridge.

Identification of the Squadron's aircraft was made easy by the introduction of an eighteen inch wide, white painted band around the fuselage, forward of the tailpane assembly.

On 30th August No.15 was moved yet again, this time to Longasvesnes, approximately six miles north-east of Peronne, where they re-newed their link with the 12th Wing and III Brigade.

The new equipment did not come up to the expectations of Fifteen's flyers, a fact they found out in combat against the enemies latest fighters when, during September, No.15 took over No.3 Squadron's duties with No.IV Corps.

No.15 Squadron's role had not changed with the introduction of the R.E.8., and both the men and machines of the Unit continued to photograph and reconnoiter the battle areas over Arras, Ypres and Messines. Raymond John Ward, who had survived many such missions, continued to produce valuable photographic information.

During a patrol on 24th September, whilst observing for the artillery, 2nd/Lts G W Armstrong and J D Steele were attacked by a German Albatros D.III fighter. As the R.E.8 replied to the attack, Steele's machine gun jammed, thus necessitating the need for immediate evasive action. As the R.E.8. went into a spin, the errant machine gun broke free from its mounting and fell overboard, rendering the 'new' English aircraft totally unarmed. The German machine followed its adversary down and forced it to land near Bertaincourt, leaving Armstrong and Steele to fight another day.

On the night of 27th September, No.15 Squadron detailed five R.E.8's to participate in bombing raids behind enemy lines. The targets were to include ammunition dumps, communication centers and an enemy headquarters at Cisy-le-Verger, and apart from No.15, were to be attacked by No's 8, 12, 13 and 59 Squadrons.

During the course of the five raids undertaken on this night, a total of one hundred and forty-nine 25lbs Cooper bombs and ten Hales bombs were dropped.

The five crews from No.15 reported good results with most of their bombs falling on, or close to, the specified target, an ammunition dump at Rumilly.

The same target was visited again the following night by five more crews led, as it had been the previous evening, by Captain J B Solomon. The ground defenses were a little more active on this second occasion, but all No.15's aircraft returned safely to base.

The fact that 2nd/Lt Schofield demolished the cookhouse during a landing accident, had no bearing on the decision to move the Squadron again. At the end of the first week of October, on Monday 8th, No.15 moved base to Lechelle, south-east of Bapaume, from where it continued to undertake its usual role. Some excellent results were achieved whilst flying from Lechelle, and the quality of the reconnaissance operations, artillery ranging and photographic missions were all carried out to the high standard being associated with the Squadron. The weather did however deteriorate towards the end of October, preventing the Squadron from carrying out artillery co-operation work for several days.

The Squadron was fortunate in that no casualties had been inflicted on it since 15th June, when Lt John de Conway and 2nd/Lt Cecil Powell failed to return from a mission. However, that was to change in dramatic form.

The Battle of Cambrai commenced during the third week of November, with an attack which took the enemy by surprise. It concluded three weeks later with the capture of part of the German's front-line trenches, and many thousands of prisoners. On 20th November, as part of the battle tactics, No.15 Squadron were detailed to co-operate with the 1st Cavalry Division, during which time their duties entailed much low-level flying. Being that much closer to the ground brought their aircraft within range of a higher than usual assortment of gun fire, and No.15 paid a price. One pilot, 2nd/Lt George Young, and two observers, 2nd/Lts Alan Wylie and Wilfred Davey were killed, whilst Lieutenants' Robertson and Creaghan and 2nd/Lt Mayoss were all wounded.

British pride was also wounded when the Germans pushed forward and advanced on Gouzeaucourt, causing No.15 Squadron to move northwest to an advanced landing ground at Dapaume. After approximately one week at this location, the Squadron returned to Lechelle, where it was re-attached to V Corps with whom it was successfully associated earlier in its short history.

Having successfully evaded an attack by four enemy aircraft, whilst engaged in an artillery operation with Lt J R Hodgkinson on 6th December, James Schofield was not quite so lucky the following day. Schofield and Hodgkinson had the misfortune to encounter eight Albatros D.III's, which closely resembled those flown by the Richthofen *'Flying Circus'*. During the inevitable fight, the observer received leg wounds, whilst the aircraft was badly damaged, necessitating a forced-landing behind British lines.

The end of the year was looming, and with it came thoughts of Christmas and the prospects for the new year. Nobody knew what 1918 would hold, but at least the personnel from No.15 Squadron knew they would endeavor to have a good Christmas.

The Mess was duly prepared with Christmas decorations. Flags and bunting brightened up the otherwise drab wooden hut, and gave the intended festive feeling. Even Major Walker, the Commanding Officer, got struck-in and lent a hand. Along with Hubert Griffith, a member of aircrew, from whom he had been taking lessons, Major Walker helped paint a back-cloth for a variety show which was to be held after the Christmas lunch. For one day at least, Christmas day itself, the Squadron could (almost) forget the war and indulge in the tradition that Christmas represents.

'Patterns in the Snow'

Lieutenant William Ritter

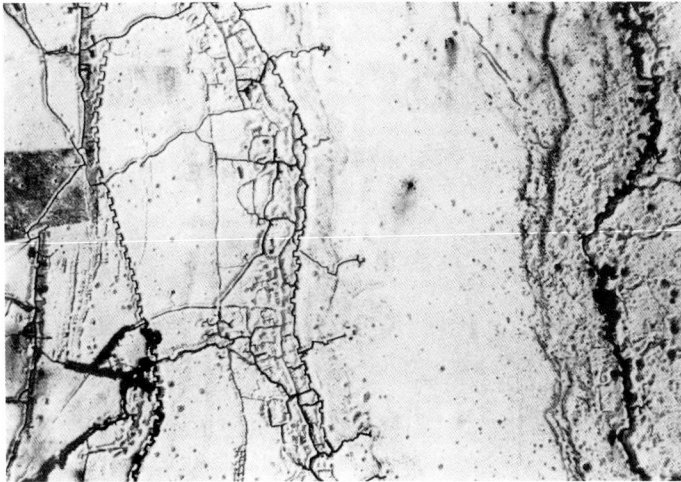

A blanket of snow defines the features of the front line for the aerial camera, as roads, trenches and shell and bomb craters, south of Arras, are clearly seen in this photograph, taken by Raymond Ward on 23rd January 1917. *Courtesy of and via Flt/Lt T. N. Harris, XV (R) Squadron.*

A poor quality photograph of Lieutenant William Henry Ritter, Observer, photographed sitting on the lower wing root of a B.E.2. *Courtesy of and via Flt/Lt T. N. Harris, XV (R) Squadron.*

In Good Hands

Four unidentified mechanics of No.15 Squadron pause for a photograph whilst working on B.E.2c, serial 2810, during March 1917. *Courtesy of and via Ray Vann.*

A B.E.12, serial C3236, coded 4, operated by No.15 Squadron. The aircraft, which was built by the Daimler Company Ltd, at Coventry, England, during 1917, was one of an original contract for two hundred B.E.1 machines. The contract was later modified to include B.E.12a/b aircraft. *Courtesy of and via Flt/Lt T. N. Harris, XV (R) Squadron.*

"If You Can Walk Away..."

'They walked away from this one'. The remains of a B.E.2e, serial A2866, of No.15 Squadron, which crashed at Courcelles-le-Court, France, on 28th May 1917. Not all crews were so fortunate. *Courtesy of and via Flt/Lt T. N. Harris, XV (R) Squadron.*

'Eagles' Eye-View

A view no doubt often seen by German Scout pilots, of an R.E.8. This unidentified aircraft of No.15 Squadron, was photographed whilst flying over France. *Via Chaz Bowyer.*

'Harry Tates'

Lieutenant James William Schofield, pilot, and Lieutenant G Gibbs, observer, prepare to start the motor of their R.E.8., prior to carrying out a reconnaissance mission. *Courtesy of and via Ray Vann*

No.15 Squadron's new R.E.8 aircraft were marked with a white painted identification band, situated at the fin root, as shown on this example wearing the serial, A3697. The photograph was taken during October 1917. *Via Chaz Bowyer.*

Battle-Scarred Landscapes

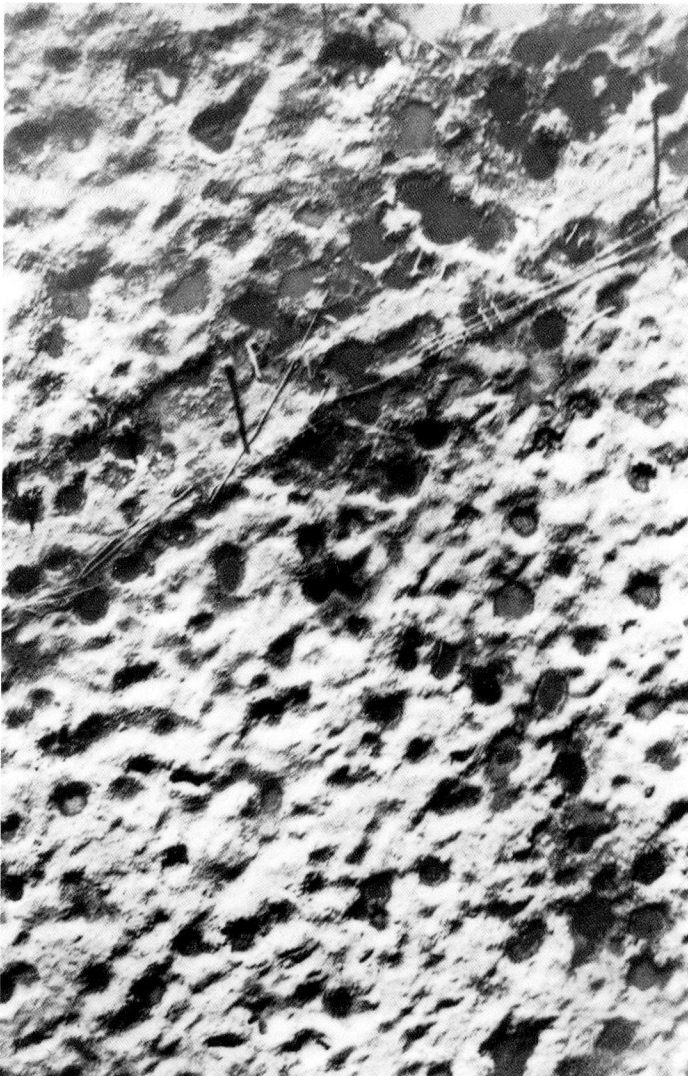

An aerial photograph, taken during August 1917, clearly defines the river, pontoon bridges, trenches and shell craters at Trendresby, France. *Courtesy of and via Flt/Lt T. N. Harris, XV (R) Squadron.*

Left: An example of Raymond John Ward's work shows the battle-scarred ground near Ypres, France, during August 1917. *Courtesy of and via Flt/Lt T. N. Harris, XV (R) Squadron.*

Teamwork

Two officers, identified only as Levy (left) and Morrison (center left), pose with two members of their groundcrew in front of a No.15 Squadron R.E.8. *Courtesy of and via Flt/Lt T. N. Harris, XV (R) Squadron.*

Christmas 1917

Christmas day preparations in the Mess at Lechelle, France, 1917. *Courtesy of and via Flt/Lt T. N. Harris, XV (R) Squadron.*

Photographed outside their billet, sometime during 1917, are: Lear, Mayoss, Casey, Morrison, unknown, Hall. *Courtesy of Don Neate.*

"B" Flight, No.15 Squadron

Members of 'B' Flight, No.15 Squadron pose for an informal photograph at Savy, France, during August 1917. Included in the group from left to right are: MacCloud; 2nd/Lt R Barrett; Lt J Vachell; Clark; 2nd/Lt B Morgan (KIA -07.10.18); Kelsell; Lt Cambridge; Rowbridge; Richard; Lt G Gibbs and Lt J Hodgkinson. Morgan, Cambridge, Gibb and Hodgkinson were all observers who flew at some time with Lieutenant James Schofield. *Courtesy of and via Ray Vann*

4

1918 - 'The Final Push'

Snow, rain, mist and low cloud all took their turn to welcome the new year and cause havoc with No.15's work schedule; although the elements occasionally relented and enabled the Squadron's aircraft to get airborne.

On 6th January Lt H Corsan and Lt D Munro, pilot and observer respectively, took advantage of one of those breaks in the weather and carried out an artillery co-operation mission. Working together, Corsan and Munro were able to range the Siege guns of the 48th and 56th Batteries and record two direct hits, resulting in the destruction of two gun pits and quantities of ammunition.

Later the same day Lt C Douglas and his observer, Lt W Haddow, flew a similar mission. After a patrol lasting nearly three and three-quarter hours, Douglas and Haddow were able to report two direct hits, with one gun pit being destroyed and severe damage inflicted on the whole of the enemy battery position.

Another crew who successfully attacked enemy positions was that comprising of Lt J Schofield and Lt B Morgan, who discharged four bombs on to units in Cateau Wood. Shortly after completing their task they were attacked by a German fighter. In the ensuing fight, the R.E.8. sustained damage to the elevator and rudder controls, but Schofield managed to evade further trouble after Morgan drove off the attacker by discharging the contents of three drums of ammunition at it.

Apart from a change in date and year there was little, as far as No.15 Squadron was concerned, to signify the arrival of 1918, although they did record a change of command two weeks into the new year.

On 14th January, Major Walker, who had led the Squadron for one month short of a year, relinquished his position as Commanding Officer, in favor of Major H V Stammers, M.C., and was posted to Home Establishment.

Although the skies over France were very overcast, 2nd/Lt L Hobbs and 2nd/Lt G Gibbs found a break in the weather on 19th January and carried out an artillery ranging patrol. By virtue of the manner in which they carried out their mission, which lasted three and a half hours, Hobbs and Gibbs enabled the 56th Siege Battery to score three direct hits, and destroy two enemy gun pits.

Excitement abounded through the Squadron during February, with the arrival of a single Bristol F.2b, two seater Army Co-operation aircraft. This gave rise to the hope that No.15 might soon convert to this type of machine. The crews fortunate enough to fly the Bristol during its two month stay with the Squadron, saw obvious advantages with the F.2b over the their current machines. However, whatever the reasons for the F.2b's issue to No.15, the Squadron was destined to continue the war with the R.E.8.

For one R.E.8, serial B2257, the war very nearly came to an end at 15.45 hours, on 5th March, whilst engaged in an artillery observation over La Vacquerie. The R.E.8 piloted by 2nd/Lt James Burchett, a former Texan Ranger and ex-cow-puncher from the U.S.A., was flying at 4,000', when it

was attacked by four enemy aircraft who had used the clouds above to conceal their approach.

As they dived down into the attack, the leader of the enemy formation opened fire, wounding the observer 2nd/Lt Robert Fear, in the shoulder. With the observer slumped in his cockpit, the guns of the R.E.8. remained silent, rendering the aircraft defenseless. Although there was no return fire, the four individual attacks made by the enemy aircraft failed to record any further hits on the British machine. Having eluded his attackers, 2nd/Lt Burchett headed back to base where Robert Fear died from his wounds that same day.

Five minutes after the attack had began on Burchett's aircraft, a patrol of five enemy machines attacked the R.E.8, serial 6508, piloted by 2nd/Lt H S Pett. Pett was carrying out a similar mission over the same area as his colleague, when five Albatross Scouts emblazoned with black and white stripes on their fuselages, used the same method of attack. Opening fire at a range of 200 yards, the hostile formation split into two sections, with the larger group containing three aircraft, continuing the attack. The observer, Lt F B Rees, returned fire and, with the help of anti-aircraft fire from the ground, drove off the attackers who turned and flew off towards the east. Fred Rees, an efficient observer and modest man, was the 1914 Champion Steeplechase jockey. He survived the war, resumed his interest in racing and went on to win the English Grand National race on a horse named *Shaun Spaduh*.

During the early morning of Thursday, 21st March, an intense enemy artillery barrage opened up along the entire front occupied by the British Third and Fifth Armies, from the Oise to the Scrape. This barrage signified the start of an offensive prepared by the German High Command, which they hoped would bring an end to the stalemate on the Western Front. By making use of the low ground mist, the enemy's infantry rapidly advanced and re-captured the area astride the Somme, the scene of heavy fighting during 1916.

As the Germans advanced a number of squadrons, including No.15, were forced to vacate their respective bases. No.15 was hastily moved from Lechelle to Lavieville on 22nd March, followed by a further move to La Houssoye three days later. The only recorded casualty during these withdrawals was the loss of one abandoned Leyland lorry which succumbed to shell fire. However, within twenty-four hours of their arrival at La Houssoye, No.15 was moved again north-west, to Flenvillers, near Doullens.

The move to Flenvillers on 26th March, was marred by the loss of 2nd/Lt Vernon Reading and 2nd/Lt Matthew Leggett, pilot and observer respectively. On this particular day they were flying R.E.8., serial B742, engaged on an artillery ranging patrol, when they were attacked by five Fokker DrI triplanes, from Jasta 1, of the German Air Service. This well-known and revered enemy unit, better known as the notorious '*Richthofen Flying Circus*', was led by German's highest scoring ace, Rittmeister Manfred von Richthofen, otherwise known as the '*Red Baron*'.

The German ace dived into the attack, firing off approximately one hundred rounds of ammunition at close range. As fire spat from the twin machine guns, mounted on the upper fuselage in front of the cockpit of his aircraft, Von Richthofen watched as his 70th victim caught fire. Although 2nd/Lt Leggett endeavored to protect and defend both himself and his pilot, the R.E.8. fell to earth in flames; killing both the occupants.

As if metaphorically rubbing salt into the Squadron's loss wound, the *'Red Baron and his Circus'* had taken-off from No.15's recent base at Lechelle, which Jasta 1 was using for operations that day.

Such was the force of the enemy advance, both on the ground and in the air, that at one stage the German offensive appeared so serious that the British High Command made plans to withdraw and render the Channel ports ineffective at the same time.

However, despite all the planning that went into the offensive, the German High Command had underestimated the resistance they would meet from the British infantry and cavalry divisions facing them. Admittedly, the British lines were forced back, but not for long. New positions were taken up from where the fight could continue; and new policies could be adopted.

Amidst all the fighting during the German offensive, a new British military force was created; the Royal Air Force. On Easter Monday, 1st April 1918, the Royal Naval Air Service squadrons (RNAS) flew into Flenvillers, where No.15 was based, to amalgamate with the five Royal Flying Corps squadrons already there, to form the RAF.

On 10th April No.15 Squadron, Royal Air Force, returned to Vert Galand airfield where it had been based two years before, and immediately resumed its allotted tasks of contact, ranging and reconnaissance patrols.

The fact the British air arm had been re-structured and renamed did not detract from the fighting spirit of the men who served in it. This was proven by a crew from No.15 Squadron the day after the unit's arrival back at Vert Galand.

The fight started when a flight of four German Pfalz scout aircraft attacked an R.E.8. crewed by 2nd/Lt Raymond Hart, pilot, and 2nd/Lt L Handford, observer. The latter were flying a contact patrol at an altitude of 3,500 feet over Bouzincourt, when the enemy machines climbed up through low cloud and opened fire at the *'Harry Tate's'* tail. As Hart turned his aircraft into face the attack, both the elevator and aileron controls were shot away, whilst the pilot received shell splinter wounds to his ankle.

However, the German pilots were not having things all their own way. One of their number plummeted to his death as the starboard planes of his aircraft broke off, after 2nd/Lt Handford had retaliated with fire from his own guns. The enemy machine crashed inside British lines at Millencourt, and was confirmed by the 47th and 63rd Divisions. In an effort to avenge his colleague's death, a second enemy scout took-up the fight and attacked the R.E.8., wounding Handford twice in the same knee. Despite the obvious pain, the observer managed to return fire and saw the German aircraft fall steeply away from the fight, before he succumbed to the darkness of pain. This enemy aircraft was also reported to have crashed in flames.

As 2nd/Lt Hart fought to get his crippled machine safely on to terra firma, another member of the German pack followed him down to within one hundred feet of the ground, as if to inflict the coup-de-grace. Then, without any further incident, the German pilot pulled up and flew away.

The weather on 13th April continued to be misty with low clouds, preventing a number of operations other than low-level reconnaissance and contact patrols. Lieutenant H Chippendale, pilot, and his observer, 2nd/Lt G Hobbs, managed to get airborne and carry out a successful ranging operation. Working in co-operation with the 56th Siege Battery, 'Chips' Chippendale and Hobbs, were able to direct forty-seven rounds of artillery fire at the Aveluy bridge. Nine of those rounds scored direct hits on the target, whilst others completely destroyed the eastern approaches to the bridge.

Although the Germans repaired the bridge, No.15 Squadron shared in its destruction again one month later, when they ranged the target for the 48th Siege battery.

At 05.00 hours on the morning of Thursday, 18th April, a fresh-faced young Scotsman arrived at Vert Galand, having been posted to No.15 Squadron from No.5 Aerial Gunnery School at New Romsey, Hampshire, England.

Robert Alexander Fraser, was born on 19th November 1898, in a very small village near Kiltarlity, to the south-west of Beauly in Invernesshire. The village was so small that the narrow country roads were not named. However, for ease of identification for the visitor, the cottages were numbered in conjunction with the name of the village; Robert Fraser being born at No.18 Culburnie.

From the age of three and a half years, young Robert attended the village primary school, a small stone building surrounded by forest and fields. His secondary education was undertaken at Inverness Royal Academy, whilst his studies were finally completed at Aberdeen University. It was whilst he was at university, sometime between March and April 1917, that Robert Fraser enlisted in the British Army. He was posted, with the rank of Private, to E Company, 39th Reserve Battalion, Seaforth Highlanders, based at Dunfermline, Fife, near Edinburgh. Towards the end of 1917, Private Fraser volunteered for transfer to the Royal Flying Corps, and applied for a commission. The former was granted and, probably for the first time in his life, he left his native Scotland. The fledgling airmen's flying training was to be undertaken in another country; England. It was whilst he was training at Reading, Berkshire, that a commission to the rank of 2nd/Lt was approved and Gazetted on 28th January 1918. Three weeks later, on 15th February, the young officer was sent to No.10 Aerial Gunnery School at Winchester, Hampshire, where he spent six days prior to going on two weeks leave.

Further training took 2nd/Lt Fraser to Lydd in Kent where No.15 Squadron had sent a detachment during the summer of 1915. On completion of his course, Robert Fraser was given a further seven days embarkation leave, prior to going overseas. He left Lydd on Easter Sunday, 31st March, and immediately headed for Scotland. His week in the Highlands went all too quickly and on Friday, 5th April, Robert said good-bye to his family and caught a train south from Inverness.

After four days at No.5 AGS at New Romsey in Hampshire, 2nd/Lt Fraser reported to the Air Board in London, where he received a posting to France. He was to take the route that so many young men before him had taken, some never to return.

The train which Robert had boarded left Victoria station, London, at 07.35 on the morning of 17th April, and headed for Folkestone, Kent, where he was to catch a ferry to Boulogne. As the train steamed towards the channel port, 2nd/Lt Fraser watched the English countryside roll past the carriage window, not knowing whether he would pass this way again.

The channel crossing he recorded as being *"quite decent"*, but also added that having reached Boulogne, he had a wait of several hours at the French port before being driven to Abbeville. Although he had not arrived at the latter until midnight that same day, the young Scot was ready to leave at 06.00 hours the next morning to continue his journey. Less than twenty-four hours later, Second Lieutenant Robert Fraser reported for duty with No.15 Squadron, Royal Air Force.

Taken on Squadron strength with almost immediate effect, the 'rookie' observer flew his first operational patrol at 09.30 hours the following day. Although he did not see fit to record any private observations in his diary, he did however write a letter to his sister Lizzie, which was dated the same day. In it he made mention of his journey, and his first thoughts on the war when he wrote:

..."What a weird sensation you feel when you listen to the continual rumbling of the guns early in the morning, for the first time."... The letter continued, ... "it makes me think how unspeakably awful it must be in the trenches. Yet nobody is discouraged. Everybody, indeed seems to be cheery (happy) including myself" ... "Yet Lizzie, please remember me in your prayers, for I need them".

Robert Fraser flew patrols every day without incident, until Monday, 22nd April, when he saw an artillery barrage, from the air, for the first time. Recording the event in his diary, he wrote:

..."Patrol 06.30 – 08.35" ... "First barrage seen, horrible and yet [a] splendid sight in the [morning] gloom"...

The next day 2nd/Lt Fraser flew a photographic reconnaissance mission. Again, he recorded in his diary:

..."Clouds at 4,000'. Archie (gunfire) galore, but no 'Hun' (German) aircraft. Got 18 photographs"...

The Squadron obtained a number of other photographs during the first week of May, but they were not aerial pictures of the trenches etc. During a quite period between missions, some of No.15s aircraft and personnel were caught on camera.

The day after the group pictures were taken, there was one photograph the Squadron would rather have not included in its annals. That of the graves of 2nd/Lt Leslie Derrick and Lt Harold Browne.

Derrick had been engaged on an army co-operation mission with Lt Browne, who was piloting R.E.8., serial C2361, when the aircraft was attacked by three enemy aircraft over Buire, France. The two Englishmen were killed when their machine fell in flames.

During this period of time, a number of the experienced crews of No.15 Squadron received special mention in Brigade reports etc, in recognition of the valuable contributions they had made to the war effort. However, one crew who received special mention became the subject of a further report, but for a different reason.

Lieutenant Felix Dymant and his observer, Lt Hubert Griffith (who after the war became a respected author and dramatic critic for the Observer newspaper), were ordered into the air on a wet Sunday afternoon. Although it was summer, the tall, fair young South African pilot did not warm up the engine of his machine to the usual extent. The result of this action was that as soon as the R.E.8. left the ground it become obvious that there was insufficient power for the aircraft to gain altitude. Felix Dymont held the biplane steady at approximately ten feet above the ground, hoping to gain height before reaching the hangers. Unfortunately, between the R.E.8. and the hangers was a road along which a truck was travelling. The flying machine and the truck both reached a 'crossing' point on the road at the same time. The outcome was that, at seventy miles an hour indicated airspeed, the lower starboard wingtip of the aircraft touched the cab of the truck. The wingtip folded back, and in almost vaudeville style, the aircraft completed a full 360 degree roll, at an approximate height of fifty feet, and came to rest having flown through the timber wall of the Power Targets Hut. Felix Dymant walked uninjured from the debris and was found, thirty minutes later, inspecting the crash site. Hubert Griffith however was slightly hurt and taken off to hospital with superficial injuries. Many years later, Griffith was to relate this story in one of his books.

Apart from the pilot and observer, the occupants of the main Squadron Office also had a lucky escape, as their hut was immediately adjacent to the Power Targets hut!

Not all No.15 Squadron's aircraft accidents were as spectacular as that of Lts Dymant and Griffith. Some occurred during take-off, others happened whilst endeavoring to land. However, it was said that any landing you could walk away from was a good landing!

A crew whose luck ran out on 18th May, comprised of 19 year old Lt Solomon Fine, pilot, and his 19 year old Scottish observer, 2nd/Lt Robert Alexander Fraser. They were engaged on an army co-operation mission, from which they failed to return. Although Robert had asked his sister Lizzie to pray for him, her prayers went unanswered, and exactly one month after his arrival on No.15, the young Scot's name was added to the Squadron's Roll of Honor.

Robert Fraser's original resting place was marked in the usual fashion of a cut-down aircraft propeller, with an inscribed metal plate at its hub. He was buried alongside his pilot, whose grave was marked in similar fashion, but with a Star of David. Although he never made the return journey across the English countryside back to his native Scotland, Robert was not forgotten in his homeland. His name was later to be recorded on the War Memorial, in the center of the village at Kiltarlity.

No.15 lost another pilot at the end of May, but this one was in happier circumstances. Having flown his last mission with the Squadron on 29th of the month, and logged over 295 flying hours, James William Schofield was transferred to Home Establishment as an instructor.

There was a lack of activity during the latter part of June as far as aerial operations were concerned, due to the inclement weather conditions. No.15 Squadron were called upon for artillery ranging missions, but these were of limited assistance to the ground forces.

At the end of June an experienced pilot joined the Squadron, when (Acting) Captain James Craig was posted in to No.15. James Anderson Craig, like Robert Fraser, was a native of Scotland and originated from Glengarnock, Ayrshire. He commenced flying training with the Royal Flying Corps in November 1916, having transferred from the Royal Fusiliers. On completion of his training, Craig was posted to No.59 Squadron in France, with effect from March 1917. He remained with that unit until his move to No.15.

It was obvious within a few days of his arrival that Capt Craig would undertake his duty with the same courage and determination of those who, by their earlier actions, had moulded the Squadron into a competent and respected unit. The evidence of this came on 1st July when James Craig, with Lt P Foot acting as his observer, undertook an artillery ranging mission.

Working with both the 56th and 431st Siege Batteries, Craig and Foot were able to direct the guns onto enemy gun positions, thus assisting the former unit to inflict serious damage to two gun pits, cause an explosion, and create two large fires. Also under Craig and Foot's direction, the latter unit was equally successful, destroying two gun pits, damaging a third and reporting two large explosions.

As it had done at the beginning of May, No.15 Squadron used the end of July and beginning of August 1918 to update its own photographic records. A number of official pictures where taken, together with a host of unofficial ones, all of which were added to the Squadron's annals.

August 1918 started very wet, very windy and with very low cloud. It began to improve at the beginning of the second week of the month, just in time for the start of the 'final push'; the offensive which would ultimately bring the war to an end. The 8th August saw the British Army, supported by American, Canadian, Australian and French forces, launch a massive assault against the Germans near Amiens. The Royal Air Force also assisted by flying fighter patrols, low-level strafing attacks and bombing raids. Overwhelmed by the force of the assault, and having had their front line penetrated, many German soldiers withdrew, whilst others gave up the fight and surrendered. As many as 30,000 prisoners were recorded as having been taken in just a few days.

No.15 Squadron was tasked with aerial reconnaissance of the battle area, with orders to report specifically on the state of the enemies defences and condition of the wire protecting his trenches etc. In general, the Squadron had to be the eyes of No.V Corps and relay back any information the Divisional Commanders could not obtain from the ground.

With evidence of a large scale German withdrawal, the intensity of the 'push' was increased during the middle of the month. However, the enemy still had some fight left in it as No.15 Squadron was to find out. On 22nd August, whilst flying a patrol over the forward lines, 2nd/Lt Walter Fox, observer, received wounds from ground fire, which were to prove fatal.

Two days later, on 24th August, Capt Craig carried out an action which generated an article in the (English) Times newspaper.

Entrusted with the task of trying to locate the position of a battalion of British troops who had advanced too far, and were thought to be cut off behind enemy lines, Capt Craig flew a low level mission over the battle front in search of them. In crossing enemy positions at such a low altitude, he was vulnerable to, and attracted, hostile gunfire. Although his aircraft was hit in many places, Capt Craig was successful in his mission and found the 'missing' troops. Whilst Craig circled above the latter, Capt Hill, the observer, signaled to them in morse-code on a klaxon horn. The British troops answered, by laying out ground strips, that they were short of ammunition. This information was relayed back to the Squadron, who sent out another aircraft which dropped the required supplies by parachute.

Further ammunition supplies were dropped the following day when the 38th Division, who were engaged in a bitter conflict with the German army between Mametz and Bazentin-le-Petit woods, called for urgent supplies. The call was again answered by No.15 Squadron who despatched an R.E.8. with the required small arms ammunition boxes, complete with parachutes, fitted to the aircraft's bomb racks. Second Lieutenant R A P Johnson piloted the aircraft, serial 3440, whilst 2nd/Lt Dalgleish was detailed to act as observer. Having reached the 'drop zone', these two men became so engrossed in their task, they failed to notice an enemy two-seater aircraft, with evil intent, descending upon them from the clouds above. Suddenly realising their dilemma, Dalgleish quickly aimed his machine-gun at the intruder and opened fire with a burst of only twelve rounds of ammunition. Initially the enemy machine was seen to turn away eastwards as though under control, but within seconds it burst into flames and spun down. As the German aircraft fell to earth, Johnson and Dalgleish continued with their task. 2nd/Lt Dalgleish was later commended for his remarkably good shooting.

The third stage of the general advance started during the early hours of the morning of 26th August 1918, for which purpose the First Army, commanded by General Horne, extended its flank of attack to the north. Support to this move was given by the Third Army, commanded by General Byng. The night of 2nd/3rd September proved the tactics employed by the First and Third Armies had been successful, when the enemy fell back from the whole of the Third Army front, and the right of the First Army front.

During the fighting on 2nd September, some small groups of British Infantry became isolated from the main force, and found it necessary to call for supplies to be dropped by aircraft. It was to these groups that No.15 Squadron, who by now had become proficient in this type of mission, dropped over 8,700 rounds of ammunition.

Another important mission entrusted to No.15 Squadron during this period, was detailed for first light on the morning of 9th September. Their task was to fly over the battle front and carry out a reconnaissance of the forward areas, to ascertain exactly where the enemy divisions were. They were also to report, and signal by use of red smoke bombs if necessary, any indication of an enemy counter attack.

This operation continued into the following day, when the Squadron reported that the enemy had occupied an African trench, and that gun fire emanated from all the trenches reconnoitered.

During the first week of September, although still based at Vert Galand, the Squadron had made use of an advanced landing ground, at Quatre Ventes Ferme. However, on the 12th of the month they relinquished the latter and set up a new ALG at Lechelle where, at the request of the 21st and 23rd Divisions of No.V Corps, two aircraft were kept in readiness for urgent reconnaissance duties. Two days later, No.15 vacated Vert Galand and relocated to a new base at Senlis, north-west of Albert.

On 15th September Lt V W Lilroe, pilot, and Lt O Izzard, observer, returned to base and filed a claim for the destruction of an enemy aircraft. They had been engaged on an artillery observation patrol when, at 17.00 hours, they encountered a German Albatross two seater aircraft at 6,000'. Following a short aerial conflict only one machine returned home.

With the tide of war definitely in the Allies favour, continuous moves forward were being made whilst the Germans naturally fell back. On 2nd October No.15 moved forward to Quatre Vents Ferme, situated to the west of Lechelle, an advanced landing ground it had used previously.

An aerial reconnaissance, on the morning of 5th October, revealed that the Germans had fallen back from the east bank of the Canal de L'Escaut, to which they had withdrawn earlier. Having taken possession of

the east bank of the canal by the same date, No.V Corps ordered an immediate advance by large patrols. However, the German forces were not totally defeated, and met the advance with a strong resistance. The ensuing 'stalemate' was finally broken three days later, with the Battle of Cambrai.

Although they had been heavily engaged in their usual tasks over the preceding days, No.15 Squadron's role in the battle started at 05.45 hours on the morning of 8th October, when Lt D McD. Northcombe and Lt E Baskcomb Harrison took-off for an artillery ranging patrol. They were successful in their allotted task and located fourteen enemy batteries, all of which were operational and active. Working in close co-operation with the Allied artillery, these two officers enabled excellent results to be achieved. They also recorded the locations of a number of other enemy batteries, together with the position of a column of German infantry, estimated to be over 100 strong.

Lieutenant R A P Johnson and Lt C E Whittaker, who had taken-off five minutes after Northcombe and Harrison, undertook a similar mission. They located a further twenty enemy artillery batteries, but unfortunately their success was only moderate due to lack of response, by the artillery, to the information they imparted.

For another No.15 Squadron crew, their mission on this same day was very nearly their final one. Captain Alfred Richard Cross, a red headed Canadian, was the Unit's expert on low-level photography. He had taken-off at 07.30 hours with his observer Lt S Alder, for a counter-attack patrol. Whilst flying in the vicinity of Walincourt, they spotted a large contingent of German troops ensconced in several orchards. Machine gun and small arms fire was soon exchanged, with Cross employing both his front and rear guns as he flew over the orchards at an altitude of 500'. As the R.E.8. crossed the British lines, the aircraft was subjected to a direct shell hit which destroyed several interplane structs. Cross had no option but to get the aircraft down, and the stricken machine made a forced-landing in (prophetically) an orchard at Ardissart Farm. Although the pilot was slightly injured, both he and the observer scrambled out of the wreckage and ran for the cover of a nearby farm building in order to take shelter from an artillery bombardment then in progress. They escaped further injury when, having reached the safety of the cellar, the building was struck by a direct hit and the structure collapsed killing six other people who were also taking shelter. Within two hours of their unscheduled landing, Cross and Alder were back with the Squadron, imparting valuable information collected during their patrol.

Further valuable information was recorded during a reconnaissance patrol undertaken by Capt James Craig, on the morning of 11th October. Flying with Lt G Wilton as his observer, Craig was instructed to reconnoiter the River Selle. The members of No.15 Squadron had always excelled in their ordained tasks and, even in these late stages of the war, they carried them out with fervency and zeal. Craig had previously shown his determination to get the best results, and this patrol was to be no different. Not only were Craig and Wilton able to confirm the content of an earlier report made by Capt C C Snow relating to the destruction of all bridges over the River Selle along the No.V Corps front, but they actually descended to very low-level in order to examine the river and its banks!! From their inspection, they ascertained the approximate width of the river, the depth of the water, and the height above water level of the banks on either side. Their report indicated that, although troops could wade across in certain places, vehicular and horse-drawn transport would require the use of bridges.

Although the end of the war may have been in sight for some, these reconnaissance patrols were still not without their elements of danger, and the British aircrews still had to be vigilant. The German Air Service was determined that the Allied forces would not have things all their own way, even at this late stage in the conflict. Whilst undertaking missions for their own retreating forces, German fighter and reconnaissance aircraft, along the No.V Corps front, would often secrete themselves in any available cloud cover, from which they would emerge and strike at any unsuspecting target; as 2nd/Lt D Mumford and 2nd/Lt E E Richardson found out. Engaged on a counter attack patrol, on Tuesday, 22nd October, 2nd/Lt Mumford was piloting his *'Harry Tate'* over Ovillers at approximately 450 feet, when it was attacked by a Hannover CL.III. This particular type of German aircraft, due to its compact design, gave the impression of being a single-seater machine. However, many an allied pilot found to his horror, when attacking from astern, that this was a two-seater with a rearward firing Parabellum machine gun! It was not however the free-firing rear gun that Mumford and Richardson had to worry about, but the fixed forward firing Spandau machine gun. Richardson immediately returned fire from the *'Harry Tate's'* .303 inch Lewis guns, managing to discharge approximately 100 rounds of ammunition before the enemy machine broke-off the engagement. Second Lieutenant Mumford gave chase, albeit at very low-level, opening fire with his Vickers gun, but he was forced to retire from the fight when enemy ground fire, towards the rear of the German lines, threatened his aircraft.

Three days later six Fokker D.VII's attempted to shoot down an R.E.8. piloted by Lt G Griffin, who was engaged on an artillery patrol with 2nd/Lt H Ecob as his observer. The British aircraft was flying over the region of Englefontaine, at an altitude of between 6,000 – 7,000 feet, when the enemy machines attacked from above in a steep dive. As Griffith banked the R.E.8. away from the attack, Ecob swung his machine guns to face the enemy and opened fire. Having failed in their endeavors the first two D.VII's pulled away, but when one of them went down in a steep spiraling glide, the main formation turned and headed east.

Second Lieutenant E E Richardson, who had been in action against the German Hannover CL.III, two seater, on 22nd October, found himself in similar circumstances on 4th November. On this occasion he was flying as observer to Lieutenant J C Holmes, engaged on a contact patrol, when they spotted an enemy two-seater. Taking the offensive, they managed to drive the German aircraft down and saw it crash through some trees. The fate of its crew was not recorded.

Two days before the end of hostilities, Lt J C Deremo, a Canadian pilot, and Lt W J Wreford, his observer, were to be on the receiving end of aggression. However, fortunately for them both men evaded injury or death. In a scenario similar to that experienced by Cross and Alder only a month earlier, Lieutenants' Deremo and Wreford had to make a forced-landing behind enemy lines having been engaged by German machine-gun fire.

Detailed to carry out a counter attack patrol on the morning of 9th November, Deremo and Wreford proved successful in their task, and gleaned much valuable information regarding enemy activities and movements. As they were flying at less than 500', they could not help but notice the many contingents of enemy troops below them; targets too tempting to ignore. However, as they flew low, strafing and causing many casualties and mayhem amongst those on the ground, they themselves became victims. A few minutes after 08.00 hours, a hail of bullets rose up into the air as ten enemy machine-gun emplacements opened fire on the British aircraft, riddling it with bullet holes and puncturing both the oil and fuel tanks. To add to their problems, the rudder control was completely shot away. Lieutenant Deremo had no choice but to put the almost uncontrollable machine down on the ground as quickly as possible. A forced-landing was made near the village of Berelles, to the east of the Foret de Mormal, close to a wood in which the crew could take shelter. Having extricated themselves from the wreck-

age of their aircraft, Deremo and Wreford gathered together the information they had acquired and buried it in the thicket where they had taken refuge. However, whilst they lay in hiding, they were able to record further information regarding enemy rearguard procedures. Concluding their 'clandestine' activities at dusk, the British airmen made their way back to Berelles, where they were assisted by French civilians. Having refused the offer of exchanging their uniforms for plain clothes, Deremos and Wreford worked their way westward and without further incident, rejoined the Squadron.

Unfortunately, Lt J C Holmes and Lt E E Richardson, who six days earlier had caused the demise of a German two-seater aircraft, met their own end on 10th November. They were shot down whilst on a contact patrol, and both killed less than twenty-four hours before the end of the hostilities.

At 11.00 hours, on the eleventh day, of the eleventh month, in the year 1918, the war came to an end. Germany had signed the Armistice at 05.00 hours that Monday morning, in a railway carriage in the forest of Compiegne. The final surrender being accepted on behalf of the Allies, by Marshal Foch.

Some of No.15 Squadron's pilots and observers were in the air when the German capitulation was announced, and witnessed the change. To their amazement, Fifteen's flyers saw groups of unarmed German soldiers standing about, whilst civilians emerged from their various locations of concealment and waved at the overflying aircraft.

Having contracted a severe bout of Spanish influenza, Major Stammers who had commanded No.15 Squadron since the second week of January 1918, relinquished his position on 27th November, in favor of Major C C Durston.

No.15 Squadron remained under the control of No.V Corps, and as such did not move forward to the German frontier, as did some squadrons under the terms of the armistice. Instead, on 3rd December, the Squadron moved back to Vignacourt situated to the north of Amiens, and some four miles north-west of Bertangles.

The end of hostilities saw a reduction in the fighting strength of the Royal Air Force in general, and No.15 Squadron in particular. By the middle of February 1919, the unit was reduced to a cadre, having relinquished its *'Harry Tates'* a month earlier. It was in this form, and around the same time, that No.15 returned to England and took up residence at Fowlmere, near Royston, in Cambridgeshire.

For eleven months, the Squadron existed under the command of No.3 (Training) Group but on 31st December 1919, the 'axe' fell and No.15 Squadron, Royal Air Force was disbanded.

During its short existence, men from all over the British Isles were posted to the Squadron. The camaraderie which built up and honed the Squadron was also enjoyed by men from the United States of America, Canada, South Africa and Australia. Together they created a Squadron whose military record was second to none, and one of which they could be justly proud of being a part. In all their tasks their aim was sure and, like the mythical phoenix rising from the ashes, they ensured that No.15 Squadron would rise again.

Major H V Stammers   **The Texan Ranger**   **The Calm Before the Storm**

A group photograph of No.15 Squadron aircrew with Lt H D White, pilot, from South Africa (seated in the foreground); Captain Roberts, pilot (wearing shorts); ? Ludlow, observer (3rd from left). Also known to be in the photograph are Fred Rees, Observer; King; Price; Fowler and Hyde. *Courtesy of and via Ray Vann.*

Left: Major H V Stammers, DFC, Officer Commanding No.15 Squadron, Royal Air Force, from 14th January 1918. The original photograph is signed and dated 7th July 1919. *Courtesy of and via Flt/Lt T. N. Harris, XV (R) Squadron.* Right: Captain James Burchett, M.C., was posted to No.15 Squadron, as a Second Lieutenant, on 12th December 1917. *Courtesy of and via Ray Vann.*

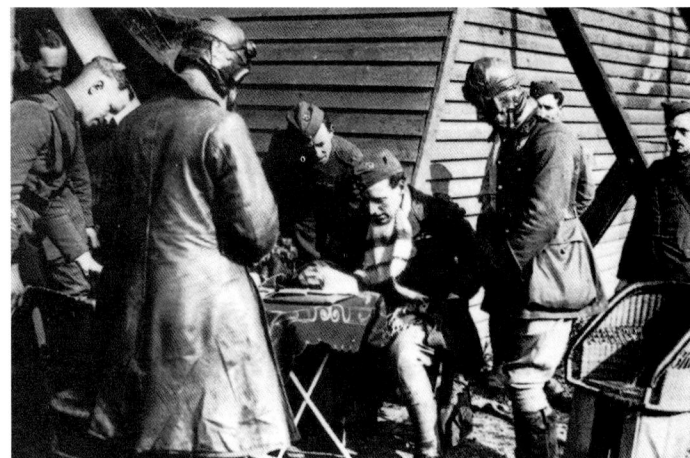

Major Stammers (seated), officer commanding No.15 Squadron, peruses a report, filed by a crew following a mission during the German offensive on 25th March 1918. *Via Chaz Bowyer*

Aerial Views

Not all the aerial photographs taken by No.15 Squadron depict battlefield scenes. These two pictures, taken on an unspecified date, show adjacent areas of Calais harbor, as seen by the pilot and observer. *Courtesy of and via Flt/Lt T. N. Harris, XV (R) Squadron.*

Vert Galand airfield, France, from which No.15 Squadron operated between April and September 1918. At least twelve R.E.8s can be seen in the foreground. *Via Chaz Bowyer.*

Aerial view of the officers' quarters at Vert Galand airfield, France. *Via Chaz Bowyer.*

"Punjab 10. Kalabagh"

Two photographs showing R.E.8., serial A3598, which carried the name 'Punjab 10. Kalabagh', of No.15 Squadron. The aircraft was struck-off-charge on 11th April 1918. *Via Chaz Bowyer*

Discussing the Mission

A fine study of Lt H A Chippendale, pilot, and 2nd/Lt G L Hobbs, observer, discussing their morning reconnaissance mission over the battlefield area, with Captain J B Solomon. Note the machine-guns, complete with ammunition drums. *Via Chaz Bowyer*

Flyer From the Scottish Highlands

The cottage at Culburnie, near Beauly, Invernesshire, where Robert Fraser was born, at 08.00 hours, on 19th November 1898. It was here that young Robert lived with his mother and father, two brothers and three sisters. *Courtesy of Moira Michie*

The Old School House at Culburnie, where Robert Fraser undertook his primary education, from the age of three and a half years old. He later attended the Inverness Royal Academy, before completing his studies at Aberdeen University. *Author's Collection*

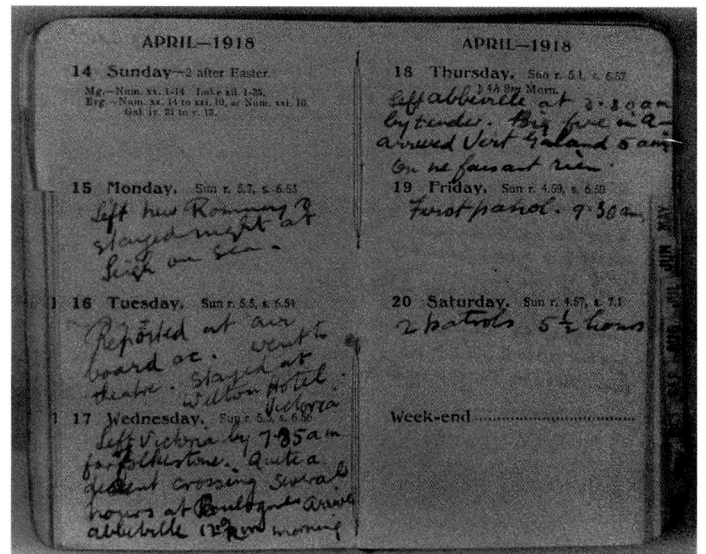

The diary of Robert Fraser, in which he records his journey across the Channel, and his arrival at Vert Galand on 18th April 1918. *Author's Collection*

Left: Second Lieutenant Robert Alexander Fraser, RFC (left), with his eldest brother John. *Courtesy of Moira Michie*.

Airmen and Aircraft

'A' Flight, No.15 Squadron, Royal Air Force, May 1918. Standing, left to right are: Collins; 2nd/Lt E Richardson; Bradley; Fowler; Hyde; Ludlow; Lt H White; Lt J Holmes; Capt Roberts. Sitting, left to right are: Foster; Moorhouse; Capt J Stodart; Lt F B Rees; Capt Stephenson. *Courtesy of and via Don Neate*

Four members of the groundcrew, pose for a photograph with their charge, R.E.8., serial B6521, coded 10. The officer wearing the flying clothing is thought to be Captain Gilbert Stephenson. *Via Chaz Bowyer*

R.E.8., serial B2260, coded 12, photographed at Vert Galand airfield, France, on 1st May 1918, with left to right: Air Mechanic I Jackson; Air Mechanic I Clunie; Unknown; Corporal Budd; Captain Hill and Captain H B Pett. *Via Chaz Bowyer*

Members of No.15 Squadron's groundcrew gather for an informal photograph in and around R.E.8., serial C2464, coded 14, possibly at Vert Galand, France. *Via Chaz Bowyer*

Loss of Two Flyers

R.E.8., serial B6661, coded 8, at Vert Galand airfield, on 2nd May 1918, with left to right: Air Mechanic I Skinner; Air Mechanic I Grey; Lt Leslie Derrick and Lieutenant Jame Schofield. Lieutenant Leslie James Derrick was killed in action the day after the photograph was taken. *Via Chaz Bowyer*

Right: The adapted propeller forming a wooden cross, with engraved center-plate, marks the graves of Lt L J Derrick, observer, and Lt H J Brown, pilot, who were killed in action, on 3rd May 1918, whilst carrying out army cooperation duties. Their aircraft, an R.E.8., serial C2361, was shot down in flames after combat with three enemy aircraft over Buire, France. *Courtesy of and via Flt/Lt T. N. Harris, XV (R) Squadron*

"Whoops !"

Another No.15 Squadron R.E.8. which attempted to 'fly' through a timber hut was (possibly) B7716, coded 15. Members of the groundcrew inspect the damage, and ponder how to remove the aircraft from its resting place. *Via Chaz Bowyer*

Lieutenant Felix Dymont, from South Africa, and his observer, Lt 'Tippy' Griffith, had a brush with death when their R.E.8. touched the cab of a truck during take-off from Vert Galand, and went through the timber wall of the Power Target Hut. Both crew members emerged from the wreckage without a scratch between them. The original photograph carries the type-written comment "When taking-off his under-carriage touched telephone wires". *Via Chaz Bowyer*

Crash Landing 1. An R.E.8, thought to be carrying the serial, 2403, coded 16, photographed after coming the grief on an unidentified airfield. *Via Chaz Bowyer*

Crash Landing 2. An R.E.8, with the readable tailfin serial, C?507, coded 9, of 'B' Flight, No.15 Squadron is minus its undercarriage and rudder following a landing accident. Note the two bombs stuck nose down into the ground in the center of the picture. *Via Chas Bowyer*

Fall of the Scottish Flyer

The original resting places of 2nd/Lt Robert Fraser (left) marked with a sawn-down propeller, and Lt Solomon Fine (right) marked with a Star of David. The center plates on both read:

In Affectionate Memory
of
A Very Gallant Airman
(2nd/Lt R.A. Fraser) (Lt S. Fine)
R.A.F
Killed in Action
18.5.18

Robert Fraser and Solomon Fine now rest together at Contay British Cemetery, France, Plot VIII, Row A, Graves 20 and 21 respectively. *Courtesy of Moira Michie*

The Cross of Remembrance, at Kiltarlity, nearly Beauly, Invernesshire, Scotland, which records the loss members of the community in two world wars. *Author's Collection*

Ready for Action

Right: Captain James Anderson Craig suitably attired, in flying helmet, goggles and fur gauntlets, as though preparing for a flight in his No.15 Squadron R.E.8. This officer never took to the skies unless accompanied by his black woolen knitted mascot doll which, in this photograph, is seen sitting on the engine nacelle. *Courtesy of Jean Campden*

Lieutenant P E Foot, observer, in full flying kit, poses with maps in hand, by the cockpit of an R.E.8., coded 8. *Via Chaz Bowyer*

Lieutenant P E Foot, in a more formal pose in the turret of an R.E.8., coded 8. The pilot is thought to be either Lt Gerald Fonseca, or Lt Landry. *Via Chaz Bowyer*

R.E.8., serial B6661, coded 8, of "A" Flight, No.15 Squadron, photographed at Vert Galand aerodrome, France, on 6th July 1918. Standing by the RAF roundel are Capt Gilbert Matthew Stephenson, pilot, and Lt Fred Rees, observer. The sergeant in the background remains unidentified. *Courtesy of and via Flt/Lt T. N. Harris, XV (R) Squadron*

At Rest

Captain Gilbert Matthew Stephenson, pilot, takes time out to relax and enjoy his pipe. Note the 'face' chalked on the nose of the aircraft. *Courtesy of and via Ray Vann*

Captain James Craig (left) and his observer, Capt Hill pose by a railway wagon converted for Squadron use. *Courtesy of and via Jean Campden*

'Action Shot'

An in-flight photograph, of 2nd/Lt R A Johnson, taken from the rear cockpit during a reconnaissance flight over France. *Via Chaz Bowyer*

Some of Fifteen's Flyers – 1918

A Lewis ammunition drum is passed up to Atkins (in the turret position) by Lt Crowther. The pilot, looking back over his shoulder, is thought to be Landry. *Via Chaz Bowyer*

Left: A poor quality photograph of 2nd/Lt R Capel-Cure, observer, in casual mood. *Via Chaz Bowyer* Right: A pilot, thought to be Lt Crowther, of No.15 Squadron, poses for a photograph at an unknown location. *Courtesy of and via Don Neate.*

Atkins, observer (left), Lt Gerald Fonseca, pilot (center) and Lt P Foot, observer, pose for a photograph by an unidentified R.E.8. Whilst the aircraft, map and flying clothing of the observers indicates an impending flight, the pilot is not suitably dressed for the occasion. *Courtesy of and via Don Neate*

A No.15 Squadron R.E.8. provides a backdrop for some of its aircrew and groundcrew. In the picture are Atkins, observer (center), Crowther, pilot (center right), and Levy. *Courtesy of and via Flt/Lt T. N. Harris, XV (R) Squadron.*

Left: Lieutenant Izzard, Observer (left) and Lt J Carlos Deremos, pilot (right). The latter was shot down by anti-aircraft fire whilst piloting R.E.8, C2796, on 9th November 1918. Lt Deremos and Lt Wreford, the former's observer during that flight, both survived. *Courtesy of and via Ray Vann*

Group Photos

Eighteen members of No.15 Squadron pose for a group photograph, utilising an R.E.8. parked in a canvas hangar as a backdrop. Standing at the back of the group are Wilton, Obs (2nd left); Barret, Obs (5th left) and Capt Richard Cross, pilot (6th left. Seated in the middle row are Jackson, pil (2nd left); Maj. Stammers, C.O. (3rd left) and Capt Dodds (4th left). *Courtesy of Don Neate*

'A' Flight, No.15 Squadron, Royal Air Force, August 1918. Standing left to right: Baddley, Obs; Harrison, Obs: 2/Lt D Mumford, Pil: Ruxton, Obs; Northcombe, Pil; Nixon, G/Staff. Seated left to right: Capt G Stephenson, Pil; Allen, I.O.; Lt H White, Pil; Capt Roberts, Pil (Flt Cmdr); Lt J Holmes, Pil; 2/Lt E Richards, Obs (K.I.A. 9.11.18); Ludlow, Obs. *Courtesy of and via Ray Vann*

'B' Flight, No.15 Squadron, Royal Air Force, France 1918. Standing left to right: G Wilton, Obs; Lt G Izzard, Obs; Lt J Deremos, Pil; Lt W Wreford, Obs; Lt G Fonseca, Pil; Evans, Obs; Barrett, Obs; Lt G Griffin, Pil; Smith, Trns; Allen, I.O.; Capt N Dobbs, R.G.A. Seated left to right: Jackson, Obs; Lt V Kilroe, Pil; Capt J Craig, Pil (Flt Cmdr); Capt A Hill, Obs. Major H V Stammers, who took command of the Squadron on 14th January, 1918, is seated on the extreme right. *Courtesy of and via Ray Vann*

'C' Flight, No.15 Squadron, Royal Air Force, 1918. Standing left to right: Unk; Unk; Levy, Obs; Maclellan, Obs; Oliver, Obs; Whittaker, Obs; Lt S Alder, Obs. Seated left to right: 2/Lt A Manders, Pil (K.I.A.); Lomberg, Pil; Pizey, Pil; Dymont, Pil; Capt C Snow, Pil (W.I.A.); Jackson, Pil; Welch, Pil. *Courtesy of and via Ray Vann*

Survivors

Some of 'Fifteen's flyers', photographed near Bapaume, France, during October, 1918. Left to right: Lt Gerald Fonseca, Pil; G Wilton, Obs; Lt G Griffin, Pil; 'Scotty' Knox, Obs; Evans, Obs; Lt G Izzard, Obs; Unknown, ? ; Allen, I.O. Note the tartan pants worn by 'Scotty' Knox, and the fur gauntlets worn by Evans. *Courtesy of and via Don Neate*

Right: Captain Richard Alfred Cross, DFC, in formal pose. This officer was slightly injured when his aircraft was shot down on 8th October 1918. Both he and his observer, Lt S Adler, survived the crash and returned to the Squadron. *Courtesy of and via Flt/Lt T. N. Harris, XV (R) Squadron*

The officers and other ranks of No.15 Squadron, RAF, possibly photographed at Selvigny, France, on 1st December 1918. Squadron Leader C C Durstan, who took command of the Squadron on 27th November 1918, is seated at the center of the second row. *Courtesy of and via Flt/Lt T. N. Harris, XV (R) Squadron.*

5

Aim Sure

On 20th March 1924, No.15 rose again when the Squadron was re-formed at Martlesham Heath, Suffolk, under the command of Squadron Leader P C Sherren, M.C., along with No.22 Squadron, as part of the Aeroplane and Armament Experimental Establishment (A&AEE).

The two squadrons were reformed to overcome the fears of the British public who were concerned about the run-down of the Royal Air Force following the end of the Great War. Although both squadrons were to be engaged in experimental, armament and performance testing work, their respective Squadron number plates expanded the Government's figures for the strength of the RAF.

During the ten years, two months and eleven days that No. 15 Squadron served with the A&AEE it undertook much experimental work and many trials involving all aspects of bomb ballistics, aerial gunnery, parachute and reconnaissance flares, photography and numerous other forms of research work. Over the aforementioned ten year period the Squadron was engaged in carrying out these various tests, it accumulated over 12,100 flying hours, flew over seventy-five different types of aircraft, constructed by sixteen manufactures. It also saw five changes of commanding officer, culminating with Squadron Leader Robert M Foster, D.F.C., A.F.C.

At the end of May 1934, the two Martlesham Heath squadrons were renamed. No.15 became the Armament Testing Squadron, whilst No.22 was known as the Performance Testing Squadron. The existing personnel of both units being absorbed into the existing establishment.

Utilising new personnel, No.15 Squadron was immediately re-formed at Abingdon, Oxfordshire on 1st June. Abingdon was a relatively new R.A.F. Station, having been constructed two years previously. It was situated to the north-west of the town from which it took its name, and was only six miles from the university city of Oxford.

No.15's new role was to be that of a day bomber squadron, flying Hawker Hart Mk.I. aircraft, under the command of Squadron Leader Thomas Elmhirst, A.F.C.

The personnel of the squadron had changed, but the pride and reputation of the Squadron had not. Whilst No.15 was being equipped with its aircraft, the new commanding officer made a decision that was to have a lasting effect and would ensure that the Squadron was instantly recognisable to all. He decided to apply the Squadron's number, in Roman numerals, to the fuselage of each aircraft. The tradition has lasted to the present day, albeit in a slightly different form. Following the Second World War, the numerals were reduced in size and relocated to the tailfin of the Squadron's aircraft. However, during times of crisis and hostility the numerals are always painted out.

The public got their first glimpse of the new markings when an aircraft, which Thomas Elmhirst had flown from Abingdon on 18th June, was exhibited at a display at RAF Hendon. No.XV Squadron would be visually known as such hereafter.

The commanding officer also gave thought to the fact the Squadron should have its own badge and, with the help of a friend in the office of the Chester Herald, set about designing one. The badge, which was eventually approved by King George V, depicted a Hart (the name of the aircraft flown by the Squadron) with Heraldic wings encircled by a garter which carried the legend "*XV SQUADRON, ROYAL AIR FORCE*". A Royal (or Tudor) Crown surmounted the design, whilst a scroll beneath the badge bore the Squadron motto "*AIM SURE*".

Throughout 1934 and into early 1935, the Squadron undertook many training and tactical exercises, a number of which entailed 'attacking' mock targets, formation bombing and formation flying for fighter interception tactics. Also around this time, and adding to their already heavy workload, rehearsals commenced for the forthcoming celebrations of King George V Jubilee Review of the Royal Air Force.

The summer of 1935 was a very busy period for the Squadron. However, its capabilities and competence had not gone unnoticed, and it was considered to be the number one light bomber squadron. Maybe because of this fact, No.XV was chosen to lead the light bomber Wing at the annual RAF Display, at Hendon, at the end of June.

Devising a dive-bombing technique for the purpose of the display, Squadron Leader Elmhirst thrilled the public by bringing his formation of nine Hawker Harts in over the airfield at 5,000' and having the aircraft pull up and wing-over into a dive, on a simulated target in front of the enthralled crowd. These tactics and set-pieces of 'exhibition' flying may well have enraptured the multitude watching, but across the channel in Europe the evolvement of another war was beginning. The flying maneuvers and demonstrations seen by the public at the air display would, in the not too distant future, be carried out for real.

Two days after his success at Hendon, Thomas Elmhirst was promoted to the rank of Wing Commander, whilst his senior Flight Commander, C H Cahill, was promoted to Squadron Leader.

The Jubilee Review of the Royal Air Force took place at RAF Mildenhall on 6th July. King George V accompanied by his wife Queen Mary, and two sons, Edward, the Prince of Wales, and George, the Duke of York, traveled to the RAF Station where thirty-eight Squadrons had drawn up a total of 356 aircraft for inspection. After the review the Royal party were driven by car to RAF Duxford, where the King watched a flypast consisting of 155 of the aforementioned aircraft from Mildenhall. Wing Commander Elmhirst, and No.XV Squadron, had the honor of (again) leading the light bomber Wing.

The Jubilee Review at Mildenhall was the only occasion where the King and his two sons, both future Kings (Edward VIII and George VI respectively), were all seen together dressed in the uniform of the Royal Air Force.

Wing Commander Elmhirst relinquished command of the Squadron on 12th August, on being appointed Station Commander at RAF Abingdon. His successor was the former senior Flight Commander, Squadron Leader Cahill, A.F.C. Charles Howard Cahill was a west country man by birth, having made his entry into this world in Bristol towards the end of the 1890s. Following his education in both Paris and Belfast, he joined the Royal Army Medical Corps. Private Cahill served with this Regiment during the first two years of the Great War, but in 1916 transferred to the Royal Field Artillery. His service as a gunner with the R.F.A. was brought to an abrupt end in 1917, when he applied for a transfer to the Royal Flying Corps. His requested being granted, Charles Cahill assumed the rank of Cadet, and began flying training. By the end of the First World War he had been commissioned to the rank of Lieutenant. He remained in the fledgling service after the war, and saw further service in both Iraq and the Mediterranean.

The unsung heroes of any squadron were the those who served as members of groundcrew, particularly during times of conflict when their task of keeping the aircraft airworthy was never ending. Many of them mistakenly saw themselves (and still do) as unimportant 'cogs' in a large wheel. Craving for more excitement, or wishing to be seen to 'do their bit', many remustered to aircrew. One such man was Thomas Grey, a Leading Aircraftman, who served as an engine fitter with No.XV between June 1933 and 27th April 1935. For nearly two years he served the Squadron without recognition, although whenever the opportunity arose he would volunteer for flying duties as an air gunner. However, following further training and conversion courses, Thomas Grey undertook training for aircrew. On completion of the latter, he qualified as an air observer/gunner and was posted to No.12 Squadron where he served as a sergeant. On his first bombing mission during World War Two, Thomas Grey was killed in an action which led to him, and the pilot of his aircraft, being posthumously awarded the Victoria Cross.

Another name which would be recorded in the annals of No.XV Squadron, was that of a young pilot officer who was posted to the unit during October 1935 and who, later in life, would reach the highest pinnacle of command in the Royal Air Force. Samuel Charles Elworthy, a New Zealander by birth, was educated in England, at Marlborough College near Swindon, and later at Trinity College, Cambridge, where he read Law. Unlike many who passed through Cambridge, Elworthy did not join the University Air Squadron, instead he took a commission with the Auxiliary Air Force and flew Harker Harts with No.600 Squadron. However, during 1935 the young 'part-time' Air Force officer gave up thoughts of a legal career and accepted a permanent commission in the Royal Air Force. Pilot Officer Elworthy's 'tour' with No.XV Squadron lasted two years, during which time he received promotion to Flying Officer, became "A" Flight Commander, was appointed Squadron Adjutant and flew both the Hawker Hart and the Hawker Hind. A couple of years later, on 22nd November 1937, Sam Elworthy was appointed Aide-de-Camp (Personal Assistant) to one of the Squadron's former commanding officers, Sir Edgar Ludlow-Hewitt, who was now Commander-in-Chief, Bomber Command. Following a distinguished career in the R.A.F, Samuel Charles Elworthy retired from the Service having attained the title Marshal of the Royal Air Force The Lord Elworthy, K.G.

Squadron Leader Cahill's appointment as C.O. lasted approximately four months but his successor, S/L F G Robinson who took over on 1st January 1936, was in office for only five weeks. Robinson was in turn succeeded by S/L Cyril Adams who took over on 5th February 1936, and held the post for just over two years.

Three weeks prior to S/L Adams taking command, the Air Ministry policy for unit badges was announced. No.XV Squadron, being to the fore, had had its own badge approved in 1934, and it was on this badge that all others were to be designed. No.XV took the opportunity to update their crest, by replacing the Hart with the image of a Hind; the name of the aircraft with which they were at that time re-equipping. The amendment was approved, along with badges for other Squadrons, by King Edward VIII in June. As with the earlier badge, the revised version was applied to the tailfins of the Squadron's aircraft.

Furthermore, the second half of the 1930's saw a rapid expansion of the Royal Air Force. No.XV played a part in that expansion on 17th February 1936, when "C" Flight was disbanded and reformed as No.98 (Bomber) Squadron.

On 21st August 1936, a young Australian pilot was commissioned into the R.A.F, having transferred from the Royal Australian Air Force. Flying Officer Hughie Edwards, who was born in Freemantle, Australia of Welsh immigrant parents, was posted to No.XV Squadron following his transfer. He flew Hawker Hind bombers with the unit before being sent to No.90 Squadron at Bicester, as squadron adjutant, in March 1937. On 4th July 1941, as commander of No.105 Squadron, Hughie Edwards led a daylight bombing attack against Bremen. Flying his twin-engined Bristol Blenheim bomber through intense flak and balloon cables, Edwards pressed home the attack. For the gallantry and determination shown during this action, he was awarded the Victoria Cross.

With the political situation in Europe giving cause for concern, the expansion and formation of new Squadrons continued and, on 18th January 1937, five officers and twenty-six airmen from No.XV Squadron, formed the nucleus of No.52 (Bomber) Squadron. The new unit remained at Abingdon for a short period before moving to RAF Upwood, Huntingdonshire on 1st March.

Squadron Leader Adams's tenure as Commanding Officer came to an end at the beginning of May 1938, when he was posted to Bomber Command Headquarters. The first duty of S/L J G Llewelyn, who was appointed to command the Squadron on 2nd May, was to train the pilots for their role of formation flying, which together with dive bombing techniques, were to be displayed at the Empire Air Days. As the displays were to be held on 26th and 28th of that same month, he had a little over three weeks in which to accomplish his task.

As the decade edged towards its closing years, the aerial exercises continued but the training was now in earnest. Although the crowds still marveled at the aerobatics and maneuvers performed at the annual Empire Air Day, the displays had a more ominous overtone.

On the first day of June 1938, No.XV saw the 'birth' of yet another daughter squadron, when No.106 Squadron was reformed from "A" Flight. Initially the new squadron operated alongside its mother unit, under the command of F/O S Hook. However, on 1st September 1938, it moved to RAF Thornaby in Yorkshire.

For No.XV Squadron, the expansion period also meant converting to a new type of aircraft, a single engined monoplane, with which it would initially go to war.

Having been born in war, No.XV Squadron rose quickly to the challenge placed upon it in time of crisis, and it emerged from the First World War with many accolades to its credit. It also rose quickly to the challenge of the work it undertook, following its peacetime resurrection, as part of the A&AEE. Although the Squadron received no official recognition for the work it carried out during this period, it could be justly proud of the part it played in the technical developments and advancements achieved. No.15/XV Squadron, R.F.C/R.A.F, had shown itself in both spheres; those of war and peace and, with the clouds of another war looming on the horizon, it was obvious the cycle had turned full circle. No.XV Squadron was to return to the task for which it had first been created. The foe would be the same enemy fought in the last war, but the men and machines would be of a different generation. However, the loyalty, steadfastness and courage that had gone before, would remain the same.

Some of the aircraft used by the Squadron at A&AEE.

Vickers Venture, serial J7282, was one of a batch of six aircraft. This particular machine served twice with No.15 Squadron, first during 1925 and again in 1927, following modifications. *Courtesy of and via Flt/Lt T. N. Harris, XV (R) Squadron.*

Westland Westbury, serial J7765, was one of a production batch of two aircraft. It was constructed with wooden wings, whereas its stablemate J7766, which also flew with No.15 Squadron, had composite wings. J7765 first flew in September 1926, at Andover, Hampshire, and joined the A&AEE in October the following year. *Courtesy of and via Flt/Lt T. N. Harris, XV (R) Squadron.*

Hawker Hoopoe, serial N237, was built to Air Ministry Specification 21/26, for trials with wheel/float undercarriage, during 1928. It was powered, in succession, by Mercury, Jaguar and Panther engines. *Courtesy of and via Flt/Lt T. N. Harris, XV (R) Squadron.*

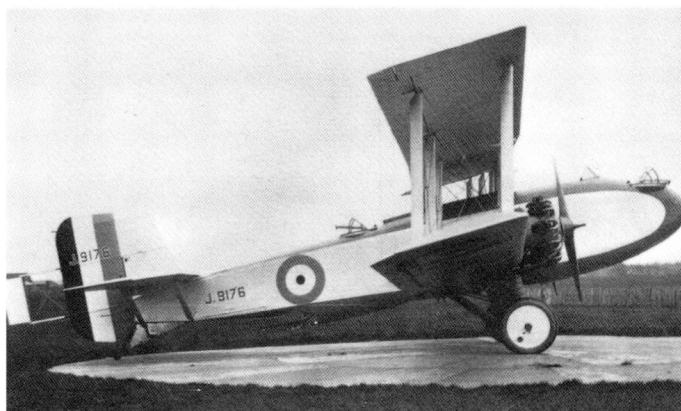

Boulton and Paul P.29A. Sidestrand Mk.II, serial J9176, was one of a batch of six aircraft, powered by two Jupiter VI engines. It served with No.15 Squadron during 1928. *Courtesy of and via Flt/Lt T. N. Harris, XV (R) Squadron.*

A Fairey Mk.IIIF, serial 5184, reputed to have served with No.15 Squadron, at Martlesham Heath. *Courtesy of and via Flt/Lt T. N. Harris, XV (R) Squadron.*

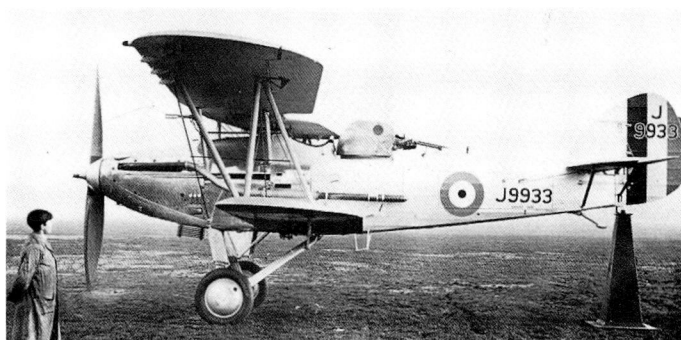

Hawker Hart, serial J9933, was one of a batch of fifteen aircraft. It joined the A&AEE during 1930, and was used for various gunnery tests and trials until 1936. Two years later, during 1938, it was used as a ground instructional frame. *Courtesy of and via Flt/Lt T. N. Harris, XV (R) Squadron.*

Hawker Hind, K3900

Left: Hawker Hart Mk.I., serial K3900, was built by Armstrong-Whitworth, and delivered between March and April 1934. The aircraft, which was flown by Thomas Elmhirst, Officer Commanding, No.15 Squadron, also saw service with the Royal Aircraft Establishment, the Central Flying School, No.503 Squadron, No.3 Elementary Reserve Flying Training School and No.2 Flying Training School. It was struck-off-charge on 1st July 1940. The aircraft, photographed at RAF Abingdon on 1st June 1934, is seen here having been adorned with the roman numerals which were instigated by Thomas Elmhirst. Note also the Squadron badge applied near the forward edge of the tailfin. Also visible is the Squadron Commander's pennant on the fuselage, aft of the rear cockpit. The original photograph carries the inscription, *"My aircraft. Thomas Elmhirst. 1934-36". Courtesy of and via Flt/Lt T. N. Harris, XV (R) Squadron.* Right: Squadron Leader Thomas Elmhirst, pilot, and Corporal Haydon, observer, fly above the clouds in Hawker Hart, serial K3900. *Courtesy of and via Flt/Lt T. N. Harris, XV (R) Squadron.*

Displaying at Hendon

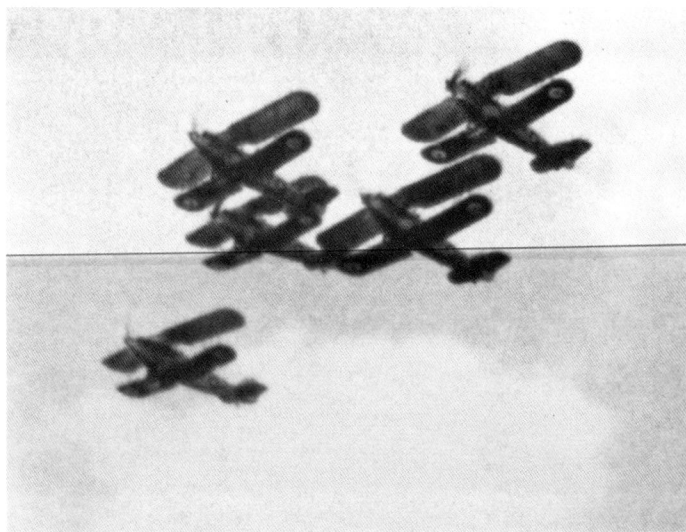

Squadron Leader Thomas Elmhirst leading nine Hawker Harts of No.XV Squadron, at the 16th RAF Hendon display on 29th June 1935. *Courtesy of and via Flt/Lt T. N. Harris, XV (R) Squadron*

Five aircraft, having broken away from the main group, fly past at low level in a tight formation. *Courtesy of and via Flt/Lt T. N. Harris, XV (R) Squadron.*

The Royal Review

A general view of some of the 356 aircraft, lined-up for the King's inspections at RAF Mildenhall. Note the No.XV Squadron aircraft to the left of the picture. *Courtesy of and via Flt/Lt T. N. Harris, XV (R) Squadron.*

On 6th July 1935, 356 aircraft, provided by 38 squadrons of the Royal Air Force, were lined up in ranks, at RAF Mildenhall, for inspection during the Jubilee Review of King George V. The photograph shows eight Hawker Hart aircraft, of No.XV Squadron, at the Suffolk airfield, prior to the formal flypast. Nearest the camera is Hart, K3040, which built by Armstrong-Whitworth. This aircraft also saw service with No.600 Squadron, No.610 Squadron and No.1 Flying Training School. *Courtesy of and via Flt/Lt T. N. Harris, XV (R) Squadron.*

Four Hawker Harts (from a formation of nine aircraft) of No.XV Squadron, form part of the Light Bomber Wing which the Squadron led during the flypast over RAF Duxford, on 6th July 1935. The aircraft in the photograph are K3957 (left), which also served with No.600 Squadron, No.1 Flying Training School, No.1 Ground Gunners School and the Air Transport Auxiliary. It was struck-off-charge on 31st October 1943. K3040 (center), saw service with No.600 Squadron, No.610 Squadron and No.1 Flying Training School. K3903 (right), went to No.603 Squadron before seeing service with No.1 Flying Training School. The camera ship remains unidentified. *Courtesy of and via Flt/Lt T. N. Harris, No.XV (R) Squadron.*

Squadron Leader C H Cahill

Squadron Leader Charles Howard Cahill, DFC, Officer Commanding, No.XV Squadron, 1st July – 12 August 1935. He is seen wearing the Full Dress uniform (as it was then), which was made from Venetian cloth. The tunic consisted of a high collar, on which the wearer's rank was indicated with gold embroidered acorns and oak leafs. The shoulder epaulettes were embroided with gold eagles and crowns. Seven highly polished brass buttons down the front fastened the tunic. *Courtesy of and via Flt/Lt T. N. Harris, XV (R) Squadron.*

Peacetime Accidents

Mishap 1. Members of the groundcrew endeavor to recover Hawker Hind, serial K5440, at RAF Abingdon, on 4th February 1937, following a taxying accident when the aircraft ran into soft ground and tipped up. The officer in the foreground, with his hands in his pockets, is Sam Elworthy, later Marshal of the Royal Air Force The Lord Elworthy, KG, DSO, DFC, AFC. He was not responsible for the mishap. *Author's Collection.*

Mishap 2. The port wing of Hawker Hind, serial K5433, dug into the ground after the pilot took evasive action in order avoid a collision, during a formation landing at RAF Abingdon on 10th May 1938. The body language of the airmen leaning on the tailfin suggests 'it was nothing to do with him'. *Author's Collection.*

The Squadron Crest

Hawker Hind, serial K5414, flying over the Oxfordshire countryside. The aircraft, which was delivered to No.XV Squadron on 27th February 1937, also saw service with No.611 Squadron, before being struck-off-charge on 9th April 1940. *Courtesy of Wing Commander Hugh George, DFC.*

The badge adopted by the Squadron, and approved by His Majesty King George V and the Chester Herald Office, in 1934, was amended in 1936 to carry the Hind's head. The badge depicted above was painted by F/O John Bruce, a Lancaster pilot with the Squadron during 1944/45. *Author's Collection.*

Hughie J Edwards

Hinds of No.XV

A line-up of six Hawker Hind aircraft, photographed at RAF Abingdon during 1937. The second aircraft in the rank is K5460, with Hind, K5421, third in line. The latter aircraft also saw service with No.609 Squadron, the Station Flight Yeadon (Yorkshire), the Station Flight Northolt (West London), No.2 Glider Training Squadron/1 Glider Training School, Glider Instructors School and No.20 (pilots) Advanced Flying Unit. It was struck-off-charge on 26th April 1943. *Author's Collection.*

Hughie Idwal Edwards (later Air Commodore H I Edwards, V.C., K.C.M.G., C.B., D.S.O., O.B.E., D.F.C.), who flew with No.XV Squadron between 1936 and 1937. *Courtesy of and via Flt/Lt T. N. Harris, XV (R) Squadron.*

Four Hawker Hind aircraft, including Hinds K5439 (left) and K5414 (right), being part of a total formation of nine aircraft, take-off from RAF Abingdon, during 1938. *Courtesy of and via Flt/Lt T. N. Harris, XV (R) Squadron.*

Four Hawker Hinds, flying as part of a nine aircraft formation, endure overcast skies as they fly over Oxfordshire, during May 1938. Represented in the photograph from left to right are K543? (extreme rear), K5421, Delivered to No.XV Squadron on 18th June 1937. The aircraft saw service with numerous other units before being struck-off-charge on 26th April 1943. Hind, K5414, possibly the most photographed aircraft of its type on No.XV Squadron. Hind, K5408, which flew with No.52 Squadron, No.82 Squadron and No.211 Squadron before being delivered to No.XV Squadron, on 1st April 1938. The aircraft crashed near Caythorpe, Lincolnshire, after hitting high tension cables, on 4th January 1940, whilst serving with another unit. The Hind was struck-off-charge, on 17th April 1940, having been deemed beyond cost effective repair. *Courtesy of and via Flt/Lt T. N. Harris, XV (R) Squadron.*

Four Hawker Hind aircraft (including the camera ship), of No.XV Squadron, fly over their 'home' airfield at RAF Abingdon (top left of picture), during 1937. *Courtesy of and via Flt/Lt T. N. Harris, XV (R) Squadron.*

6

"Oxford's 'Own'"

The new aircraft with which the Squadron re-equipped, was the Fairey Battle Mk.IV, single-engine, three seat, light bomber. Conversion to this monoplane commenced on 13th June 1938, with the aid of a dual control aircraft. Instruction continued through to the end of July, by which time the Squadron's pilots were deemed 'operationally ready'.

Towards the end of August the Squadron welcomed a young pilot officer from the Antipodes whose name would, in due course, be added to its annals. Leonard 'Len' Trent, from New Zealand, who was to learn his 'trade' and be initiated into war with No.XV, would later distinguish himself and become the recipient of a Victoria Cross.

Due to the international situation caused by the Munich Crisis in September 1938, the Squadron was forced to make yet another change. Much to its chagrin, it was instructed to remove its beloved roman numerals from the fuselage of each aircraft. Instead it was to adopt the policy applied to all squadrons, whereby the aircraft were adorned with Squadron code letters. In place of the Roman numerals, the letters "EF" were applied to one side of the new red and blue R.A.F roundel. The individual aircraft letter was applied to the other side of the new national marking. Apart from the squadron code and aircraft call-sign letter, each machine still carried its own serial stenciled on the fuselage, forward of the tail assembly. Although the roman numerals reappeared when the tension eased, it was only a short time before they were removed for a second time.

A change of year brought with it a change of command when, on 21st March 1939, W/C Wingate took over from S/L Llewelyn. The latter being named Deputy Commanding Officer on Wingate's promotion.

In an endeavor to generate public interest in squadrons of the R.A.F, the Air Ministry introduced a scheme whereby squadrons could be affiliated to major cities and towns of the United Kingdom. This scheme, which was put into practice in April 1939, was also seen as a way of cultivating the airmen's 'esprit de corps'. With the airfield at Abingdon being so close to the 'city of spires', it inevitably followed that No.XV was associated with Oxford. As No.XV did not embody the name of the city in its official title, it was unofficially known as "Oxford's 'own' Squadron".

The summer months of 1939 saw an increase in the number of low and medium-level bombing exercises undertaken, together with a new defensive formation flying tactic. The latter involved an exercise lasting five and a half hours, during which W/C Wingate led a formation of nine aircraft over France. These exercises however, were not only for the benefit of the aircrew. Being the top Light Bomber Squadron of the Royal Air Force required team effort, and that included the input of the groundcrews.

Following a Home Defense exercise during early August, many members of the Squadron were given leave. However, due to a declared state of emergency, this was cut short on 27th August when all personnel were recalled to base. During this period of unrest the Squadron code letters were changed from "EF" to "LS". The latter were to adorn the Unit's aircraft throughout the war, and remained in use until 1951.

With Germany's invasion of Poland, Britain mobilised her fighting forces. At 10.00 hours on the morning of 2nd September 1939, sixteen Fairey Battle bombers of No.XV Squadron deployed to Bethenville in the heart of the champagne country of France, as part of No.1 Group. Ground staff and equipment were taken to Southampton on the south coast of England, where the steam ship *S.S. Isle of Thanet* was waiting to convey them to France. The night before their departure the Squadron personnel were lined-up in the M.O's office for the necessary injections. Pilot Officer Hugh George, who had been accepted for a Short Service Commission in the R.A.F during 1937, watched in anticipation as some of his fellow officers 'dropped' like flies having been inoculated. Although he did not succumb to fainting, the following morning P/O George found his arm had swollen-up like a balloon. He found it very difficult to load the possessions he needed to take to France into his Fairey Battle, and even more difficult to fly the aircraft one-handed across the Channel. However, on his final approach to Bethenville, Hugh George found his problems were not yet over. Some of the British Expeditionary Troops sent out to France went by air. An Armstrong Whitworth Ensign aircraft, which had been used for this purpose, had landed just before the Fairey Battle pilot. Having disembarked its passengers, the Ensign was parked in one corner of the unprepared airfield, which ran down hill. As he landed one handed, feeling extremely fatigued, Hugh George found he was heading straight for the Ensign. His tired mind wondered if he was going to stop in time!. Fortunately he did.

No.1 Group was at this time made up of ten squadrons, all of whom were equipped with Fairey Battle aircraft. Collectively, they were to form the first echelon of the A.A.S.F (Advanced Air Striking Force).

The following day Britain declared war on Germany. This action followed the lack of response by the German Government, to Prime Minister Neville Chamberlain's ultimatum for German forces to withdraw from Poland. As no such undertaking had been received by 11.00 hours on Sunday, 3rd September, Neville Chamberlain announced to the Nation, by radio, that a state of war existed between the two Countries.

The Squadron carried out its first war operation on 6th September when the C.O. led "B" Flight on a reconnaissance mission over Metz, where light flak was encountered.

Five days later, on 11th September, No.XV received a signal from Headquarters, A.A.S.F, instructing the Squadron to move with all speed to a new war station at Conde-Vraux. Although it would still be part of No.1 Group, the Squadron was to report to, and be under the control of, No.71 Wing.

Conde-Vraux was an open airfield, with only a red-roofed house to one side and no other amenities. Therefore, all manner of tasks had to be carried out before the facilities were deemed acceptable. Latrines needed to be dug, field kitchens set-up, aircraft camouflaged and numerous other works completed. Due to the status of his rank, Pilot Officer George and six of his fellow officers, took up residence in the Red House. Almost immediately problems were encountered with the sleeping arrangements, the solution to which nearly cost Hugh George his life. Camp beds were set up around the room with the proviso that the last man turning in for the night would extinguish the light. However, Ronnie Clarke liked to read in bed before going to sleep, thus creating the problem of who was going to switch off the light. The problem was solved when each man acquired a camp set and a candle, which they set up by their respective beds. Unfortunately, the candle situation was inadvertently aggravated by another member of aircrew. Brian Readhead owned a gramophone on which he continually played the one solitary record in his possession. After two weeks of a constantly burning candle, and listening to a tune called *'Sympathy'*, P/O Bassett's patience snapped. One night after Ronnie Clarke had fallen asleep with his candle still alight, and with Brian Readhead's gramophone still playing, Tom Bassett decided to put an end to the situation once and for all. He grabbed a rifle and opened fire, extinguishing the candle and shattering the record with one shot. Instinctively Hugh George, who was trying to sleep, sat bolt upright to see what was going on. As he did so the bullet whizzed over his head, almost parting his hair, and slammed into the wall!

The Royal Air Force was restricted by higher authority, at this early stage of the war, with regard to the type of offensive operations it was allowed to undertake. Strict orders were issued preventing the dropping of bombs over land; only German Naval targets could be attacked when instructed. As its airfield was not far from the Siegfried Line, most of the Squadron's flying activities centered upon reconnaissance and photographic missions. However, between these operations, it also continued to fly practice missions. On one occasion, whilst flying on one of the latter exercises, P/O Tom Bassett, P/O Ronnie Clarke and P/O Hugh George decided to see if it was possible to loop three Fairey Battle bombers in formation. Their first attempt was a near disaster, as they went into the loop too slowly and all spun off the top. Their second try having proved satisfactory, the three intrepid airmen decided to show off their skills over a nearby airfield occupied by a fighter squadron of the French Air Force. Later that same day the calm over Conde-Vraux was shattered by the roar of aircraft engines, as the aforementioned French squadron arrived over the RAF airfield in their single-engined Curtis fighters. Led by their C.O. in his brand new twin-engined Potez 63, they flew over in impeccable inverted formation at zero feet!. Following an impressive aerobatic display the French flyers returned to their base. However, a day or two later the French C.O. reappeared in his Potez 63 and, watched by the RAF, came in for a text book landing. Unfortunately, and to his total embarrassment, the French commander forgot to lower the aircraft's undercarriage. Although the Potez was badly damaged the pilot was unhurt, physically anyway.

The expected immediate aerial attacks by either air arm, following the declaration of war, did not materialise (certainly not as far as No.XV was concerned) and, although this period of quiet and dubious calm became known as the 'phoney war', it did not stop the Squadron's aircraft being shot at!. Flying an early morning reconnaissance patrol at 30,000' over Saarbrucken on 29th September, P/O George was surprised to see one of the aircraft knocked out of formation by the first salvo of a burst of flak from a heavy ack-ack battery.

Between the operational and specified training flights, to relieve the tedium of the 'phoney war' period, some of the pilots would take their Fairey Battles into the sky over the airfield to practice dive bombing tactics. These tactics usually consisted of climbing to an altitude of approximately 6,000', before plunging 'Stuka fashion' towards the ground. On one occasion, it was reported that 'someone' had been seen to fall from an aircraft, without a parachute, as the Battle went into its dive. The report was puzzling as, in the main, these 'exercises' were normally carried out solo. However, a widespread search was initiated involving all available members of the Squadron, but it drew a blank. A few hours later a pair of coveralls were found in a field, where they had landed, having 'escaped' from the air gunner's cockpit. The reported 'body', seen falling from the aircraft, was in fact the coveralls floating down realistically inflated. Unfortunately, earlier in the year, on 5th April, P/O Peter Shennan, a 23 year old pilot from New Zealand, had met his death in similar fashion. He fell from Fairey Battle, K9359, when the aircraft piloted by F/O Paul Chapman manoeuvered for a practice bombing run over their (then) home base at Abingdon. Those who witnessed the incident involving the coveralls, obviously feared history had repeated itself.

Knowing hostilities could flare-up at any moment, and without warning, the Squadron ensured that both the aircraft and the aircrews were always ready for the conflict which could arise at any time. However, their call to action was to be delayed. On 1st December 1939 orders were received by No.XV to return to the United Kingdom, where the Squadron was to re-equip with twin-engined, three seater, Bristol Blenheim Mk.IV bombers.

Arrival of the Fairey Battle.

A Fairey Battle Light Bomber, escorted by Hawker Hind, K5401 (nearest the camera) and Hind, L7177, arrives at Abingdon, during June 1938, on commencement of the Squadron's re-equipment with the type. Apart from a being a much larger aircraft, the new bomber came complete with a retractable undercarriage and an enclosed cockpit. *Courtesy of and via Flt/Lt T. N. Harris, XV (R) Squadron.*

Fairey Battle Mk.I., light bomber, K9226, banks away into the overcast skies. This aircraft also served with No.4 Bombing and Gunnery School, before being struck-off-charge on 13th October 1943. *Courtesy of and via Flt/Lt T. N. Harris, XV (R) Squadron.*

The Young Antipodean

The young Pilot Officer from New Zealand, Len Trent, poses for a photograph leaning against the trailing edge of the wing, of a No.XV Squadron Fairey Battle aircraft. He is wearing white prestige overalls emblazoned with the No.XV Squadron insignia, over his RAF uniform. *Courtesy of and via Flt/Lt T. N. Harris, XV (R) Squadron.*

Battles Over Abingdon

Fairey Battle Mk.II, K9228, EF-C, flies low over the town of Abingdon. The curve of the River Thames, and the tower of 14th century St. Helen's Church, near the right-hand river bank, can clearly be seen under the leading edge of the aircraft's wing. *Courtesy of and via Flt/Lt T. N. Harris, XV (R) Squadron.*

Formation Landing. Three Fairey Battle aircraft, of No.XV Squadron, fly low over the Oxfordshire countryside, as they prepare to touch-down at RAF Abingdon. Nearest the camera is Battle, K9311, EF-O. On completion of its service with No.XV, this aircraft went to the Royal Canadian Air Force where, from 14th October 1940, it served as No.1753. In the background is Battle, K9358, EF-V. This aircraft also saw service with the RCAF, during 1941, as No.2039. The camera ship remains unidentified. *Courtesy of and via Flt/Lt T.N. Harris, XV (R) Squadron.*

Aircraft and Aircrew at Rest

Aircraft at rest. Nine Fairey Battle light bombers, of No.XV Squadron, their daily tasks completed, are dispersed inside a hangar at RAF Abingdon. *Courtesy of W/C Hugh George, DFC.*

Aircrew at rest. Making the most of the sunshine, outside the Officers' Mess, at RAF Abingdon, are F/L Lawrence, F/O Bert Oakley, Jess Oakshott, and Ronnie Clarke. *Courtesy of W/C Hugh George, DFC.*

Paul Chapman (left), 'Red' Eames (center) and Bert Oakley, RAAF (right) relax and chat outside the No.XV Squadron hangar, at RAF Abingdon. Oaklay was killed in action, on 12th May, 1940. 'Red' Eames was wounded on the same day, whilst Paul Chapman was killed in action a week later, on 18th May 1940. *Courtesy of W/C Hugh George, DFC.*

A Fairey Battle, K9233, EF-J is photographed at RAF Abingdon, during the winter of 1938/39. The aircraft also saw service with No.4 Bombing and Gunnery School, and No.4 Air Observers School, before joining the Royal Canadian Air Force, on 2nd March 1942. The Squadron member remains unidentified. *Courtesy of and via Flt/Lt T. N. Harris, XV (R) Squadron.*

Some Rest, Some Play

Some of *"Oxford's 'Own'"* prefer to relax ... Photographed outside the Sergeants' Mess at RAF Abingdon are, left to right: Sgt Raymond Stone (seated), Sgt Readhead, Sgt Philip Camp, Sgt Warren, Sgt Piff. Both Raymond Stone and Philip Camp were to be decorated with Distinguished Flying Medals, for actions during 1940. *Courtesy of S/L P. J. Camp, DFM.*

... Some prefer to engage in sports. The RAF Abingdon Tug-o-War team, complete with two future recruits in the background, were photographed at RAF Uxbridge during 1939. Philip Camp is standing to the extreme right. *Courtesy of S/L P. J. Camp, DFM.*

The Groundcrews Practice for War

Practicing for the real thing' ... A group of No.XV Squadron armorers prepare a number of 8lb Stannic (tin) practice bombs. At least thirty bombs are in evidence. *Courtesy of S/L P. J. Camp, DFM.*

Right: Mobilisation week, and geared up to go. Sergeant Hall (right), with a colleague at War Station Weston on the Green, Oxfordshire, during June 1939. *Courtesy of S/L P. J. Camp, DFM.*

The Aircrews Practice for War

A poor quality photograph, taken in the pilots and navigators crewroom at RAF Abingdon. Left to right are: Unknown, Sgt Hall, Sgt Philip Camp, Unknown, Sgt Perkins, Unknown, Sgt Day. *Courtesy of S/L P. J. Camp, DFM.*

Discussing tactics outside the Squadron hangar, at RAF Abingdon. Left to right are: F/O Peter Douglass; F/O Bert Oakley, RAAF; F/O Webster; F/L Lawrence; Jess Oakshott and 'Red' Eames. Douglass and Oakley were both killed during an attack against bridges over the River Meuse and Albert Canal, at Maastricht, Netherlands, on 12th May 1940. Lawrence was killed a week later, on 18th May, whilst attacking the same targets. 'Red' Eames was wounded as was Webster, the latter being shot through the toes. *Courtesy of W/C Hugh George, DFC.*

And the Squadron Practices for War

The Squadron's Fairey Battle bombers being prepared for take-off from RAF Abingdon, during August 1939. The aircraft nearest the camera is wearing the code F-EF, and is therefore possibly Battle Mk.II, K9228. *Courtesy of S/L P. J. Camp, DFM.*

At least twelve Fairey Battle bombers can be identified on the all grass airfield at RAF Abingdon. *Courtesy of S/L P. J. Camp, DFM.*

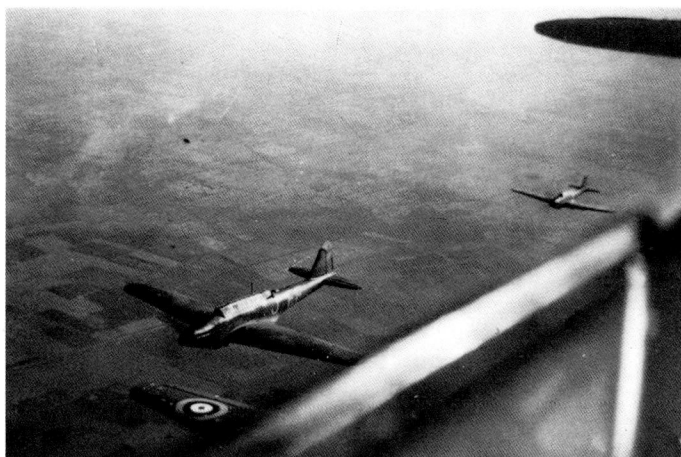

Part of a nine aircraft formation, led by W/C Wingate, flying over the English countryside on return from an exercise, during August 1939. The aircraft left center of the picture is coded J-EF. *Courtesy of S/L P. J. Camp, DFM.*

Four Fairey Battle bombers fly in formation, as the port side section of a nine aircraft configuration. The aircraft nearest the camera is Battle, K9311, EF-O, which was delivered to No.XV Squadron on 18th September 1938. Following service with two British Maintenance Units, K9311 went to the RCAF, on 14th October 1940, as 1753. *Courtesy of and via Flt/Lt T. N. Harris, XV (R) Squadron.*

Fairey Battle Mk.II, K9228, EF-F. The aircraft, which was delivered to No.XV Squadron on 23rd June 1938, also saw service with No.4 Bombing and Gunnery School and No.4 Air Observer School, before joining the Royal Australian Air Force, on 8th November 1942. It was finally struck-off-charge on 13th March 1945. *Courtesy of and via Flt/Lt T. N. Harris, XV (R) Squadron.*

Sixteen Fairey Battle bombers of No.XV Squadron, photographed at RAF Abingdon, prior to departing for France. *Courtesy of and via Flt/Lt T. N. Harris, XV (R) Squadron.*

One Down

The result of light flak or a bad landing? Bearing the scars of the elements and possibly conflict, an unidentified Fairey Battle incurs further damage following a wheels-up landing. *Courtesy of S/L P. J. Camp, DFM.*

A profile view of the Fairey Battle following its forced-landing. Note the absence of any markings, and the poor attempt to conceal the RAF Roundel aft of the gunners cockpit. *Courtesy of and via Flt/Lt T. N. Harris, XV (R) Squadron.*

The Airfield at Conde-Vraux

Conde-Vraux airfield, as seen from a Fairey Battle bomber, whilst flying in the landing circuit. To the right is the Red House, which was used as the Squadron Headquarters, Officers Mess and dormitory. *Courtesy of W/C Hugh George, DFC.*

The Red House at Conde-Vraux airfield, near Chalons-sur-Marne, France. Following the withdrawal of the Royal Air Force, the house was taken over by the Luftwaffe. Before their own retreat from France, the latter set fire to the building. *Courtesy of W/C Hugh George, DFC.*

During the Second World War, Conde-Vraux airfield was used respectively by the Royal Air Force, the Luftwaffe and the United States Army Air Force. This aerial view of the airfield, taken towards the end of the war, shows the Red House in a dilapidated state (bottom left), with four USAAF aircraft parked nearby. *Courtesy of W/C Hugh George, DFC.*

A permanent memorial in the form of a buckled propeller blade surmounted on a stone cairn, erected outside the Red House, commemorates the site of the once much used airfield. A plaque, mounted on the stonework records its use as, Royal Air Force, 1939-40; Luftwaffe, 1940-44; United States Air Force, 1944-45. *Courtesy of T. K. Garrett.*

Service and Refuel the Aircraft

The aircraft having landed, members of the Squadron's groundcrew immediately prepare them for the next sortie. Fairey Battle, K9228, EF-F, receives attention at Conde-Vraux. *Courtesy of and via Flt/Lt T. N. Harris, XV (R) Squadron.*

A Fairey Battle, wearing its protective engine and canopy covers, appears to be receiving a "top-up" from a fuel bowser trailer!! *Courtesy of S/L P. J. Camp, DFM.*

Prepare and Test the Airfield Defenses

A young airmen looks worried as his colleague 'demonstrates' how to use machine guns. The airfield defenses at Conde-Vraux consisted of twin machine guns, of World War One vintage, mounted on a rough wooden plinth. The gun emplacement was a pit dug near the perimeter of the landing ground, which also provided 'shelter' for the gunners. The whole structure being reminiscent of a first world war infantry trench. *Courtesy of and via Flt/Lt T. N. Harris, XV (R) Squadron.*

The art of camouflage also helps in the defense of an airfield. Situated near the edge of a road on Conde-Vraux airfield a tent, used by 'B' Flight of No.XV Squadron, is disguised to resemble a haystack. Riding past on an RAF issue motorcycle are "Red" Eames (driver) and F/O Webster. *Courtesy of and via Flt/Lt T. N. Harris, XV (R) Squadron.*

Billet the Crews

Officers to the Red House, other ranks with the local population. An unknown elderly French lady poses for a photograph with the members of No.XV Squadron who will be billeted in her home. Left to right are: Sgt Douglas Avent, Obs; Sgt Sidney Readhead, Obs; Sgt Philip Camp; Sgt Robert Moffat, Obs; Sgt Raymond Stone, Obs. Sgt Avent and Sgt Moffat were both killed in action, the former on 12th May 1940, whilst Sgt Moffat died a month later on 8th June. Sgt Camp and Sgt Stone were both later to be awarded the DFM. *Courtesy of S/L P. J. Camp, DFM.*

Right: A wall in the Red House is used as a blackboard to record the names of No.XV Squadron aircrew as follows:

DAY	DAY
STONE	FARMER
SHORTLAND	
FARMER	HARVEY
PERKINS	PERKINS
RYVRS	
PHILLIPS	PHILLIPS
HARVEY	PEPPER
REDHEAD	
PEPPER	RYVRS
	REDHEAD
	SHORTLAND
	STONE

The bullet hole in the wall between the two columns of writing is, according to Hugh George, the result of an effort by Brian Readhead to extinguish a candle by use of gunfire. *Courtesy of W/C Hugh George, DFC.*

En-suite accommodation 1939 style. An unidentified airmen poised ready to acquire cold water from a hand operated pump in an outhouse where he is billeted. Amongst the squalor, note the two champagne bottles on the shelf, one of which appears unopened. *Courtesy of and via Flt/Lt T. N. Harris, XV (R) Squadron.*

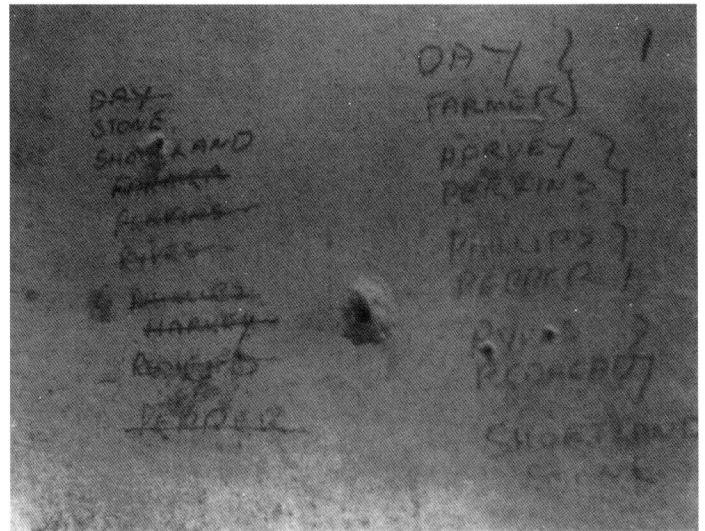

Then Take the Photographs to Send Home

One for the Squadron album. No.XV Squadron, at Vraux, 1939. Seated in the front row, left to right: F/S Nightingale; W/O ? ; P/O Robinson; P/O George; Red Eames; F/O Douglass; Jess Oakeshott; F/L Paul Chapman; S/L Llewellyn; W/C Wingate; F/O Webster; F/O Dawson-Jones; F/O Oakley; F/O Clarke; Tom Bassett; F/O Trent; P/O Frankish; F/S MacDonald. *Courtesy of and via W/C Hugh George.*

Another picture for the album. Standing left to right: Warrant Officer ? ; P/O Hugh George; Red Eames; F/O Ronnie Clarke; Tom Bassett; F/O Len Trent; P/O Claude Frankish; F/S Nightingale; F/S MacDonald. Seated left to right: F/O Peter Douglas; Jess Oakeshott; F/L Paul Chapman; S/L Llewellyn; W/C Wingate; F/O Peter Webster; F/O Francis Dawson-Jones; F/O Bert Oakley; P/O Charles Robinson. *Courtesy of and via W/C Hugh George, DFC.*

A group of unidentified groundcrew personnel pose for a picture in front of a Fairey Battle bomber. Note the covering over the engine nacelle and cockpit canopy, giving protection to the aircraft from the elements. Note also the bomb trolley, under the port wing to the bottom right of the picture. *Courtesy of and via Flt/Lt T. N. Harris, XV (R) Squadron.*

Leading Aircraftman Moorhouse (left), Sgt Coburn (center) and F/S Cleiffe pose for a photograph with their Fairey Battle bomber (possibly K9227). Although the sun was shining when the photograph was taken, the state of the aircrew's boots gives an impression of the conditions which the groundcrews had to endure to keep the aircraft fully serviced. *Courtesy of S/L P. J. Camp, DFM.*

Aircraft Ready, Aircrews Ready

A Fairey Battle, only identified by its code 'N', gets the fully service treatment from the instrument and airframe fitters, whilst the armorers wait to load the bombs. *Courtesy of and via Flt/Lt T. N. Harris, XV (R) Squadron.*

A Fairey Battle, named "Jolly Roger" gets a final check, before being classified ready for action. *Courtesy of and via Flt/Lt T. N. Harris, XV (R) Squadron.*

"Morning Prayers". Sergeants, of No.XV Squadron aircrew, assemble in casual fashion for the daily briefing at Conde-Vraux ... *Courtesy of S/L P. J. Camp, DFM.*

... Whilst the Leading Aircraftmen form-up in more formal ranks for their photograph. *Courtesy of S/L P. J. Camp, DFM.*

An unidentified pilot waits at readiness, sitting high on the engine nacelle of his Fairey Battle bomber, EF-R, at Conde-Vraux. *Courtesy of and via Flt/Lt T. N. Harris, XV (R) Squadron*

Pilot Officer Shennan (left), P/O Hugh George (center) and F/O Ronnie Clarke, pose for a photograph by the tailplane of Fairey Battle Mk.I, K9358, EF-V, of No.XV Squadron. Whilst Hugh George and Ronnie Clarke wear the RAF issue "Irving" jacket over their uniforms, P/O Shennan opts to wear leather flying pants and boots. As Peter Shennan was killed in a flying accident on 5th April 1939, the photograph was taken prior to the Squadron's move to France. *Courtesy of W/C Hugh George, DFC*

Flying Officer Len Trent in the cockpit of his Fairey Battle, wearing his field service cap and white flying overalls. Pilot Officer Hugh George relates a true story regarding his friend Len Trent, as follows: "It was our custom to practice dive bombing with either 8.5lbs practice bombs on a target such as a white sheet pegged out on the airfield, or without bombs but going through the motions as if for real, using the Red House as the target. On one occasion Len Trent, having taken off solo for such an exercise, threw the whole place into a state of alarm when it was realised his aircraft was fully bombed-up. Moreover, it was a nice day and Len had taken-off without his flying helmet, and therefore could not be contacted. I did the only thing possible and leapt into an aircraft, with the intention of heading him off. Fortunately he was taking his time climbing and I managed to catch him up and contrived to thwart him every time he showed signs of peeling off, which made him very cross. Eventually he got the message and landed, and was persuaded to look in his bomb bay where he found four 25Olb high explosive bombs"! *Courtesy of and via Flt/Lt T. N. Harris, XV (R) Squadron*

Strolling by the lock on a French canal are, from left to right: Unknown; F/O Ronnie Clarke; F/O Francis Dawson-Jones; P/O Hugh George; P/O Claude Frankish. *Courtesy of W/C Hugh George, DFC*

7

The Blenheim Era

As a farewell gesture to No.71 Wing H.Q., the Squadron held a dinner, followed by a concert, on the evening of 7th December. The guest of honor for the evening was Group Captain Hugh Walmsley, O.B.E, M.C, D.F.C, the Officer Commanding, No.71 Wing.

With formalities complete, No.XV flew back to England with the aircrews in their Fairey Battle aircraft, whilst the groundcrews were airlifted in De Havilland DH89s and Armstrong Whitworth Ensign aircraft. However, inclement weather over the British Isles meant that only three of the Fairey Battles actually landed at Wyton, the rest were forced to land at various airfields in the south of England. A change in the conditions allowed the dispersed No.XV Squadron aircraft to arrive back at Wyton in ones and twos, between 11.15 hours and 13.40 hours, on 10th December. Unfortunately, the aircraft carrying the Squadron's equipment landed at Shoreham on the south coast of Sussex, and therefore took a couple of days longer to get back.

On its return to England No.XV officially deployed to the R.A.F Station at Wyton in Huntingdonshire, where it came under the command of No.2 Group. The airfield, which was approximately three and a half miles from the county town of Huntingdon, was constructed in 1937. It boasted three bitumen covered concrete runways, in the adopted "A" configuration, C-4 type hangers and permanent accommodation.

Conversion to the Mk.IV twin-engined bomber began almost immediately, with the crews being given instruction on a dual-control, Mk.I. version of the aircraft. With the change of equipment came a change of leadership when, on 21st December 1939, S/L Ralph Lywood assumed command of the Squadron. The new C.O.'s promotion to W/C was announced eleven days later on Monday, 1st January 1940.

The second month of the new year saw the start of a new partnership, after Ronnie Clarke had introduced Hugh George to an aspiring young actress by the name of Beatrice Moor. The occasion of the latter's meeting was as a result of a double date, set up by Ronnie Clarke and his lady friend. The encounter was such a success, the dashing young pilot and the aspiring actress deemed it agreeable to see a lot more of each other. During the following weeks Betty, as she was known to her friends, gave up her place at Royal Academy of Dramatic Art (RADA) in order to spend as much time as possible with the new love of her life. However, the Squadron and the war had first priority on Hugh George's time.

For the first three months of 1940 the Squadron undertook numerous training exercises which involved everybody, including members of the groundcrew. The knowledge gained during the training exercises, both in the air and on the ground, was about to be put to the test for real.

Following reports of the German occupation of Norway, on 9th April, the Squadron was put on standby and deployed to RAF Alconbury, which was to be Wyton's satellite airfield.

The waiting was over. A month later, on 10th May, further reports were received imparting the news that German forces had invaded Holland and Belgium. When these reports proved correct all personnel on leave were recalled to the Squadron, including F/O Hugh George who had gone home to visit his family at Haverfordwest in south-west Wales.

On this same day, No.XV Squadron was ordered to detail two of its aircraft for reconnaissance duties. Both Blenheims returned safely from their missions, albeit with a number of flak holes to indicate the reception they had received from the invading forces. Later that day, at 15.00 hours, the Squadron detailed nine crews to bomb Waalhaven airfield, just south of Rotterdam. Although all the Blenheims returned safely, it was evident from the damage inflicted on some of the aircraft, that the German paratroopers holding the airfield had treated the main force to the same reception as the two reconnaissance machines which had ventured out earlier.

Two days later, on 12th May, twelve Blenheims from No.XV Squadron, were detailed to bomb bridges across the Albert Canal at Maastricht, in order to try and stop the German advance. When the Blenheims arrived over the target area they were again subjected to very heavy and intense groundfire resulting, on this occasion, in the loss of six Squadron aircraft.

Of the six aircraft that returned to base, only two were deemed serviceable. P/O Robinson's Blenheim landed with no flaps and the undercarriage retracted, due to the hydraulics system being severely damaged by enemy fighter attacks during the mission. Flying Officer Tom Bassett, F/O Bert Oakley, P/O Claude Frankish and Sgt Hubert Hall, and their respective crews, were all killed. Bassett's aircraft crashed three kilometres north of Maastricht, Oakley came down at Munsterbilzen, Frankish crashed at Genk, whilst Hall's aircraft sank in the Albert Canal.

Although F/O Douglass and his observer, Sgt Shortland, were also killed, Sgt Davies, their air gunner survived and was taken prisoner of war. Other PoWs included Sgt Pepper, Sgt Booth and LAC Scott.

Over the next few days the groundcrews worked feverishly to repair the aircraft, in an effort to make them airworthy. However, the Squadron was only able to detail three Blenheims for an attack against targets in the Sedan on 15th May. For the purpose of this operation, these three aircraft were instructed to operate with a flight from No.40 Squadron, who were also based at RAF Wyton.

The target was successfully attacked through a heavy and intense curtain of exploding flak shells, some of which obviously found its mark. No.40 Squadron lost two aircraft, complete with their crews, whilst P/O Harriman of No.XV Squadron, endured a dicey moment on his way home. As his Blenheim bomber, serial L8856, was approaching the Belgian coast, the port airscrew detached from its mounting. The pilot, who managed to keep control of his machine, force-landed the aircraft in Holland close to the Belgium border. Pilot Officer Harriman and his air gunner, LAC Moorhouse, were able to return to England the following day, but their observer, Sgt Stanford, who was injured during the incident, was detained in a Belgian hospital.

With its numbers at Alconbury seriously depleted, the remnants of No.XV Squadron were ordered to operate again from RAF Wyton, with some replacement aircraft.

The nightmare re-occurred for No.XV on 18th May, when the Squadron was ordered to detail six aircraft to attack enemy armored columns massing at Le Cateau. Accompanied by Blenheims from No.40 Squadron, the attack was to be made under the protection of a French fighter escort, which both No.40 and No.XV were supposed to meet at Abbeville. However, when the fighters did not appear S/L Lawrence and F/L Chapman, leading "B" Flight and "A" Flight respectively, decided to go on unescorted. They approached the target from the west at an altitude of 6,000' and, having peeled-off, attacked in a fairly steep dive. Flying Officer Hugh George's Blenheim flying as No.6 in the formation, was the last aircraft to attack. As he went down in the dive, he was aware of the thick 'curtain' of flak and tracer rising up to meet him and his colleagues. He felt it impossible to fly through it, but that is exactly what he decided to do. Knowing the emergency boost for the Blenheim bomber should not be used for more than three minutes at any one time lest the engines explode, Hugh George 'pulled the plug' and the aircraft went down like a rocket. He was aware of passing some other aircraft whilst in his dive, and then releasing two delayed action bombs at extremely low altitude. He completed the tactic by pulling up, straight into the path of two waiting Messerschmitt Bf109s! Believing them to be members of Adolf Galland's *'Yellow Nosed'* Squadron, F/O George threw the Blenheim into every conceivable maneuver in an effort to evade the enemy fighters. As they gained a position on the tail of the Blenheim, the rear gunner opened fire, but after only a couple of rounds a cartridge case slipped and jammed the gun. Heaving back on the 'stick', and throwing the Blenheim round in a stomach wrenching turn, Hugh George managed to get a close-up view of the rear end one of the Bf109s. He could not miss, but when he pressed the gun button nothing happened, the front guns refused to fire. Realising that discretion was the better part of valor, F/O George decided to try and extricate himself from the deadly situation he now found himself in. Throwing the Blenheim around the sky, with compass and instruments spinning wildly, he headed back down to the 'deck'. When the instruments began to settle down, the pilot realised to his horror that he was actually flying eastwards, towards Germany!!! Again, he pulled the twin-engined bomber round into a tight turn in order to bring the aircraft onto a correct heading. Although they had not managed to shoot down the Blenheim, the two Bf109s had not given up the chase. Flying as low as he possibly dare, F/O George 'hugged' the ground and followed the contours of the French countryside. As he cleared the brow of a hill, he became aware of a village ahead of him, with its main street directly in line with his flight path. Realising the Blenheim was not going to increase altitude, one of the enemy fighters pulled away but the other pilot, being of a more determined nature, continued the chase at rooftop height, down the main street. Engrossed in the task at hand, Hugh George gave little thought to the enemy aircraft behind, and was therefore surprised when the air gunner informed him that the Bf109 was no longer there. Its sudden disappearance was a mystery to both the Blenheim pilot and the air gunner. When he was finally able to collect his thoughts, F/O George found the emergency boost was still on. The specified three minutes had been extended to the best part of three quarters of an hour.

Intent now on getting back to base, Hugh George took stock of the situation. Scanning the skies around him, the Bomber pilot recognised the shape of another Blenheim aircraft flying above and ahead of him. Remembering the old adage 'safety in numbers', he endeavored to catch up with the aircraft ahead of him. However, try as he might, he could not close the gap between the two machines. It seemed as if the pilot of the other machine was intent on keeping it that way. Due to a shortage of petrol, Hugh George was forced to reduce speed and land at Poix, an airfield still operated by the R.A.F where he was able to refuel. He was surprised to find the other Blenheim, a machine from No.40 Squadron, had also landed there to take on fuel. On hearing the other pilot relate how he had been chased all over France by a Messerschmitt 110, Hugh George could not resist replying, *"Look sonny, that Me110 was me"!* The former pilot's pow-

ers of aircraft recognition were obviously in some doubt, as the Junkers Ju.88 resembled a Blenheim more closely than a Me.110.

For the second time that week heavy and intense flak bursts peppered the sky, forcing the Blenheims to again fly through a curtain of steel. The plan had been for the bombers to land at Poix, re-arm with a new load of bombs and return to the fray but, apart from P/O George, only two other aircraft emerged from the flak barrage and landed as instructed. The two other Blenheims, piloted by F/O Len Trent and P/O Robinson, were deemed by the engineering officer as unfit to fly. Indeed, 'Robbie' Robinson's aircraft was so full of holes, it was beyond repair. Bristol Blenheim, serial L8848, piloted by F/O Hugh George, was the only No.XV Squadron aircraft to return to Wyton. Flying Officer Trent remained overnight with his machine at Poix, where repairs were carried out the following day. Under cover of darkness that second evening Len Trent took-off and, evading both searchlights and flak, managed to return to England, landing at RAF Mildenhall, at 20.30 hours. 'Robbie' Robinson and his crew 'hitched' a ride back to Wyton in a No.40 Squadron Blenheim.

In total the Squadron lost four aircraft, with eight aircrew killed. Amongst those who perished were F/O Dawson-Jones and F/L Paul Chapman, together with their respective crews. Squadron Leader Lawrence and his observer, Sgt Hopkins, were also killed. Lawrence's air gunner, LAC Thomas, survived the onslaught, was captured and taken prisoner of war.

Flight Lieutenant Webster gave cause for concern when he failed to return from an attack against enemy armored fighting vehicles, in the vicinity of Boulogne on 21st May. However, he was back with his Unit just over a week later.

The Squadron was down, but certainly not out. On this same day four new pilots, all from No.101 (Bomber) Squadron, were posted to No.XV as replacements for those previously lost in action. Three of the 'new boys', Masters, Henderson and Gilmore all wore the rank of pilot officer, whilst the fourth member, Roland Megginson, wore the three stripes of a Sergeant. Gilmore and Megginson would fly both Blenheim and Short Stirling aircraft with the Squadron, but Masters and Henderson were to meet quick and untimely ends only a week after their arrival on the Unit. James Masters was shot down on 23rd May whilst piloting Blenheim, serial L9403, during an attack against German troop columns south of Arras. Henderson was killed the following day when his aircraft's port engine seized whilst landing at RAF Alconbury. The aircraft, Bristol Blenheim, serial R3614, coded LS-D, was returning from an attack against German troops who were surrounding the British Garrison at Aa Canal. Over the target area the Blenheim had been the recipient of an explosive shell which struck the aircraft between the leading edge of the wing and the port engine nacelle. The machine spun into the ground at 21.00 hours killing not only the pilot, but also the observer, Sgt Arthur Holmes, and the air gunner LAC Ronald Austin. Having been born in December 1919, Sgt Holmes was always known as Noel, his second name.

Flying Officer Robertson, P/O Bamber and P/O Roberts, three more replacement pilots all arrived at Wyton on 23rd May, the same day that promotions were made to existing members of No.XV Squadron. F/L Webster was promoted to the rank of Squadron Leader, and given command of "B" Flight, whilst both Flying Officers Len Trent and Jess Oakshott were promoted to Flight Lieutenant. Jess Oakshot celebrated his promotion by leading his Flight on a bombing raid against a target in the Bois du Boulogne area, but the raid was marred by the loss of P/O Masters and his crew.

Flight Lieutenant Oakshott led another attack, on 25th May, against an enemy motor transport column which was located two miles north of Calais. This mission was also marred by the loss of another crew when P/O Douglas Harriman's Blenheim was shot down.

Due to the nature of their friendship, and as a result of the events on 18th May, Hugh George and Betty Moor decided to make their relationship a permanent one. The wedding was set for the 25th May at St. Mary's Church, Houghton, near Wyton airfield. Unfortunately, due to the loss of P/O Harriman and his crew that same morning, the Squadron felt it imprudent to hold any festivities to celebrate the marriage even to the point no photographs were taken. The following morning, knowing of the latter circumstances, Len Trent rectified the situation by restaging the event. He was a very keen photographer and, having instructed the young couple to dress in their wedding finery, recorded the occasion on film; albeit twenty-four hours late. Unfortunately, Ronnie Clarke would never know that his match-making led to a partnership which would last well over fifty-nine years.

On 29th May the Squadron participated in a mission against the German forces, who had encircled the British army at Dunkirk. Three experienced pilots, F/L Len Trent, F/L Oakshott and P/O Robinson each led a section. The mission was successfully carried out but, due to very heavy morning mist upon their return, the participating aircraft were forced to land at various airfields in the south and east of England.

During the first few days of June, W/C Joseph Cox arrived at RAF Wyton to take command of No.XV Squadron. He was an experienced officer who had joined the Royal Air Force during May 1928 as a probationary Pilot Officer, with a Short Service Commission. He had seen service in both the Middle East and India, interspersed with a period as flying instructor at the Central Flying School. Wing Commander Lywood, from whom Joe Cox was taking over, was posted to command No.40 Squadron. Unfortunately, Lywood was lost during the first raid undertaken with his new command.

The 11th June was another sad day for the Squadron, when one of its long serving pilots was killed. Flying Officer Ronnie Clarke was leading a flight on a bombing raid against enemy concentrations in the region of Boisement-St. Clair-Vernon-Venables. Prior to reaching the target area, the formation flew into a heavy bank of cloud. Whilst concealed in the cloud, Blenheim, serial L8851, piloted by P/O R Werner, somehow unknowingly overtook the formation leader. Whilst executing a turn to port, Werner's aircraft struck the Blenheim piloted by Ronnie Clarke. A wing broke-off Werner's machine and his aircraft crashed to the ground, killing him and his crew. Flying Officer Clarke managed to retain control of his badly damaged Blenheim, but shortly after the starboard engine broke free. Sergeant Maloney and Sgt Piff, the observer and air gunner respectively, managed to bale out before Blenheim, L9024, crashed at Freneuse-sur-Risle.

Squadron Leader Singer, D.S.O., reported for duty with No.XV Squadron on 14th June, only to be informed he had been reposted to the Operational Training Unit at West Raynham, Norfolk, with immediate effect. However, two weeks later, on 30th June, S/L Singer was back with No.XV Squadron where he took over command of "A" Flight.

The summer months of 1940 saw a continuation of bombing raids against targets in Holland, Belgium and France. Enemy occupied airfields in the low countries, together with the introduction of attacks against industrial plants, such as oil refineries in Germany, were now considered fair targets. Occasionally the Squadron reverted to its original role and undertook reconnaissance missions.

Sergeant Philip Camp's pilot, F/L Jess Oakshot, received further promotion during the second week of July when he was made up to Squadron Leader. Sergeant Hunter and Sgt Davis were both decorated on 29th July by Air Marshal Portal, D.S.O., M.C., with the award of a D.F.M., in recognition of the missions they had respectively undertaken. Hugh George was to receive promotion to the rank of Flight Lieutenant at the end of the following month.

Three French airfields, Abbeville, Evere and Hingene, all occupied by the German Luftwaffe, were attacked by No.XV on 2nd August. Although resistance was encountered, all Squadron aircraft returned safely to base. Unfortunately, the Squadron was to lose an aircraft, with a crew of four, two days later on 4th August, when Blenheim, serial R3771 crashed. The aircraft had been engaged in a fighter affiliation exercise when it dived into the ground.

Amongst the new pilots who joined No.XV at this time was a young Irishman from Dublin, by the name of William 'Bill' Garrioch. He had an aeronautical background, held the rank of Sergeant and, at the time of his arrival on the Squadron, had flown a number of both single and twin-engined aircraft. Although he was educated in Dublin, Bill Garrioch had been fortunate in securing an apprenticeship with the Armstrong-Whitworth Aircraft Company. Based at Coventry in the Midlands, where he was to produce parts for Whitley bombers and Ensign aircraft, Bill commenced his engineering training in January 1937. The weekly salary of an apprentice not amounting to very much, Bill decided to look for a way of increasing his income. This was achieved during 1938, by joining the Royal Air Force Volunteer Reserve where, apart from learning to fly, he was paid for his services. By the time war was declared in September 1939, Bill Garrioch had accumulated forty five hours flying time on Avro Cadet aircraft. On completion of his training, and conversion to Blenheim Mk.I, Sgt Garrioch was posted to No.XV Squadron at Wyton, where he reported for duty on 24th August 1940.

The formalities dealt with, the young Irishman started to find his way around the Squadron. Upon introducing himself to the groundcrew chief, Bill was informed there was no aircraft readily available for him. However, he was informed that Blenheim, serial R3704, which had previously been badly shot-up, was being repaired and would shortly be returned to Squadron strength. Sensing the new pilot's dismay, 'Chiefy' Wright informed Bill that this particular aircraft was one of the best, having been built by the parent company at Bristol, plus the fact it could outclimb the other Squadron aircraft.

Another initial worry for Sgt Garrioch was that, six weeks after his arrival on the Squadron, No.XV converted to night bombing operations. Unfortunately, one very important factor in this type of aerial warfare was overlooked; the aircraft had not been repainted according to the dictates of their new task. Therefore, when they were caught in the glare of enemy searchlights, the duck-egg blue undersurfaces made the aircraft easy to see! The Irishman's fears were overcome when the aircraft were hurriedly painted black.

August and September 1940 saw a massive build-up by the Germans of barges and assorted sea craft, in the Channel Ports in preparation for their proposed invasion of Britain. They also increased the number of aircraft at airfields within striking distance of London and other desired targets. During these two months, No.XV Squadron was heavily engaged in attacking enemy occupied airfields, shipping and harbor installations. It was against one of the latter targets that Bill Garrioch undertook his first operational mission, on 7th September, when three aircraft were detailed to attack the harbor at Dunkirk. For the raid Bill was allocated Blenheim, serial R3704, an action which initially caused him a moment of concern, but as the aircraft headed out over the English Channel he realised the truth of 'Chiefy' Wright's words. The other two Blenheim aircraft detailed for the attack were T2231 and R3494, piloted by F/L Mahler and F/O St John respectively. These two battle experienced pilots located and bombed the target. However, the glare of searchlights precluded Sgt Garrioch from identifying his aiming point.

October saw an increase in the attacks against harbor installations and the Channel Ports. One such mission, undertaken on 10th October, was a combined operation with the Royal Navy. Although twelve Blenheims were detailed to bomb Cherbourg, only eleven were recorded as having taken-off. All but two of the Blenheims were to bomb the town, creating a diversion which would enable the Royal Navy warship, *HMS Revenge*, to position itself, so that the salvoes fired by the ship's guns would be within range of the installations. Apart from creating a diversion, the exploding bombs would also start fires which would give the ship's gunners an aiming point. Meanwhile, the two remaining Blenheims were detailed to attack any shore batteries which endeavoured to return fire at the ship. The two remaining Blenheims, serials R3594 and T2227, were piloted by W/C Joe Cox and S/L Stewart 'Paddy' Menaul respectively. At one point, as they were circling separately over the target area, both pilots realised they were on a path to destruction; a head-on, mid-air collision. Both took avoiding action, and the two aircraft skimmed past each other, a hair's breadth from disaster.

Recovering his composure, when the shore batteries responded to the *Revenge's* gunfire, W/C Cox dived into the attack and unleashed his bombs on the offending gun emplacement. Following his commander's example, S/L Menaul struck at the same position, adding to the destruction and fires already started.

The spectrum of color which prevailed as bombs burst, shells exploded, fires burned and flak traced illuminated trials up into the sky, was a scene to behold. Behind the facade of death and destruction was a fairy tale illusion, as the colors twinkled and glowed, rose in intensity and then faded, only to rise again. From his viewing point 6,000' over the target area, Bill Garrioch became engrossed. It was the greatest 'firework' display he had ever seen. His reverie was broken by a warning from 'Taffy' Reardon, the wireless operator/air gunner, that a brilliant light emanating from a Messerschmitt Bf110 behind them was illuminating the Blenheim. Without waiting for further instructions, Sgt Garrioch threw his aircraft into a power dive, which recorded 340 mph on the air speed indicator. Thinking the Blenheim to be in trouble, Sgt Reardon prepared to evacuate the plunging machine, but was spared the jump when the aircraft leveled out.

During the flight home, Paddy Menaul's wireless operator tuned his set to the British Broadcasting Corporation wavelength and heard a dance band playing a tune called *'Fools rush in where angels fear to tread'*, which seemed very apt following the recent near-miss. At the de-briefing back at Wyton, Stewart Menaul related this fact to W/C Cox in front of an audience of participating aircrew and invited press reporters. Many weeks were to pass before S/L Menaul was allowed to forget the incident. So successful was the operation that, not only was it recorded in the newspapers of the day, the Royal Navy also recorded its thanks to the Royal Air Force for the excellent co-operation it received from the latter.

The final operations undertaken by No.XV Squadron, using the Bristol Blenheim bomber, took place on 29th October when seven aircraft attacked Hamburg, one aircraft attacked Schipol airfield and one other aircraft attacked Flushing docks.

During its time at RAF Wyton, whilst operating with the Bristol Blenheims, No.XV Squadron carried out thirty training sorties before 10th May 1940 and 578 operational sorties between that date and the end of October. The Squadron was initiated into night operations on 12th August and, over a two and a half month period, carried out 236 sorties of this type. A total of 373 tons of high explosive bombs and nearly 28 tons of incendiaries were recorded as having been dropped. Furthermore, in keeping with its original role, a total of 4,100 operational aerial photographs were taken. A total of twenty-four aircraft were lost on operations during this period.

The Squadron, always ready to adapt, was about to move into a new era.

Conversion to the Blenheim

Pilot Officer Red Eames (left) and a fellow officer photographed in front of Bristol Blenheim, Mk.I, L1196. The aircraft was a dual-control version of the Blenheim, and was used for pilot training during December 1939. *Courtesy of W/C Hugh George, DFC*

The ever-dependable groundcrew strive to keep the aircraft flying, whatever the conditions. Here, an unidentified Bristol Blenheim, Mk.IV, receives attention on the snow covered airfield at Wyton, during January 1940. *Courtesy of S/L P. J. Camp, DFM*

One to send home for the family album ? Whilst the groundcrew continue to make minor adjustments to the port engine, Sgt Philip Camp stands proudly in front of his Blenheim at RAF Wyton. *Courtesy of S/L P. J. Camp, DFM*

A fine study of Sgt Philip Camp, observer, at his crew position in the glazed nose section of a Blenheim, Mk.IV. Note the flying clothing, including the zip-fronted, one-piece, fur flying suit. *Courtesy of S/L P. J. Camp, DFM*

Blenheim Groundcrew

Bristol Blenheim, Mk.IV, P6917, LS-H, provides the background for a photograph of Ned Wiltshire (left) and Tommy Moore (right), both members of groundcrew. The aircraft was lost on 18th May 1940, when it was shot down during an attack against enemy columns at Le Cateau, France. *Courtesy of W/C Hugh George, DFC*

Aircraft mechanic Tommy Moore, a member of No.XV Squadron groundcrew, in posing for a photograph by the starboard engine of 'his' aircraft, unknowingly provides some close-up detail of the motor and the 'prop'. The aircraft on which the engine is mounted is Bristol Blenheim, Mk.IV, P6917, LS-H. Following service with No.XV Squadron, the aircraft went to No.264 Squadron, but later returned to No.XV. *Courtesy of W/C Hugh George, DFC*

In order to bring the aircraft of 'A' Flight to a state of combat readiness Aircraftman Bailey (with an ammunition belt draped round his neck) prepares to re-arm the Blenheim, whilst Aircraftman Davies, from Canada, sees to matters in the cockpit. The photograph was taken during August 1940. *Courtesy of Ken James*

Left: Another member of the Blenheim groundcrew was Ken James, an engine fitter, who later remustered to aircrew. Following aircrew training Ken returned to No.XV Squadron, with whom he flew as a flight engineer on the Short Stirling four-engined bomber. The white flash in his forage cap indicates the photograph was taken during his aircrew training period. *Courtesy of Ken James*

Blenheim Aircrew

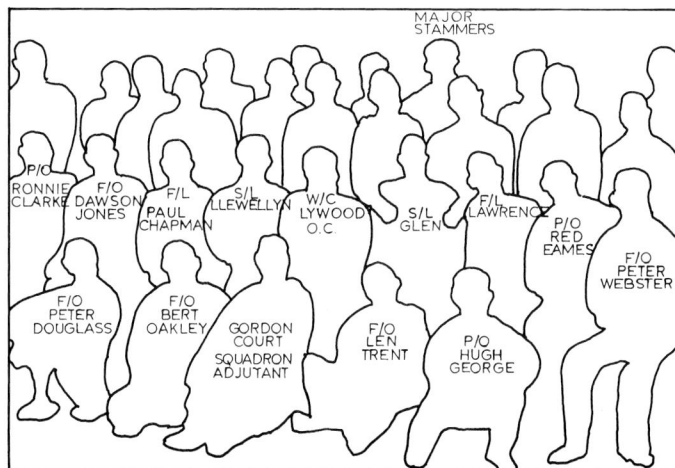

MAJOR STAMMERS

P/O RONNIE CLARKE — F/O DAWSON JONES — F/L PAUL CHAPMAN — S/L LLEWELLYN — W/C LYWOOD O.C. — S/L GLEN — F/L LAWRENCE — P/O RED EAMES — F/O PETER WEBSTER

F/O PETER DOUGLASS — F/O BERT OAKLEY — GORDON COURT SQUADRON ADJUTANT — F/O LEN TRENT — P/O HUGH GEORGE

During the first week of April 1940, a dinner was held at the Old Bridge Hotel, at Huntingdon, near Wyton. Apart from officer pilots serving with the Squadron, the guest list included a number of Fifteen's flyers from the First World War. *Courtesy of W/C Hugh George, DFC*

One Up, One Down

Bristol Blenheim, Mk.IV, N3627, LS-L, following a 'wheels-up' landing. The aircraft was one of a production batch of 100 machines built and delivered by Messers A. V. Roe & Co. Following service with No.XV Squadron, it went to No.139 Squadron. On 8th August 1941, N3627 caught fire during a training flight and spun into the ground at North Barningham, Norfolk. *Courtesy of and via Flt/Lt T. N. Harris, XV (R) Squadron*

An in-flight photograph of P/O Hugh George at the controls of his Blenheim bomber, taken by his observer, Sgt Box. *Courtesy of W/C Hugh George, DFC*

At Readiness – Alconbury

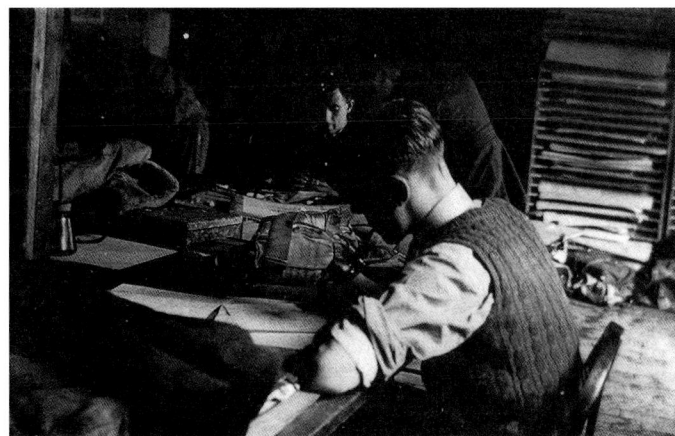

The aircraft were ready and so too were the aircrew. However, even during a friendly game of cards, the tension shows on the faces of these young men who wait in the crewroom at Alconbury, to respond to the call to action. *Courtesy of S/L P. J. Camp, DFM*

Some play cards while others, like Sgt "Pirks" Perkins, pass the time smoking and writing letters home. Note the map and chart storage unit to the right of the photograph. *Courtesy of S/L P. J. Camp, DFM*

"Aim Sure". Some try to ensure their aim is sure and true by playing darts in the crewroom. *Courtesy of S/L P. J. Camp, DFM*

Trying to look relaxed whilst on 'standby', some take time out for the inevitable group photograph. For some, it was to be their last photograph. *Courtesy of S/L P. J. Camp, DFM*

Sergeant Edwin George Roberts, observer (left), appears very relaxed as he makes a model of the German warship *Scharnhorst*. It is not known whether the model was ever finished, as Sgt Roberts was killed in action on 12th May 1940. His aircraft, P6912, was shot down during an attack against the bridges at Maastricht, and crashed at Genk, Belgium. His pensive looking colleague is thought to be LAC Ernest Cooper, wireless operator/air gunner. Cooper was killed in the same action, along with the crew's pilot P/O Claude Frankish. *Courtesy of S/L P. J. Camp, DFM*

Sgt Edwin Roberts (extreme left) leaves his model making to join in a group photograph. Also in the picture, from left to right are: Sgt Edwin Roberts, Obs; Sgt Readhead, Obs; Sgt Cobourn, Obs; Sgt Avent, Obs; Sgt Prior, Obs; Sgt Philip Camp, Obs; Unknown; Sgt Hopkins, Obs; Unknown; Flight Sergeant, groundcrew. Kneeling at the front is F/S Cleiffe, pilot. Note the camouflaged building in the background. *Courtesy of S/L P. J. Camp, DFM*

Relieving the Tension

Right: To relieve tension some took-up other pastimes, such as lizard fishing ...*Courtesy of S/L P. J. Camp, DFM*

... gardening ...*Courtesy of S/L P. J. Camp, DFM*

... or just giving the impression of walking tall ... *Courtesy of S/L P. J. Camp, DFM*

The 'Balloon Goes Up'

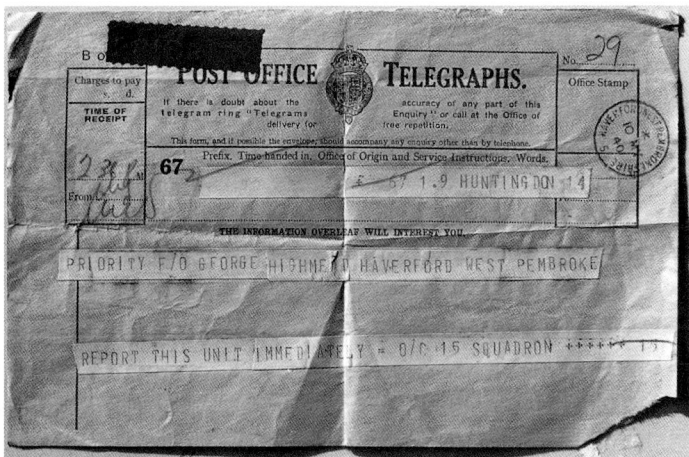

Post Office Telegraph, sent to F/O Hugh George who was on leave when the Germans invaded the Low Countries. It read, "Report this Unit immediately = O/C 15 Squadron". *Courtesy of W/C Hugh George, DFC*

The cost to No.XV in attacking the bridges at Maastricht and over the Albert Canal was high. The Squadron lost seven aircraft shot down, fourteen crew members killed, one injured and four taken prisoners of war. The damaged section of this arched bridge was replaced with a metal span construction. *Author's collection*

The 'new' road bridge across the Albert Canal on the approach to the village of Gellik, Belgium. From this point, on the original bridge on 12th May 1940, German gunners shot down a No.XV Squadron Blenheim. The aircraft flew over the bridge, and crashed onto the canal bank to the left of the picture. *Author's Collection*

The Albert Canal at Gellik, Belgium, photographed from the 'new' road bridge. The picture shows the bank where No.XV Squadron Blenheim, P6914 crashed, on 12th May 1940. The bomber came down on the right-hand embankment and slid down into the canal, taking two of the crew with it. The wounded pilot, Sgt Hubert Hall, died on the path alongside the canal. Some time was to pass before the bodies of the observer, Sgt Edward Perrin, and the air gunner, LAC Patrick 'Joe' McDonnell, were recovered from the wreckage. All three men were originally buried on the canal bank, but were removed to a military cemetery, by the American liberators of the area, later in the war. *Author's Collection*

Left: 56340 Sergeant Hubert H. Hall, pilot, No.XV Squadron, killed on 12th May 1940, is remembered, along with his two crew members, on a purpose built memorial, at Gellik, Belgium. *Author's Collection* Right: 581058 Sergeant Edward R. Perrin, observer, No.XV Squadron, as remembered on the Gellik memorial. *Author's Collection*

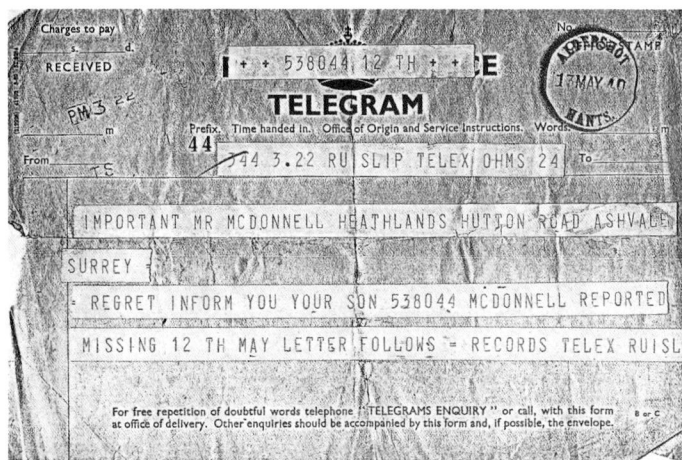

The day after LAC McDonnell was shot down and killed, a telegram was sent to his parents. Due to the fact his body was concealed in the wreckage of the aircraft, below the surface of the water in the canal, the air gunner was listed as 'Missing'. *Courtesy of Mrs Irene Murphy (nee McDonnell)*

528044 Leading Aircraftman Patrick Joseph McDonnell, air gunner, No.XV Squadron, was known to family and friends as Joe, and is still remembered by that name. *Courtesy of Mrs Irene Murphy (nee McDonnell)*

The original cross and nameplate marking the last resting place, at Hotton War Cemetery, of Sgt Hubert Hall, Sgt Edward Perrin and LAC Patrick 'Joe' McDonnell. *Courtesy of Mrs Irene Murphy (nee McDonnell)*

The brick built memorial at Gellik, Belgium, erected to the memory of the three No.XV Squadron airmen who were killed on 12th May 1940. Their names, Sgt Edward Perrin, observer(left), Sgt Hubert Hall, pilot (center) and LAC Patrick (Joe) McDonnell, air gunner (right) are engraved on marble tablets mounted on the brickwork. *Author's Collection*

In Some Foreign Field

Broken trees and scattered wreckage marks the crash site of Blenheim Mk.IV, P6911, of No.XV Squadron. The aircraft, piloted by F/O Albert Oakley, crashed at Munsterbilzen, thirteen miles west-north-west of Maastricht, at 09.10 hours on the morning of 12th May. *Courtesy of and via Flt/ Lt T. N. Harris, XV (R) Squadron*

A buckled and broken propeller lies in the shrubs, having 'cartwheeled' away from the main crash site of Blenheim, P6911. The aircraft was shot down during an attack against bridges at Maastricht and the Albert Canal. *Courtesy of and via Flt/Lt T. N. Harris, XV (R) Squadron*

A poor quality photograph of the original resting places of the crew of Bristol Blenheim, P6911. From left to right: 551566 LAC Dennis Victor Woods, Observer. 566960 Sgt Douglas John Avent, Observer. 36132 F/O Albert Edward Oakley, Pilot. *Courtesy of and via Flt/Lt T. N. Harris XV (R) Squadron*

The final resting place of F/O Oakley and his crew, at Munsterbilzen Communal Cemetery, Belgium, where the simple wooden crosses have been replaced with the standard Commonwealth War Graves Commission headstones. *Courtesy of W/C Hugh George, DFC*

The Attacks Continue

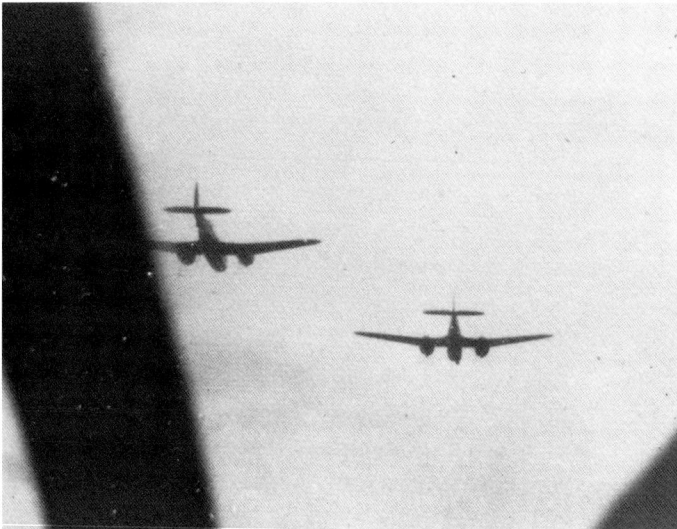

Three Bristol Blenheim Mk.IV bombers fly in formation to a target in France. *Courtesy of S/L P. J. Camp, DFM*

A Bristol Blenheim, Mk.IV, of No.XV Squadron, identified only by its Squadron code of LS-R. *Courtesy of and via Flt/Lt T. N. Harris, XV (R) Squadron*

Flight Lieutenant Paul Geoffrey Chapman, Officer Commanding, "B" Flight, between 1938 and May 1940. He was killed in action on 18th May 1940, whilst piloting Blenheim bomber, P6917. The aircraft was shot down and crashed at Landrecies (Nord), France. Sergeant Cecil Colbourn and LAC John Fagg, observer and air gunner respectively, were also killed. *Courtesy of and via Flt/Lt T. N. Harris, XV (R) Squadron*

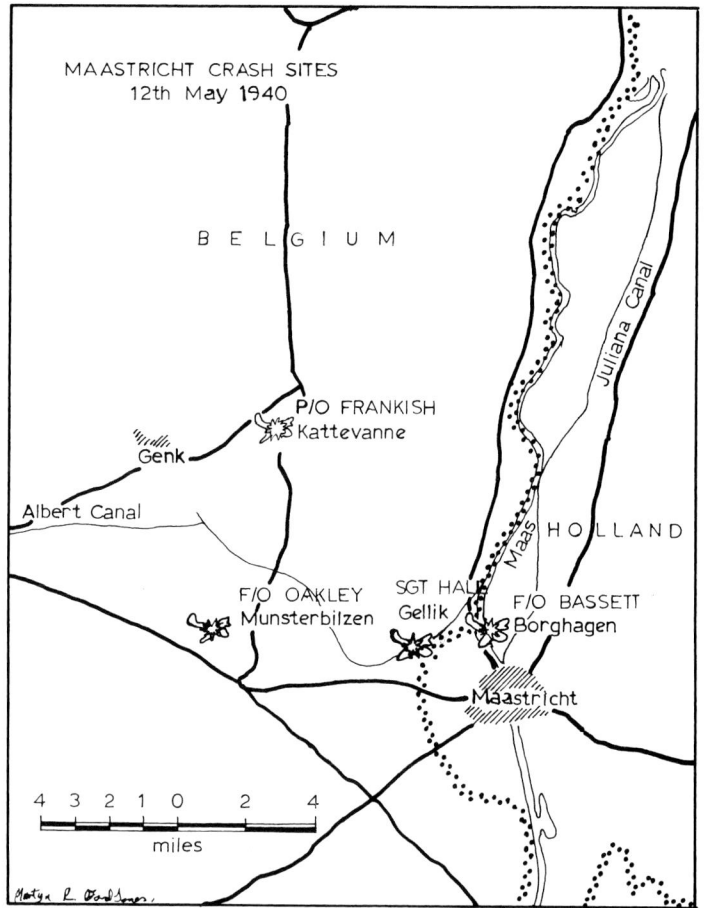

MAASTRICHT CRASH SITES
12th May 1940

Two views of the wreckage of Bristol Blenheim, Mk.IV, R3614, LS-D, which crashed at RAF Alconbury, at 21.00 hours, on 24th May 1940. The aircraft was returning from an attack against German troops who were surrounding the British Garrison at the Aa Canal, and was about to land when the port engine seized. The Blenheim spun into the ground killing P/O Henderson, the pilot, Sgt Holmes, the observer, and LAC Ronald Austin, the air gunner. Prior to joining No.XV, the aircraft, which was built by Rootes, had previously been in service with No.40 Squadron. *Courtesy of and via Flt/Lt T. N. Harris, XV (R) Squadron*

Squadron Leader Jess Oakshott, wears a British Army issue 'tin' helmet, over his leather flying helmet, whilst piloting a Bristol Blenheim bomber. *Courtesy of S/L P. J. Camp, DFM*

'First tour photograph' – Sergeant Philip Camp (center) with his pilot, S/L Jess Oakshott (right), and air gunner, Sgt Trehearne (left). *Courtesy of S/L P. J. Camp, DFM*

A poor quality photograph of a Bristol Blenheim, which force-landed on 29th May 1940, after being hit by flak over Dunkirk whilst covering the evacuation of British troops. *Courtesy of S/L P. J. Camp, DFM*

Hugh George – Then and Now

Pilot Officer Hugh George (right), in light-hearted mood, strolls across the dispersal area at RAF Wyton, on his return from a mission. With him is Jess Oakshott (left) and Peter Webster (concealed). Note the flying clothing and the British issue 'tin' helmet carried by P/O George. *Courtesy of W/C Hugh George, DFC*

Right: Fifty years on. Wing Commander Hugh George, DFC (left), and Martyn R Ford-Jones, at a Bomber Command Association Reunion Dinner, held at the Park Lane Hotel, London, on Saturday, 27th April 1991. *Author's Collection*

Wing Commander Joe Cox, D.F.C

Musical Interlude

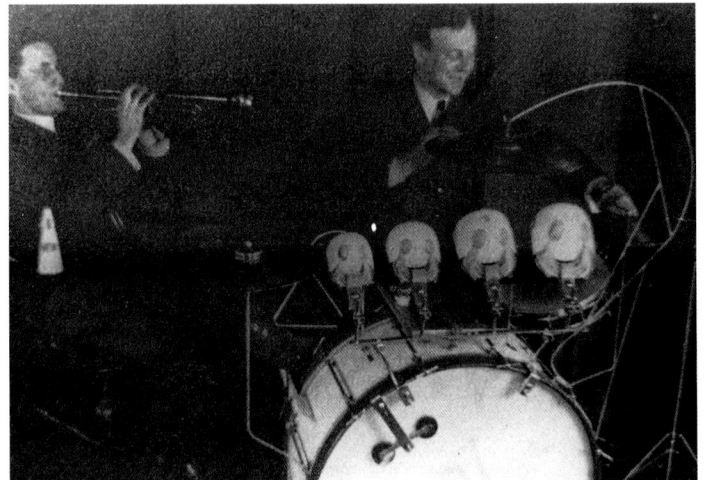

Some like to make a noise. The drum bashing, jazz playing, F/O Ronnie Clarke, as his friend Hugh George (left) wishes to remember him. The two pilots make music during an off-duty 'jam' session. *Courtesy of W/C Hugh George, DFC*

Some like peace and quiet. Wing Commander Joe Cox, DFC, who commanded No.XV Squadron between June and December 1940, takes a few minutes away from the war for a quiet smoke. *Courtesy of Patricia Ann Banks (nee Cox)*

'Ops' On

Squadron Leader Jess Oakshott starts-up the port engine of his Bristol Blenheim, Mk.IV, at RAF Wyton, prior to the start of a mission during the summer of 1940. *Courtesy of S/L P. J. Camp, DFM*

A No.XV Squadron Bristol Blenheim in the air over France during 1940, photographed from the gunner's turret of a similar aircraft of the same Squadron. *Courtesy of S/L P. J. Camp, DFM*

The 'Flying Twins'.

Sergeant 'Bill' Watson (left), with Ken Walker, at an unidentified, snow-covered airfield. William dedicated the photograph to his Mother and Father, with the inscription, "The Flying Twins – Two Battle Scarred Airmen". *Author's Collection*

Preparations Before Take-Off

Sergeant Bill Garrioch, perched on top of his Bristol Blenheim, Mk.IV, bomber. The "B" Flight aircraft and its pilot were caught by the camera during a quiet moment at RAF Alconbury, during 1940. *Courtesy of William "Bill" Garrioch*

Poor quality photograph of Bristol Blenheim, Mk.IV, LS-Z, of No.XV Squadron, at RAF Wyton, circa 1940. Whilst the groundcrew make final adjustments to the aircraft, the pilot holds a last minute briefing with his crew by the tailplane. *Courtesy of and via Flt/Lt T. N. Harris, XV (R) Squadron*

Left: A poor quality photograph of Sgt Bill Garrioch at the controls of Blenheim bomber, R3704. *Courtesy of William "Bill" Garrioch* Right: A poor quality photograph showing Sgt Bob Beioley, observer/navigator, in the bomb aimer's compartment of Blenheim bomber, R3704. Note the glazing bars of the aircraft's 'greenhouse' nose. *Courtesy of William "Bill" Garrioch*

'Second Tour Photograph' – Sergeant Philip Camp (right) with his second pilot, W/C Joe Cox (center) and Sgt Trehearne, at Wyton 1940. As a sign of affection for his wife, Joe Cox dedicated his aircraft to his wife Dorothy, which therefore carried the code letter "D". *Courtesy of Patricia Ann Banks (nee Cox)*

Wing Commander Joe Cox (left) talking to his observer, Sgt Philip Camp, on 11th October 1940. The air gunner, Sgt Trehearne, can by seen the turret uncovering the gun. The aircraft's serial, R3594, and Squadron code, LS, can clearly be seen in this photograph. *Courtesy of Patricia Ann Banks (nee Cox)*

Wing Commander Joe Cox (second from right) receives assistance putting on his parachute, from members of the groundcrew. Sergeant Philip Camp, whose parachute can be seen on the ground in front of the corporal (center), is shielded by Joe Cox. Meanwhile, Sgt Trehearne climbs up onto the aircraft to gain access to his turret. *Courtesy of Patricia Ann Banks (nee Cox)*

Right: Watched by the groundcrew corporal (center), W/C Joe Cox and Sgt Philip Camp make final adjustments to their parachute harnesses. Two other members of the groundcrew await the departure of the aircraft, whilst a fourth is busy on the leading edge of the port wing. Having gained entry to the aircraft, Sgt Trehearne's head can be seen protruding from the entry hatch, forward of the turret. *Courtesy of Patricia Banks (nee Cox)*

8

The Wellington Era

Although the Squadron was to continue to use Wyton as a base, No.XV was to move to No.3 Group, into which the RAF Station was incorporated on 1st November. With this 'paper' move came a change of equipment to the geodetically designed, metal-framed, fabric-covered, twin-engined, Vickers Wellington bomber. Conversion to the 'Wimpey', as the new aircraft was affectionately to become known, began on 7th November when Wellington, serial N2871, arrived at Wyton, together with the required support equipment. This machine was soon joined by two others.

Squadron Leader Mahler and F/L Morris were the first to benefit from instruction on the new bomber. They in turn passed on their newly acquired knowledge to two other members of the Squadron, who repeated the process until each pilot had received basic instruction. Sergeant Garrioch flew a forty minute 'circuit and bumps' exercise with F/O Manahan, prior to the latter sending him on a solo flight.

As each Wellington bomber required a crew of five, compared to the three man crew of a Bristol Blenheim, the beginning of November also saw the arrival of additional aircrew personnel. Both aircraft and crews were put through a full training program during that same month. The crews were detailed for daylight and night flying cross-country exercises, practice bombing and air firing trials, whilst the aircraft were subjected to fuel consumption tests, altitude tests and loaded take-off trials. Apart from getting to learn the technicalities of their new aircraft, the training period also gave the airmen time to get to bond as an efficient crew.

Having served as No.XV's Commanding Officer for the last six months, Joe Cox was declared tour expired in December. The following month, in January 1941, he was posted to No.33 Flying Training School at Carberry in Manitoba, Canada, where he assumed the position of Chief Instructor. On his departure, the aircrew of No.XV presented Joe Cox with a silver cigarette case on which was inscribed their signatures. Unfortunately, most of those whose names were recorded in silver, were to be killed later in action during the course of the war. The cigarette case survived, and was presented to the Squadron by Mrs Dorothy Cox on 22nd May 1987, after the death of her husband. Joe Cox, who retired from the RAF during 1958, rose to the rank of Air Vice-Marshal. His decorations and awards included the C.B., O.B.E. and D.F.C.

Wing Commander Herbert Dale was appointed to take over as C.O., and assumed command of No.XV on 14th December 1940. As he settled in to the task ahead of him, W/C Dale planned a mission for two days hence, but the weather deteriorated and caused the cancellation of the operation. It was to be a further four days before a crew managed to get airborne and complete its allotted task. Wellington bomber, T2702, LS-H, piloted by F/L Morris, attacked a petrol concentration at Antwerp. Two runs were made over the target, in the face of little enemy opposition. Of the ten 250lbs general purpose bombs dropped, only five from the first pass were seen to explode in the target area. Results of the second pass were not observed.

As 1940 gave way to the new year, the weather continued to impede operations. Not only did the snow which had fallen hamper movement, but the freezing conditions which prevailed caused icing-up of the aircraft. However, even in the insufferable conditions confronting them, which made the simplest task an arduous operation, the groundcrews continued to work.

During the morning of 12th January 1941, Sgt Garrioch almost became a victim of the cold weather whilst flying an air test. He was piloting Wellington bomber, R1169, LS-D, when a pipe to the hydraulic fluid line severed. He was able to land the aircraft, but did not have the use of flaps or brakes. Unfortunately the airfield was covered with snow and slush, which was turning the landing area into a quagmire. Although Bill Garrioch came into land as slowly as possible, the bomber traversed across most of the airfield sliding on the wet surface. The aircraft came to a halt at 10.50 hours, when it struck a boundary fence. This action caused the nose of machine to dip down into the snow and the tail to rise into the air. The bomber held this position for a moment, then fell back onto its undercarriage. Squadron Leader Menaul took a dim view of the situation, and stated that the pilot should have been able to land his aircraft easily in the given conditions! The former decided to show the 'errant' sergeant pilot how the landing should have been carried out, and nearly 'pranged' his own aircraft on the same spot.

The cold weather continued into February 1941 and although some raids were canceled, a number of operations were undertaken. One mission during this period was to see the loss of F/Lt Morris's usual 'mount' in dramatic fashion, and with tragic results. Wellington bomber, T2702, was one of ten aircraft from No.XV Squadron detailed to participate in an attack against Hannover, on the night of 10th/11th February. On this occasion T2702 was to be piloted by Sgt W. Garrioch, with Sergeant Bill Jordan who was on his second familiarisation trip, flying as 2nd pilot. Having undertaken the pre-mission ritual of carrying out an air test, attending the operational briefing and partaking of the aircrew meal of eggs and bacon, the crews were driven out to their respective aircraft in preparation for take-off. As the Bedford truck carrying Sgt Garrioch and his crew neared their aircraft dispersal point, 'Taffy' Reardon made a request which was to cost him his life. He had, since converting to the Wellington bomber, usually flown as wireless operator, but asked if he could undertake this mission in the front turret. He was fed-up being ensconced inside the fuselage, and wanted to see what was happening outside. As there was no objection from the captain or crew, the request was granted.

At the allotted departure time of 17.30 hours precisely, the first Wellington bomber, accompanied by the roar of its twin Bristol Pegasus engines, rolled down the runway and lumbered into the air. At regular intervals, on receipt of the green Aldis light from the control caravan, the remaining nine aircraft carried out the same procedure.

The attack, which was led by S/L Stewart Menaul, was carried out on a very cold, clear night, lit by a full moon which reflected off the snow covered landscape below. Patches of white cumulus cloud scurried along, pushed by the very strong westerly winds, which gave the aircraft a ground speed of nearly 200 miles an hour.

Out over the North Sea, the rattle of machine gun fire was heard above the drone of the smooth running engines, as the gunners tested their weapons, ensuring the coldness of the night had not caused the guns to freeze. The Dutch coast lay ahead, and so too did the enemy fighters; if they were not already in the vicinity. In fact, the outward journey to the target was surprisingly quiet, with no fighter opposition and no flak.

Navigator/bomb aimer Bob Beioley's calculations were spot-on, and Sgt Garrioch needed no final course corrections for his run-in to the target area. Having got T2702 to Hannover, Sergeant Beioley clambered down to the bomb aimer's compartment and prepared to unleash the cargo of two 500lb bombs, one 250lb bomb and six canisters of incendiaries. As the bomb aimer guided the aircraft towards the aiming point by instructing the pilot to bring the Wellington to the right, left or hold it steady, Sgt Garrioch witnessed the first of No.XV's bombs exploding on the target; dropped by the aircraft ahead of him. Suddenly, as if awake to the fact their city was under attack, the German gunners responded with a barrage of light flak. The bomber maintained its position and, as the aiming point came into view in the bomb sight, Bob Beioley pressed the 'tit' and called *"Bombs gone"*. The light flak, which arced lazily up into the night sky, was now joined by heavy ack-ack. A *"task completed"* signal was transmitted back to base at 21.35 hours, as Sgt Garrioch pulled on the control column and hauled the Wellington bomber into a steep turn to port away from the target, leaving the angry bursting anti-aircraft shells to find another victim.

With a total of 222 aircraft participating in the attack, fires quickly developed and engulfed Hannover, which John Hall the rear gunner, reported were visible forty miles away.

The strong winds, which had been in their favor on the outbound journey, were now against the attacking force on the homeward leg. Flying against a strong headwind, T2702's ground speed was reported as 85 knots and, to add to their problems, the patches of cumulus clouds were increasing in intensity; it was going to be a rough ride home. As the aircraft crossed the eastern coast of the Zuider Zee, over Kampden, to the west of Zwolle, the rear gunner yelled a warning of an enemy fighter coming in low from the rear. Sergeant Garrioch responded by initiating a steep turn to starboard, in order to face his attacker, and saw a Messerschmitt Bf110 turning to meet him. A violent turn to port by the Wellington enabled Sgt Hall to open fire with a long burst, but the enemy pilot ignored the fire and continued to attack. As strikes hit the bomber in the fuselage and port engine, Sgt Garrioch lowered the flaps of T2702, causing the Bf110 to overshoot its quarry. However, the respite was short lived and the enemy fighter attacked again, wounding Sgt Hall who was still in the rear turret. The radio equipment exploded into fragments, as cannon fire raked the fuselage, missing the wireless operator and the navigator by inches. The Bf110, piloted by Hauptmann Walter Ehle, commander of 11./NJG1, turned away over the Wellington's port wing and circled for yet another attack. By this time the bomber was full of smoke, with very little fight left in it. As Sgt Garrioch instructed the crew to prepare to bale out, he heard a faint cry for help from 'Taffy' Reardon, who was asking to be helped out of the front turret. Bill Jordan, the 2nd pilot, opened the forward escape hatch, and then went down to assist Sgt Reardon. As the enemy fighter came round again, intent on inflicting the 'coup de grace', Sgt Garrioch raised the flaps and put the aircraft into a dive. The fighter attempted to follow the maneuver, but again overshot when the bomber yawed violently to port after the pilot lowered the flaps for a second time. The bursts of machine gun and cannon fire from the Bf110 missed the bomber, as the latter sideslipped with the pilot holding the throttles back. The enemy aircraft roared over the top of the bomber, placing itself in the front gunner's sights. Sergeant Garrioch called for 'Taffy' Reardon to open fire, but the guns in the front turret remained silent. As the bomber twisted and turned in an effort to evade the enemy nightfighter, Bill Jordan tried desperately to open the front turret doors, but failed in his endeavors.

With fire raging across the port engine and inner wing, and inside the fuselage, Sgt Garrioch knew time was running out. Apart from being past the point where the crew could safely jump, the Wellington bomber was going down in a steep dive, at a speed of over 300 knots. Through the flight deck window Sgt Garrioch could see the frozen surface of the Zuider Zee rapidly coming up to meet the crashing aircraft. With the help of Sgt George Hedge, Wop/AG, RNZAF, Sgt Garrioch hauled back on the control column in a desperate effort to level out, but both men found the elevators sluggish in their response. The aircraft struck in a nose down attitude, and skidded for a great distance across the ice before finally coming to rest. However, their nightmare was not over. As the crew made their escape from the wreckage, the ice under the front of the aircraft began to crack, and water flooded into the bomb aimer's compartment. The severity of the situation was increased when Bob Beioley shouted a warning of fire within the fuselage, creating a further need for the crew to get out as quickly as possible. Sgt Garrioch made his escape through the hatch in the cockpit roof, closely followed by George Hedge. Jock Hall, who was both wounded and burned, needed assistance from Bob Beioley and Bill Jordan. Having gained egress through the same hatch onto the top of the fuselage, Bob Beioley and Bill Jordan lowered the injured rear gunner down to the pilot and Wop/AG, who were already on the surface of the ice. Sergeant Garrioch was concerned that the burning wreckage would explode, but there was a more imminent danger facing the crew. As the fire raged, the ice beneath the wreckage was beginning to melt due to the heat. In turning around to move away, Sgt Garrioch glanced at the front turret. It appeared undamaged, but was pointing down towards the melting ice and slushy water. It was then he realised it contained the unmoving body of Sgt Reardon. In desperation the crew tried to break the plexi-glass cover on the turret with their bare hands in an effort to save their colleague, but to no avail. With the ice melting rapidly, and water rising up to their knees, the situation became desperate as the ammunition in the rear of the bomber started to explode. The survivors of the crew had no choice but to move away, as the front of the Wellington bomber slipped forward and the turret sank below the water line, taking Sgt Glyndwr 'Taffy' Reardon with it.

The five airmen now faced another ordeal, that of a trek of approximately 15 kilometres across the frozen surface of the Zuider Zee. It was only at this point the extent of Jock Hall's injuries was fully realised, and it was obvious he was not going to walk anywhere. Apart from bullet holes in his clothing, and burns to his face and head, the rear gunner had also sustained a severed foot. After rendering what medical attention they could, the crew utilised two parachute harnesses as a makeshift sledge on which to pull the injured flyer across the ice. Some twelve hours later, having been to the limits of human endurance, in appalling arctic like conditions, the five flyers were 'rescued' by a group of German soldiers. In true Hollywood film style they were told, *"For you the war is over"*, and it was. Sergeant Hall was taken to the Queen Whilhelmina Hospital in Amsterdam, where his leg was amputated just below the knee whilst, after the usual interrogation period, Sgt's Garrioch, Beioley, Hedge, and Jordan were interned in PoW camps. During 1969, when areas of the Zuider Zee (Ijsselmeer) were being drained for reclamation, the wreckage of T2702 was found. The remains of Glyndwr Reardon were recovered from the turret, by the Royal Netherlands Air Force, and taken to Nijmegen for burial in the Jonkerbos War Cemetery.

On 3rd March, F/Lt Morris led an attack against Cologne. Eight Wellington bombers took-off, but Sgt Bernard Kelly piloting Wellington, T2961, LS-S, was forced to turn back with engine trouble. However, nearly a month later, Sgt Kelly and his crew were destined not to return. On 31st March, they were participating in an attack against Bremen when their aircraft, Wellington bomber, T2703, LS-A, was attacked by an enemy nightfighter piloted by Feldwebel Karl-Heinz Scherfling, of III Gruppe/ Nachtjagdgeschwader 1. Cannon and machine gun fire raked the bomber killing the bomb aimer Sgt Thomas McWalter, and wounding the observer, P/O Hargrave McCosh. The aircraft, which was forced down, landed on the railway line near the station at Harens-Em, Germany.

Bernard Kelly, together with three members of his crew, was captured and taken prisoner of war. Sergeant McCosh, who was found near Osterfeld, succumbed to his injuries and died a week later, on 6th April.

Targets in Germany, France and Holland were attacked during March and April, a number of the raids being led by S/L Menaul. During the second week of March, Stewart Menaul adopted Wellington bomber, T2961, as his own aircraft, and it was in this machine he flew the last few missions before No.XV Squadron relinquished these bombers. He had the aircraft adorned just below the cockpit, with the figure of a scantily clad young lady named Jane. Jane was a cartoon character from an English newspaper who, together with her pet Dachshund dog, got into many scrapes which invariably led to Jane losing her clothes. In a scroll, above the figure, S/L Menaul had written the motto, *'No Bono Panico'*.

Although No.XV had only commenced operations with Wellington bombers during the last month of 1940, news began to circulate during April 1941, to the effect that the Squadron was to convert to Stirling bombers. The Stirling, which was a very large aircraft requiring a crew of seven, was the first of the four-engined heavy bombers to see service with RAF Bomber Command. However, until the news of another change of equipment became official, operations were to continue to be flown on the 'Wimpeys'.

A few days after the rumors started, the mighty bulk of a four-engined aircraft descended out of the sky and landed at Wyton; the first of the Stirling bombers had arrived. It was a dual-control aircraft from No.7 Squadron, and had made the short flight from Oakington, Cambridgeshire. The aircraft come complete with an instructor by the name of F/L Best. However, due to the fact there were many checks to be made, plus the fact the Stirlings were not scheduled to arrive all at the same time, the Wellington bombers were deemed to carry on.

'Robbie' Roberts was a member of No.XV Squadron's groundcrew who had serviced both Bristol Blenheim and Wellington bombers. When he saw the approach of the first Stirling to arrive at RAF Wyton, he stood and gazed in awe at this monster descending from the skies. When he saw it at close quarters, after it had landed, he wondered if he had a head for heights; he soon found out.

During mid-April, seven Wellington bomber crews from No.XV, were ordered to proceed to the Middle East immediately, with three more crews following before the end of the month.

The seven selected crews all arrived safely in Malta on 17th April. Unfortunately, of the crews posted later in the month, one failed to arrive. Sgt Walsh and his crew were later reported missing.

On 24th April, two Wellington bombers, R1498 and R1218, piloted by Sgt Smith and Sgt Jones respectively, took-off for an attack against Kiel. Both aircraft reached and bombed the target, then turned for home. The mission was uneventful until the aircraft reached England, when Sgt Jones and his crew were forced to abandon their aircraft due to lack of fuel. Their machine, R1218, crashed in Yorkshire, whilst Wellington bomber R1498, made it safely back to Wyton.

Another aircraft which came to grief was Wellington bomber, R1169, LS-D, which was 'run over' by a Stirling bomber. The Stirlings commenced operations on the night of 30th April, when five of the type were detailed to bomb Berlin. The raid was led by W/C Dale who, due to engine problems, bombed a concentration of searchlights at Hamburg. It is not known whether the engines continue to give trouble after the aircraft had landed but, as the Stirling taxied back to its dispersal, it collided with a Wellington bomber. The Wellington's back was broken, at a point adjacent to the tailplane assembly.

Squadron Leader Menaul, who also participated in the raid, abandoned his mission due to the failure of both starboard engines. He jettisoned his bomb load into the sea, and landed at Wyton without incident.

However, an incident occurred which 'Robbie' Roberts was never to forget. In those early days at Wyton the runways were not suitable for use by fully laden Stirling bombers, and it was necessary to fly the aircraft to Alconbury, where they were bombed-up, refueled and prepared for operations. On one occasion, having completed his duties, Robbie Roberts was recalled to his aircraft where an irate flight engineer made it known to Robbie his instrument panel was not functioning. Robbie made it quite clear that the panel had been checked, and left in full working order. However, when the rear of the instrument panel was checked, it was found that the wiring had been cut!! This of course rendered the aircraft unserviceable. Fortunately for Robbie, he never heard anymore about the incident and assumed it had been 'hushed-up'. The airfield at Alconbury was by the side of a busy road, with only a hedge dividing it from the public highway. Robbie remembers how small groups of people would congregate and watch the activities on the airfield, especially when the aircraft were being prepared for 'ops'. He often wonders!!

The Wellington era, for No.XV Squadron, was a relatively short period, but as the war progressed Bomber Command was planning to take the war to Germany in force. No.XV, being only the second Squadron to re-equip with the Stirling bomber, was to be a major part of that force.

The Geodetic Bomber

An unidentified Vickers Wellington bomber, of No.XV Squadron, photographed at R.A.F Wyton. *Courtesy of Don Clarke, MBE, The Mildenhall Register*

Stalag Luft 1, Germany,

Photographed outside the latrines at Stalag Luft 1, by a German photographer, are the crew of Wellington bomber, T2702, which was shot down on 10th February 1941. From left to right are: Bill Garrioch, pilot; Bob Beioley, obs; Bill Jordan, 2nd pilot; and George Hedge, w/op. *Courtesy of William Garrioch*

Photographed against a backdrop on the stage of the camp theater at Stalag Luft 1, on 2nd January 1942, are: Dicks; Flower; Bill Garrioch; George Hedge; Unknown; Bill Jordan; Brooks; Barber; Bristow; Bob Beioley; "Bonnie". *Courtesy of William Garrioch*

Higher Office

Joseph Cox, who relinquished command of No.XV, in December 1940, a month after the Squadron began to re-equip with the Wellington bomber. He retired from the Royal Air Force in 1958, with the rank of Air Vice-Marshal. His Decorations and Awards included the C.B., O.B.E. and D.F.C. He is seen in this photograph wearing the dress uniform of an Air Vice-Marshal, together with his Decorations and awards. *Author's Collection*

Out in the Cold

Vickers Wellington, Mk.Ic, bomber, T2624, LS-B, photographed at RAF Wyton, during the winter of 1940/41. The aircraft, which was delivered between November 1940 and February 1941, also saw service with the Czech Training Wing and No.311 (Czech) Squadron. It was lost on 23rd October 1941, when it crashed into the Irish Sea, during a cross-country exercise involving a flight to the Isle of Man. *Courtesy of Don Clarke, MBE, The Mildenhall Register*

Reunited

Left: After twenty-six years submerged in the Zuider Zee, Wellington bomber, T2702, 'surfaced' from the reclaimed land in 1967. The wreckage, along with the remains of 'Taffy' Jordan, the front gunner who perished in the crash, was found by a local farmer. In August 1971 William Garrioch and members of his family visited the area of the crash site, where he was present with the control wheel of his old aircraft. From left to right in the picture are: Col. A. De Jong, RNethAF; Bill Garrioch; Mrs Doreen Garrioch, wife; Doreen, cousin; and Mr Aalderink, farmer. *Courtesy of William Garrioch* Right: Forty-eight years after being shot down, four of the crew of Wellington bomber, T2702, meet for a reunion in a London hotel. From left to right are: George Hedge, W/Op; Bob Beioley, Obs; John Hall, R/G; and Bill Garrioch, pilot. *Courtesy of William Garrioch*

Demise of T2703

'On track'. On the night of 31st March/1st April 1941, Vickers Wellington, Mk.Ic, bomber, T2703, LS-A, was attacked by an enemy nightfighter, whilst participating in a raid against Bremen, Germany. The bomber, which was forced-down, landed almost intact on the railway line, near the station at Harens-Em, Holland. Following its removal from the railway line, the bomber, seen in the center of the photograph, was deposited in Werkhalle F, at Rechlin. A variety of other aircraft types can be identified, including a complete Wellington bomber outside the hangar doors (top left). *Courtesy of and via Flt/Lt T. N. Harris, XV (R) Squadron*

A 'Wellington' named 'Jane'

Squadron Leader Stewart 'Paddy' Menaul looks down from the cockpit of Vickers Wellington Mk.Ic bomber, T2961, LS-S. Above the scantily clad figure of *'Jane'*, is a scroll bearing the motto *"NO BONO PANICO"*. *Author's Collection, via the Late Air Vice-Marshal Stewart Menaul*

Vickers Wellington Mk.Ic, bomber, T2961, LS-S, photographed at RAF Wyton, during March/April 1941. The aircraft was one of a production batch of 300 machines delivered between June 1940 and February 1941. It was 'adopted' by S/L Stewart 'Paddy' Menaul as his personal aircraft, during the second week of March 1941. On the fuselage, under the cockpit, it carried the artwork form of *'Jane'*, a cartoon character from the Daily Mirror newspaper. The aircraft went onto serve with No.57 Squadron and No.18 OTU. It was 'struck-off-Charge' on 14th March 1942, after it hit some trees whilst taking-off from Bramcote. *Author's Collection, via the Late Air Vice-Marshal Stewart Menaul*

'Wellington' at Rest

Vickers Wellington Mk.Ic, bomber, R1279, LS-L, photographed at RAF Wyton. The aircraft, which was delivered between August 1940 and May 1941, went on to serve with No.9 Squadron. It was reported missing in action following an attack against Genoa, on 29th September 1941. *Courtesy of and via Flt/Lt T. N. Harris, No.XV (R) Squadron*

Squadron Leader 'Paddy' Menaul, pilot (center) poses with members of his crew under the forward fuselage of Wellington bomber, T2961. From left to right the crew members are thought to be: F/O Downs, F/E; Sgt Cossar, Wop; S/L Menaul, pilot; F/O Crochane, Nav; and Sgt Warner, RG. *Author's Collection, via the Late Air Vice-Marshal Stewart Menaul*

One of the Groundcrew

A studio photograph of 'Robbie' Roberts, who serviced Blenheim, Wellington and Stirling bombers, whilst serving as a member of No.XV Squadron's groundcrew. *Courtesy of 'Robbie' Roberts*

R1218 and R1498

Photograph of a painting of Vickers Wellington, Mk.Ic, bomber, R1218, LS-H, being prepared for operations. The aircraft, which was piloted by Sgt A W Jones, RAFVR, was delivered between August 1940 and May 1941. This aircraft crashed at Sand Hutton, seven miles north east of York, on 25th April 1941, after running short of fuel whilst returning from an attack against Kiel. The pilot circled beacons and searchlights using the distress procedure without response. Due to the fuel state, the crew abandoned the aircraft. Sergeant Gallaway, the rear gunner, damaged his leg, but the rest of the crew escaped injury. *Courtesy of and via Don Clarke, MBE, Mildenhall Register*

Vickers Wellington Mk.Ic bomber, LS-W, thought to be carrying the serial, R1498, was photographed at RAF Wyton after a mishap. The aircraft was one of a production batch of 550 bombers delivered between August 1940 and May 1941. Apart from being on-charge with No.XV Squadron, the machine also saw service with No.1504 Beam Approach Training Flight, No.22 Operational Training Unit and No.81 OTU. *Courtesy of Don Clarke, MBE, Mildenhall Register*

"D" for Damaged !!

The tailend of Vickers Wellington bomber, R1169, LS-D, after it was 'run over' by a four-engined Stirling bomber, on the night of 30th April 1941. The Stirling, piloted by W/C Dale, was landing back at Wyton following an attack against Hamburg, when it collided with the smaller static twin-engined aircraft. *Courtesy of and via Flt/Lt T. N. Harris, XV (R) Squadron*

Wellington bomber, LS-D undergoing repairs, following its encounter with a Stirling bomber. *Courtesy of Don Clarke, MBE, Mildenhall Register*

9

The Stirling Era

The first week of May 1941 saw the twin-engined Wellington bombers operating alongside the Squadron's newly arrived larger four-engined Stirling bombers.

On the night of 2nd May, the Squadron despatched one Wellington bomber to participate in an attack against Rotterdam. The aircraft, R1498, was flown by Sgt Leggate. Although the raid was successful, there was a tense moment for the pilot as he made his bombing run. With bomb doors open the Wellington approached the aiming point, but when the bomb release button was pressed only the incendiaries tumbled out. Five 500lbs H.E. bombs, which the aircraft was also carrying, had failed to release. With fifteen to twenty searchlights probing the night sky, and slight heavy flak bursts endeavoring to find a victim, Sgt Leggate made another attempt to release the bombs. On this occasion he was rewarded by the sight of three bomb bursts in the target area; the other two overshooting.

Two nights later the situation was reversed, when a single Stirling bomber from the Squadron joined an attack against the docks at Brest. The operation went without incident and the aircraft returned safely to base.

A further two nights later, Sergeant Leggate had the opportunity to fly with the 'big boys', when the two types of aircraft operated together on what proved to be the twin-engined bombers' last operational mission with No.XV Squadron. Three Stirlings accompanied Wellington, R1498, to Hamburg as part of a total force of 115 bombers.

The attack went without incident, apart from Sgt Leggate failing to locate his specified target and returning with his bomb load. On landing back at Wyton, he taxied the aircraft back to its dispersal area and shut down the engines. For No.XV Squadron, this action brought to a close the Wellington era. However, a new one was already underway. The 'big boys' had taken over; the Stirling era had begun.

On the night of 10th/11th May, Wing Commander Herbert Dale took-off at 22.35 hours for an operation against Berlin. His aircraft, N3654, LS-B, was one of two Stirling bombers detailed by the Squadron, to join a main attacking force of twenty-three aircraft. Shortly after midnight the C.O.'s aircraft was attacked by an enemy nightfighter, piloted by Oberleutnant Prinz Egmont Zur Lippe-Weissenfeld, of I./NJG1. The Stirling succumbed to the German's gunfire and crashed at Hoog-Spanbroek, between Hoogwoud and Opmeer, Netherlands, at 00.20 hours, killing the seven man crew. Although W/C Dale was buried at Bergen General Cemetery, his crew were listed as missing. They have no known graves and are commemorated on the RAF Memorial at Runnymede, Surrey.

Flight Lieutenant Raymond's aircraft, Stirling bomber, N6018, LS-C, was severely damaged whilst bombing from 18,000'. The aircraft was hit in the port inner engine by intense heavy flak, causing the airscrew to shear off and the motor to burst into flames. As the propeller cartwheeled away, it struck the port outer airscrew, causing further damage. Unfortunately, F/L Raymond's troubles were not over. His aircraft, whilst being held in the blinding glare of searchlights, was attacked by a single-engined enemy

fighter. Raymond's rear gunner, Sgt Jack Bushell, responded to the machine gun fire aimed in his direction by replying in a like manner. As the bomber 'corkscrewed' down into a dive, the crew released a flare which the searchlights swung onto. This respite from the intense light enabled the crippled bomber to gain the sanctity of the night sky. Having managed to evade both the fighter and the searchlights, N6018 limped home safely on two engines.

For their respective parts in the action, F/L Cuthbert 'Bobby' Raymond, RNZAF, was awarded a Distinguished Flying Cross, whist Jack Bushell, RAFVR, was awarded a Distinguished Flying Medal.

Wing Commander Patrick Ogilvie arrived on the Squadron on 16th May, to assume command of No.XV following the loss of W/C Dale. The former, who had been based at RAF Oakington, had only received his promotion to the rank of acting Wing Commander four days earlier. Further down the RAF 'chain' of command was a young man who held the lowly rank of Leading Aircraftman. During the month of May this young LAC was in Blackpool, Lancashire, but he too was soon to join No.XV. Although he was a groundcrew armorer, Don Clarke would, in later years, earn the affection and respect of many of his contemporaries for his work on behalf of the Squadron.

A quiet period between 16th-26th May when no operations were undertaken, should have given the 'hard-pressed' groundcrews a brief respite from repairing damaged aircraft. However, the mighty Stirling bomber was prone to problems with its undercarriage, and flying air tests and practice flights did not prevent some of the Squadron's aircraft from incurring 'self-inflicted' damage.

On one occasion Robbie Roberts was to gain first hand experience of the undercarriage problem, and came to know how the aircrew felt when the 'undercart' gave cause for concern. A No.XV Squadron Stirling had landed at a Yorkshire airfield with a technical problem. Robbie was sent north with instructions to inspect and repair the fault, and then fly back to base on the errant aircraft. Shortly after take-off an oil leak developed on one of the engines, which meant the pilot had to shut down the defective motor. Later, as the aircraft approached Wyton, it became obvious the undercarriage was not functioning properly. After circling the airfield for forty-five minutes, both the undercarriage legs were finally in the landing configuration, but it was not known if they were locked down. The pilot brought the mighty aircraft in for a 'touch and go' landing, bumping the wheels hard on the ground to see if they remained locked. A safe landing ensued on the second attempt, with the emergency vehicles following the Stirling across the airfield.

Amongst the bombers known to have succumbed to troublesome undercarriages around this period of time were Stirlings, N3638, LS-S and N6004, LS-F.

A Pilot Officer, who was already an experienced pilot with a completed tour of operations on Wellington bombers to his credit, arrived at

Wyton on 7th June and reported for duty with No.XV. His name would forever be synonymous with the Squadron, and he would serve it with honor, dignity and pride. Apart from his exploits when serving with the Squadron, the name of Peter James Somerville Boggis would be recorded in later years, when asked to officiate or represent No.XV at many official functions and events.

Pilot Officer Boggis, who had been accepted for a Short Service Commission in June 1937, had mixed feelings about flying the Stirling bomber. However, he recorded that once it was airborne, he found the Stirling pleasant to fly, and added it could take a lot of punishment.

As if to confirm the point made by Peter Boggis, Stirling, N6004, which had sustained damaged to its port undercarriage a couple of weeks before was, thanks to the groundcrews, back in service in time to participate in a mission against Dusseldorf on 11th June.

On 13th June, S/L Menaul led an attack against the German pocket battleship *Lutzow*, which was off the coast of Norway. Approximately one hour after take-off the formation, which included aircraft flown by F/L Gilmour, P/O Thompson and F/O Campbell, was recalled to base. However, on their return, these four crews found they were not allowed to stand down. Instead, they were detailed to participate in an attack later that same day against three cruisers anchored at Brest.

Other operations throughout the rest of June included missions against Kiel, Hannover, Wilhelmshaven, Bremerhaven and Hamburg. It was during an attack against the latter target, on the night of 29th/30th June, that the Squadron lost two more crews. The day itself was bright and sunny and, whilst the groundcrews toiled in the heat of the day to prepare the aircraft for the forthcoming operation, some of the aircrew took the opportunity to relax and sunbathe. One of the latter was Sgt Thorkilsen, wireless operator to Sgt Ronald Smith and his crew. As he rested in the warm sunshine, little did Theo Thorkilsen realise this would be the last time he would feel English soil against his back for four years. Later in the day, following the mission briefing, the nineteen year old wireless operator wrote a letter to his mother. He also prepared a small parcel for his two year old sister, to whom he sent chocolate which he had saved from his ration. Both of these items he sent off, prior to joining his crew for their pre-operational meal.

The Stirling bomber stood silhouetted against the waning sun as the aircrew bus which had driven them from Wyton to Alconbury, drew up along side it. Receiving an expression of good luck from the driver, the crew disembarked from the vehicle and were last seen climbing into the fuselage of the aircraft. Stirling bomber, N6016, LS-G, one of six machines detailed for the mission, took-off from Alconbury at approximately 23.00 hours; it was neither seen nor heard from again.

Many miles distant from the target, the crews of the No.XV Squadron aircraft saw the signs of battle. Fires in Hamburg were burning fiercely, flak was bursting around the attacking bombers and searchlights swept back and forth across the moonlit sky. As they continued on towards the target, one of the beams suddenly locked onto Stirling bomber, N6016, and no matter how much evasive action the pilot took, the brilliant shaft of light stayed with the aircraft. The flak continued to explode around the Stirling as other searchlights swung in its direction and formed a cone. Within seconds of the pyramid of light being formed the flak barrage stopped, and the bomber crew knew they were in serious trouble. Before they had time to think about their situation, 'Jock' Storie, rear gunner, shouted a warning of fighter attack. Theo Thorkilsen left his station in the fuselage, and endeavored to take a look out of the astrodome. Being short in stature, the wireless operator found it necessary to climb up and balance with one foot on his radio plotting table and the other foot on a stanchion. As he viewed the situation in this precarious position, with his legs wide apart, a hail of machine fire passed between his lower limbs. The firing ceased as the nightfighter, which he had not seen, turned away. Jumping down, he scurried back to the safety of his radio compartment and looked out of the small side window. To his horror, Sgt Thorkilsen saw the two port engines were on fire. The seriousness of the situation was brought home to him when he heard the pilot issue the order to get out. Without hesitation, he destroyed all secret paperwork and then set the explosive

charge inside the I.F.F. set, which transmitted a call sign to identify the aircraft as a friendly machine. These tasks completed, the redundant wireless operator grabbed his parachute and clipped it on. Having helped Sgt Dave Rees, the Welsh flight engineer, with his parachute, the pair made their way towards the forward escape hatch. In the confusion of trying to evacuate the stricken bomber, one of the crew members stood on the escape hatch, thereby losing precious seconds. A hefty kick from Sgt Thorkilsen quickly resolved the situation, as the offender moved and the hatch cover was discarded. One by one the men around the hatch jumped from the aircraft into the night sky. On clearing the aircraft, their parachutes billowed open into white 'mushrooms' above them and lowered them gently to earth. As Theo Thorkilsen floated down, he watched Stirling, N6016, LS-G, spinning and gyrating wildly in a downward spiral to destruction.

Although five of the crew survived and were taken prisoners of war, Sgt Ronald Smith, the pilot, known as D.A. to his crew, was killed when the bomber crashed and exploded at Ellerbeck, 14kms from the target. Flight Sergeant Richard 'Jock' Storie, the rear gunner, is thought to have been killed in the initial attack by the nightfighter, as no further instruction were received from him after his warning of the presence of the enemy aircraft.

The second No.XV Squadron aircraft lost that night was Stirling bomber, N6015, LS-A, piloted by P/O Richard Renshaw. This aircraft crashed into the sea off Kiel, having been shot down by flak. All seven members of the crew were killed.

At the beginning of 1941, the Commander-in-Chief, Fighter Command, Air Marshal Sir William Sholto Douglas, devised the concept of 'Circus' operations. A 'circus' was a daylight attack, in which a large force of fighters would escort a small formation of bombers to the target area. It was hoped that the bombers, acting as bait, would entice the Luftwaffe fighters into the air, where they could be 'pounced on' by the RAF fighters. One such attack, Circus No.33, took place on Saturday, 5th July 1941.

Whilst the main formation of this operation attacked the Five-Lille steel works, F/O Marshall attacked the marshaling yards at Abbeville. Although he had fighter escort, Marshall reported no enemy fighter opposition, and very little flak. The raid was not considered successful as most of the bomb load overshot the target and straddled the railway lines, creating some damage to the tracks.

The following day the same target was attacked by the same formation, when S/L Menaul led F/L Gilmore and F/O Thompson, as the first of two formations to, again, bomb the steelworks. The second formation, led by S/L Tim Piper, followed in and dropped their bombs on those already laid. Enemy fighters attempted to intercept the bombers but Sgt Ward, front gunner on F/L Gilmour's crew, was ready and opened fire as the hostile aircraft swung into the attack. Sergeant Ward later claimed the enemy machine as damaged. The rear gunner on Sgt Needham's crew, who were flying in the second formation, was not as successful as Sgt Ward had been. As he was about to open fire at an incoming enemy aircraft, two escorting RAF fighters flew into his sights, forcing him to abort his action.

On 7th July S/L Menaul led a three aircraft formation in an attack against a chemical factory at Choques, France, whilst F/O Boggis led another three aircraft to attack Frankfurt. The former attack was reported as successful with every stick of bombs hitting the target, sending large plumes of smoke skywards. The following day Stewart Menaul read about his own exploits, when the success of the raid was reported in the national press. He was later awarded a Distinguished Flying Cross for his part in the mission. At this same time, having completed thirty-two operational missions with the Squadron, Stewart 'Paddy' Menaul was declared 'tour expired' and posted to No.3 Group Headquarters, on 19th July.

The attack against Frankfurt was not as successful as the raid on Choques, due to the fact only two of the three aircraft detailed reached the target. Unrecorded circumstances led Peter Boggis to unleash his bomb load on Hocheim, fifteen miles to the west of the target.

Having previously seen service as a peacetime officer on No.148 Squadron, F/L T H 'Brian' Tayler, who had been a pilot instructor with

No.15 Operational Training Unit for approximately the last sixteen months, decided to apply for a posting to an operational squadron. His application was approved and on 22nd July 1941, he reported to No.XV at RAF Wyton. Knowing that one of his closest friends, S/L Tim Piper, was already on the Squadron, Brian Tayler was more than happy with the posting.

Unfortunately, unbeknown to Brian, S/L Piper had failed to return from a mission three days earlier, and was listed as missing. The news of the loss of his friend depressed F/L Tayler, and the feeling 'deepened' slightly when he was informed he would be taking over S/L Piper's Flight.

Initial reports implied Tim Piper had died along with his crew. However, later reports confirmed six of the crew had perished although the pilot, along with Sgt Armstrong one of the air gunners, had survived and were both prisoners of war. Their aircraft, Stirling bomber, N6018, LS-C, was shot down during an attack against the Lille power station, on 19th July. Flying with this crew as rear gunner, and recorded as one of those on the casualty list, was P/O Bushell. Thirty-three year old Jack Bushell, who had been awarded a D.F.M for his courage, had only recently been commissioned to the rank of Pilot Officer.

For his first operation with No.XV, F/L Tayler was detailed to fly as second pilot to Sgt Needham and his crew. The mission, on 23rd July, was a daylight attack against the German pocket battle-cruiser *Scharnhorst* which was berthed at La Pallice. However, due to the fact the aircraft's undercarriage could not be raised after take-off, Sgt Needham had no option but to jettison the bomb load, dump the fuel and return to base.

Two days later F/L Tayler was detailed to fly with F/O Thompson, on a raid against Berlin. The mission was carried out without incident and, although searchlights scanned the sky and flak burst around the bombers, Stirling, N6029, emerged unscathed. However, navigational problems on the homeward run caused the aircraft to run short of fuel and when the engines stopped due to fuel starvation, F/O Thompson was forced to ditch the aircraft in the North Sea. Although he did not know it at that precise moment, having more urgent matters on his mind, F/L Brian Tayler would soon be re-united with his friend.

After only four days on the Squadron, and five days in a survival dinghy, F/L Tayler's operational career ended. He, along with the rest of the crew, was captured by the Germans and made a prisoner of war.

Two sergeant pilots who had served the Squadron well were rewarded during early July when, on the 8th day of the month, Sgt Needham and Sgt Leggate were both commissioned to the rank of Pilot Officer. Further promotion was granted to P/O Frank Needham nearly a month later when, on 6th August, he was elevated to the rank of Flying Officer. Unfortunately, his aircraft was shot down two days later, during the early hours of 8th August, whilst participating in an attack against Essen. Stirling bomber, N3658, LS-E, was attacked by a Bf110 nightfighter piloted by Leutnant Kurt Loos, from I Gruppe, Nachtjagdgeschwader 1 (I./NJG1). The bomber crashed in flames on a farm 8kms south-south-west of Nijmegen, at 02.45 hours, with disastrous consequences. Apart from killing the entire crew, the 34 year old farmer, his 32 year old wife and their 6 month old son were also killed. Rescue of either the crew, or the farmer and his family, was out of the question due to the intensity of the fire. However, permission was granted by the Germans on Sunday, 10th August, for the bodies to be removed from the wreckage of the house and buried. Pilot Officer Leggate was more fortunate, he became 'tour expired' on 11th August and, having served his time on No.XV Squadron, was posted to No.3 G.T.F, at Newmarket.

On the night of 19th/20th August, Bomber Command despatched a total force of 108 aircraft, comprised of Hampdens, Halifaxes, Wellingtons and Stirlings, for an attack against Kiel. Amongst the latter type of aircraft were three Stirlings detailed by No.XV Squadron.

Apart from their usual crew compliment that night, each of the Stirling bombers detailed by No.XV was to carry an additional observer. The three 'passengers' consisted of a Short Brothers test pilot, an army colonel and a naval officer, all of whom were to gain first hand experience of an RAF bombing mission. The Royal Navy observer was First Lieutenant Peter Scott, son of Robert Falcon Scott the British explorer who lost his life in 1912, whilst attempting to reach the South Pole.

The rain, which had accompanied the crew as they journeyed out to the aircraft's dispersal area, ceased just before the scheduled take-off time. As the heavy clouds broke, a stream of sunlight bathed the airfield in a golden mellow light, causing Peter Scott to reflect on the forthcoming mission and what it stood for; the salvation of the British Empire.

Taking his place in the fuselage of Stirling bomber, W7431, LS-A, Peter Scott prepared for take-off. As the aircraft piloted by P/O Coran climbed into the darkening sky, the young naval officer watched with interest everything that was going on around him. During the outbound flight he was able to chat with the navigator and the wireless operator, both of whom shared the same section of the aircraft, aft of the flight deck of the bomber. There was little to report during that period of the mission, other than the appearance of flak whilst crossing over Heligoland. A heavier belt of flak was later encountered as the aircraft crossed the coast over Schleswig-Holstein, north-west of Kiel. This in turn was joined by the probing beams of searchlights, which swept back and forth across the sky as the aircraft flew steadily towards the target area. The flak grew in intensity, although most bursts appeared to occur below the bomber. Far from being afraid, the battle hardened naval officer found the experience edifying and informative. He was aware that nothing he could do would affect their chances of survival one way or another. However, the most heart-stopping moment of the attack occurred on the homeward leg of the mission. Pilot Officer Coran was holding the Stirling in level flight over the sea when, without warning, the engines lost power and the aircraft entered a stall. Fearing the worst, and pre-empting the pilot's command to *"stand by to abandon the aircraft"*, Peter Scott made a grab for his parachute. However, in endeavoring to regain control of his aircraft, P/O Coran's action of adjusting the propeller blades to fine pitch saved both the aircraft and its occupants. Gradually the bomber, which had fallen 4,000', pulled out of its dive, thankfully, before the order to abandon the machine had been given. Following recovery to level flight, an assessment of the situation was made, resulting in the conclusion that a build-up of ice could have caused the incident. Taking careful note of the reading on each instrument, P/O Coran nursed the bomber over the English coast north of Lowestoft.

Another heart-stopping moment occurred as LS-A was preparing to land, when a flare illuminated the night sky. The Stirling bomber was number two in the airfield circuit when the flare, dropped from above, signified the presence of an enemy 'intruder' nightfighter. Although the flare went down on the opposite side of the airfield. the crew of LS-A still felt vulnerable, but landed safely and without further incident.

None the worst for his aerial experiences with No.XV Squadron, First Lieutenant Peter Scott returned to the sea and his duties aboard ship. Those duties, combined with his skill and courage earned him many accolades. By the end of the war he had been promoted to the rank of Lieutenant-Commander, and received a D.S.O, an M.B.E. and three Mentions in Despatches. In later years he was to become Britain's most eminent ornithologist, work for which he was to receive a Knighthood.

At approximately 21.00 hours, on the night of 7th September 1941, Stirling bomber, N6045, LS-U, took-off from RAF Wyton to participate in an attack against Berlin. The bomber piloted by F/L Tarry, was one of seven machines detailed by No.XV Squadron as part of the main attacking force of 197 aircraft. N6045 reached the target at 00.54 hours, and transmitted a *'task completed'* signal. Nothing more was heard until 01.28 hours, when another signal was sent stating that the aircraft had been hit by flak and was badly damaged. It also stated the bomber was returning to base, but it never arrived. Not only were the crew captured by the Germans when the bomber crash-landed in a Dutch field, but so too was the almost intact Stirling.

Sergeant Richard Pape, the aircraft's observer, did not adapt to the idea of incarceration, and adopted an extraordinary policy for escape. The crash which led to his capture, and his subsequent escapades, are recorded in detail in his book, *'Boldness Be My Friend'*. Richard Pape later rose to the rank of Warrant Officer, and was awarded the Military Medal.

Another pilot who served with great distinction, and whose name was justly recorded in the annals of No.XV Squadron, was James Fraser Barron.

Born in New Zealand in 1921, Fraser Barron applied for pilot training with the RNZAF in October 1939. Having been awarded his 'wings' in January 1941, he was sent to England where following further training, he was posted to No.XV Squadron. The arrival at Wyton, in mid-June 1941, of this quiet, well-dressed, well-mannered young man, gave no indication as to the courage contained in his five feet six inch high frame. Sergeant Barron undertook his first operation on 7th July, when he participated in an attack against Frankfurt, flying as second pilot to F/O Peter Boggis. Another member of this crew had been Sgt Richard Pape.

By the middle of September James Fraser Barron had gained much valuable experience which, fortunately, would prove an asset at the end of that same month. Flying as captain of his own aircraft, Sgt Barron participated in an attack against Genoa, on the night of 28th/29th September. As his Stirling bomber, N6040, LS-C, crossed the French coast it was hit by flak. Barron elected to continue with the mission, unaware that shell splinters had severed part of the bomb-release cable. It was not until the aircraft was over the target, on its bombing run, that both the pilot and his crew realised that approximately 5,000lb of bomb load was refusing to budge from the fuselage bay. Unable to dislodge the offending bombs, Sgt Barron had no choice but to turn for home with the load still hung-up. To add to his problems, a detour from the planned route had to be made due to a severe deterioration in the weather. This detour had the adverse effect of consuming a substantial amount of fuel which in turn gave further cause for concern. As the bomber flew over the Channel Islands, the pilot was informed that the aircraft could remain airborne for only another ten minutes. Following a hurried consensus of opinion with the crew, it was decided to press-on rather than attempt ditching the aircraft in the sea. Fortunately, the fuel held out and allowed the aircraft to cross the Hampshire coast and effect a safe landing at Thorney Island. As the Stirling rolled down the runway, three of the engines spluttered and died through fuel starvation; the tanks were dry. The damage to N6040 was repaired, and aircraft and crew were reunited for a mission together on 14th October.

Climbing Out

A setting sun reflects on the clouds, as three No.XV Squadron Stirlings climb in 'Vic' formation into the evening sky. *Courtesy of Don Clarke, MBE, Mildenhall Register*

Groundcrew Armorer

Leading Aircraftman Don Clarke (later Corporal Clarke) photographed at Blackpool, Lancashire, during May 1941. LAC Clarke was an armorer with No.XV Squadron at RAF Wyton, RAF Bourn and RAF Mildenhall. He also served with No.622 Squadron, which was formed from "C" Flight of No. XV Squadron, in August 1943. *Courtesy of Don Clarke, MBE*

Line-Up of Heavies

A line-up of seven Stirling bombers, of No.XV Squadron, photographed at RAF Wyton, circa 1941. *Courtesy of S/L P.J.S. Boggis, DFC*

One of the 'Big Boys'

Short Stirling, serial N6018, coded LS-C, from a painting by the aviation artist, Keith Aspinall. The aircraft was severely damaged, on the night of 10th/11th May 1941, during an attack against Berlin. *Courtesy of Keith Aspinall*

'Belly Landing'

Undercarriage Problems 1. A sequence of three poor quality photographs showing a Short Stirling bomber, identified only by the Squadron code, S-LS, flying at almost zero feet over the runway, moments before making a wheel-up landing. The presence of the twin-engined Vickers Wellington bomber in the background implies the airfield could be Wyton. The second picture shows the aircraft moments before the four-engined bomber makes a belly-landing. In the third picture, the Stirling makes contact with the runway and slides to a halt with a minimum of damage to both aircrew and airframe. *Courtesy of and via Flt/Lt T. N. Harris, XV (R) Squadron*

"It was the undercarriage's fault, Sir"

Left: *Undercarriage Problems 2*. The first Short Stirling bomber taken-on-charge by No.XV was delivered to RAF Wyton on 10th April 1941. It carried the serial N3638, and was piloted by F/L Best. Apart from delivering the aircraft, F/Lt Best was charged with the task of instructing and training members of aircrew in flying the new four-engined bomber. The Stirling came to grief when, having landed from a training flight, it was taxying back to its dispersal area when the undercarriage collapsed. The presence of engine covers over the four motors indicates the photograph was taken some time after the incident, which occurred at 16.30 hours, on 21st May 1941. *Courtesy of and via Don Clarke, MBE, Mildenhall Register* Right: *Undercarriage Problems 3*. Short Stirling bomber, coded LS-F (possibly serialled N6004), photographed following a collapse of the port undercarriage leg; circa 1941. *Courtesy of and via Flt/Lt T. N. Harris, XV (R) Squadron*

At Dispersal

Stirling Bomber, N6004, LS-F, photographed in late May 1941. During its 'career' this aircraft saw service with No.7 Squadron, No XV Squadron, No.1427 Flight, No.1657 Conversion Unit and RAF Northolt. It was struck-off-charge on 13th September 1946. *Courtesy of and via Flt/Lt T. N. Harris, XV (R) Squadron*

Theo Thorkilsen – P.O.W.

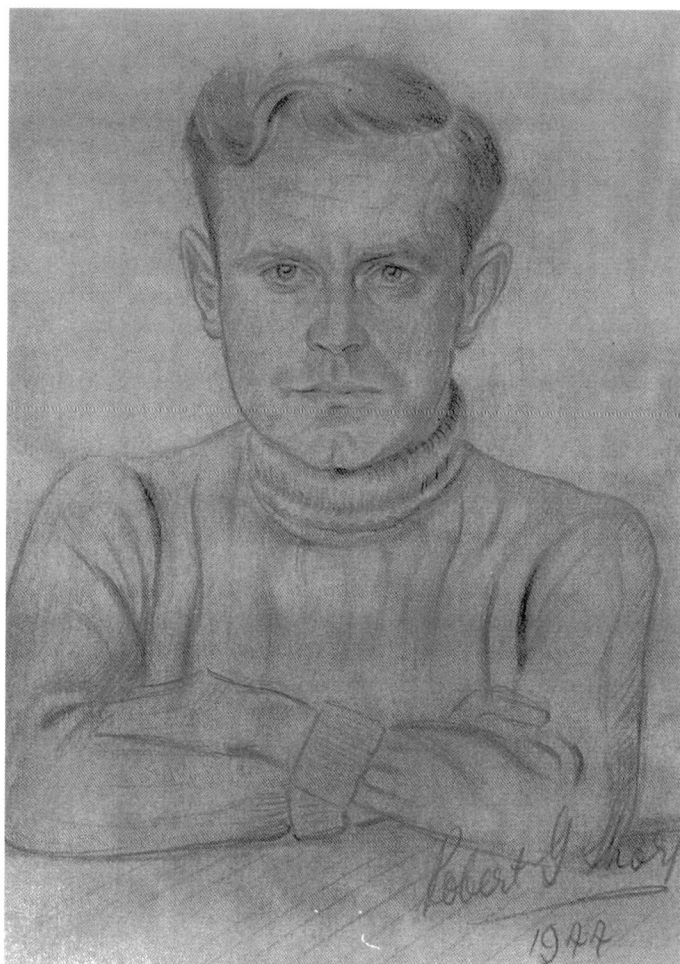

The first photograph Theo Thorkilsen was able to send home from Dulag 'A', Stalag IXc, to his mother, was a group picture with Theo sitting fourth from left in the front row. *Courtesy of Theo Thorkilsen*

Copy of a pencil sketch of Theo Thorkilsen in somber mood. The original was drawn by fellow prisoner of war Robert Thorpe, whilst at Stalag Luft IV, at Kiepheide, during 1944. *Courtesy of Theo Thorkilsen*

Circus Operation, No.33

'Circus' photo 1. On Saturday, 5th July 1941, S/L 'Paddy' Menaul was detailed to lead F/L 'Bobby' Gilmore and F/O Thompson on 'Circus' Operation No.33, against the Five-Lille Steelworks, France. The day was bright with good visibility and S/L Menaul had no difficulty joining up with the fighter escort (who were from Northolt and Kenley), at 10,000 feet over the east coast of Essex. The Biggin Hill and Tangmere Wings were flying as Target Support, with W/C 'Sailor' Malan leading the operation. The target support aircraft arrived over Lille ahead of the bombers as planned and circled the target until the Stirlings arrived. Messerschmitt Bf109 fighters could be seen some distance away forming into small groups, getting ready to pounce. *Author's Collection, via the late AVM Stewart Menaul*

'Circus' photo 2. Squadron Leader Menaul led F/L Gilmore and F/O Thompson over the French coast at Dunkirk, where the flak immediately burst into life, peppering the blue sky with shrapnel. As the bombers approached the target area, the Bf109s dived into the attack. Wing Commander Malan raced after two of the German fighters and opened fire at one of them, from a range of 300 yards. The German fighters endeavored to draw the supporting Spitfires away, by diving between the Stirlings. The Czech No.312 Squadron gave chase and claimed a Bf.109 as it made a stern attack on the bombers, whilst No.610 Squadron stayed in position to protect its charges. Owing to the target being obscured by 5/10ths cloud, S/L Menaul declared the first run over the aiming point a dummy run. Leaving their escort to take care of the German fighters, the three Stirlings ran the gauntlet of the intense flak a second time. *Author's Collection, via the late AVM Stewart Menaul*

'Circus' photo 3. Stirling Bomber, N3658, LS-E, receives a lot of attention from the German ground gunners, as flak bursts around the aircraft. The bomber, which was delivered around May 1940, survived the above action but failed to return from a raid against Essen, on 8th August 1941. *Author's Collection, via the late AVM Stewart Menaul*

'Circus' photo 4. With exploding steel peppering the sky around them, the three bombers dropped their loads, comprising of 15 x 1,000lbs and 30 x 500lbs bombs. The steelworks shook as the bombs detonated on impact with the target, which was left severely damaged. The sequence of photographs were taken from S/L Stewart Menaul's aircraft, and show Stirling bomber, LS-E, piloted by F/L 'Bobby' Gilmore. *Author's Collection, via the late AVM Stewart Menaul*

LS-J

In flight. Stirling bomber, W7429, LS-J, possibly piloted by F/O Marshall, flies low over the English countryside. The photograph was taken whilst flying in close formation with S/L Menaul's aircraft. *Courtesy of and via Don Clarke, MBE, Mildenhall Register*

In print. A page from a German issued aircraft recognition manual, shows Stirling bomber, W7429, LS-J (center), as one of three photographs depicting that type of aircraft. *Author's collection, via John and David Booth*

Stirling Formation

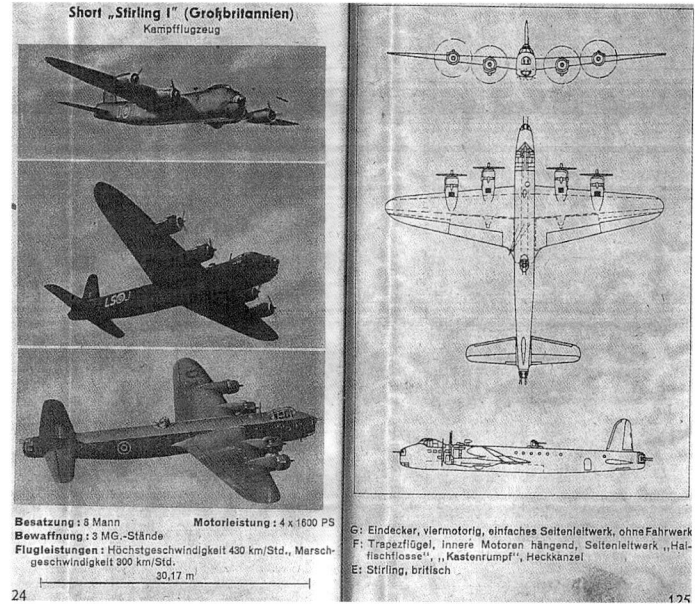

Left and two below: A sequence of three photographs taken during a daylight mission against a chemical factory at Choques, France, on 7th July 1941. The attack, on 7th July, was led by S/L 'Paddy' Menaul whose Stirling bomber, LS-H, was photographed from F/O Thompson's aircraft. The three aircraft involved in the mission were Stirlings, N3656, coded LS H, piloted by S/L Menaul, N3658, coded LS-E, flown by F/L Gilmore and N6029, coded LS-K, piloted by F/O Thompson. *Courtesy of and via Flt/Lt T. N. Harris, XV (R) Squadron*

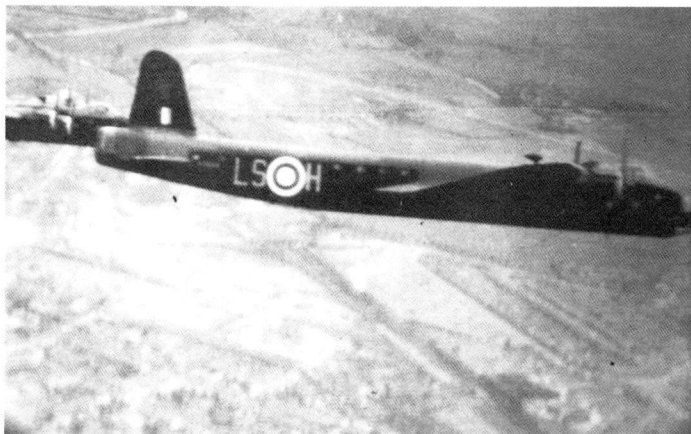

From in Service with the RAF to...'In the bag'

In the R.A.F. Photographed with some colleagues outside "B" Flight hangar (possibly of No.148 Squadron), is F/L 'Brian' Tayler, who joined No.XV Squadron on 22nd July 1941. He was posted missing three days later when Stirling bomber, N6029, LS-K, ditched in the North Sea, whilst returning from a raid against Berlin. From left to right in the photograph are: Robin ? ; Peter Rhodes; John Llewellin; T H 'Brian' Tayler. *Author's Collection, via G/Capt. T.H.B. Tayler*

'In the Bag.' Flight Lieutenant T H 'Brian' Tayler (second from left) photographed with his friend S/L Tim Piper (2nd from left) and two other 'inmates' at Offlag XXIB, Prisoner of War Camp, Schubin, Poland, during November 1942. Brian Tayler had hoped to fly with Tim Piper on No.XV Squadron, but on arrival at the unit, found the latter had been shot down. Brian Tayler was eventually reunited with his friend in a P.o.W Camp. *Author's Collection, via G/Capt. T.H.B. Tayler*

A Tail of Two Stirlings

Two of a kind 1. A Stirling bomber, identified as a No.XV Squadron machine by its LS-X code, photographed at RAF Wyton, during 1941. *Courtesy of and via Flt/Lt T. N. Harris, XV (R) Squadron*

Two of a kind 2. A Stirling bomber, coded LS-N, photographed at RAF Wyton. Three other No.XV Squadron Stirlings can be seen in the background. *Courtesy of and via Flt/Lt T. N. Harris, XV (R) Squadron*

A Prize for the Luftwaffe

A series of six photographs showing a 'prize' for the German Air Force. A Luftwaffe recovery team takes time out from their task, to pose for a photograph in front of Short Stirling bomber, N6045, LS-U, of No.XV Squadron. The aircraft crash-landed near Hengelo, Holland, at 04.05 hours, on the morning of 8th August 1941, whilst returning from an attack against Berlin. *Author's Collection via Jan Uithol*

With wings and tailfin removed, the fuselage of the Stirling resembles a beached whale. Note the open crew entry door adjacent to the Squadron code letter 'S'. *Author's Collection via Jan Uithol*

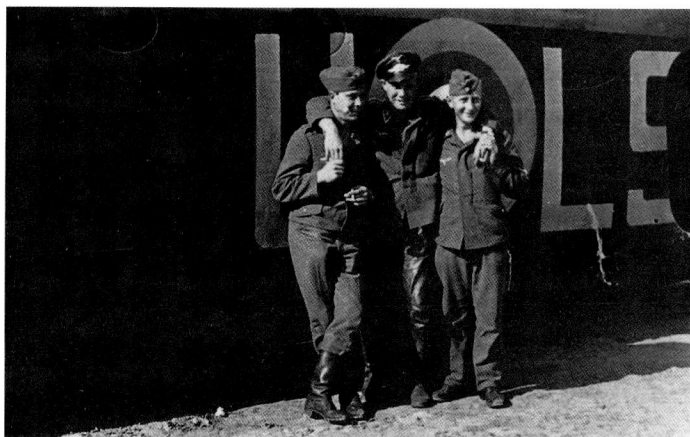

During the course of dismantling the aircraft, three unidentified members of the Luftwaffe recovery team find time to be photographed by the RAF roundel and Squadron identification code. *Author's Collection via Jan Uithol*

The fuselage lies battered and forlorn after the wings and four Bristol engines have been removed. A member of the recovery team stands guard by the tarpaulin covered motors, prior to their removal from the crash site. *Author's Collection via Jan Uithol*

Having been removed from the crash site, the fuselage is towed by a tractor unit through the local streets, accompanied by a group of interested Dutch children. *Author's Collection via Jan Uithol*

At its final destination, Stirling bomber, LS-U, becomes airborne again at the Luftwaffe compound, but only whilst being moved by heavy duty crane. The damage to the underside of the forward fuselage can clearly be seen in this photograph. *Author's Collection via Jan Uithol*

James Fraser Barron

Right: James Fraser Barron (extreme left) photographed with his crew (later in his career), at the tail end of their No.7 Squadron Lancaster. *Courtesy of S/L C. Lofthouse, OBE, DFC*

10

'MacRobert's Reply'

Short S.29 Stirling Mk.I. bomber, N6086, LS-F (Freddie), was a special aircraft. It carried not only a name, a coat-of-arms and a bomb load, but also a mother's revenge.

The German Luftwaffe had brought the fight to the British public by way of the Battle of Britain and the indiscriminate bombing of London.

Far from breaking the morale of the population, the ordinary citizens were showing their 'fighting spirit' by donating money, in varying amounts, to numerous fighting and aircraft funds. In the main, the money raised for the latter fund went towards the purchase of a fighter aircraft, usually a Supermarine Spitfire or a Hawker Hurricane. However, during 1941, a mother who had lost two of her three sons in action, revealed her 'fighting spirit' in a more determined way. She provided the total amount of money for the purchase of a Stirling bomber, in order that the fight might continue in the name of her sons.

Lady Rachel MacRobert, an American by birth, was the daughter of Dr and Mrs William Workman, of Massachusetts, U.S.A. It was in 1909 whilst she was travelling across Europe with her parents, both well-known explorers and authors, that Rachel Workman met Alexander MacRobert. At the time of the meeting Alexander MacRobert, Rachel and her parents were all on board a ship returning from India. Although there was a thirty year age gap between Alexander and Rachel, they were attracted to each other. After a very discreet courtship they married in York, on 7th July 1911, a year after Alexander had been knighted for his public services in both Britain and India. Naturally, his young wife become Lady MacRobert.

For Alexander MacRobert, a Scottish gentleman of some means, it was a second marriage; his first wife having died in 1905, following an illness of a cancerous nature.

Following the creation of his British India Corporation, Sir Alexander was given a further honor in 1922, when he was created a Baronet, and given the full title of Sir Alexander MacRobert of Cawnpor and Cromar. The latter name in his title came from the Cromar Estate, which he had acquired from Lord Aberdeen in 1918, although the latter remained in residence on the Estate until his death. The House at Cromer, which was built in 1904, was later transferred into a Trust set up by Lady MacRobert and renamed Alastrean House. This estate, comprising of some nine thousand acres, abounded the hundred and twenty-two acre estate, now known as Douneside, which Sir Alexander had acquired in 1888.

During their eleven year marriage (Sir Alexander died at Douneside on 22nd June 1922), the couple were blessed with three sons; Alasdair, Roderic and Iain.

Alasdair, being the eldest son, succeeded to the baronetcy on the death of his father. He developed an interest in aviation and formed a company which provided aviation related services. His life was cut short on 1st June 1938, when he was killed in a plane crash near Luton, Bedfordshire.

Roderic, the second son to whom the title passed, had been commissioned into the Royal Air Force, and was flying with No.6 Squadron in Palestine. Three years later, having been promoted, F/L Sir Roderic MacRobert was in Iraq, commanding a detachment of Hawker Hurricanes. On 22nd May 1941, he was engaged in a strafing attack against the airfield at Mosul, which was host to many German aircraft. Having made a diving attack in which he set light to two petrol laden lorries, Sir Roderic dived again for a second attack. He was seen to complete the action and pull away, but after that his aircraft was lost from view. The young pilot's remains were recovered, and laid to rest in the War Cemetery at Mosul.

The third son, aged 24, became the fourth baronet. Like his brother, whom he had just succeeded, Sir Iain was a commissioned officer in the Royal Air Force. Following Roderic's demise, the R.A.F gave Sir Iain a short period of leave, which he spent with his mother at Douneside. It was to be his last trip home.

Approximately five weeks after his brother's death, Sir Iain was reported missing whilst carrying out a search for a bomber crew known to have ditched into the North Sea. No trace of the young Baronet, or his aircraft, was ever found. His name, like so many others, was later added to the Runnymede Memorial.

Having lost her family, speculation arose as to whether Lady MacRobert would return to her native America. In reality, she had no reason to stay in Scotland, but the memory of her sons and what she could do in their name, gave Lady MacRobert cause to remain in Britain.

Her reply to those who wondered what she would do, came less than a month after the loss of Sir Iain. Enclosed in a letter to Sir Archibald Sinclair, the Secretary of State for Air, was a cheque for £25,000, with which to purchase a Stirling bomber. In making this donation, Lady MacRobert asked for the aircraft to carry the inscription, *MacRobert's Reply*. She also requested that it bore the MacRobert family crest, and be piloted by a Scotsman.

As if this was not sufficient, Lady MacRobert donated a further £20,000 for the purchase of four Hawker Hurricane fighters. Three of the fighters were to bear the title and name of each of her sons respectively, whilst the fourth aircraft carried the name, *MacRobert's salute to Russia*.

Stirling bomber, N6086, LS-F, was duly selected to become *MacRobert's Reply*, whilst No.XV Squadron was chosen as the Unit to take charge of the aircraft. Unfortunately, Lady MacRobert's wish for a Scotsman to fly the aircraft was not acceded to. However the pilot chosen, F/O Peter Boggis, who came from Barnstaple, Devon, was a distant relative of Lady MacRobert, through the 'Workman' side of the family.

The aircraft, which had been taken-on-charge by No.XV Squadron on 15th September, was officially handed over to F/O Boggis by W/C Ogilvie, following an official ceremony at RAF Wyton on 10th October 1941. With the bomber came a letter from Lady MacRobert, in which she wished the crew good luck. Flying Officer Peter Boggis and his crew formed up in front of the Stirling, whilst W/C Ogilvie read the letter aloud before it was presented to F/O Boggis. The letter also stated that she knew the crews who flew the aircraft would always strike hard, sharp and straight to the mark. In conclusion, Lady MacRobert gave thanks to those who had care of her *'Reply'* and prepared the aircraft for its flights.

The first mission undertaken by *MacRobert's Reply*, and her crew, occurred two days later, on 12th October, when seven aircraft from No.XV Squadron were detailed for an attack against Nuremberg. When the Stirling arrived over the target, Peter Boggis noticed a considerable number of fires burning. He circled the area for thirty-five minutes, before commencing his bombing run. Five 1,000lb and seven 500lb bombs tumbled out of the aircraft's bomb bay, at 01.15 hours. It was with a sense of pride, on behalf of Lady MacRobert, that the crew watched as the bombs erupted into clumps of billowing black smoke as they exploded in the sea of flame below. By a strange coincidence, as if representing Lady MacRobert, Major Egan, an American military officer, flew as an observer on this operation. He flew with S/L Gilmour and his crew, on Stirling bomber, W7443.

The mission was considered successful, even through one aircraft failed to return to base. Stirling, N6047, piloted by P/O Colbourne, was shot down by a nightfighter, and crashed at Mariembourg, Belgium. *MacRobert's Reply* was more fortunate, and landed at Wyton, at 03-50 hours.

During the rest of October and November, *MacRobert's Reply* and her crew carried out attacks against targets in Germany, Czechoslovakia and France.

A daylight attack, scheduled for 15th December against the docks at Brest where the German warships *Gnisenau* and *Scharnhorst* were berthed, had to be abandoned due to severe weather conditions. However another attempt, made three days later, was to prove very successful.

An attacking force of forty-seven aircraft, nine of which were detailed by No.XV Squadron, headed for the docks. The rest of the force was made up of aircraft from No's 7, 10, 35 and 76 Squadrons. As the 'invaders' reached the target area they found enemy fighters awaiting their arrival. Cannon and machine gun fire streaked back and forth across the sky, and three Stirlings from No.7 Squadron went down in flames.

No.XV Squadron, who were detailed as the second wave, went into the attack. As they dived from 17,000' down to 14,000' heavy flak burst around the sky. Enemy fighters above peeled off and joined the fray, each endeavoring to get a bomber in its gun sights.

Peter Boggis informed his crew he was commencing the bombing run, and held the Stirling in straight and level flight. At 12.37 hours, *MacRobert's Reply* lifted slightly in the air as the weight of the cargo was released from the bomb bay. Black columns of smoke rose up into the air just short of the dry dock installation, but still within the prescribed target area. F/O Boggis also reported black smoke rising from *Gnisenau* after the bombing. During the raid *MacRobert's Reply* was attacked by three Bf.109 enemy fighters, one of which was later claimed as damaged.

Whilst F/O Boggis and his crew were lucky to evade the wrath of the three enemy fighters, two other crews from No.XV Squadron were not so fortunate. Stirling, W7428, piloted by F/O Bunce, was last seen over the target area, going down in a vertical dive towards the sea, with fire emanating from the port wing. Stirling, N3665, piloted by F/L Heathcote, MiD, also failed to return. This aircraft was last reported going down under control, but with smoke trailing from one engine. Although both aircraft were clearly out of the fight, enemy fighters continued to attack both bombers.

The following day a message was received from The Chief of the Air Staff, Air Chief Marshal, Sir Charles Portal, K.C.B, D.S.O, M.C, congratulating the crews who took part in the raid.

Sir Charles stated he had the greatest admiration for the skill with which the attack was planned, and the gallantry with which it was carried out against fierce opposition.

Four members of No.XV Squadron were awarded immediate decorations for their respective courage during this mission. Wing Commander Patrick Ogilvie, F/O Peter Boggis and F/O Cyril Vernieux each received a D.F.C., whilst F/S Richard Hardy was awarded a D.F.M. The latter, a member of F/O Vernieux's crew, had shot down a Bf109, whilst the pilot took evasive action following the fighters attack.

Lady MacRobert could be justifiably proud, not only of her Stirling, but also the crew who flew it. On a Christmas card dated 1941, she wrote, *"My thoughts are always with you"*.

Having been included in the collective congratulations from the Chief of the Air Staff on the success of the raid against Brest, F/O Boggis received personal congratulations nine days later. On 28th December he received a postagram from the Commander-in-Chief, Bomber Command, Air Marshal Sir Richard Pierse, following the award of the D.F.C.

Wing Commander J C MacDonald arrived on the Squadron during the first week of January 1942, prior to officially taking command of No.XV, on 7th of the month. Wing Commander Ogilvie, being tour-expired, was posted to Farnborough, Hampshire on 11th January.

Having completed his second tour of operations, Peter Boggis was posted, on 26th January, to No.15 Conversion Flight. A new unit, formed on that same date, No.15 C.U. was established for the operational training of Stirling pilots and crews. It was to be based at Alconbury and commanded by F/O Cyril Vernieux.

MacRobert's Reply was also destined to leave the Squadron at the end of the month, unfortunately, in a somewhat undignified way. A member of the aircraft's groundcrew was to experience first hand the episode which led to the Stirling's demise.

Sergeant Stanley P Smith, who had joined the R.A.F as an Instrument Maker Apprentice at RAF Cranwell, on 10th January 1939, was posted to No.XV Squadron on 2nd September 1941. He was placed in charge of the Squadron's Instrument Section, in conjunction with a 'Class E' Reservist by the name of Sgt Iggleston. It was in this capacity that Sgt Smith flew north to Scotland on *MacRobert's Reply*, when ten aircraft from the Squadron were deployed to Lossiemouth, on 28th January, for what transpired to be an abortive attack against the German battleship *Tirpitz*.

The ten Stirlings left RAF Wyton at 10.30 hours, led by W/C McDonald; each bomber armed and carrying a bomb load. There were no reported incidents on route, and all aircraft had arrived and landed safely at Lossiemouth, by 16.00 hours.

The following day, eight crews were detailed for an attack against the *Tirpitz* moored in Trondheim Fjord. The take-off time was set at 00.30 hours on the morning of 30th January. Although the weather was not particularly good over the Scottish base, it was stated to be excellent over Norway. Eventually, only five Stirlings took-off for the attack, but the promised good weather never materialised forcing the aircraft, including *MacRobert's Reply*, to return to Lossiemouth.

Upon the return of the five bombers, the groundcrews worked feverishly to prepare their charges in readiness for another attempt at the German warship. Stanley Smith was attending to some details aboard one of the Stirlings with three of his colleagues, when somebody yelled a warning of *"Fire"*. The aircraft, which was still 'bombed-up' with eight 2,000lb S.A.P's, had caught fire whilst being refueled. The four men on board did not need to be told twice, and scrambled out of the fuselage without delay. As he made his escape, the electrician cut off the power at the accumulator trolley, but the armorer called for it to be restored. As the electrician did not retrace his steps to the acc. trolley Sgt Smith restored the power, thereby enabling the armorer to 'jettison' the bombs. This action caused the runners to accelerate their pace, but they need not have worried as the bombs were 'safe'. As Stanley Smith and the armorer made their escape to a safe distance, the crash tenders arrived and dealt quickly with the situation.

The inclement weather precluded any operations from taking place during the following week and, with snow and rain aggravating the situation, a decision was made for the Squadron to return to Wyton on 6th February. The first Stirling bomber to take-off headed for Wyton, but the remaining aircraft which followed were diverted to Peterhead due to becoming iced-up shortly after getting airborne.

The next morning during its take-off run, *MacRobert's Reply* slew to port off the runway and careered through a dispersal area. It collided with a stationary Spitfire, which was crushed against the rear fuselage. The port undercarriage of the bomber dug into the soft ground and was ripped off, due to the sudden deceleration of speed. The port mainplane broke away from the wingroot as the aircraft slid to a halt. Sergeant Smith and the aircraft's rear gunner, who were at the aircrew rest position, released the upper escape hatch and made good their escape whilst checking for signs of fire. At the same moment an unknown person 'inadvertently' released the crew dinghy, an action which was to have an unexpected bonus for Stanley Smith. As they scrambled out of the aircraft, one of the aircrew retrieved the half-bottle of rum contained in the dinghy rations and passed it around. They all had a good swig from the bottle, including Sgt Smith, who was experiencing the taste of rum for the first time in his life.

Although the crew of *MacRobert's Reply* returned to Wyton aboard another Stirling, Stanley Smith was left at Peterhead with instructions to recover certain items from the crashed bomber. Included amongst the items, which were to be taken back to Wyton, were the aircraft's bombsight and clock. An airman was also left at Peterhead to assist Sgt Smith in his allotted tasks. In order to keep the image that Lady MacRobert's aircraft was still flying, Sgt Smith was also ordered to remove the family coat-of-arms crest panel from the nose of the aircraft and return it to RAF Wyton for application to another Stirling. However, the instruction to paint out the name *MacRobert's Reply* was overlooked, and was seen by many a bystander as the dismantled aircraft headed back to Cambridgeshire, on the back of a "Queen Mary" low-loader, RAF transport vehicle.

Once back at Wyton, the MacRobert family coat-of-arms crest panel was applied to Stirling bomber, W7531, which was also given the code LS-F. The intention had been for this second Stirling to continue to act as Lady MacRobert's '*Reply*', but unfortunately the aircraft failed to return from a mission on the night of 17th/18th May 1942. It had taken-off from RAF Wyton at 21.40 hours, as one of eight aircraft detailed for mining operations in the Baltic and Heligoland Bight. As W7531 flew over the target area, enemy ships below opened fire with a barrage of flak which hit the port inner engine. Illuminating the night sky around it with the fire which inevitably followed, the Stirling flew over the Little Belt Bridge, inducing the ground defenses to open fire. The stricken aircraft crashed in the forest at Galsklint, on the Island of Funen, Denmark at 02.10 hours. Sergeant Duncan Jeffs, the mid-upper gunner, was thrown out of the crashing aircraft as it impacted with the ground. He survived, was captured and made a prisoner of war The rest of the crew were killed, including two of the original *MacRobert's Reply* crew, together with F/L Neville Booth who had gone along 'for the ride'. Initially, the Germans did not remove the bodies from the wreckage of the aircraft, an insensitive action which angered the local people. However, sensing trouble could arise, the German Commandant ordered the local fire brigade to amend the situation. The fallen airmen of No.XV Squadron were duly removed from the debris of *MacRobert's Reply* and taken to the chapel at Odense cemetery, approximately fifteen miles from the crash site.

Although no longer on the Squadron, Peter Boggis was saddened to learn of the loss of this aircraft and its crew, especially as F/O John Ryan, the navigator, had flown in the same capacity on the original '*Reply*'. He was also a good friend and had been invited to Peter's wedding, which took place the day before the mission. Unfortunately, due to pressure of operations, John Ryan had to decline. In an almost prophetic letter to Peter Boggis's mother, dated 12th May 1942, declining the invite, John Ryan wrote:

"...we are putting on maximum efforts every night, so I won't be able to be present in person, but I most certainly will be present in spirit ...

...[Peter] is a wonderful person and a great character ... He carried me through the major part of my operations and I was very proud to be part of his crew – as were all the other members ...

... my sincerest wishes to Peter and Kay for a long, happy and holy life. God will bless them both".

The wedding took place at St John's Church, Torquay on Saturday, 16th May 1942. Unbeknown to the bride and groom, whilst they were inside the church taking their wedding vows, a Guard of Honor, made up of cadets from the local Initial Training Wing, was forming up outside; a gesture which surprised the couple as they emerged from the church. Another surprise which awaited them was a telegram in verse, from Lady MacRobert, in which she wished them a happy future.

Less than one month after the occasion of his marriage, Peter Boggis was engaged in another important event and, like his wedding day, it was one he would never forget. On 12th June 1942, whilst at No.1651 Conversion Unit, F/L Peter James Somerfield Boggis, D.F.C, R.A.F, was formally presented to the King and Queen. The occasion was an official visit to RAF Waterbeach, Cambridgeshire by His Majesty King George VI and Queen Elizabeth.

Two weeks later F/Lt Boggis received an invitation to attend an Investiture at Buckingham Palace on 14th July, where His Majesty King George VI would officially present the award of the Distinguished Flying Cross.

Due to his having been the original pilot of *MacRoberts Reply*, and having remained in contact with her, Peter Boggis and his wife Kay, became good friends with Lady MacRobert. Kay Boggis, a nursing sister, eventually became a companion and confidante of Lady MacRobert, and was to spend a great deal of time at Alastrean House.

It was at Alastrean House, which had been set aside by the MacRobert Trust in 1943 as an RAF and Commonwealth Air Forces recreation and leave center, that Lady MacRobert was able to provide a fitting and more permanent memorial to the memory of her sons. Carved in the portico over the entrance, the memorial could not fail to be seen by those who visited the estate.

Lady MacRobert loved her 'adopted' home and chose not to return to her native America. Following her death from heart failure on 1st September 1954, she was laid to rest in consecrated ground at Douneside. Her passing was mourned by many, including Peter and Kay Boggis. At the funeral the Royal Air Force paid homage to its generous benefactor, and saluted Lady MacRobert with a formation flypast of nine Gloster Meteor twin-engined jet fighters.

The Baronetcy had been extinguished in 1941 with the death of Sir Iain, and although this generous hearted American 'British' Lady had gone, her legacy was to live on. Lady MacRobert's wealth was not entirely due to the provisions made by her late husband, a large inheritance from her American family made her a very wealthy lady in her own right. At Rachel MacRobert's request, her trusted friend and confidante, Kay Boggis, was appointed a Trustee, and went on to serve the fund for over forty-one years. During that period Kay, with the other Fund Trustees, ensured that Lady MacRobert's wish to help others was carried out.

Portrait of a Lady

A post war photograph of Lady Rachel MacRobert, the American born benefactor of the Royal Air Force, taken in the gardens at Douneside. *Courtesy of and via S/L P.J.S. Boggis, DFC*

Country Estate

Alastrean House, nestling in the Scottish hills, with the estate of Douneside in the background. *Courtesy of and via S/L P.J.S. Boggis, DFC*

Alastrean House, photographed before it was destroyed by fire in 1952. *Courtesy of and via S/L P.J.S. Boggis*

Sterling Wings

A gold 'Sweetheart' brooch, given to Lady MacRobert, by her son Sir Iain, during his visit home on leave, following his brother's death. Five weeks later, Sir Iain was killed in action whilst flying with the Royal Air Force. Following the death of Lady MacRobert, her friend and confidante, Kay Boggis was allowed by the MacRobert Trusr to wear the 'wings' in memory of her close relationship with Lady Rachel MacRobert. *Author's collection*

A Mother's Reply

Flying Officer Peter Boggis and his crew listen attentively to an address by W/C Patrick Ogilvie, O.C., No.XV Squadron (right), during the official handover ceremony of the Stirling bomber, on 10th October 1941. The aircraft was purchased by Lady MacRobert in memory of her three sons, and named, *MacRobert's Reply. Courtesy of and via S/L P.J.S. Boggis, DFC*

Having previously read out aloud its content, W/C Patrick Ogilvie hands out copies of the letter from Lady MacRobert, following presentation of the Stirling bomber purchased by her. Included in the photograph are from left to right: W/C Ogilvie, C.O.; Sgt Wilson, R/G; F/O Boggis, Pil; Sgt Watson, F/E; P/O King, 2/Pil; Sgt Clark, F/G; P/O Ryan, Nav. *Courtesy of and via S/L P.J.S. Boggis, DFC*

Flying Officer Peter Boggis (left) gives the 'thumbs-up' sign from the pilot's seat of Stirling bomber, N6086, LS-F. *Courtesy of and via S/L P.J.S. Boggis, DFC*

Flying Officer Peter Boggis (wearing the white flying suit) and his crew take a final glance at a flight map, before boarding the aircraft. *Courtesy of and via Flt/Lt T. N. Harris, XV (R) Squadron*

Pilot Officer Ryan, navigator (left), P/O King, 2nd pilot (center) and F/O Peter Boggis, aircraft captain (left), look out from the starboard side of the flight deck of the Stirling named, *'MacRobert's Reply'. Courtesy of and via S/L P.J.S. Boggis, DFC*

Sergeant Wilson, rear gunner, Sgt Allen, wireless operator, and P/O Ryan, navigator, pause in the crew entrance doorway of Stirling, N6086, prior to taking the aircraft into the air. *Courtesy of and via Flt/Lt T.N. Harris, XV (R) Squadron*

MacRobert's Reply banks round at low level to fly over Stirling, W7439, LS-N (left) and Stirling, N3665, LS-S (right). Both aircraft, having been serviced by the groundcrews, are left in a state of readiness. Cockpit and engine nacelle covers have been applied to both aircraft. *Courtesy of and via S/L P.J.S. Boggis, DFC*

The starboard inner engine of LS-F, bursts into life, as F/O Peter Boggis starts-up the first of the four motors. Although this photograph was taken at RAF Alconbury, at the start of *MacRobert's Reply* 12th mission, it is reminiscent of the start of the aircraft's inaugural flight from Wyton. *Courtesy of and via S/L P.J.S. Boggis, DFC*

With all four motor running smoothly and cockpit drills completed, F/O Peter Boggis prepares to taxi the Stirling bomber out from its dispersal area. *Courtesy of and via Don Clarke, MBE, The Mildenhall Register*

With F/O Peter Boggis at the controls, *MacRobert's Reply* lifts into the air. *Courtesy of and via S/L P.J.S. Boggis, DFC*

With a roar of thunder from the four Bristol Hercules engines powering *MacRobert's Reply*, F/O Peter Boggis performs a low-level banking maneuver, over Stirling bomber, N3665, LS-S. *Courtesy of and via S/L P.J.S. Boggis, DFC*

The 'flying display' over, Peter Boggis returns *MacRobert's Reply* to the allocated dispersal area, where the crew leave the aircraft. Walking towards the camera, in line-abreast, from left to right are: Sgt Watson, F/E; Sgt Wilson, R/G; Sgt Soned, MUG; P/O King, 2/Pil; F/O Boggis, Pil; P/O Ryan, Nav; Sgt Clark, F/G; Sgt Allen, W/Op. *Courtesy of and via S/L P.J.S. Boggis, DFC*

With the sun low in the sky, and its duty for the day complete, Stirling bomber, *MacRobert's Reply* is left at readiness for the next mission. At least five other Stirling bombers are visible in the line-up. *Courtesy of and via Don Clarke, MBE, The Mildenhall Register*

The setting sun provides an interesting background to nine Stirling bombers, of No.XV Squadron, Photographed at RAF Wyton, during late 1941. The aircraft second from left is thought to be *MacRobert's Reply*. *Courtesy of and via Don Clarke, MBE, The Mildenhall Register*

At work ... The groundcrew of *MacRobert's Reply*, utilise a loaded bomb trolley under the nose section of the aircraft, as a seating arrangement. Standing center rear of the picture is 'Robbie' Roberts, who was concerned about losing his head for heights whilst working on the Stirling. *Courtesy of 'Robbie' Roberts*

... And at play. 'Robbie' Roberts (extreme right) takes time out for some liquid refreshment at the Rising Sun Public House, Bourn, Cambridgeshire. With him are two friends from the Squadron, Fred 'Lofty' Batchelor, standing in the doorway, and Bill 'Appy' Eaton. *Courtesy of 'Robbie' Roberts*

Right: Flying Officer Peter Boggis (left) and Wing Commander Patrick Ogilvie, D.S.O (Officer Commanding No.XV Squadron), photographed a few days after the daylight raid against the German battleships in Brest Harbor, on 18th December 1941. For the valor displayed during the attack, a number of members of the Squadron were to receive awards. Wing Commander Ogilvie, received a D.S.O, F/O Boggis and F/O Vernieux were each awarded a Distinguished Flying Cross, whilst F/S Richard Hardy was awarded the Distinguished Flying Medal. *Courtesy of and via S/L P.J.S. Boggis, DFC*

'Warship Raiders'

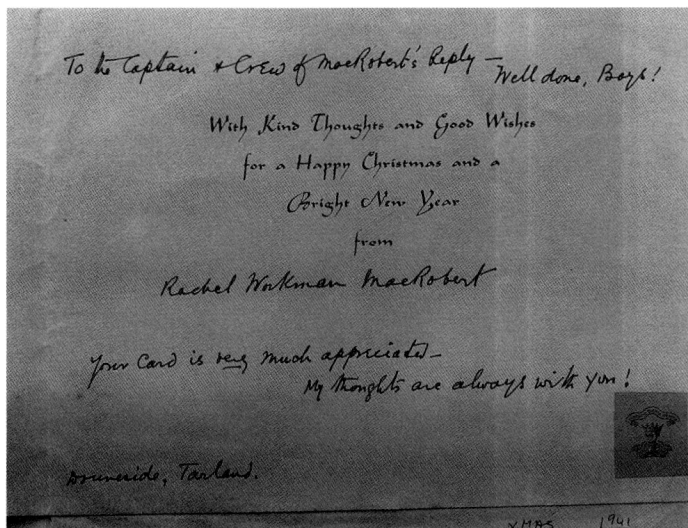

A message from Lady MacRobert, to Peter Boggis and his crew, written in a Christmas card dated 1941. To the bottom right of the card is the MacRobert Family Coat-of-Arms. *Courtesy of S/L P.J.S. Boggis, DFC*

Squadron Photograph

The officers and other ranks of No.XV Squadron, photographed with an unidentified Stirling bomber, at RAF Wyton, on a cold, misty day in January 1942. *Courtesy of S/L P.J.S. Boggis, DFC*

Words of Appreciation

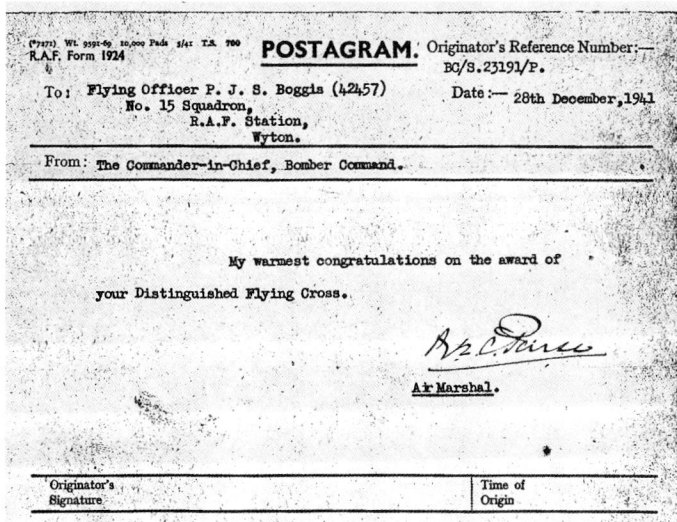

Photo-copy of a Postagram received by Peter Boggis, on the award of his D.F.C, from Air Marshal Sir Ralph Peirse, the Commander-in-Chief, Bomber Command. The text reads, *"My warmest congratulations on the award of your Distinguished Flying Cross". Courtesy of S/L P.J.S. Boggis, DFC*

The MacRobert Crest

The MacRobert Family Crest, as displayed on Stirling bomber, Serial N6086, coded LS-F. *Author's Collection*

The Loss of the Second MacRobert's Reply

Left: Duncan Jeffs, the sole survivor of the second *MacRobert's Reply*, photographed standing at the former German anti-aircraft gun positions, at Little Belt Bridge. The picture was taken after the capitulation in 1945. *Courtesy of and via 'Robbie' Roberts* Right: Flying Officer John Ryan, navigator on *MacRobert's Reply* and good friend S/L Peter Boggis. Due to pressure of operations, the former had to decline an invitation to his friend's wedding. Two days after Peter was married, F/O Ryan was killed on one of the operations which had prevented his attendance. *Courtesy of S/L P.J.S. Boggis, D.F.C.*

The preserved wooden crosses which marked the original resting places of those killed when the second *MacRobert's Reply* crashed, still bear the names of the crew. They are from left to right: F/L N G R Booth (passenger), Sgt A Spriggs, Sgt R Nicholson, Sgt J B Butterworth. The cross on the extreme right is not that of one of the crew. It bears the name Sgt H Davies, with a date of death given as 1 October 1942. *Courtesy of and via 'Robbie' Roberts*

Squadron Leader Peter Boggis, D.F.C., represents the MacRobert Trust at the commemoration of the 50th anniversary of the end of the war at Assistens Cemetery, at Odense, during November 1995. The Danish Guard of Honor present arms as S/L Boggis marches forward in order to lay a wreath from the MacRobert Trust, to the memory of twenty-five allied airmen buried in the cemetery. *Courtesy of and via S/L P.J.S. Boggis, D.F.C.*

The memorial stone, in Galsklint Forest, near Odense, Denmark, marks the crash site of Stirling bomber, W3571, LS-F. The bronze plaque on the lower portion of the stone records the names of the crew members. *Courtesy of and via "Robbie" Roberts*

Flying Officer Peter Boggis, D.F.C., married Kay at St. John's Church, Torquay, on Saturday, 16th May 1942. Cadets from the local Initial Training Wing, identified by the white flashes in their forage caps, formed-up outside the church whilst the ceremony was in progress. Neither Peter or Kay had any idea the Guard of Honor was there until they emerged onto the sidewalk. *Courtesy of Kay Boggis*

A Most Important Occasion – One

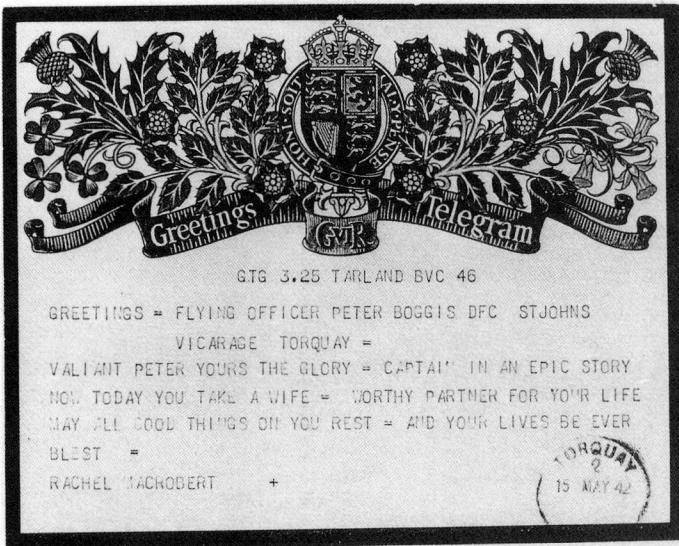

Meeting his King. Flight Lieutenant Peter Boggis, D.F.C., bows to formality as he shakes hands with His Majesty King George VI, during a visit by the Monarch to No.1615 Conversion Unit at RAF Waterbeach, on 12th June 1942. *Courtesy of and via S/L P.J.S. Boggis, D.F.C.*

The greetings telegram (wire) sent by Lady MacRobert, to the Peter and Boggis, on the occasion of their wedding. It read: *"Valiant Peter yours the glory = Captain in an epic story, Now today you take a wife = Worthy partner for your life, May all good things on you rest = and your lives be ever blest. Courtesy of Kay Boggis*

A Most Important Occasion – Two

Meeting his Queen. During the Royal visit to No.1615 Conversation Unit, at Waterbeach, Peter Boggis (2nd from right) had the double honor of also being formally introduced to Her Majesty Queen Elizabeth. *Courtesy of and via S/L P.J.S. Boggis, D.F.C.*

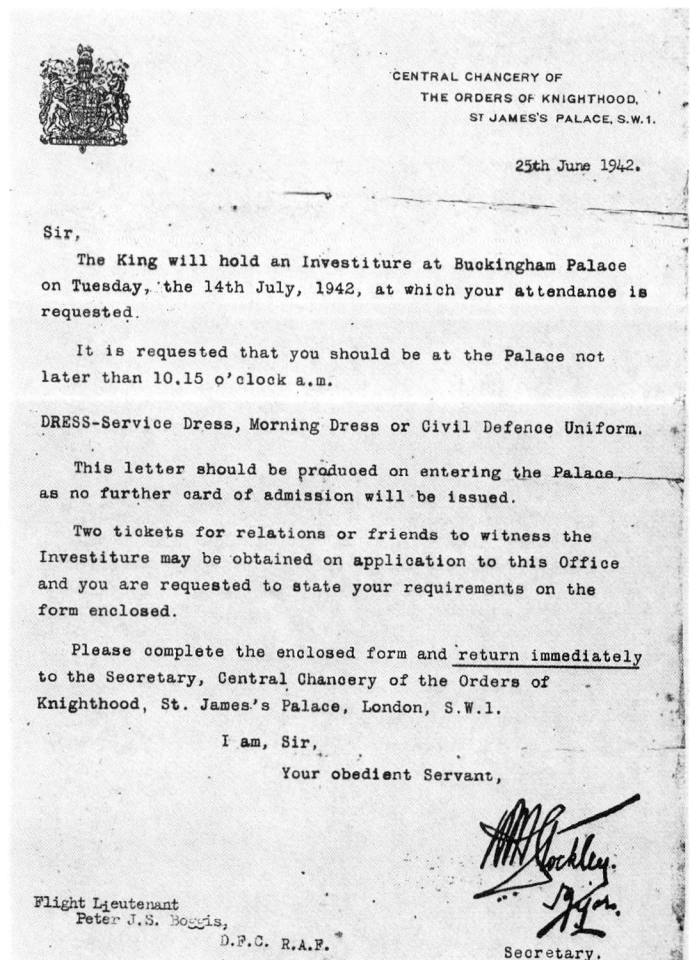

A copy of the formal invitation, dated 25th June 1942, requesting F/L Peter Boggis to attend Buckingham Palace, London, on 14th July 1942, for the Investiture of his Distinguished Flying Cross. *Courtesy of S/L P.J.S. Boggis, D.F.C.*

A Most Important Occasion – Three

Looking happy and relaxed, Kay and Peter Boggis enjoy an evening out in the company of Lady MacRobert. *Courtesy of and via S/L P.J.S. Boggis, D.F.C.*

Carved in Stone

Lady Rachel MacRobert's permanent memorial to her three sons, carved in stone, in the portico over the entrance to Alastrean House. The memorial, which also pays tribute to other airmen who died, including those of the Battle of Britain reads:

VIRTUTIS GLORIA MERCES

ALASTREAN HOUSE

IN MEMORY OF MY SONS

SIR ALASDAIR SIR IAIN SIR RODERIC
1912-1938 1917-1941 1915-1941

MACROBERT

R.A.F
COMRADES OF ALL THE BRAVE
THE FAITHFUL AND THE TRUE
AND
IN GLORIOUS MEMORY OF
"THE FEW"

Courtesy of and via Kay Boggis

At Rest

Lady Rachel MacRobert died of heart failure on 1st September 1954. Although born in the United States of America, Rachel MacRobert never returned to her native homeland. She was buried in consecrated ground at Douneside, Scotland, where the Royal Air Force paid homage to its benefactor. *Courtesy of and via 'Robbie' Roberts*

The MacRobert Banner Goes Home

The tradition lives on. At an informal presentation at RAF Laarbruch, West Germany, during May 1988, 'Robbie' Roberts offers the MacRobert family banner into the safe keeping of No.XV Squadron. The occasion was witnessed by a number of ex-No.XV Squadron wartime personnel. Included in the photograph are: Back row: Len Miler, George Mead, Peter Youde, 'Chick' Chandler. Front row: Unknown, 'Robbie' Roberts, Hugh George, Betty George, Ken Pincott, W/C Mike Rudd, O.C. XV Sqdn, Rodney Pope. *Courtesy of and via 'Robbie' Roberts*

11

"The Thousand Plan"

The inclement weather, which had created so many problems during the first couple of months of 1942, abated in early March allowing operations to increase in intensity.

On 8th March, a maximum effort attack against Essen was carried out by a total force of 211 aircraft. The leading machines were fitted with the Gee navigational aid, which made identification of the German city easier than on previous occasions. The attacking force included twenty-seven Stirling bombers, of which ten were from No.XV Squadron. The latter were led by W/C MacDonald piloting Stirling, N6065, LS-G. The only aircraft of the type lost during this mission was Stirling, N3673, LS-D, piloted by F/Lt Nicholson. The No.XV Squadron aircraft, which had taken-off shortly after midnight, was shot down by flak and crashed in Holland at 03.05 hours. The crew of nine, three of whom were qualified pilots, were all killed.

Essen was attacked again over the following two nights, and on the second of these two missions, No.XV Squadron nearly lost its commanding officer. Having arrived over the target early, W/C MacDonald circled the city for approximately twenty minutes. His aircraft, Stirling, W7463, LS-B, attracted the attention of the German flak gunners, who attempted to shoot down the bomber. Although they failed in their endeavors, the Stirling was hit in many places by shrapnel from bursting flak and, as a result, a considerable amount of fuel was lost. However, both the aircraft and its crew returned safely to base, where they landed at 00.26 hours.

Flight Sergeant Fraser Barron participated in all three attacks against Essen and on completion of the third raid (his 29th mission) was informed he had been awarded a Distinguished Flying Medal.

The weather turned again on 11th March. Thick fog lingered during the day, increasing in intensity as night fell, preventing any major operations taking place until the night of 25th March. However, mine-laying operations were carried out on 23rd and 24th March.

James Fraser Barron, D.F.M, who had been promoted to F/S in December 1941, received a commission to Pilot Officer on 23rd March; the same day he had another encounter with enemy flak. It was his 30th, and almost final, mission. Barron's aircraft, Stirling bomber, N3669, LS-C, was one of three machines detailed for the first of the two mining operations. As he was preparing to drop the mines from an altitude of 850', Stirling N3669 was struck by bursting flak. Apart from the outer skin of the aircraft being peppered with nearly one hundred holes, elevator trim cables were severed and damage was sustained inside the forward fuselage. It was a long and arduous journey home but, with the assistance of F/S Cowlrick the second pilot, Fraser Barron managed to get the bomber back to England.

Two nights later, on 25th March, the Squadron detailed three experienced crews to participate in an attack against Essen, whilst one 'rookie' crew was sent on a mine-laying operation in the Bay of Biscay. The newcomers, or 'Freshman' crew as they were known, were ordered to plant their mines off Lorient. All was quiet as they reached the dropping zone. Sergeant Lucas, the captain of the 'rookie' crew, piloted Stirling, N6065, LS-G, low over the water. The mines were released in the 'Gardening' area, at 01.00 hours from an altitude of 800'. Suddenly and without warning, the German flak ships made their presence known. Shrapnel from exploding flak bursts hit the aircraft and severed the hydraulic pipes serving the rear turret, rendering it u/s. Mindful of the condition of his aircraft, Sgt Lucas diverted to Predannock airfield in Cornwall, where he landed at 04.00 hours. The Stirling was repaired, and two nights later was back in action from its home base.

On the night of 5th April, No.XV detailed seven aircraft to participate in a raid against Cologne, two of which returned early. Shortly after take-off, F/O Brian Ordish began to experience problems with the port inner engine of Stirling bomber, W7505, LS-V. By the time he reached the French city of Lille, the defective engine was u/s, forcing F/O Ordish to abandon his mission. Although the bomber was carrying five 2,000lb high capacity bombs, only two of these were jettisoned into the North Sea during the return journey. The other early return was S/L Matthew Wilson who, having taken-off late, would have to have made the return journey from Cologne in daylight. A thought neither he, or his crew, relished.

The mission, which was led by W/C MacDonald, was made in bright moonlight. Although this assisted the Luftwaffe nightfighter pilots in their task of finding the bombers, it also meant that any alert gunner would see the approach of the fighters. The mid-upper gunner on Stirling, N3674, LS-T, was one of the latter. He saw a Bf109 positioning itself for an attack and yelled a warning to his pilot, W/C MacDonald. Flames spat from the fighter's guns as the mid-upper turret turned to face the aggressor. The Bf109 peeled off and away, giving the bomber crew the impression the fighter had been hit. However, no claim was made against the enemy aircraft.

Of the five aircraft which took-off to bomb Hamburg on the night of 8th April, only two actually made it to the target. Flying Officer Ordish was forced to return because the 2nd pilot, F/S Rigby, RCAF, had forgotten his oxygen mask! Jimmy Rigby was to redeem himself some time later, by being granted a commission. Flight Lieutenant Barr returned because his gyro instruments went u/s after take-off, and F/O Bennitt endured undercarriage trouble. The two crews who reached Cologne, P/O Barron and S/L Wilson, found the target obscured by 10/10ths cloud, but nevertheless released their respective bomb loads in the appropriate area. Although flak was both heavy and intense, it was the elements which caused the only injury of the night, when Sgt Gould the mid-upper gunner on S/L Wilson's crew, suffered frostbite. However, two nights later on 10th April, fate was to deal a cruel blow to both S/L Matthew Wilson and Sgt Edgar Gould. Squadron Leader Wilson, F/O Ordish and F/O Bennitt were all detailed for an attack against Essen, whilst F/Sgt Shepherd and Sgt Lucas were order to carry out mine-laying operations at Heligoland off the German coast, east of the Frisian Islands. Owing to elevator control problems, coupled with a faulty intercom system, F/O Ordish was again forced to return early. Flying Officer Bennitt and S/L Wilson however continued with their allotted task. Over the target area flak was again heavy and accurate, and both bombers succumbed to shrapnel damage. Squadron Leader Wilson's aircraft, Stirling, N3703, LS-G, was severely hit, but fortunately remained airworthy. However, tragedy struck at 04.35 hours, when the bomber was approximately four miles from its home base. The aircraft became difficult to control, dropped from the sky and crashed at Godmanchester, one mile south-south-east of Huntingdon. Both S/L Wilson and Sgt Gould were killed in the crash; the rest of the crew survived.

Cyril Vernieux, who had been awarded the D.F.C for his part in the attack against Brest in December 1941, returned to the Squadron on 12th April with the rank of Squadron Leader. A week later F/O Brian Ordish, having been on active duty with No.XV since July 1941, was declared 'tour expired'. He was posted, as an instructor, to No.214 Conversion Unit on 19th April 1942, but like Cyril Vernieux, he too would return to the Squadron.

The night of 25th/26th April saw ten aircraft from No.XV Squadron detailed for an attack against Rostock. Three of the bombers were forced to return early, including Stirling, W7505, LS-V, piloted by W/C MacDonald. This aircraft had taken-off at 22.05 hours and was flying over the island of Sylt, on the outward leg, when was it attacked by a German nightfighter. Fire spat from the cannons as the enemy aircraft made a firing pass at the Stirling, inflicting damage upon the bomber. Whilst the damage was being assessed the ground gunners opened up with a barrage of flak, which also struck the aircraft causing further impairment to its aerial stability. W/C MacDonald abandoned his task and immediately returned to base. As he nursed the aircraft gently across the North Sea, he took the opportunity to jettison the bomb load. The crew, each aware of the situation in which they found themselves, worked as a team and urged the Stirling to keep flying and get them home. A landing, albeit with the undercarriage retracted, was made at RAF Alconbury at 03.04 hours, without injury to any of the crew. F/O Conran and his crew also had the misfortune to encounter an enemy nightfighter, but unfortunately they were not quite so lucky. A Messerschmitt Bf110 from V Gruppe/Nachtjagdgeschwader 3 (V./NJG3) attacked Stirling, W7514, LS-B, at 01.05 hours, killing Sgt David East the rear gunner. Although he was mortally wounded in the stomach, Sgt Gordon Surridge the wireless operator/air gunner, managed to bale out of the stricken aircraft with the rest of the crew. Sadly, he died of his wounds three days later in Tonder Hospital, whilst undergoing treatment. The rest of the crew were captured and made prisoners of war. On this same night, F/S Rigby flew his second mission as captain of his own aircraft, Stirling Bomber, W7515, LS-Q. Having carried out a 'Nickel' (leaflet dropping) raid over Lille the previous night, he was now detailed to attack Dunkirk with twenty 500lb general purpose bombs. Although the defenses were reported as fairly strong, he was able to successfully complete his task and return to base.

The first week of May 1942 saw some changes to the Squadron with regard to personnel. Flight Sergeant Jimmy Rigby was granted a commission and promoted to the rank of Pilot Officer. Being 'toured expired', both S/L Vernieux and P/O Wright were posted non-effective to Wyton, whilst A/S/L Bobby Gilmore made a welcome return to the Unit. Operationally, the month of May was a fairly quiet period for No.XV Squadron, with a minimal number of aircraft participating in mine-laying sorties and attacks against German cities. However, losses were still incurred. Flight Sergeant Ivan Charbonneau, RCAF, and his crew failed to return from an attack against Warnemunde, on the night of 8th/9th May. The entire crew were killed, when their Stirling bomber, W7528, LS-G, crashed at Brodersby, Germany. Sergeant Doyle and his crew were lost on the night of 29th/30th May, when their Stirling bomber, W7515, LS-Q, failed to return from a mine-laying operation off the Frisian Islands.

Since his installation as Air Officer Commanding-in-Chief Bomber Command on 22nd February 1942, Air Marshal Sir Arthur Harris had been 'fighting' against the politicians who wanted to direct resources to other areas of the war effort, rather than to Bomber Command. It became obvious to Air Marshal Harris that in order to show the worth of the men and machines under his command, he needed to undertake an operation of great importance and magnitude. By the middle of May 1942, Harris had formulated a plan for an operation involving the maximum number of bombers available, to attack one selected German town on the same night. The operation became known as *'The Thousand Plan'*, as that was the number of aircraft intended to participate. Originally Hamburg was to be the recipient of this massive aerial attack, but in the event of unfavourable conditions on the chosen night, Cologne was to be substituted. Two sets of dates, covering the full moon period, were also set aside for the attack. The nights of 27th/28th May and 30th/31st May. The first date passed without incident, due to unfavorable weather conditions over both cities. However, at approximately 09.30 hours on the morning of 30th May, having received weather reports for that night, Air Marshal Harris looked at a map of Europe and pointed to Cologne. Just over twelve hours later, 1,047 aircraft from Bomber Command, Coastal Command and Flying Training Command were despatched to Germany. Included in the total figure of assorted machines heading for the cathedral city, were eighty-eight Stirling bombers, of which twelve were detailed by No.XV Squadron. Due to exactor throttle control problems, F/O Phillips and F/S McMillan were both forced to return early, but the remaining ten aircraft continued on to the target. Locating the aiming point presented no difficulties to the aircrews, as clear skies and bright moonlight illuminated the target area from above. Wing Commander MacDonald, who was leading the Stirlings of No.XV, reported later that both the streets and railways could be clearly identified; as were the Cathedral and the River Rhine.

Sergeant Melville piloting Stirling, N3756, LS-O, had identified the aiming point by locating the Hohenzollern Bridge, across the Rhine to the east of the Cathedral. Having dropped his bomb load and obtained the aiming point photo, Sgt Melville received a course for home from his navigator. However, the pilot of an enemy twin-engined nightfighter had other ideas, and endeavored to shoot down the bomber, but found he could not finish the task he had started. Each time the fighter attacked, the gunners on the bomber responded with equal vehemence, whilst Sgt Melville threw the Stirling into evading corkscrew actions. After the fourth attack, the mighty four-engined aircraft successfully evaded its pursuer, and returned to base without further incident.

Two nights later, on 1st June, another 1,000 bomber raid was scheduled; on this occasion the target was to be Essen. Again, No.XV Squadron detailed twelve aircraft to participate in the attack, but W/O Cowlrick's aircraft, Stirling, W7561, LS-F, failed to take-off. Only nine of the Squadron's aircraft actually bombed the target due to the fact both W/C MacDonald and S/L Gilmour returned early. The former experienced problems with an electrical fault to the bomb release gear, whilst the latter arrived too early over the target and could not definitely locate the aiming point. Weather conditions over the target were not as accommodating as they had been over Cologne. Eight-tenths cloud, combined with industrial haze, hampered identification of the aiming point, but most crews bombed on existing fires having followed the River Rhine to a point north of the Rhur.

The following night Essen was again the target for Bomber Command, but on this occasion only 195 aircraft participated. No.XV Squadron detailed seven Stirlings for the attack, one of which failed to return. Stirling bomber, N3728, LS-T, piloted by W.O. Andrew Cowlrick, RNZAF, was attacked by a nightfighter flown by Feldwebel Paul Gildner, of II./NJG2. The bomber, which was shot down, crashed between Melick and Herkenbosch near Roermond, Holland at 03.10 hours.

On 6th June W/C Lay, who had arrived at RAF Wyton four days earlier, formally took over command of No.XV from W/C MacDonald, the latter having been posted to Horsham St. Faith, Norfolk. That night the Squadron detailed seven aircraft for an attack against Emden. The raid, which was led by the new C.O., was very successful and all crews returned safely.

During the second week of the month, a number of mine-laying 'Gardening' operations were carried out by the Squadron in the Frisian Islands. The following week on 19th and 22nd June, further attacks against Emden were detailed. From the Squadron's point of view, the attack on 19th June was unsuccessful. Of the ten aircraft detailed to participate, two failed to take-off, one did not reach the target, two bombed the secondary target and one aircraft failed to return. The second raid fared little better. Pilot Officer Miles, piloting Stirling, N3759, LS-Q, and Sgt Robert Melville, flying Stirling, N3756, LS-C, were both forced to return due to exactor throttle problems. Although P/O Patterson was one of the eight remaining crews to reach the target, he experienced problems of a more serious nature. As he held Stirling bomber, R9351, LS-R, straight and level for his bombing run, his aircraft was caught in a cone of searchlights. Violent, evasive action failed to shake of the blinding beams, forcing the pilot to order the bomb aimer to jettison the bomb load. Enemy fighters, seeing the plight of the four-engined bomber, swung into the attack and opened-up with bursts of cannon and machine-gun fire. The rear turret of the Stirling was rendered u/s, damage was sustained to the starboard mainplane and main landing wheel which was holed in several places. P/O Patterson eventually managed to evade his attackers and get both his aircraft and crew home.

A return to *'The Thousand Plan'* was made on the night of the 25th of the month, when the selected target was Bremen. This time only 960 aircraft were mustered for the attack. Included in the total number were thirteen Stirlings detailed from No.XV Squadron. Although all thirteen aircraft took-off, Sgt Melville was again forced to turn back as his Stirling flew over Alkmaar, Netherlands when N3756, LS-C, once more developed exactor throttle problems. The remaining twelve Stirlings continued on course to the target, over which was found 10/10ths cloud cover. By the time the aircraft of No.XV Squadron started their respective bombing runs, a number of large fires were already burning. Mushrooms of fire, smoke and debris erupted into the air as the bombs dropped by the Squadron rained down and exploded amongst the sea of flames; the reflection of which was later reported as seen on the clouds. Although W/C Lay reported the sighting of three bombers being shot down, all No.XV's aircraft returned, including two Stirlings which sustained damage when hit by flak. Sgt Bebbington, piloting Stirling, W7525, LS-E, was forced to land at Bircham Newton, Norfolk due to a shortage of fuel.

The month ended with the loss of S/L Richmond and his crew, on 29th June. They were participating in a raid against Bremen, when their Stirling, N3757, LS-G, was shot down by flak. Fortunately the crew were able to bail out before the aircraft crashed at Hartward, Esens, Germany. Unbeknown to the Squadron, who despatched Sgt Russell-Collins on an air-sea rescue search off the Dutch coast, S/L Richmond and his crew had become 'guests' of the enemy.

During the latter part of June, P/O Fraser Barron became 'tour expired' and was posted to No.1651 Conversion Unit for a rest. Three months later, in September 1942, he returned to operational duties with No.7 (PFF) Squadron, with whom he completed a second tour. In December 1943, following another rest period, Fraser Barron volunteered for further operational duties. His request was granted and he rejoined No.7 Squadron. Unfortunately, on the night of 19th/20th May 1944 his aircraft failed to return, and he was listed 'missing'. James Fraser Barron, who started his operational career with No.XV Squadron as a Sergeant, rose to the rank of Wing Commander. His decorations included a Distinguished Service Order and Bar, Distinguished Flying Cross and Distinguished Flying Medal.

Apart from a number of mine-laying sorties, the month of July saw No.XV Squadron participate in attacks against harbor and port installations, as well as industrial targets in Germany, with attacks against Bremen, Wilhelmshaven, Lubeck, Vegesack, Duisberg, Hamburg and Dusseldorf. The Squadron recorded the loss of two aircraft during this period. Stirling bomber, W7524, LS-D, piloted by P/O Robert Melville, RAAF, was lost on the night of 16th July during an attack against Lubeck. The aircraft, which was shot down by flak, crashed into the tidal area of the River Sneum, near Esbjerg, Denmark at 21.30 hours. Six members of the crew were killed, including Robert Melville, whose commission had been granted that same

day. Sergeant Donovan and Sgt Masfen, from Southern Rhodesia, Africa, survived the crash, were captured, and took-up residence at PoW camp Stalag 344, at Lamsdorf. The other aircraft lost was Stirling bomber, W7576, LS-G, piloted by P/O Wilbert Shoemaker, D.F.C, RCAF. The aircraft, which was participating in an attack against Duisberg, was shot down by a nightfighter piloted by Leutnant Reinhold Knacke, of I./NJG1 on 26th July. It crashed beside the Tienraysweweg at Horst, 12kms north west of Venlo in Holland, at 01.32 hours. As the aircraft impacted with the ground, it broke in two pieces scattering nineteen containers of 90 x 4lbs incendiaries around the crash site. Some of these incendiaries ignited, causing fires at three nearby farms. The occupants, however, were able to extinguish the flames before too much damage was created. Three crew members were killed in the crash, their bodies being found in the neighborhood of the aircraft. They were recovered and buried, but later removed, and now rest in the Jonkerbos War Cemetery.

No.XV Squadron spent the first week of the new month training crews in general, and navigators in particular, with a number of training flights, each lasting nearly four hours. The training was put to good use on the night of 6th August, when No.XV detailed nine aircraft for an attack against Duisberg. Although engine problems prevented two of the Stirlings from taking-off, six of the remaining seven aircraft went on to locate and bomb the target.

By a strange twist of fate two pilots with very similar sounding surnames, were killed on successive nights, along with five members of their respective crews, following attacks on the same city. Flight Sergeant William McCauseland's Stirling, N3756, LS-C, was attacked by two Junkers Ju.88 nightfighters during an attack against Mainz, on the night of 11th/12th August. Although seriously damaged, the bomber limped home and crossed the English coast near Harwich. As F/S McCauseland prepared for an emergency landing at RAF Wattisham north west of Ipswich, Suffolk, the stricken bomber fell from the sky. It crashed into a pond at Potash Farm, Brettenham at 03.37 hours, and burst into flames. The only survivor was the rear gunner, Sgt Egri, RCAF, who was pulled from the wreckage by Jim and John Hammond, and Stan Arbons. The following night, 12th/13th, Stirling bomber, BF329, LS-A, piloted by F/S Kenneth McAusland, suffered the same fate. Shot down by an enemy nightfighter, F/S McAusland's aircraft crashed 2 kilometres south east of Romedenne, Belgium at 01.15 hours. The crew, with the exception of Sgt Goodbourne, were all killed.

Later that same day, 13th August, in accordance with orders received, the Squadron moved base from Wyton to Bourn, seven miles west of the university city of Cambridge. The airfield, which stood by the side of the main A45 Highway, was constructed on the A-plan configuration, with three concrete runways. It was able to accommodate approximately two thousand personnel of all ranks, and had hardstanding for thirty-six bombers.

Operations from the new station commenced on the night of 15th August, when seven aircraft were detailed for an attack against Dusseldorf. From the Squadron's point of view, the raid was not a success. Four crews reported bombing on scattered fires, whilst two aircraft returned with their bomb loads, one of which jettisoned two 500lbs bombs over the sea. The seventh Stirling landed at RAF Mildenhall, Suffolk, because of engine failure.

The remainder of the month was fairly quiet with regard to missions undertaken by No.XV. Although only four targets in Germany (Osnabruck, Frankfurt, Kassel and Nuremberg) were attacked over a ten night period between 17th and 28th August, the Squadron paid a high price. Two aircraft crash-landed and three others failed to return. Two of the latter aircraft were lost during the raid against Kassel, on the night of 27th August. Stirling bomber, BF327, LS-D, piloted by P/O John Thornton, was shot down by an enemy nightfighter at 23.55 hours. The aircraft crashed at Beeusichen on the south bank of the River Lek, Holland. Flight Sergeant George Barton-Smith's Stirling, W7624, LS-E, was also attacked and shot down by a nightfighter. The enemy pilot, Oberleutnant Victor Bauer of III./NJG1, despatched the bomber from the sky, causing it to crash at Bentelo near Hengelo, also in Holland. The following night one aircraft, Stirling, R9153, LS-U, piloted by P/O Eric Patterson, RCAF, was lost. The aircraft, which failed to return from the attack on Nuremberg, crashed at Mesmont, France. Twenty-one young men from No.XV Squadron were killed, whilst four others were injured.

September 1942 saw an increase in the number of missions undertaken by No.XV, with the Squadron operating against targets in Germany virtually every day during the first three weeks of the month.

A cause for celebration arose on 9th September when Stirling bomber, N3669, LS-H, piloted by F/S 'Porky' Halkett, completed its fiftieth mission, a mine-laying operation in the Baltic. Unfortunately, the celebrations were slightly marred by the crash-landing at Waterbeach of Stirling, BF352, LS-U, piloted by Sgt Russell-Collins. Due to severe engine problems encountered on the outbound leg of the operation, the pilot decided to return early. He ordered his crew to bale out of the defective bomber, before landing at the Cambridgeshire airfield. Just before the aircraft touched-down, forty-five minutes after taking-off, the port inner engine failed. As the bomber landed, it swung off course and the undercarriage collapsed. Due to the pilots actions, no injuries where reported. Sadly, just ten days later, Charles Russell-Collins and his crew failed to return from an attack against Munich, Germany. The entire crew were killed when their aircraft crashed at Noyers Le-Val, France. The pilot had only recently been promoted to the rank of Flight Sergeant.

During the early hours of 13th September, a new aircraft flying its first operation with the Squadron had to abandon its mission due to engine failure and unserviceability of the rear turret. The Stirling, BF356, a replacement machine for LS-D and bearing the same code, landed away from its home base six nights later due to a fuel shortage, following an attack against Munich.

Apart from an aborted attack against Vegesack, on the 23rd of the month, only two mining-laying operations were detailed during the remainder of September. The rest of the time was spent undertaking fighter affiliation exercises, air firing exercises and compass swinging. The lack of operations during the latter period of September was a 'blessing in disguise' for two new Squadron 'Erks'. Les Butcher and John Pratt were both experienced members of groundcrew who had recently arrived at Bourn from No.1 Air Gunnery School, Pembrey, South Wales. Although the aircraft still had to be maintained, the fact there were no operational directives adding to their daily duties, allowed the men time to get to know their new environment.

The airfield at Bourn was much like any other RAF Station, but there was one hanger situated away from the rest, which aroused the curiosity of the two 'Erks'. The hanger was actually used by Sebro (Short Brothers Repair Organisation), which repaired and re-assembled damaged Stirling bombers. One day, whilst standing on the airfield fairly close to the hanger, they watched the approach of an unfamiliar looking Stirling bomber. Appearing small in size, they deemed the aircraft to be some distance away from the airfield, but were amazed to see it descend towards the threshold of the runway. After the aircraft had landed and taxied past them, Les Butcher and John Pratt realised they were looking at a half-scale replica of the huge four-engined bomber. Known as the S.31, the plywood constructed aircraft powered by four Pobjoy motors, continued towards the isolated hanger. It was then maneuvered into the structure, before the hanger doors closed, rendering it lost to view. Their curiosity satisfied, both Les Butch and John Pratt were to see the machine a few more times before it was destroyed in a take-off accident at Stradishall, at the end of 1942.

The first week of October saw a return to operations, with attacks against Lubeck, Krefeld, Aachen and Osnabruck. During this period only one aircraft was lost when Stirling bomber, W7634, LS-G, failed to return from the raid against Lubeck. It crashed into the Baltic Sea near Peenemunde, with the loss of the entire crew. Sergeant George Stark, the rear gunner, whose body was subsequently washed ashore, was the only member of the crew to be recovered.

Another aircraft and crew were lost on 17th October whilst participating in a 'Gardening' mission off the French coast. Three aircraft took-off from Bourn at approximately 18.15 hours on the night of 16th, with orders to drop mines in the Bayonne area. Although two of the crews discharged their duty and returned without incident, Stirling bomber, R9312, LS-C, piloted by Sgt Arthur Tanner, crashed at Pont du Caens, north-west of Nantes. Five of the crew were killed whilst two, Sgt Edgar and Sgt Daly, were taken prisoners of war.

The Squadron came close to losing its commanding officer on 23rd of the month when W/C Lay crash-landed at Bradwell Bay airfield, following an attack against Genoa, Italy. The mission, which was carried out in very poor weather conditions by a force of ten aircraft, was led by the C.O. who had taken-off from Bourn at 18.00 hours. The cloud base was low and hampered the crews, not only on the outward journey but also over the target. Apart from the weather, Stirling bomber, W7635, LS-V, piloted by F/L Baigent, also had to contend with being attacked by two Ju.88 nightfighters. However, the gunners defending the bomber were alert and, although the mid-upper was slightly wounded, one of the Ju.88s was believed to have been shot down. The elements took their toll on the returning bombers, forcing a number of them to land at airfields along the east coast of England, due to lack of fuel. One of those bombers being Stirling, W7633, LS-P, flown by W/C Lay. The C.O., his aircraft and crew, had been airborne for just over nine hours as they approached Bradwell Bay on the Essex coast, near the estuary of the River Blackwater. The fuel situation was critical as W/C Lay struggled with the aircraft, and attempted to keep the airfield in sight. The cloudbase was down to 500', visibility was decreasing and of course it was dark. The bomber turned onto finals and, with the undercarriage still retracted, the engines coughed and splattered as the last of the fuel was used. At 03.02 hours, the four-engined bomber belly-landed on the airfield, fortunately without injury to the crew. The total lack of fuel on the aircraft, which was later put down to extremely inaccurate fuel gauges, probably accounted for the fact there was no fire following the crash-landing.

Similar circumstances as those which faced W/C Lay, were to dictate that the stay of an experienced officer and aircraft captain, who reported for duty with No.XV Squadron towards the end of the month, was to be of a fairly short duration. Squadron Leader Michael Wyatt was posted from No.1651 Conversion Unit to No.XV (Bomber) Squadron, on 26th October. Less than one month later, his aircraft would also fail to return from an attack against a target in Italy.

October ended on a tragic note for the Squadron, when eleven personnel were killed in one crash, on 29th of the month. Stirling bomber, BF386, LS-Q, was flying on an air test with S/L Charles Fisher at the controls. Also on board the aircraft were Sgt Walter Hood, RNZAF, acting as second pilot, S/L Fisher's aircrew and some members of the groundcrew. The aircraft had been airborne for approximately forty minutes when it was seen to emerge from cloud cover in normal flying attitude. It executed a steep turn and immediately went into a steep dive, from which it did not recover. It impacted with the ground at 11.50 hours, at Salter's Lode, five miles south-west of Downham Market, Suffolk, and burst into flames. The names of all eleven men were added to the Squadron's Roll of Honor.

The Wing Commander's 'mount'

Members of "A" Flight groundcrew pose for a photograph by Short Stirling, N6065, LS-G, at R.A.F Wyton. This aircraft, which was delivered to the Squadron in February 1942, was flown by a number of pilots including W/C MacDonald, the Unit's C.O. The aircraft also saw service with No.149 Conversion Flight and No.1657 Conversion Unit. *Author's Collection*

Brian Ordish

Charles 'Brian' Ordish, together with an unidentified fellow sergeant, takes time out for refreshment. Apart from a bar of chocolate in his left hand, Brian appears to have a bread roll tucked into the sleeve of his tunic. *Courtesy of Mrs Elizabeth Park*

Return of a Crippled Stirling

Short Stirling, W7505, LS-V, photographed on 26th April 1942, after its crash-landing at R.A.F Wyton. The aircraft, which was piloted by W/C J C MacDonald, had been seriously damaged by enemy fire over Sylt during an attack against Rostock. The damage incurred forced the aircraft's immediate return. *Author's Collection*

Operational Loss Card

The Operational Loss Card for Short Stirling, N3673, LS-D, which was shot down on the night of 8th/9th March 1942, during an attack against Essen. Stirling bombers normally flew with a crew of seven or eight men, but on this occasion the card records the names of nine crew members lost. *Authors Collection*

One of a 'Thousand'

Short Stirling, W7524, LS-D, was flown by F/S Lucas on the first 1,000 bomber raid on 30th/31st May 1942. The aircraft, seen here at rest at R.A.F Wyton, was lost on the night of 17th July 1942. It was shot down by flak and crashed into the North Sea, off Esbjerg, Denmark. *Courtesy of and via Flt/Lt T. N. Harris, XV (R) Squadron*

A poor quality photograph, taken over Cologne Cathedral at the end of the war, shows the result of continuous bombing on the German City of Cologne. The city was subjected to the first of the 1,000 bomber raids, on the night of 30th/31st May 1942. *Author's Collection*

'Right on the Nose'

Hit by flak whilst flying over Antwerp, during the first 1,000 bomber raid on Essen on the night of 1st June 1942, Stirling bomber, W7513, LS-R, sustained damage to the fuselage and port tyre. Due to the latter damage, the aircraft swung on landing and tipped over onto its nose. The photograph is reputed to have been taken two days after the event. *Courtesy of and via Don Clarke, M.B.E., The Mildenhall Register*

Members of No.XV Squadron groundcrew, two of whom are perched high up by the tailfin, wonder how they are going to get LS-R, back onto the ground. Apart from the damage to the aircraft, the rear gunner sustained injuries to his neck and head. *Courtesy of and via Flt/Lt T.N. Harris, XV (R) Squadron*

Three from Fifteen

Three Stirling airmen from No.XV Squadron, photographed between March and April 1942 are, from left to right: F/L Fred Raefaelli, DFC, F/L Ronnie Bar, DFC, F/L R A Strachan. *Courtesy of and via Don Clarke, MBE, The Mildenhall Register*

The aircrew and groundcrew of Short Stirling bomber, W7513, LS-R, with their aircraft as a backdrop. The aircraft also saw service with No.149 Squadron and No.75 Squadron, before being lost during a mine-laying operation, in the Fehmarn Bely, on 29th April 1943. Standing left to right are members of the aircrew: Sgt R Quartly, F/G; Sgt R Milne, W/Op; F/L D Parkin, Pil; P/O R Strachan, Nav/B.A.; Sgt Drake, F/E; Sgt E Lancashire, F/E; P/O Haines, 2/Pil; Sgt A Birch, R/G. Members of the groundcrew sit together in the front row. *Courtesy of and via Don Clarke, MBE, The Mildenhall Register*

'Ops' is Tiring Work

Right: Pilot Officer Bill Prune, the Squadron's Bulldog mascot, photographed with P/O Jimmy Rigby, RCAF, at the tail end of a No.XV Squadron Stirling bomber. The aircraft is inscribed with the name of the target which had been the subject of the second 1,000 bomber raid. It is not known if Bill Prune participated in the raid, but he appears to ready to 'hit the sack'. *Courtesy of and via Don Clarke, MBE, The Mildenhall Register*

Officers and Gentlemen of No.XV

Identified amongst the Officers of No.XV Squadron are: Back Row (7th, 8th & 9th) from left: P/O J Sinclair, P/O R Strachan, P/O D Burgess. Centre row: P/O Roy Lair, F/L F Raffaelli, F/O D Parkin, S/L R Gilmour, W/C Day, W/C J MacDonald, F/O Wright, S/L Newman, F/L R Barr, F/O N Bennitt, P/O D Wright. Front Row: W/O Smith, F/O Phillips, F/O Putt, Unknown, Unknown, F/O R Munns. *Courtesy of and via Don Clarke, MBE, The Mildenhall Register*

The Gentlemen of No.XV Squadron, utilise one of their aircraft for a photograph, at RAF Wyton during 1942. *Courtesy of and via Don Clarke, MBE, The Mildenhall Register*

Stirling Bomber, N3756

Air Training Corps Cadets lend a hand and help push a bomb trolley into position under Stirling bomber, N3756, LS-C, during the summer of 1942. The aircraft, which had previously seen service with No.214 Squadron, crashed at Potash Farm, near Ipswich, on 12th August, whilst attempting an emergency landing. Sergeant Egri, the rear gunner, was the only survivor. The rest of the crew were killed in the crash. *Courtesy of and via Flt/Lt T. N. Harris, XV (R) Squadron*

'Nearly Home'

A Stirling bomber, which was nearly home, lies crumpled and forlorn at an unknown location. *Courtesy of and via Flt/Lt T. N. Harris, XV (R) Squadron*

The Stirling bomber which nearly made it home, attracts the attention of two airmen. The one on the left (with the bicycle) appears to be RAF, whilst the second airman on the right appears to be wearing USAAF clothing. *Courtesy of and via Flt/Lt T. N. Harris, XV (R) Squadron*

Planning the Next Mission

Members of aircrew make notes, as W/C D.J.H. Lay, D.S.O., D.F.C. (center of the picture) gives a crew briefing for a forthcoming operation, at RAF Bourn. The walls of the room are adorned with maps and aircraft recognition posters, whilst models of British, American and German aircraft hang suspended from the ceiling. *Courtesy of and via Flt/Lt T. N. Harris, XV (R) Squadron*

Cause for Celebration

Flight Sergeant Alexander Morgan 'Porky' Halkett, D.F.M., RCAF (standing 2nd from right), and his crew give the "V" for Victory sign in front of Stirling, N3669, LS-H. The reason for their gesture being, having landed safely at Bourn, their aircraft had just 'notched-up 62 operational missions against enemy targets. *Courtesy of and via Don Clarke, MBE, The Mildenhall Register*

Having divested themselves of their flying clothing, the crew of Stirling, N3669, LS-H, returned to their aircraft and joined the members of their groundcrew for a photograph to celebrate the safe completion of the Stirling's 62nd operational mission. The bomb tally, recording all the operations can be seen to the left of the picture. The groundcrew sergeant (standing 3rd from left, back row) had the honor of painting the 62nd bomb insignia. The aircraft, which was placed on display outside St. Paul's Cathedral, London, went on to record a further five sorties, before being struck-off-charge in February 1943. *Courtesy of and via Don Clarke, MBE, The Mildenhall Register*

12

Sojourn in Spain

Upon his arrival at Bourn on 26th October 1942, S/L Michael Wyatt found operational life on No.XV very carefree, and at times extremely rowdy. The spirit on the Squadron was very good, and he was aware that everybody thought highly of the Commanding Officer. It was the C.O., W/C Lay, who greeted the experienced pilot and informed him he would be taking over as "A" Flight Commander.

The 'newcomer', who was first commissioned in the RAF on 23rd March 1934, experienced his first mission with the Squadron the day after his arrival when he flew as second pilot to W.O. Sayers, on a mine-laying operation in the Gironde river west of Bordeaux. The flight was uneventful, apart from a small amount of flak encountered whilst crossing the French coast both on the outbound and homeward journeys.

Being in a position of some authority, Michael Wyatt included his own name on the Battle Order for an attack against Genoa in northern Italy, on 7th November. On this occasion, he flew as second pilot to F/L Baigent and his crew, who were operating in Stirling, W7635, LS-V.

Seven aircraft were detailed to participate in the attack, but two returned early due to engine problems. One of them was Stirling, BF395, LS-F, piloted by the experienced bomber captain, F/S 'Porky' Halkett. Much to their frustration, the aircraft was within ten miles of the target when F/S Halkett decided, at 20.23 hours, that discretion was the better part of valor, and informed his crew he was abandoning the mission. The timing of 'Porky' Halkett's decision proved reassuring to the crew when the starboard motor, which had been the source of their problems, finally cut out on landing. The remaining five aircraft continued with the mission, and reported successful results on their return.

Squadron Leader Wyatt included his name on a number of other Battle Orders, and flew on a leaflet dropping 'Nickel' operation on 8th November, before taking his own crew on a mine-laying mission in the Frisian Islands two nights later. This was followed by a similar sortie, to the estuary of the Gironde river on the night of 16th of the month.

The Squadron detailed eleven aircraft for an attack against the FIAT Works at Turin, on the night of 18th/19th November. However, during the course of the day, five of the allotted bombers were withdrawn by Group Headquarters. In the event, only two aircraft managed to get airborne for this mission, following a take-off accident involving Stirling bomber, BF384, LS-R. Battling against a heavy crosswind, the bomber was endeavoring to gain airspeed when it developed engine failure. With the resulting loss of power and gusting winds, the aircraft veered to one side. The pilot, Sgt Frank Millen, an American from Rhode Island, attempted to check the swing, but the Stirling failed to respond. As the bomber continued on its chosen course, the undercarriage collapsed causing severe damage to the airframe. The only member of the crew hurt in the crash was Sgt Ken Dougan, RCAF, the bomb aimer, who received injuries to his left leg and right ankle. He was quickly assisted into the 'blood wagon' and transferred to the Station Sick Bay, at RAF Oakington, where he was admitted

for treatment. Although he was still unable to walk, he was discharged after only a few days. The problem of getting around was soon overcome with the aid of a stout stick, which enabled him to hobble about.

The first opportunity for S/L Wyatt and his crew to operate together as a team against a major enemy target, occurred on the night of 20th/21st November, when 232 aircraft made another attack against the FIAT works at Turin. In checking the Squadron's aircraft availability list, Michael Wyatt again added his own name, and those of his crew, to the Battle Order and allocated himself Stirling bomber, BK595, LS-A. He also detailed a further eight crews from No.XV to participate in the operation. Having completed the cockpit checks, taxied out to the threshold of the runway and received the green signal from the control caravan Aldis lamp, S/L Wyatt took-off at 18.02 hours. The eight other aircraft followed him at regular intervals, with the last machine lifting into the night sky at 18.47 hours. Squadron Leader Wyatt set course for the French coast, which the bombers from No.XV Squadron were instructed to cross at Fecamp, between Dieppe and Le Havre. Upon their arrival over enemy occupied territory, the Stirlings received the usual welcome from the German ground gunners. However, the flak bursts did little to impress the aircrews who, on this occasion, did not feel threaten by the uninvited attention. Of more concern to the air gunners was the visibility. As Stirling, BK595, flew on over France the full moon rose high in the sky, lighting up the French countryside and illuminating everything in sight. Surprisingly the bombers flew on unhindered by either flak or fighters, and all seemed calm and serene. However, for Michael Wyatt and his crew, who were approaching the Alps, all that was about to change.

Momentarily, the captain of BK595 had been lost in the natural beauty of the sight which lay before him as he looked out of the cockpit. He was admiring the beautiful vista of moonbeams dancing across, and reflecting off, the snow capped mountains, when suddenly the starboard inner engine cut out. Michael Wyatt's immediate reactions was that of an ignition failure, as though the switches had suddenly been turned off. Both Sgt Amies the flight engineer, and P/O Crich the second pilot, were put to work trying to locate the source of the problem. When both reported they were unsuccessful in their endeavors, S/L Wyatt decided to feather the propeller, but much to his consternation it would not feather fully. Following a restart of the offending motor, a second attempt at feathering it also failed. Although they held their course for a while on three engines, the drag created by the defective motor would not allow the aircraft to maintain height. The pilot realised the Stirling would not be able to cross the Alps on the planned course. The mountains, which a few minutes beforehand had induced a feeling of beauty, peace and serenity, now became a formidable enemy.

Hoping to find a way over some of the lower mountain peaks to the west of the original crossing point, S/L Wyatt asked his navigator for a new course. Next, he called up the crew and informed them of the nature of the problem, explaining that once the bomb load had gone the aircraft should be able to maintain height. As they were only one hundred miles from the target, S/L Wyatt reasoned it would be better to drop their bombs in the proper manner, rather than just jettison them. The crew agreed with their captain's decision, and trusted in his judgment.

Needless to say, by the time Stirling, BK595, reached the target the other aircraft had bombed and were on their way home. Lying at his station in the nose of the aircraft, the bomb aimer guided his pilot towards the target which could be identified by the remains of the target indicators dropped by the Pathfinder Force. The aiming point however was not fully visible, due to the smoke rising from fires created by the previously dropped incendiary bombs.

With their task completed, Michael Wyatt asked P/O Daborn for a reciprocal course back over the Alps. The navigator, when he responded to the call, sounded very defeatist. He advised the pilot that the flight engineer was concerned about the fuel load, and added that the latter's calculations indicated there was insufficient petrol to cover the homeward journey. Squadron Leader Wyatt immediately instructed the navigator and second pilot to check all figures and details relating to the fuel state, and report back as quickly as possible. The answer, when it came, was not good. The calculations proved correct, and the aircraft was carrying insufficient fuel to get back to the English coast, let alone its home base. A discussion ensued between the pilot and navigator, during which all the possibilities were examined. Having made his decision, Michael Wyatt informed his crew he was going to make for Spain, explaining it was much nearer than England. He also advised them that if they reached their objective, he would either ditch the aircraft in the sea or try to locate a suitable beach on which to crash-land. Having once ditched a Wellington bomber in the North Sea, the second pilot, who was not keen to repeat the experience, immediately responded over the intercom, and pleaded with his captain to reconsider the implication of such an action. Not knowing what sort of terrain they would encounter at the end of their journey, S/L Wyatt promised his crew he would not ditch the four-engined bomber if it was possible for them to either bale out or crash-land. Without hesitation, and given the choice, the crew all stated their preference for a crash-landing.

Although it was not the one he originally asked for, the pilot received a course from the navigator which would take them on a south-westerly heading to the Franco-Spanish border, just to the south of Port Bou. Being mindful of the fuel situation, S/L Wyatt kept the aircraft out over the sea, making it easier to ditch if the need arose. Furthermore, he was also aware that the whole of the South of France had, two days earlier, been handed over to the Germans by the Vichy Government, and did not relish the idea of crash-landing there. However, for one heart stopping moment, it seemed as though the pilot would be forced to carry out one of the two options when, about halfway through the journey whilst flying off the coast of Marseilles, the port inner engine began to falter. As this motor kept the only remaining generator operational, Michael Wyatt carefully 'nursed' the engine by reducing the revolutions as much as possible.

Although at this point in time the aircraft was flying at 8,000' and gradually losing height, S/L Wyatt was not too concerned, as he wanted to be down to about 1,500' when the Stirling crossed the Spanish coast. The crossing of the frontier was recorded on the navigator's log at about 02.30 hours, on 21st November. The crew, having previously received instruction from their pilot, knew what to do now the moment for a crash-landing had arrived. As each crew member took up his respective crash position, the bomb aimer activated the explosive charges on the GEE and IFF sets. The noise of the explosions drowning out all other sounds, and reverberating down the fuselage, whist giving off an overpowering smell of cordite. Before taking up his crash position, the wireless operator had rendered the radio set inoperative by smashing it. Whilst his crew were taking the necessary actions further down the aircraft, S/L Wyatt scanned the terrain from his seat on the flight deck, trying to locate a suitable beach on which to set down his four-engined charge.

He quickly spotted a suitable location, a strip of beach with plenty of sand and not too many rocks. Turning on a final circuit, he ordered the crew to brace themselves, and brought the Stirling in for a wheels-up land-ing. The bomber skimmed across the beach, leaving a sandstorm in its wake, but just before it came to rest the port wing glanced a blow against some rocks. The impact was sufficient to swing the aircraft round in the direction of the offending rocks with such force that the fuselage broke in two at a point just aft of the wing root. The crew, uninjured, but looking bewildered and shaken, quickly made use of this new, and unexpected, exit point from the shattered aircraft.

Assembling his crew on the sand near the wreckage, S/L Wyatt issued instructions for the total destruction of their aircraft by setting fire to it. He told the crew to make use of the two incendiary devices carried on board specifically for the purpose, which were still in the shattered fuselage. Two volunteers entered the wreckage, located the incendiaries and set about igniting them. However, although the instructions for use printed on them were followed to the letter, neither device would function. Realising that something had to be done quickly, P/O Crich, the second pilot, volunteered to re-enter the aircraft and make a bonfire on the flight deck using maps, charts and other inflammable items. By the time he returned to his colleagues on the beach, a good fire was burning in the front section of the aircraft, with ammunition from the front turret exploding in all directions. The flight engineer had, in the meantime, also tried to play a part by opening the fuel jettison valves in the hope that any petrol left in the tanks would seep out onto the sand below the wings. Any hope the crew had of setting fire to the center section of the Stirling were dashed when S/L Wyatt became aware of four figures approaching the scene from out of the shadows. He realised they were wearing uniforms of some description and were carrying rifles which they started to discharge over the Englishmen's heads.

Calling them to order, Michael Wyatt instructed his crew to abandon their attempts at destroying the aircraft and walk slowly, with their arms raised, towards the four oncoming figures. The armed figures, who turned out to be members of the Guardia Civil, were all talking at once in a guttural tongue which the pilot later realised was Catalon. The tense situation was relieved momentarily for Michael Wyatt, when he noticed the black patent leather hats they where wearing, which reminded him of typewriters. Obviously, he kept this observation to himself. With two of the guards walking ahead of them, and the other two behind them, the crew were ushered along the beach, away from the crashed bomber. However, it amused the crew when their captors repeatedly stopped and looked back, with puzzled expressions on their faces, every time gun fire was heard emanating from the burning Stirling. The Guardia Civil obviously thought that somebody was still in the wreckage taking pot-shots at them!

At the end of the beach was a steep cliff, up which the crew were signaled to climb. On top of the cliff was a cafe which, at the unearthly hour of 03.30 in the morning, and much to the crew's surprise, was still open. However, whist their captors downed copious amounts of black coffee and brandy, the captives were not allowed to partake of any refreshment. Neither were they allowed to freshen-up or restore their personal comfort. Whilst one of the guards made a number of telephone calls, his three colleagues seemed quite happy drinking, chatting and smoking foul smelling cigarettes. Occasionally, they passed furtive glances at the Englishmen who by now just wanted to get some sleep.

Eventually a Spanish Army Officer who spoke good English arrived. He was obviously annoyed at having been woken from his slumbers, and was consequently rather terse towards the English aircrew. He explained to them in no uncertain terms, that because they had entered the country illegally, without papers and uninvited, they would all be sent to an international internment camp at Miranda, near Gerona. This statement horrified Michael Wyatt, as he had heard all manner of stories relating to this internment center. During lectures on escape and evasion back on the squadron, the Intelligence Officer had repeatedly warned of the dangers associated with the camp at Miranda. The conditions were reported as appalling, with ill-treatment of internees, a lack of rations and filth everywhere, all of which contributed to a very high death rate. Any objections raised by S/L Wyatt were quickly dismissed by the Army Officer, who also disallowed any request to communicate with the British Consul General in Barcelona.

Michael Wyatt's protestation at being told he and his crew would be required to march the 25 miles to Gerona also fell on deaf ears. The obnoxious Army Officer informed the internees that as there was no available transport, they had no choice but to walk. S/L Wyatt continued to protest, pointing out that most of his crew were wearing flying boots which were totally unsuitable for the exercise. This comment was met with a disinterested shrug of the Army Officer's shoulders. Very much against their will and escorted by the armed guard, the crew set off in a southerly direction towards S. Feliu de Guixols. The ill-fitting flying boots did not take long to exact a toll on the crew members wearing them, with sore feet and blisters soon becoming evident. One or two members of the crew started to complain but the guards, who did not understand a word of English, totally ignored their charges. The words of complaint and abuse did not cease as they entered the suburbs of S. Feliu, in fact they got louder as the group marched through the streets. Lights were switched on in several properties, and at one house a man wearing a dressing gown appeared at the front door and called out something which Michael Wyatt did not understand. Realising the crew were talking amongst themselves, the newcomer approached them and asked them, in perfect English, if they were British airmen. With the fact confirmed, Michael Wyatt quickly explained their predicament. It transpired the man in the dressing gown was a solicitor who had read Law at Oxford, and had practices in both Gerona and S. Feliu. To the crews' relief, he informed the members of "Oxford's 'Own'" Squadron that the Guardia Civil had no right to deny their request to make contact with the British Consulate, or to make them walk to Gerona. Furthermore, he instructed the Guardia Civil to take the crew to the council offices in the town where he said the latter could spend the remainder of the night, whilst he arranged for their proper treatment according to International Law. It was due to the intervention of this unknown Spanish Solicitor that the fortunes of S/L Wyatt and his crew changed, and ultimately led to their repatriation to England in April 1943.

At approximately the same time that S/L Wyatt and his crew flew over the Franco-Spanish border, the last two Stirlings of No.XV Squadron, who had participated in the Turin raid, landed back at Bourn. All crews reported a successful attack with fires, which the rear gunners and mid-upper gunners reported, were visible up to forty miles away on the homeward journey.

Turin was to receive further attention from Bomber Command before the month ended. On 28th November, seven Stirling bombers from No.XV Squadron were detailed to participate in another attack against the Fiat Works. Unfortunately, two of the Stirlings were withdrawn prior to take-off, whilst two others jettisoned their respective bomb loads and returned early. Of the three remaining aircraft which continued on towards Italy, two bombed the City and one attacked the primary target. All returned safely to base. The following night two more machines from No.XV were detailed to join a force of twenty-nine Stirling bombers and seven Lancasters from No.3 Group, Bomber Command, for another attack on the Fiat Works. On this occasion the weather conditions were extremely poor, and only eighteen aircraft were recorded as having crossed the Alps and reached the target. Four of those aircraft were Stirlings, one of which came from No.XV Squadron. Flying Officer Hank Tilson, piloting Stirling, R9168, LS-T, was forced to abandon the mission at Dijon due to engine trouble, and return early. As the sole representative of No.XV, Sgt Frank Millen, completed his task and bombed at 03.35 hours, from an altitude of 1,500 feet. However, Sergeant Millen's aircraft did not come away from the target unscathed. As he held Stirling, BF350, LS-O, steady for its bombing run, light flak rose into the night sky. It found its target and struck the aircraft, causing damage to a fuel tank. By careful coaxing, and without the misfortune of further incidents on the homeward leg, Sgt Millen nursed the crippled bomber back to England. Unfortunately, due to a shortage of fuel the pilot was forced to land his bomber at Bradwell Bay, a fighter airfield near the south bank of the River Blackwell, in Essex. Although it was only eleven days since their accident, and he was still not fully mobile, Sgt Dougan had decided to fly on this operation with his crew. However, it was only with their help that he could get in and out of the aircraft. Had they been obliged to leave the Stirling in a hurry, the bomb aimer would have been in serious trouble.

Whilst Hank Tilson and Frank Millen were heading for Turin, Brian Ordish was re-establishing contact with some of his former colleagues back at Bourn. Not only had the latter been reposted back to his old Squadron, he had also been promoted to the rank of Acting Flight Lieutenant the same day.

Operations during the first week of December were hampered by adverse weather conditions, with a number of missions being canceled. Some crews were given leave, including that of P/O Millen. The latter crew headed north, to Liverpool, to the home of Jim Perring. It was whilst they were visiting the family of their rear gunner that Ken Dougan's legs started to give cause for concern. A visiting doctor took one look at the patient, and immediately had him admitted to a civilian hospital. Although it was never confirmed to him, Ken Dougan suspected that he had blood poisoning and was subsequently grounded for some weeks. One thing which was confirmed to him, was his application for a commission which was granted on 6th December.

It was on this same Sunday, 6th December, that Brian Ordish finally undertook the first operational mission of his second tour. His aircraft, Stirling bomber, R9168, LS-T, was one of eight Squadron machines which took-off to attack Mannheim. Although the target area was covered by 10/10ths cloud, the experienced pilot dropped his bomb load on the marker flares put down by the Pathfinder Force. Due to the prevailing conditions, the participating crews considered the raid to be a wasted effort.

The following day the Squadron despatched two aircraft for a mine-laying operation in the Frisian Islands. One of the two crews involved had arrived on the Squadron on 25th October, but to date had not undertaken a mission. Under the leadership of their pilot Sgt John Monteith, RCAF, they had carried out a number of flying training exercises, including fighter affiliation, bullseyes, and cross-country 'bombing' exercises. Their first mission was carried out without incident, and the rear gunner, Sgt John Greenwood, reported the successful 'planting' of their six mines in the allocated position. That same day another familiar face returned to No.XV, also with promotion. Stewart Menaul arrived back from No.1651 Conversion Unit to take up his third posting with "Oxford's 'Own'". On this occasion however, he came with the rank of Acting Wing Commander, to assume command of the Squadron. Wing Commander J H Lay, having been posted to H.Q., No.3 Group, officially handed over to W/C Menaul and left Bourn on 13th December. The 13th December also saw promotion for John Monteith, who was commissioned in the rank of Pilot Officer.

On 16th December, No.XV detailed three experienced crews for an attack against Deipholz; their target was to be a German aircraft Depot. The crews selected were those of F/O Cooke, F/S McMonagle and P/O Frank Millen. Flying Officer Cooke, piloting Stirling bomber, BF411, LS-A, took-off at 17.22 hours closely followed by the two other aircraft. Two and a quarter hours after leaving Bourn, the bombers approached their target. Flying Officer Cooke, flying at 7,000', made a low-level bombing run at 19.37 hours. He was followed five minutes later by F/S McMonagle who, unbeknown to the latter, was being stalked by two Bf110 nightfighters. He knew of their presence, when cannon and machine-gun fire spat forth from the guns of the enemy aircraft, illuminating a path in the night sky towards the bomber. As the Stirling took evasive action, the gunners swung their turrets towards the aggressors and returned the fire. As the mid-upper gunner continued firing at his target, he had the satisfaction of seeing the Bf110 go down. The second nightfighter took up the challenge and made a firing pass, scoring hits on the bomber and wounding the mid-upper gunner. Surprisingly, it did not return to complete its task, and Stirling, BF355, LS-F, survived to fight another day. Stirling bomber, R9168, LS-T, flown by P/O Frank Millen, was not so fortunate. The aircraft was shot down by a nightfighter, and crashed at Epe, Holland where the whole crew, with the exception of Sgt J Perring, perished in the crash. Although Sgt Perring survived, it was recorded he lost a leg as a result of incident. Of the six men whose names were added to the No.XV Squadron Roll of Honor, two were Americans, two were Canadians, one was a New Zealander and one was

British. Frank Millen was a Sophmore (Second year student in a four year course) of Norwich University, at Northfield, Vermont. Pilot Officer Emerson Kieswetter, RCAF, an observer, was also an American who came from Seattle, Washington. The two Canadians, Hugh Hill and Russell Holmes, air gunner and wireless operator/air gunner respectively, had both been commissioned to the rank of Pilot Officer only seven days earlier. The New Zealander was Sgt Robert McKillop, who had replaced Ken Dougan as bomb aimer on this trip. The sixth member of the crew who died was Sgt Grantley Hutton, the flight engineer. They were all buried together in Epe General Cemetery.

When P/O Ken Dougan returned to Bourn he found a number of changes. Many of his friends were either dead or had been posted and, for a while, he was treated very much as a 'newcomer'. However, during mid-December P/O Dougan's feelings changed, when he was accepted into the crew led by F/S 'Dicky' Craddock. The latter was an experienced pilot who was to be commissioned in the rank of Pilot Officer on the 19th of the month.

Ten days after he took command of No.XV, W/C Menaul resumed operational flying, but it was very nearly his last trip. The mission was a mine-laying sortie in the Frisian Islands, for which purpose he flew with S/L Faulkner and his crew. The aircraft took-off at 16.45 hours, and headed out over the North Sea. The trip was uneventful, and the crew were able to plant the six mines without hindrance from flak ships. It was not until Stewart Menaul was preparing to land Stirling bomber, BF356, LS-D, that things started to go wrong. The aircraft, which during its career had aborted many missions, was on final approach when the port outer engine cut out, due to the exactor going u/s. Unable to throttle back, W/C Menaul opened up the power to go round again, but the port inner motor cut, giving the pilot no option but to land. As the huge bomber began to settle, Stewart Menaul reduced the power to the two remaining engines, but the aircraft overshot the runway and hit a tree before bursting into flames. Fortunately, all eight members of the crew were able to evacuate the wreckage safely, with only one minor injury being recorded.

The rest of the month was a very quite period for No.XV Squadron, with only two operations being scheduled. Seven aircraft were detailed for an attack against Duisberg on 20th December, but only five carried out the raid due to two of their number returning early. The second operation detailed for eight days later, was canceled before the target was named, due to adverse weather conditions.

The year was to end on a tragic and sad note, when Stirling bomber, W7585, LS-U, crashed at Bassingbourn killing the four persons on board. The pilot, Brian Ordish, had lifted the bomber into the air at approximately 12.00 hours for an air test. All the other machines having been passed serviceable, Stirling, LS-U, was the last to be tested. Also on the aircraft with F/L Ordish were S/L The Reverend Denis Guy Ashill, the Squadron Chaplain, Sgt Francis Jackson, D.F.M, flight engineer and LAC James Hunt, a member of the aircraft's groundcrew. As the flight drew to a close, and the aircraft made an approach towards the airfield, the starboard inner engine failed. Although the aircraft was very low, Brian Ordish feathered the defective motor and made a sweeping turn in order to 'go round again'. As the Stirling bomber circled to the south of its home airfield of Bourn, it was seen to reduce speed, causing the pilot to lose control of the aircraft. At such low altitude, and with no hope of recovery, the Stirling, which had been airborne for twenty minutes, crashed to the ground. It came to rest on the east side of Main Road, on the boundary of Bassingbourn airfield, home of the USAAF 91st Bombardment Group. Rescuers from the 91st BG, or 'The Ragged Irregulars' as they were known, were quickly on the scene, but both S/L Ashill and LAC Hunt had been killed outright. Brian Ordish, who was badly burned and had sustained a broken thigh, was extricated from the wreckage and taken to Ely Hospital. Unfortunately, two days later, he succumbed to his injuries and passed away.

An unconfirmed report relating to the accident, stated that during the flight the pilot had reported the problem of a jammed undercarriage, but due to the condition of the airframe after impart, neither this nor the actual cause of the crash, could be ascertained.

The funeral of twenty-three year old F/L Charles Brian Ordish, RAFVR, was held on Monday, 4th January 1943. It took place at Harpenden Cemetery, and was attended by his parents, his sister Elizabeth, members of the family and other relatives. The Squadron was represented by W/C Stewart Menaul, DFC, AFC, S/L Faulkner, F/L Raffaelli and F/L Cope. Many friends from the R.A.F, including S/L Charles Lofthouse, also paid their last respects as they followed behind the Union Jack covered coffin. As this popular young man was laid to rest, many reflected on the fact he had done more than his duty. He had risen from Aircraftman 2nd Class to Flight Lieutenant in three years. He was the veteran of over forty operational missions over Germany and Italy, including the thousand bomber raid on Cologne, and had completed a tour as a flying instructor between operational postings. Just over a month after the funeral, Brian Ordish's record of achievement was recognised by the posthumous award of a Distinguished Flying Cross.

There was a mixture of weather during the first month of the new year, with both heavy rain and snow showers rendering operational flying almost impossible until the first week of February. Apart from a small number of mine-laying sorties being undertaken, the only other flying carried out during this period was a number of air tests. The latter being necessitated by the arrival of new equipment in the shape of Stirling, Mark.I bombers. One of the aircraft delivered during mid-January was Stirling, BF439, LS-D, which was to become the aircraft usually flown by Sgt Geoff Ware. When not in the air, LS-D was the responsibility of 'Chiefy' Warne and his groundcrew team. Les Butcher, who had been surprised by the sight of the half-scale Stirling arriving at Bourn, was a member of the team and flew on LS-D, with Sgt Ware on numerous air tests.

Also around this time Sergeant Renner, RZNAF, who had been with the Squadron since October 1942, was given his own aircraft. It was a Mk.I. Stirling, BK611, LS-U, which he promptly named '*Te-Koote*'. On 15th January, the day after Sgt Renner took receipt of his aircraft, W/C Stewart Menaul took it on a mine-laying operation to Lorient.

Apart from taking-on-charge new aircraft, the Squadron also benefited from the arrival of a number of new 'Freshmen' crews, including a pilot officer by the name of Dennis Haycock.

In order to encompass the new crews into the Squadron, No.XV was officially 'stood down' on 21st January, with a directive to participate in an intensive flying training program. As part of that program, three Supermarine Spitfire fighter aircraft were attached to the Squadron for three days, to provide fighter affiliation practice. The exercises, which also included formation flying, low level bombing practice and both warload height and air tests, gave valuable knowledge not only to the 'Freshmen' crews, but also proved profitable to the more experienced crews.

The first maximum effort raid of the new year in which No.XV Squadron participated, occurred on the night of 3rd/4th February. The target was Hamburg and No.XV detailed eleven aircraft, all of which took-off. At least two 'Freshmen' flew as second pilots on this mission, gaining first hand experience of 'Ops'. Pilot Officer Haycock flew with Sgt Irwin on Stirling, BF436, LS-E, whilst Sgt Smale flew with P/O Monteith on Stirling, BF411, LS-A. The first machine away was Stirling, BF448, LS-T, piloted by F/L Chave. This aircraft was, however, to return just over an hour and a half later due to the starboard outer motor going u/s. Unfortunately for the Squadron, F/L Chave was the first of five pilots to return early. Flying Officer V. Harris also suffered a defective starboard outer engine, whilst P/O Monteith and S/L English suffered excessive icing and excessive fuel consumption respectively. Pilot Officer 'Dicky' Craddock, who was also returning early due to a defective starboard outer engine, jettisoned his bomb load when attacked by an enemy nightfighter over Holland. With the aircraft, Stirling, R9193, LS-S, relieved of its 'cargo', P/O Craddock was able to take evasive action and lose the fighter. He landed safely and without further incident, at 20.28 hours. Unfortunately, Sgt Forbes and his crew were not so lucky. Their Stirling, R9274, LS-B, flew into the same patch of sky as the Bf110 piloted by Hauptmann Wilhelm Dorman, of III Gruppe/Nachtjagdgeschwader 1 (III./NJG1). The bomber succumbed to the impact of fire from the enemy aircraft's guns, and crashed at Renkum, Holland, at 20.58 hours. The whole crew managed to bale out before the stricken aircraft struck the ground, but the parachute used by Sgt Frederick Lax, the flight engineer, failed to deploy properly and he was killed.

No doubt working hard throughout the day, the tireless groundcrews repaired the u/s engines, and prepared the aircraft for the next scheduled operation. This came the following night, when eight aircraft were detailed for an attack against Turin. All designated aircraft took-off and, on this occasion, there were no early returns. The mission was considered a success, with all crews claiming to have bombed the target. Flight Lieutenant Chave's aircraft was attacked by enemy nightfighters on two separate occasions during the mission, but returned to Bourn undamaged.

Wing Commander Menaul led a force of fifteen Stirlings from No.XV Squadron, on a maximum effort attack against Lorient on 7th February. A total force of 323 assorted bombers participated in this raid, which was considered highly successful. Unfortunately, most of the crews from No.XV were diverted to RAF Oakington upon their return from the mission, due to one of their number being bogged down on the runway at their home base. The same target was attacked again two weeks later, on 13th February, with similar results.

Cologne received further attention from Bomber Command, on 14th of the month, when a total of 243 aircraft attacked the city. No.XV Squadron provided twelve bombers, one of which failed to return. Stirling, BF448, LS-T, piloted by F/L Owen Chave, was attacked by a Messerschmitt Bf110, flown by Oberfeldwebel Fritz Schellwat, of V Gruppe/Nachtjagdgeschwader 1. The enemy nightfighter caught the bomber flying at 4.500m over Belgium, and opened fire, causing the British aircraft to explode in the air. The remains of the Stirling crashed at Helchteren, a few minutes before 21.00 hours. Flight Lieutenant Chave, who was the son of Sir Benjamin Chave, K.B.E., and Lady Chave, was killed along with the rest of his crew. Sergeant Arkinstall, whose aircraft had been experiencing engine trouble and would therefore have been 'easy pickings' for any nightfighter, crossed the English Channel unmolested and landed at RAF Manston on the Kent coast.

Whilst No.XV Squadron had continued to play its part in the war against the Nazi oppression, S/L Michael Wyatt had been busy in Spain. Being the senior British internee in that Country, he was asked by the Spanish Authorities to help set up an internment center, run on service lines. The center, which was to house British internees at present scattered in various locations around the area, was to be located in an hotel at Alhama de Aragon. Seeking some form of remuneration for his work, S/L Wyatt requested he and his crew be repatriated to England, as soon as the center was running to his satisfaction. A deal was struck and Michael Wyatt undertook the task. He virtually took over the hotel, drafted rules and regulations for the internees to abide by, set up sleeping quarters for Officers and N.C.O's, arranged dining facilities along the same lines, and even arranged lounges for them to relax in. By the time he had completed his task, Michael Wyatt had all but given the internees a servicemen's club. True to their part of the bargain, the Spanish Authorities arranged for S/L Wyatt and his crew to be repatriated. On 15th January 1943, the British airmen crossed the frontier into Gibraltar where, after a couple of days stay, they boarded a ship for Scotland.

Following a long and tedious de-briefing at the Hotel Metropole in London, Michael Wyatt was sent on leave until 18th February, on which date he was to report back to his Squadron.

Needless to say, when he arrived at Bourn the pilot found a number of changes had been made. Apart from new faces on the Squadron generally, there was also a 'new' C.O. whom, much to Michael Wyatt's amazement, did not know much about the latter's trip to Turin, or his subsequent sojourn in Spain. Furthermore, it seemed as if the C.O. was not the least bit interested, as he made no effort to question S/L Wyatt about the latter's experiences. A further indignity the returning officer had to endure was the loss of his Flight, which had been given to somebody else.

The day after Michael Wyatt's return, the Squadron detailed thirteen aircraft for an attack against Wilhelmshaven but, due to engine problems, only twelve Stirlings took-off. The mission was a disaster as far as No.XV

was concerned, with three aircraft returning early and three failing to return. Above the bombers a full moon shone brightly, casting its silver rays across a cloudless night sky, allowing them nowhere to hide, whilst below, considerable haze and smoke from the bomb loads covered the target area. The night hunters of the Reich took full advantage of the prevailing conditions, and plied their trade in force. Oberleutenant Hans-Joachim Jabs, the nightfighter ace and Kommandeur of IV./NJG1, was one of those hunters, and it was he alone who was responsible for the loss of the three No.XV Squadron aircraft. Oberleutenant Jab's first encounter was with Stirling bomber, BF457, LS-B, piloted by F/O David Hopson, RAFVR. The German pilot applied all his knowledge and experience to the situation, and stalked his prey carefully. When fire spat from the guns of his twin-engined Bf110 nightfighter the shells found their target. The Stirling, the first No.XV Squadron Mark III aircraft to be lost, crashed at Nijlantsrijt, east of Buren, on the island of Ameland at 21.00 hours. There were no survivors. Fourteen minutes later the same scenario was to be re-enacted. Oblt Jabs located and shot down Stirling bomber, BF378, LS-T, flown by F/O Bernard Crawford, RNZAF. This machine went down into the North Sea 10kms north-west of Schiermonnikoog, Holland. As before, the entire crew perished. Finally, thirty minutes later, Oblt Jabs struck again, when the shape of Stirling bomber, BF411, LS-A, filled his gun sight. The target aircraft, piloted by P/O John Monteith, RCAF, shuddered under the impact of incoming enemy shells, before crashing at Terschelling, Holland, at 21.45 hours. In the short time span of forty-five minutes, No.XV Squadron had lost three aircraft and twenty-one crew members. Twenty-one young men whose names would initially be listed as 'missing', but ultimately added to those already recorded on the Squadron's Roll of Honor.

Flight Lieutenant Patrick Brennan, for some unrecorded reason, decided on that fateful night to fly as rear gunner on P/O Monteith's crew. His decision cost him his life, but saved that of Sgt John Greenwood, the crew's usual rear gunner. However, an entry was made in the latter's Log Book implying that he had been lost when the aircraft failed to return. In fact John Greenwood, who had flown seven operations with P/O Monteith, was to undertake another seven missions with another crew before events overtook him nearly two months later.

The weather during the third week of February was poor, and very few operational sorties were detailed for No.XV Squadron. The time however was put to good use and, apart from the usual flying training exercises which occurred during such periods, a number of lectures were arranged. The first was a talk by W/C Donaldson of No.7 Squadron, on 'Pathfinder Force target marking techniques'. This was followed by talks on 'Night Vision', presented by F/L Raffaelli, and 'Fighter Tactics', hosted by F/L Collins. Both of the latter lectures were accompanied by films on the subject. On the morning of 21st February a lecture, aptly titled 'Escape Tactics', was given to all members of aircrew by S/L Michael Wyatt and F/O Hartley.

Towards the end of the month, the general public had the opportunity to meet a 'heroine' from No.XV Squadron, and raise money for the war effort at the same time. The 'heroine', Stirling bomber, N3669, LS-H, which had completed sixty-nine bombing operations, a record for this type of aircraft, was put on display outside St Paul's Cathedral, in the City of London. The machine, which generated a lot of interest, remained at this location into the following month.

The night of 25th February saw a return to operations for No.XV, when the Squadron detailed eleven aircraft for an attack against Nuremberg. All the aircraft took-off, but two returned early due to the effects of severe icing. Unlike the Squadron's last mission, this raid was carried out in a moonless sky, with thick haze over the target area. Upon their return, the crews reported a successful mission, with several heavy explosions and a number of concentrated fires. One Stirling, BK632, LS-V, piloted by Sgt Arkinstall, was hit by flak over the target, but returned safely to base.

The next day the Squadron scheduled eleven aircraft for a raid against Cologne, whilst a further three were detailed for a mine-laying operation in the Frisian Islands. One of the main force machines however, was withdrawn from the Battle Order prior to take-off. Six of the ten remaining aircraft were loaded with incendiary bombs, whilst the other four carried a mixed cargo of high explosive bombs. Although the target area became obscured by haze and smoke, all crews reported bombing on the Pathfinder Force marker flares. Some crews reported good concentrations of fires, the glows of which were visible from the Dutch coast on the homeward journey. One crew however, were destined not to see the glow of those fires, their Stirling bomber, R9279, LS-J, piloted by F/L Victory Harris, RCAF, disappeared without trace. Victory Harris's promotion in the rank of Flight Lieutenant had been granted only ten days before his demise. The entire crew, including F/S Richard Ashdown, who had been seconded to the Squadron from No.1651 Conversion Unit for operational experience, were never heard of again.

The previously mentioned posthumous award of the D.F.C to F/L Brian Ordish, was recorded in the Squadron's Operational Record Book at the end of February, along with two others awarded to F/O Thompson and F/O Bob Munns, the latter being on his second 'tour' with No.XV Squadron. Amongst those receiving promotion at this time were P/O 'Dicky' Craddock and P/O Dennis Haycock, both of whom were appointed to the rank of Flight Lieutenant.

Two more crews failed to return from an attack against Berlin, on the first night of March. The Squadron had detailed twelve aircraft for the mission, but due to engine trouble one bomber failed to take-off. Sergeant G Ware, piloting Stirling, BF439, LS-T, led the Squadron away at 18.08 hours. The remaining eleven machines followed at regular intervals, until the last Stirling was airborne at 19.16 hours. They headed up towards the coast of Norfolk, and crossed out over the North Sea at Cromer. The long haul to the Reich Capital took the Stirlings over Hamburg, where they received attention from the ground defenses. Searchlights swept back and forth across the night sky seeking the intruders, as the sirens wailed in city below. Suddenly, a bomber was caught in the blinding glare of one of the menacing beams. Stirling, BK658, LS-K, piloted by P/O Irwin, was the unfortunate victim against whom the German heavy flak gunners now directed their attention. Hot steel arced up into the sky and exploded around the bomber but, by either good luck or fortune, did not succeed in bringing the aircraft down. Nevertheless, the pilot was forced to jettison his bomb load and take violent, evasive action. Having successfully evaded the beams and the bursting flak, P/O Irwin abandoned his mission and returned to Bourn. His colleagues, who were more fortunate, continued their journey unmolested towards the 'Big City'. The first of XV's flyers to make his bombing run over Berlin was F/S Shiells. His bomb load consisting of both 4lbs and 30lbs incendiaries, were recorded as having burst three miles west of the aiming point, at 22.05 hours. As F/S Shiells flew over the target area his bomber, Stirling, BK657, LS-C, was hit by heavy flak, but fortunately damage was minimal and the aircraft returned safely to base. Another aircraft hit by flak was Stirling, BF376, LS-N, piloted by P/O Moffat. This machine was struck by light flak whilst flying over the region of Amsterdam, on the homeward leg of the mission. It too made a safe landing at Bourn. Luck ran out for F/S Harold Howland and his crew, who had completed their mission and were on their way home when they encountered a nightfighter. The enemy aircraft, operating out of Leeuwarden and piloted by Oberleutnant Wolfgang Kuthe, of IV./NJG1, stealthily crept into the attack and opened fire. The bomber, Stirling, EF347, LS-T, shuddered under the impact of the enemy gunfire, peeled over and fell to earth. It crashed at Mantgum, Holland, at 00.58 hours, killing all of the crew. Just two minutes later, another No.XV Squadron aircraft went down in flames, also having been attacked by an enemy nightfighter. Stirling bomber, W7518, LS-G, piloted by F/S Arthur Tilley, crashed at Muye Polder, at Sint Maartendijk, Zeeland, Holland, at 01.00 hours. The only survivor from this crew was Sgt G E Williams the rear gunner, who was captured in a wounded condition, and made a prisoner of war. Oberleutnant Kuthe, the victor in the first attack, was himself shot down and killed six weeks later, on 14th April 1943, with a score of eight victories.

Pilot Officer William Moffat's luck was to run out the following night, 3rd March, when the aircraft he was piloting, Stirling bomber, EF333, LS-X, was shot down during a raid against Hamburg. The aircraft, carrying eight crew members, crashed into the North Sea, west of Texel, Holland, killing everyone on board. The extra crew member was Pilot Officer Adolph Meijer, RAFVR, a No.3 Group, Technical Officer. At 36 years of age, Adolph Meijer was the oldest member of the crew. A graduate of the Delft University, P/O Meijer was also holder of the Dutch Cross of Merit. The Squadron had detailed eight aircraft for this attack, but three of them returned early; two due to unserviceable fuel lines and one due to erroneous navigational fixes! The latter aircraft, Stirling bomber, BK654, LS-W, was flown by P/O I Renner, an experienced pilot who had been with the Squadron since the previous October. When it was found that the aircraft was well off track, he had no option but to abandon his mission, due to the uncalculated fuel consumption. The bomb load was 'dumped' in the North Sea before returning to base.

With the advent of March, and the coming of spring, Sir Arthur Harris decided that the time was right for a major concentrated effort against specific targets in Germany, starting with the Reich's industrial area. To this 'campaign' he applied the name *'Battle of the Ruhr'*, with Essen being selected as the first target. The 'home' of Krupps, the arms manufacturer, was attacked by a total force of 442 aircraft on the night of 5th March; seven of which were detailed by No.XV Squadron. However, on the outward journey, F/L Haycock was forced to abandon the mission due to the port inner engine overheating. He jettisoned his bomb load and returned to base. Flight Sergeant J Shiells had just dropped his bomb load on the aiming point, at 21.11 hours, when his aircraft was hit by flak. The Stirling BK657, LS-C, sustained damage but remained airworthy and returned to base.

Nuremberg was attacked on 8th March, followed by a raid against Munich on the following night. P/O Jack Ripley, RCAF, and his crew were lost during the first mission when their Stirling, BK697, LS-P, was shot down. The aircraft crashed at Campneuville, France. The second raid was more successful with all aircraft, with the exception of that flown by S/L Wyatt, returning to base. Michael Wyatt, who had led the attack, was forced to land his machine, Stirling bomber, BK694, LS-O, at Manston, Kent.

Flight Lieutenant Dennis Haycock had a lucky escape on the night of 12th/13th March, when his aircraft was attacked by two Bf110s during a raid against Essen. His aircraft, Stirling, BK656, LS-A, was one of six bombers detailed by No.XV Squadron for the mission. Having taken-off from Bourn at 19.41 hours, F/L Haycock's aircraft had just overflown the Dutch/Germany border south-east of Arnhem, near Emmerich, when it encountered two enemy nightfighters. As both the mid-upper and rear gunner prepared to open fire, the pilot order the release of the bomb load. With turrets traversing to face the incoming attackers, the Stirling started evasive action and went into a 'corkscrew' maneuver. As the enemy aircraft positioned themselves for the attack, fire spat from the guns of the rear and mid-upper turrets. One of the twin-engined nightfighters replied by closing on the bomber and returning the fire, slightly injuring Sgt Bromley the mid-upper gunner. Urged on by self-preservation and anger, the gunners on board the Stirling kept up a fusillade of gun fire, which struck its target and sent one of the attackers down in flames. The second attacker either gave up the chase or was 'shaken-off' as Dennis Haycock, the pilot, continued to fly evasive tactics. Having received a damage report on the aircraft and established the seriousness of the injuries to the crew, F/L Haycock returned to Bourn, where he landed at 23.06 hours. On arrival back at base, Sgt Bromley was transferred to the Station Sick Quarters, where the nature of his injuries were ascertained and treated.

Weather conditions during the middle of the month precluded the undertaking of any missions, including 'Gardening' operations and cross-country exercises. An attempt to launch an attack against St. Nazaire on 22nd March was canceled as the main force crossed the French coast. Although a recall signal was sent out, one of the nine aircraft detailed by No.XV Squadron, Stirling bomber, EF339, LS-Y, piloted by P/O Renner, actually bombed the primary target. Flight Sergeant J Shiell's aircraft encountered problems on the way home, due to two engines failing in flight. Five of the crew baled out of the aircraft, prior to the pilot crash-landing the Stirling, BK667, LS-H, at Clyffe Pypard, south of Swindon in Wiltshire.

Unfortunately, the huge four-engined bomber overshot and ploughed through the boundary on the north-east side of the airfield, before finally crashing into trees. The impact with the trees caused some of the incendiary cannisters aboard the Stirling to burst open, whereupon the contents ignited. Flight Sergeant Shiells and Sgt James, the bomb aimer, escaped from the burning wreckage unhurt, but Sgt Compson, the second pilot, was initially trapped in the cockpit with leg injuries. Sergeant Taaffe, the flight engineer, who had baled out of the stricken bomber was also injured. He was found close to the perimeter fence in a state of unconsciousness. Sergeant Clive Perring, navigator, and Sgt Gould, wireless operator, who had also baled out, were both found near the airfield, hanging by their parachutes from some trees.

Five days after the encounter with the two Messerschmitt Bf110s, F/L Haycock's crew were still minus an air gunner, due to the fact Sgt Bromley had not been passed fit for flying duties. The problem was overcome when Sgt John Greenwood, whose own crew were lost on 19th February, joined F/L Haycock to fly as Bromley's replacement. Apart from having John Greenwood as 'stand-in' gunner, Dennis Haycock also had to contend with another addition to his crew; he was to take Sgt McLeod along as second pilot. By the time his name was recorded on the Battle Order for the night of 27th/28th March, Sgt Greenwood had not flown operationally for nearly six weeks. Instead of an easy target to start with as he may have hoped for, John Greenwood had to face the daunting task of a raid against the 'Big City', Berlin. However, failure in the fuel tank cocks forced F/L Haycock to jettison the bomb load and return early. Flight Lieutenant Haycock's aircraft, Stirling bomber, BK619, LS-F, was repaired and air tested in readiness for a second attempt at Berlin two nights later. On checking the Battle Order, John Greenwood found his name had been retained on the crew list as an air gunner. Unfortunately, on this occasion, the weather became the enemy of the aircrews and produced severe icing conditions. Ten No.XV Squadron aircraft were detailed for this raid, half of which returned early due to the adverse conditions. Having, again, jettisoned his bomb load into the sea, F/L Haycock was forced to turn for home.

John Greenwood never did carry out an operation against Berlin.

Portrait of a Bomber Pilot

Anyone for Spain ?

Destined for a sojourn in Spain, the crew of Stirling bomber, BK595, LS-L, pose for a photograph at Bourn, prior to their enforced 'holiday. From left to right are: P/O H Daborn, Nav; Sgt G Clary, W/Op; Sgt A Ames, F/E; S/L Wyatt, Pil; Sgt E Henry, ? ; Sgt J Heal, ? ; Sgt T Kemp, ? . *Courtesy of Group Captain M. Wyatt, D.F.C.*

Wing Commander (later Group Captain) Michael Wyatt, D.F.C., MiD, joined No.XV Squadron (as a Squadron Leader) on 26th October 1942. Nearly four weeks later, on the night of 20th/21st November he and his crew participated in a raid against Turin, from which they did not return. Their aircraft, Stirling bomber, BK595, LS-A, ran short of fuel on the return journey. Michael Wyatt altered course for Spain, where he force-landed his aircraft on a beach. The bomber came to rest, with the forward fuselage on the beach and the tail end in the sea. The pilot and his crew were arrested and interned, but were repatriated in February 1943. Squadron Leader Wyatt returned to England and rejoined No.XV Squadron. *Courtesy of Group Captain M. Wyatt, D.F.C.*

Left: The twisted tangle remains of the Bristol Hercules engines, which once powered Stirling bomber, LS-A, lie strewn across the beach at Playa de Aro. *Courtesy of Group Captain M. Wyatt, D.F.C* Right: The tailplane and rear turret of BK595 appear in remarkable good condition, following the forced-landing on the Spanish beach. *Courtesy of Group Captain M. Wyatt, D.F.C*

Spanish mechanics inspect the shattered remains of Michael Wyatt's aircraft, which he force-landed on a beach at Playa de Aro, at 01.15 hours, on the morning of 21st November 1942. *Courtesy of Group Captain M. Wyatt, D.F.C*

The Return of Brian Ordish

Pilot Officer Brian Ordish (3rd left) abstains from taking a drink, whilst his colleagues enjoy a small glass of beer. The Officer to the extreme left is thought to be P/O J A Sinclair, who flew with Ordish between December 1941 and April 1942. *Courtesy of Mrs Eliabeth Park*

A Stirling Bomber, coded LS-N, of No.XV Squadron, photographed by P/O Brian Ordish, during late 1941. The aircraft is thought to be W7439, which saw service with No.XV Squadron, No.26 Conversion Flight and returned to No.XV before being taken-on-charge by No.1651 Conversion Unit. *Courtesy of Mrs Elizabeth Park*

The proud sister Elizabeth Park, photographed at her home in Gloucestershire, in 1998, displays the Distinguished Flying Cross, posthumously awarded to her brother F/L Brian Ordish in February 1943. *Author's Collection*

New Crews ?

Sergeant John Greenwood (5th from left) flew with P/O Monteith and his crew between December 1942 and February 1943. On the night of 19/20th of the latter month, P/O Monteith's aircraft was shot down and the crew were killed. John Greenwood, who was not operational that night, later joined the crew led by F/L Haycock. *Courtesy of June Watkins*

Right: A photograph, found amongst the papers of Sgt John Greenwood, of an unidentified crew standing by their aircraft, typifies many such wartime pictures. Men of No.XV Squadron, who currently remain nameless. *Courtesy of June Watkins*

Aircraft and Groundcrew

A 'hive' of activity. Stirling bomber, BF439, LS-D, being prepared for 'ops' at Bourne, during 1943. The lower section of the port inner engine nacelle has been removed, and placed at the bottom of the work trestle, whilst work is undertaken on the motor. A small group congregates around the crew entry door at the rear of the fuselage, whilst armorers load canisters of incendiaries into the bomb bay. *Author's Collection via Les Butcher*

The groundcrew of Stirling bomber, BF439, LS-D, find other chores to deal with. Included in the group cutting down the trees and clearing shrubbery are: Sgt McCann; Charlie Hawker; Harry MacClannen; Jack Head and Les Butcher (extreme left). *Author's Collection via Les Butcher*

A Stirling Named 'Te Koote'

The original crew of Stirling, BK611, *"Te Koote"*, from left to right are: Bob Renner, RNZAF, Pil; Doug Mepham, F/E; Jimmy James, R/G; Ian McColl, B/A; Norman Southern, seated in doorway, WOP/AG; Eric Siegel, Nav; Danny Millership, MUG. *Courtesy of Arthur Edgley*

Three times Fifteen !

Wing Commander Stewart 'Paddy' Menaul, D.F.C., Officer Commanding No.XV Squadron, 7th December 1942 to 7th May 1943. After a long and distinguished career with the Royal Air Force, he retired from the service with the rank of Air Vice-Marshal. His awards, both civil and military included the C.B., C.B.E., D.F.C., A.F.C. *Author's Collection via the late AVM Menaul*

Whilst the aircraft are at rest, groundcrew chief Sgt Bobby Gault takes the opportunity to pose for a photograph on the port inner engine of Stirling, BK611. Note the artwork below the cockpit. *Courtesy of Arthur Edgley*

Right: Close-up of the artwork and name *"Te-Koote"* on Stirling, BK611. The emblem, painted on the post side of the fuselage only, has New Zealand Maori tribal connections. *Courtesy of Arthur Edgley*

A glint of sun reflects off the winching mechanism during the bombing-up of Stirling, BK611. Pilot Officer Ian Renner (center right) checks the operation, whilst Sgt Norman Sorthern (2nd left) and members of the groundcrew look on. *Courtesy of Arthur Edgley*

New Arrivals, Old Hands

Found amongst the papers of John Greenwood, this photograph shows the original crew of F/L Dennis Haycock. Standing left to right: Sgt Nornam Hobden, F/E; Sgt Peters, A/G; Sgt Herbert Fortune, WOP/AG; Sgt Thomas Bromley, A/G. Seated left to right: Sgt Blackburn, Nav; Dennis Haycock, Pil; Sgt Henry Fiddes, B/A. The photograph is believed to have been taken on Wednesday, 23rd December 1942, approximately one month prior to the crew being posted to No.XV Squadron. *Courtesy of June Watkins*

Short Stirling, BF436, LS-E, served with No.XV Squadron between late 1942 and early 1943. The aircraft was usually flown by either Sgt W Irwin or Sgt J Smale. P/O Dennis Haycock flew his first mission with the Squadron on this aircraft, on 3rd February 1943, with Sgt Irwin as captain. *Author's Collection*

P/O Craddock and Crew

The aircrew of R9193 consisted of, left to right: P/O Ken Dougan, B/A; Sgt Arch MacDonnell, Nav; P/O Richard 'Dicky' Craddock, Pil; ? F/E; Sgt W Pengelly, MUG; Sgt Pete Devine, R/G; ? WAG; P/O Bill Prune, Squadron mascot. The two unidentified crew members are Sgt Dodds and Sgt Booth. *Courtesy of and via Ken Dougan*

A poor quality photograph of the groundcrew of Stirling, R9193, LS-S. Note the thirty mission bomb tally by the crew access door. *Courtesy of Don Clarke, MBE, The Mildenhall Register*

'Bombing Up'

Pilot Officer Ken Dougan, B/A (left) and P/O 'Dicky' Craddock, pilot, check out the bomb load, whilst Bill Prune appears totally disinterested in the whole procedure. *Courtesy of and via Ken Dougan*

Corporal 'Rocky' Stone (left) watches as Corporal Armorer Don Clarke (unseen in the bomb-bay) winches a bomb covered with British stamps up into the fuselage of Stirling, R9193, LS-S. Note the unidentified Stirling in the background. *Courtesy of and via Ken Dougan*

Armorers and Electricians

The Armorers and Electricians of No.XV Squadron pose for a photo-call at RAF Bourne, on a wet day during February 1943. *Courtesy of Don Clarke, MBE, The Mildenhall Register*

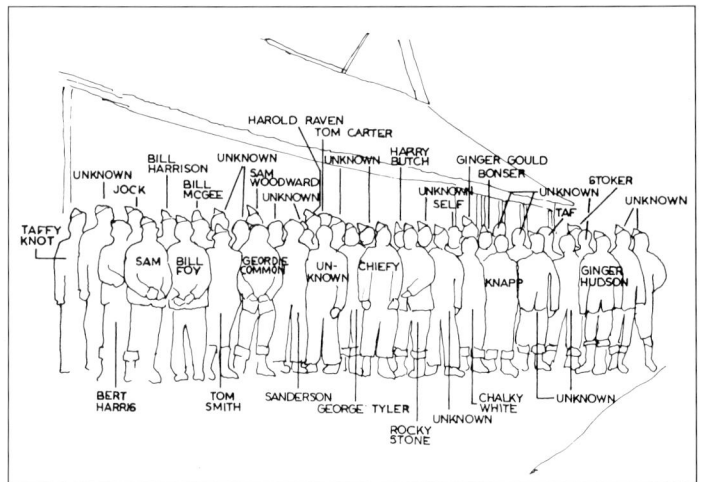

Key to above photograph. *Author's Collection*

A 'Heroine' Prepares to Meet Her Public

Short Stirling, N3669, "H"- Harry, being re-assembled for public display outside St. Paul's Cathedral, London, on 25th February 1943. *Courtesy of Don Clarke, MBE, The Mildenhall Register*

Whilst Stirling bomber, N3669, LS-H, was being shown-off to the British Public in London, a new 'Harry' (Stirling BK667) had taken to the skies. Unfortunately the second machine, piloted by F/S Shiells, crash-landed at Clyffe Pypard, in Wiltshire, a month later. The observer/navigator of the second aircraft, Sgt Clive Perrin, sent a photograph of the original machine home to his family. On it he wrote, 'A "Stirling" in which Clive "serves" – March 1943' and "H" Harry of 15 Squadron – now pensioned off. We fly the new "H" for Harry'. *Courtesy of and via Steve Smith*

Members of the Womens Auxiliary Air Force (WAAF) assist in the re-assembling of Stirling bomber, B3669, which was displayed in the City of London during February 1943. *Courtesy of Don Clarke, MBE, The Mildenhall Register*

One to send home to Mom. An unidentified member of No.XV Squadron groundcrew is photographed admiring the 62 mission bomb tally, painted on the port fuselage of Stirling bomber, N3669, adjacent to the crew entry door. *Courtesy of Don Clarke, MBE, The Mildenhall Register*

A 'Heroine' meets her public. Stirling, N3669, of No.XV Squadron is proudly exhibited for the public to inspect, in the hope of raising funds during 'Wings for Victory' week. *Courtesy of and via Flt/Lt T. N. Harris, XV (R) Squadron*

Flying Officer Bob Munns, DFC

Groundcrew with Their 'Lady Love'

Wing Commander Bob Munns, D.F.C, Signaler, who first flew as a Sergeant, wireless operator with No.XV Squadron, between May and November 1941. He was commissioned to the rank of Pilot Officer in 1942, but was promoted to Flying Officer on 1st October the same year. The award of a Distinguished Flying Cross was made on 9th February 1943, whilst on his second tour with No.XV Squadron. Flight Lieutenant Munns was posted to No.1657 Conversion Unit, with effect from 16th March 1944. He was later promoted to Wing Commander, the rank with which he saw the end of his service. *Courtesy of Wing Commander Bob Munns, D.F.C.*

As far as the members of groundcrew were concerned, the aircraft to which they were assigned became their charges. The aircrew only 'borrowed' the machine. Stirling bomber, W7518, was delivered to No.XV Squadron during April 1942. Although it flew as LS-U, LS-C, LS-W and LS-G, it was still their aeroplane. W7518 failed to return from a raid against Berlin, having been shot down during the early hours of 1st March 1943. Six members of the crew including the pilot, Sgt A Tilley, were killed. The only survivor, Sgt Williams, the rear gunner was wounded. The groundcrew, including Peter Hartley (sitting at the rear in the open doorway of the aircraft), were saddened by the lost of their aircraft. *Author's Collection*

13

The Battle of the Ruhr

The Battle of the Ruhr, which had commenced on the night of 5th/6th March, was to continue through until 24th/25th July. During that period many of the men posted to No.XV Squadron would, against all odds, complete a 'tour' of operations. There were some, however, who fate had determined, would not complete their first mission.

The first operations undertaken in April 1943 occurred during the evening of the second day of the month. Five aircraft set out to attack the French port of St. Nazaire, whilst two others carried out mine-laying sorties. One of the main force machines, Stirling bomber, BK611, LS-U, piloted by Sgt Johnson, returned early with both the mid-upper and rear turrets unserviceable. The four remaining aircraft continued onto the target where they were greeted with a barrage of flak, some of which struck Stirling, EF354, LS-Q, causing slight damage. Although the pilot, Sgt VanEupen, managed to keep control of the aircraft, he was forced to land at Moreton-in-Marsh, an Operational Training Unit airfield in Gloucestershire.

Two nights later a maximum effort raid was carried out against Kiel. No.XV Squadron detailed eighteen aircraft for the mission, all of which took-off. Unfortunately, three were to return early due to mechanical problems. Flight Lieutenant Haycock, with Sgt John Greenwood now flying as a regular crew member, returned due to an unserviceable port engine. Sergeant Curtis experienced problems trying to get his aircraft, EF348, LS-N, to gain height, whilst Sgt Trezise, who was piloting Stirling bomber, EF351, LS-L, found his aircraft would not respond to the controls. The remaining crews, who reached the target, reported bombing on the Pathfinder Force marker flares. Although the crews were unable to observe the results of their respective attacks, many also reported seeing large fires reflected on the 10/10ths cloud covering the target.

A maximum effort attack against Duisburg on 8th April, was almost a repeat of the raid on Kiel three days earlier. Fifteen aircraft took-off for the operation, but encountered severe icing conditions which forced four of them to return early. Two other machines returned early due to mechanical problems. As with the previous raid, 10/10ths cloud blanketed the target area, but this did not preclude the remaining aircraft from completing their task. Stirling bomber, EF359, LS-B, piloted by Sgt John Gurr, RAFVR, failed to return. The aircraft, which possibly exploded in mid-air, crashed at Woltershoff, on the west bank of the River Rhine, approximately ten kilometres from the target. John Gurr's application for a commission was granted three weeks after his death, allowing him, eventually, to be buried with the rank of Pilot Officer.

Two nights later another crew was lost when Sgt Eric Trezise's aircraft, Stirling bomber, BF475, LS-T, was shot down. The aircraft, which was participating in a raid against Frankfurt, crashed at St. Genevieve, France. The crew were all buried together in the Military Cemetery, at Montcornet; five kilometres north-east of the crash site. No.XV Squadron had detailed eighteen aircraft to participate in the attack, all of which took-

off. F/Sgt O'Conner was forced to return early, due to failure of the starboard outer engine, of Stirling bomber, BK704, LS-Z. His aircraft had almost reached Abbeville, where he jettison the bomb load, before he aborted the mission. The raid was not considered successful by the returning crews, due to 10/10ths cloud over the target which caused the bombing to be scattered.

The Squadron was officially 'stood-down', as from 11th April, pending a move to the newly re-opened R.A.F Station at Mildenhall. As each section of the Squadron packed away its equipment ready for transportation to the new base, the groundcrews continued to work on their charges, using the time to bring the aircraft up to full serviceability.

The airfield at Mildenhall had been constructed during the early 1930's and although an advanced party of R.A.F personnel arrived during the preceding August, the new station was not officially opened for military use until October 1934. The first unit deployed to Mildenhall was No.99 (Bomber) Squadron, who arrived there the following month, equipped with Handley-Page Heyfords.

During the pre-war years a number of notable events occurred at RAF Mildenhall. The Royal Aero Club supervised the start of the England to Australia air race from there, during October 1934. On 6th July 1935, the Royal Silver Jubilee review of the Royal Air Force was held at Mildenhall and, during October 1937 a delegation of top German Luftwaffe officials, including General Erhard Milch, carried out an official visit of the station. The members of the German mission were allowed to inspect aircraft and equipment, and to watch a fly-past, before being entertained to lunch. Later that same year, on 21st December, the Air Officer Commanding-in-Chief, Air Chief Marshal Sir Edgar Ludlow-Hewitt (No.15 Squadron's second official Commanding Officer), arrived at RAF Mildenhall to present No.99 Squadron with its Unit Badge.

During the final years of peace a number of squadrons served, or were formed, at Mildenhall. Amongst them was No.149 (Bomber) Squadron, who were still based at the station when war was declared. No.149 Squadron remained at Mildenhall until the end of 1942, when the base was closed temporarily for expansion and modernisation. When it re-opened in April 1943, following its refurbishment, the airfield had been equipped with a new control tower, concrete runways and Drem lighting.

Personnel of the new Base Headquarters, RAF Mildenhall, formed on 24th March, devoted much time and energy in preparing for the arrival of No.XV (B) Squadron. Part of No.XV's own preparations for the move from Bourn included sending an advanced party of ground staff to Mildenhall, on 13th April. Use was made of two Airspeed Horsa Gliders from No.38 Wing, which arrived at Bourn that same day, to move both personnel and ground equipment. The following day, the gliders were towed by two Whitworth Whitley Mk.V bombers to the 'new' station, which was to be No.XV's home for the duration of the war. The Stirlings naturally flew the short distance to Mildenhall during the day, and could be seen lined-up in the circuit waiting to land, looking as though they were returning from a daylight operation. Later that same day, the A.O.C. Bomber Command, Air Vice-Marshal A P M Sanders, C.B.E, visited the station to ensure all had gone well.

An example of the camaraderie which existed on the Squadron came at the time of the move. Aircraftman 2nd Class Ken Rogers had joined No.XV at Bourn in March, a month before the move. At the time of the move, AC2 Rogers was in hospital and therefore not 'officially' informed of the situation. The groundcrew team of which he was a member was a close knit group, and they packed his kit and transported it, along with their own, to Mildenhall. The next time they visited him in hospital they told Ken of their action, of which he was extremely grateful. However, in the interest of security, and to protect his friends, when he was released from hospital AC2 Rogers reported for duty at the old base. Needless to say he was officially informed of the Squadron's move, and instructed to report to RAF Mildenhall.

A settling-in period was permitted on 15th April and although no operations were detailed, several air tests were carried out. Whilst the Administrative Section settled into their new offices and the groundcrews prepared the aircraft for operations, some of the aircrews carried out a reconnaissance mission of the immediate area. Their task was to locate and map local landmarks such as the Bird in Hand Hotel, the Rose and Crown public house, the Judes Ferry Inn and the White Hart Hotel. The night of 16th April 1943, witnessed the first operational use of RAF Mildenhall in just over five months, when eighteen aircraft from No.XV lined-up for take-off, from one of the new concrete runways; their destination was Mannheim. Stirling bomber, EF348, LS-N, piloted by Sgt E Curtis was first away from Mildenhall, at 21.14 hours. The rest of the Squadron followed, and within twenty-nine minutes the other seventeen aircraft were all airborne and setting course across Suffolk towards the east coast.

One Stirling, BK611, LS-U, flown by Sgt Forbes, returned early after both port engines cut out whilst overflying Cayeux. Having jettisoned the bomb load off the French coast, Sgt Forbes landed back at Mildenhall at 00.01 hours. The seventeen remaining aircraft droned on towards Germany. The sky was cloudless, although there was slight ground haze in the target area. Flight Sergeant O'Conner, piloting Stirling, EF351, LS-L, arrived ahead of the e.t.a but, having seen the red and green target indicators floating down, and identified the target, he started his bomb run.

As the marker flares cascaded down to illuminate the aiming point, the probing fingers of the enemy searchlights stabbed up into the night sky in an attempt to illuminate the bombers. The searchlights found a victim in the shape of Stirling, BK654, LS-W, piloted by Flying Officer Norris. The probing beams held the aircraft for ten minutes, an eternity for the crew, but by much evasive action, BK654, managed to slip back into the safety of the night. In a gesture of defiance, the gunners aimed their machine guns at the base of the blinding ray and opened fire. Much to their amazement and satisfaction the tenacious beams went out.

Flying Officer Bowyer was one of three pilots who took his crew 'train busting'. The other two pilots were W/O Wilby, piloting Stirling, BF482, LS-R, who shot-up two trains in a marshaling yard, and P/O Hugh Wilkie, flying Stirling, BF470, LS-G, who attacked a train in a siding. Hugh Wilkie's navigator, Sgt Frank Diamond, later recorded that it was on this raid that LS-G was christened by flak.

His bombing run completed, F/S O'Conner turned his aircraft onto a course for home. He was flying at an altitude of 9,000' and had just begun to lose height, when the front gunner, Sgt Price, reported a nightfighter 600 yards off the port quarter front. Receiving the instruction to 'corkscrew', F/S O'Conner hauled the Stirling over into a turn, bringing the Messerschmitt Bf109 into full view of Sgt Shearer, the rear gunner. Simultaneously Sgt Shearer and the enemy pilot opened fire at a range of approximately 500 yards. The Bf109 scored hits on the starboard inner engine and the rear turret, putting it out of action. The nightfighter passed over the bombers starboard beam, some two hundred yards away, and seemingly broke-off the attack. Within minutes Sgt Gaylor, the mid-upper gunner, yelled a second warning of fighter attack, as the Bf109 re-appeared

coming down from high onto the port beam. To add to the bomber crews' problems, a second Bf109 made a head-on attack. F/S O'Conner again threw the Stirling into a 'corkscrew' maneuver. The first nightfighter passed 500' beneath the bomber, scoring hits on the mainplane as it dived. Sergeant Gaylor fired a short burst at the enemy aircraft as it came into view on the starboard side on the Stirling. The second Messerschmitt, attacking from head-on, scored hits on three airscrews and the front turret. Sergeant Price returned fire and was himself hit though not wounded, when a bullet was deflected by coins in his pocket! Breaking-off again, the two enemy fighters were seen to be positioning themselves for another, identical, attack. Forewarned is forearmed, and Sgt Price in the front turret was ready. As the first aggressor came in with cannons firing, Sgt Price took careful aim. He opened fire, but only one of his machine-guns responded; the other being put out of action in the previous attack. Nevertheless, Sgt Price and F/S O'Conner both saw strikes on the Bf109, which dived under the Stirling and was not seen again. The second aggressor attacked from the port beam and, like his comrade, passed under the bomber and disappeared from view; but the fight was not over. The Stirling had lost much height, and was flying on three engines when the third attack came. Not able to take evasive action, F/S O'Conner turned in to meet the attacker as it came in from the front starboard quarter, closing from 600 yards to 300 yards and firing its cannon continuously. Sergeant Gaylor fired 400 rounds in reply, and saw the enemy aircraft peel-over and go down in a steep dive. The fighter is reported to have exploded on impact with the ground. Flight Sergeant O'Conner nursed the crippled bomber back to England, where it landed at the fighter airfield of Tangmere, near the south coast. A close inspection of the Stirling revealed that, apart from damage to the inner starboard motor and a defective rear turret, the aircraft was found to have damage to the bomb-aimer's panel and the mainplane. The tailwheel had been shot away, and the airframe had over three hundred bullets holes. F/Sgt O'Conner and Sgt Gaylor were both awarded Distinguished Flying Medals.

Sergeant Johnson, who had bombed Ludwigshaven, was forced by the combination of a lack of fuel and flak damage to both wings, to land his Stirling bomber, BF339, LS-Y, at another fighter airfield; that of West Malling, in Kent. Flight Lieutenant Lyons crashed Stirling bomber, BK659, LS-X, on the runway at Mildenhall, as a result of damage caused by enemy action during the raid. The wreckage caused the diversion of a number of Squadron aircraft to other bases. Stirling bomber, BK476, LS-D, piloted by P/O Watson, landed at Lakenheath, as did Stirling, EF348, LS-N, flown by Sgt Curtis. Two other aircraft, Stirling BK704, LS-Z, and Stirling, BK648, LS-J, piloted by F/O Bowyer and Sgt Hunt respectively, were diverted to Newmarket. Flying Officer Norris, having recovered from his searchlight ordeal, landed at Tangmere.

At the de-briefing F/O Norris and his crew reported the sighting of a friendly aircraft being shot down, and three parachutes emerging from the stricken bomber. Flying Officer Bowyer and his crew reported not only the sighting of a bomber going down, but also what appeared to be an intact Stirling in a field! The attack was generally reported as good, with large well concentrated fires seen by the crews. A total of 98,000lbs of bombs and incendiaries were dropped in return for a loss of seven aircraft, two of which came from No.XV Squadron. Fate caught up with Sergeant John Greenwood when Stirling bomber, BK691, LS-F, piloted by F/L Dennis Haycock, DFC, crashed at Hetzerath, Germany, killing six members of the crew. Sergeant Norman Hobden, the flight engineer, was the only survivor. John Greenwood had, erroneously, been presumed killed in action some weeks before, but on this occasion there was no error. Many members of aircrew adopted a custom, or ritual, that they would undertake prior to setting off on a mission, and John Greenwood was numbered amongst them. Finding a few minutes to be alone before boarding his aircraft, Sgt Greenwood would stand in the darkness enveloping the airfield and whistle the Twenty Third Psalm. It is not known whether his usual custom was practiced on this particular night.

The second crew lost on this raid was that of P/O James Shiells, D.F.M, whose aircraft, Stirling bomber, BF474, LS-H, crashed at St. Erme, France. This aircraft, "H" – Harry, was the replacement machine referred to on the newspaper cutting, showing the previously coded aircraft outside St. Paul's Cathedral. Also flying with P/O Shiells as second pilot, was Pilot Officer Kenneth Piche, RCAF, who was the adopted son of Alfred and Arlina Piche, of Lake Linden, Michigan, U.S.A.

Following the attack against Mannheim, there was a brief respite from operations whilst the Squadron re-grouped and repaired its battle damaged aeroplanes. The quiet period only lasted four days before No.XV returned to the fray on the night of 20th April, when it attacked the Heinkel aircraft works at Rostock. No.XV detailed fifteen aircraft to join a main force of eighty-six Stirling bombers, but two of the Squadron's aircraft failed to take-off. The attacking crews found the area covered by a smoke screen, which caused the bombing to become scattered. Due to the heavy smoke, Sgt Smale piloting Stirling, BK656, LS-A, dropped his bomb load at a point north of the town, on e.t.a. He was on the homeward leg of the journey when his aircraft was attacked by a nightfighter. In the ensuing fight, which was indecisive, the mid-upper gunner was wounded in the foot. Stirling bomber, EF348, LS-N, piloted by Sgt Curtis had its port wing holed by an exploding cannon shell after its bombing run, but managed to return to Mildenhall safely. One aircraft which did not return safely was Stirling, BF476, LS-P, which was shot down over Denmark. The bomber was flying at the exceedingly low altitude of 500 feet on the homeward leg of its mission, when the ground defenses opened up. The flak found its target, and caused the bomber to crash-land at Kragelund, 15 kms north west of Vejle, at 02.46 hours. Apart from the whole crew being captured and made prisoners of war, two members of the crew were reported as being injured. Sgt Howland receiving a flak wound to the foot and burns to the face, whilst Sgt Huggett was wounded in the thigh by flak. The severity of the latter's wounds necessitated attention by a field hospital at Schleswig, across the border in Northern Germany.

Due to adverse weather conditions no operations were planned for No.XV Squadron until the night of 26/27th April, when sixteen aircraft were detailed to attack Duisberg. However, at the time of take-off, one aircraft aborted. Stirling bomber, BK652, LS-V, piloted by F/S Arkinstall, failed to leave the ground. The fifteen machines which did take-off set course from Mildenhall, and crossed the east coast of England at Southwold. They flew out over the North Sea and headed for Holland, where the flak and nightfighters were waiting. One aircraft which encountered the latter was Stirling, BK657, LS-F, piloted by P/O Watson. The bomber had crossed the Dutch coast and was approximately thirty miles inside enemy-occupied airspace, when it was attacked by a Ju.88 nightfighter. Although the enemy machine was approximately 1,000' below them, F/S Bearnes, the rear gunner, had seen its approach. He warned the pilot that the Ju.88 was passing underneath the bomber and lost to view. As P/O Watson initiated a diving turn, cannon and machine gun fire spat forth from the guns of the attacking aircraft, and ignited some of the incendiaries carried by the bomber. At least two of the crew are known to have baled out, a fact the rear gunner reported to the pilot. P/O Watson asked for a damage report, to which F/S Bearnes replied that the aircraft was on fire. Unsure as to whether the mid-upper gunner had been hit in the attack, and knowing the nightfighter was possibly still in the vicinity, F/S Bearnes stayed in his rear turret. Flames issued from the Stirling as the enemy fighter continued to attack. Now, with no hope of saving the bomber, the pilot ordered the rest of his crew to bale out. Sergeant Bearnes went out via the rear escape hatch, and had to endure the pain of a broken thigh bone, the left hip being dislocated, and of being dragged through water by his parachute after landing in a canal. Sergeant Gordon Whittaker, the mid-upper gunner, was less fortunate. He had apparently baled out of the aircraft, but his body was found in a field with a bullet wound to the head. On closer inspection, it was also found that Sgt Whittaker's collapsed parachute was full of bullet holes. It is not known whether Gordon Whittaker died during his descent, or on impact with the ground!

The first man out of the aircraft had been Sgt Cyril Mora, RNZAF, who had crawled along the floor of the bomber to reach the escape hatch. He landed in a greenhouse, in the garden of a property at Breukelen, north-west of Utrecht. Remembering the drill instilled in him at escape lectures,

on those non-operational flying days back at base, Sgt Mora made good his escape from his point of impact on Dutch soil. Luck was certainly on his side as he was quickly able to establish contact with the local Dutch Resistance unit and, with their assistance, was back in England by 6th May 1943. The five remaining members of the crew were all captured and made prisoners of war. The stricken Stirling, continued its diving turn and was heading on a westerly course when it crashed at the edge of a dyke at Portengen, two kilometres south-west of Breukelen.

Stirling bomber, BK611, LS-U, developed a fault in the inner starboard engine but, after jettisoning some of the incendiary bombs, P/O Renner the aircraft's captain, decided to carry on to the target. Having identified the aiming point, and seen the T.I's, the crew bombed at 02.33 hours, from a height of 13,000 feet. Their mission completed, the pilot turned for base, where they landed at 05.10 hours.

Over the target area two aircraft were caught by searchlights and held in cones but, fortunately, both were eventually able to escape the beams. The blinding glare of a searchlight beam invaded the darkness on the flight deck of Stirling bomber, EF345, LS-M, as the aircraft started its bombing run. The pilot, P/O Jim Stowell, continued his task with flak bursting all round the aircraft, some of which punctured the skin of the bomber with approximately sixteen holes. Sergeant Bill Jennings, the mid-upper gunner, experienced a heart-stopping moment when a large chunk of shrapnel shattered the plexi-glass of his turret and narrowly missed injuring him. Although the Stirling survived the experience, it was destined to fight only two more battles. The second aircraft to be caught and held in the searchlight beams was Stirling, EF348, LS-N, flown by Sgt Curtis. The aircraft was coned for approximately ten minutes whilst over the target area, during which time it was subjected to intense flak. Although flak damage was inflicted on the bomber, and the rear turret was rendered u/s, the aircraft returned safely to base.

Apart from a 'Gardening' operation in the Bay of Biscay on the night of 27/28th April, when F/L Bowyer was forced to return early due to a technical problem with his aircraft, no further operations were carried out until the first week of May.

The first week of May also saw a number of postings and promotions. On 3rd May, S/L Michael Wyatt was promoted to the rank of Wing Commander and given command of No.75 Squadron, based at Newmarket. Five days later, on 7th May, W/C Stewart Menaul was appointed to the rank of Acting Group Captain and given command of R.A.F Station Bourn, the base from which he had recently moved. On the same day, W/C John Stephens arrived at Mildenhall to assumed command of No.XV Squadron.

The night of 4/5th May saw a return to operations for the Squadron, when No.XV detailed twelve aircraft to participate in an attack against Dortmund. Although all the designated aircraft took-off, one returned early due to artificial horizon instrument failure. The raid was led away by P/O Renner at 22.20 hours, flying his usual aircraft, Stirling, BK611. It took twenty-six minutes for all the Squadron's participating aircraft to take-off and set course. Unfortunately three of them, together with their respective crews, would not be returning.

The flight to the target, which was made in clear skies illuminated by the light of a new moon, was uneventful. It was as the bombers approached the target area that the German ground defenses made their presence known. Searchlights weaved back and forth across the night sky in search of a victim, and the gunners opened up with a barrage of flak. Flying Officer Norris saw the target indicators go down, and prepared to make his run in over the aiming point. However, before he could release his bomb load, the aircraft, BK654, LS-W, shuddered under the impact of an exploding flak shell. A crescendo of noise rose above the known sounds of battle. Metal tore against metal as one of the aircraft's four propellers broke away from its shaft; the huge three-bladed airscrew collided with the blades of the adjoining propeller, before cartwheeling down into the night, and the burning fires of the city below. Flying Officer Norris jettisoned his bomb load over the city, in an area fairly close to the aiming point, at 01.14 hours, before returning to base. The flak turned its attention to another aircraft caught in the blinding glare of the searchlights. Stirling bomber, EF345, was coned a couple of minutes before reaching the aiming point. Although the pilot, P/O James Stowell, RNZAF, took violent evasive action, he could not shake off the tenacious beams. The brilliant shafts of light followed the aircraft down as it dived from 16,000' to 4,000' in an effort to lose them. Angry black bursts of flak erupted all around the twisting, turning bomber. Suddenly, without any warning to the pilot, a shell burst under the wing of the Stirling, igniting the port inner motor and wingroot. At the moment of impact, the control column was wrenched from the hands of the pilot, rendering the aircraft out of control. Jim Stowell gave the order to abandon the stricken bomber, but only two crew members managed to take to their parachutes. The blazing machine went down, and penetrated deep into the ground, on the outskirts of Anholt, with the rest of the crew still on board. The two crew members who jumped were Sgt Frederick 'Steve' Stevens, rear gunner, and Sergeant Tom Malcolm, a new pilot, who was on his first 'second dicky' trip. Frederick Stevens, who vacated the aircraft via the rear escape hatch, remembered nothing of his descent. The last memory he had before regaining consciousness whilst hanging by his parachute in a tree, was that of steadying himself by the escape hatch with one hand, whilst firmly gripping the parachute release handle in the other. Tom Malcolm evacuated the blazing bomber from the forward escape hatch, and floated down to land near the Dutch border. Both men were immediately captured and placed in captivity for the duration of the war.

The other two losses endured by No.XV that night, were as a result of attacks by nightfighters. The first to be shot down was Stirling bomber, BK658, LS-K, piloted by Sgt Bill McLeod, RNZAF. The aircraft was over north Holland, heading south towards the target. The crew knew there was plenty of fighter activity in the area; the mid-upper gunner, Sgt Harry Flowerday, having previously given warning of the presence of at least two enemy aircraft. Suddenly all hell broke loose, as bursts of cannon and machine-gun fire ripped through the bomber, from astern, with the rear gunner's turret taking the main force of the attack. The shells flew through the length of the aircraft and out through the nose of the Stirling, passing between the pilot and bomb aimer, wounding the latter in the shoulder. Some of the incendiaries and oxygen bottles ignited, shooting pressurised tongues of flame from the necks of the latter into the fuselage. The aircraft being uncontrollable, and with the loss of the intercom, the pilot indicated to the crew to bale out. Sergeant Eric Willis, wireless operator/air gunner, wasted no time in reaching for his parachute. Unfortunately, he was unaware that a fellow crew member who had carried the parachute aboard the aircraft for him, had accidentally dropped the 'chute into its stowage point the wrong way round. As Sgt Willis pulled it out by the release ripcord instead of the carrying handle, the 'chute canopy billowed out around him. Gathering up armfuls of silk, he staggered to the escape hatch and dropped out into the night.

Eric Willis opened his eyes; it was daylight and the sun was shining directly on his face. He could remember nothing of his descent, and did not know where he was. An attempt to sit up caused him great anxiety, as for a moment he thought he was paralysed. Then, gradually, he remembered that on leaving the aircraft he had struck the trailing aerial, a great length of which had wound itself round his body and legs. Many hours were to pass before he was eventually found and taken into custody by the Germans.

The Stirling, which had been attacked by a Ju.88 nightfighter piloted by Unteroffizier Karl Pfeiffer, of IV./NJG1, crashed at Midwolda, Holland, at 01.05 hours. Sergeant Peter McNulty, the rear gunner, was killed during the initial attack by the nightfighter, and went down with the bomber. He was the only major casualty from this crew.

The Squadron also lost Stirling bomber, BK782, LS-X, to nightfighter action. Unfortunately, when this aircraft, which was piloted by P/O Thomas Emberson, crashed at 01.45 hours, in the Houten/Schalkwijk area of Holland, there were no survivors.

Flight Sergeant Arkinstall's aircraft, Stirling bomber, BK652, LS-V, was lucky to escape with only slight damage following an engagement with a Ju.88 nightfighter.

With the loss of fourteen members of aircrew, the loss of three aircraft and a further two machine damaged, the Squadron was not considered fully operationally. Furthermore, the adverse weather conditions were not favourable for operations. Although one mine-laying (Gardening) mission was undertaken on 5th May, the Squadron did not return to maximum effort raids until 12th May. On that day, the Squadron detailed twelve aircraft to participate in an attack against Duisberg. However, before take-off, three crews were withdrawn. Of the nine which took-off, two returned early. Stirling bomber, BK648, LS-J, piloted by P/O Hunt, developed a fault in the starboard outer motor, whilst Stirling, EF348, LS-N, flown by P/O Curtis, had a failure of the port outer engine. The remaining seven aircraft all completed their tasks and returned safely to Mildenhall.

The next day the Squadron detailed another twelve aircraft for a raid against Bochum. On this occasion, three aircraft returned early and one failed to return. Sergeant Gabel, piloting Stirling bomber, BK354, LS-L returned due to the failure of the port inner engine. Sergeant Johnson, flying Stirling, BK649, LS-O, lost the use of the rear turret, whilst Sgt VanEupen, captain of Stirling, EF354, LS-Q, could not get his aircraft to maintain height.

An example of the caliber of the men who served with No.XV Squadron was exemplified by the courage of Squadron Leader Clifford Bowyer, RAFVR, during this attack. When his aircraft, Stirling bomber, BK704, LS-Z, was hit by flak and severely damaged whilst flying over Dusseldorf at 11,000', S/L Bowyer ordered his crew to bale out. Although the aircraft was on fire, S/L Bowyer stayed at the controls of his aircraft so as to ensure the safe evacuation of his crew. The bomber exploded as Clifford Bowyer was about to make good his own escape from the flying inferno. The inscription on his headstone, in the Reichwald Forest War Cemetery reads, *"No greater love hath man than this than to lay down his life for his friends"*. In this case a true and fitting epitaph.

As had happened so often in the past, No.XV welcomed back another ex-Squadron member, when Robert Megginson returned on 17th May. As a Sergeant, Megginson had flown twin-engined Bristol Blenheim bombers with the Unit during 1940, and been awarded the Distinguished Flying Medal. He was now a commissioned officer, with the rank of Flight Lieutenant, although this was to change the day after his arrival when he was promoted to Acting Squadron Leader. He had also converted to four-engined Stirling bombers, and returned complete with a full crew which included another officer, P/O Desmond Mitchell. However, as A/S/L Megginson was about to start his second tour of operations, P/O Ken Dougan, RCAF, was declared 'tour expired' and posted home to his native Canada.

The next major raid in which No.XV Squadron participated occurred on 23rd May when fifteen aircraft were detailed for an attack against Dortmund. Flight Lieutenant Norris was forced to return early when the electrical circuits operating the bomb doors on his Stirling failed. Stirling Bomber, EH875, LS-S, also found it impossible to continue with the mission after a generator seized up and filled the aircraft with smoke. The pilot, Sgt Cornell, jettisoned the bomb load into the Zuider Zee at 01.39 hours, before returning to Mildenhall. However, when Sgt Smale's aircraft lost power to the rear turret, the crew elected to continue with the mission, which they completed successfully.

Sergeant Johnson, RCAF, was killed in action, together with three members of his crew, when his aircraft Stirling, BF482, LS-R, was shot down by flak over the target area. One of those who was killed in the crash was Sgt Henry Waite who, at only eighteen years of age, was one of the youngest men to die in the service of Bomber Command, during 1943.

Two nights later, on 25th May, a further ten young men from the Squadron were to perish, whilst four more would be made prisoners of war, when their respective aircraft crashed in enemy held territory. These fourteen airmen were aboard two of the eighteen aircraft detailed for a raid against Dusseldorf that night.

The first machine to be lost was Stirling bomber, BK611, LS-U, which was hit by flak at 01.32 hours, whilst flying at approximately 12,300'. The salvo struck the starboard engines, both of which lost their respective propellers, causing the aircraft to lose height. The starboard inner motor also lost its cowling, whilst the outer engine burst into flames. The bomb load was immediately jettisoned and, although the aircraft remained airborne and had turned for home, the wireless operator/air gunner, Sgt Seabolt, RCAF, thought the Stirling was going down. His request for permission to bale out was granted by the pilot.

Sergeant Arthur Edgley, the rear gunner, was momentarily trapped in his turret, due to the loss of all power, following the demise of the starboard motors. Having disconnected his oxygen supply, heated flying suit and intercom, he used the hand control to manually turn and centralise the turret in order to effect his own escape. Gaining the relative safety of the rear fuselage, Sgt Edgley collected his parachute from its storage position, clipped it on and made his way to the escape hatch. As the release handle of the escape hatch was operated, the slipstream caught the panel and sent it tumbling down into the darkness. With a howling gale blowing around him, the rear gunner lowered himself to the floor of the aircraft, and prepared to bale out. As he started to ease himself through the aperture, still clinging on to sides of the opening, the same slipstream which sucked out the escape hatch panel, caught Sgt Edgley and endeavored to pull him out in the same manner. Even in the nightmare scenario of swinging about in the terrifying force of the slipsteam, the rear gunner was able to ascertain the aircraft was still flying in a reasonably level attitude. Quickly working on the principle the aircraft might well continue in this manner for some time, he summoned up all his strength and hauled himself back into the aircraft. Climbing back into his turret, Arthur Edgley reconnected the intercom and informed the pilot of his recent escapade, and explained he had decided to stay with the aircraft.

Remembering the rear turret was out of action, Sgt Edgley took-up station in the mid-upper turret, where he found everything functioning properly. On rotating the cupola to the starboard side of the aircraft, he was able to see the damage inflicted on the two engines. It was a sight which did not impress him. Swinging the turret back he cast a glance at the port engines and saw, to his relief, they were functioning properly. As a visual interpretation of the problems the pilot was enduring, he also noticed the port wing was slightly low and the rudder was hard to one side. This action however, could not stop the aircraft from continually losing height. When the Stirling got down to 5,000', Sgt Wilson, the Australian pilot, ordered all excess equipment to be discarded through the hatches, but this did little to rectify the problem. After the loss of a further 1,500' of altitude, the pilot ordered the crew to abandon the stricken bomber. Once more Sgt Edgley made his way through the fuselage, on this occasion going forward, towards the cockpit. There he witnessed the struggle that Sgt Wilson and Sgt Arnott were experiencing holding the aircraft steady. Their only chance of escape would be through the forward escape hatch, which Sgt Edgley endeavored to open. Unfortunately, as he pushed the handle to unlock the panel, it broke off leaving the hatch firmly in place. Holding the broken lever aloft, as a signal to his colleagues, the rear gunner gestured to the other crew members to go back to the rear of the fuselage. As he passed the pilot, still struggling to maintain air speed and altitude, Arthur Edgley noticed the aircraft was down to 1,500'. The latter raised a gloved hand and pointed towards the rear of the aircraft, he waited only long enough to see Sgt Wilson raise his own hand in acknowledgment. Time was running out.

Standing by the open rear escape hatch, Arthur Edgley signaled P/O Cooper, the navigator, to jump. As the latter disappeared into the night, Sgt Pittard appeared by the hatch.

Not fearing for his own safety, and without hesitation, Sergeant Edgley stood aside and gestured for the flight engineer to go first. Unbeknown to both of them, there was insufficient height for anyone to bale out, as a result of which Sgt Ronald Pittard was killed when his parachute failed to fully deploy. Unaware of the tragedy which had just occurred outside the aircraft, Sgt Edgley crouched down by the open hatch. As he prepared to leave the aircraft there was a violent impact.

Forty-five minutes after Sgt Seabolt's departure from the mid-upper turret, the aircraft did go down. Sergeant Jack Wilson, RAAF, fought with the controls, as the bomber got closer to the ground. Neither he, nor Sgt Patrick Arnott, the bomb aimer, gave up hope, and remained at their sta-

tion until the end. The Stirling struck the ground to the torturous screams of wrenching, tearing metal as the airframe twisted, turned and broke into many pieces. Burning wreckage was scattered over a vast area, with exploding ammunition causing further mayhem.

Instinctively, at the moment of impact, Arthur Edgley had covered his head and face with his arms in order to protect himself. Dazed, but otherwise unhurt, he crawled out of the wreckage into the silence of the night, to look for other survivors. A shout was quickly answered by Sgt Maxted, the wireless operator who, apart from an injury to his leg, was able to get out of the shattered remains of the Stirling.

The Stirling, which had been christened 'Te-Koote' by its original pilot back in January, and was known to everybody on the Squadron, crashed near Grubbenvorst, north-west of Venlo, Holland, at 02.15 hours.

After evading capture for six weeks, due mainly to the help rendered by Dutch, Belgian and French resistance units, Arthur Edgley and Sgt Maxted were betrayed and arrested. Originally, they were taken to Fresnes Prison, south of Paris, but were later moved to Stalag Luft IVb, a PoW Camp situated near Muhlberg.

The second aircraft which failed to return, was the victim of a cruel stroke of fate. Stirling bomber, BF354, LS-L, piloted by F/O Thompson, crashed as a result of being caught in the blast when a Halifax bomber from No.77 Squadron blew-up, following an attack by a nightfighter. Apart from the No.XV Squadron Stirling being lost, a similar machine from No.7 Squadron was also destroyed, presumably allowing the German pilot to claim three four-engined bombers in one attack.

Although eighteen aircraft had been detailed for this operation, one aircraft failed to take-off. Of the seventeen machines which did get airborne, two were forced to return early. The first home was Stirling, EF391, LS-M, which suffered intercom failure at altitude. The pilot, P/O Lown, elected to jettison his bomb load and return to base. At 03.20 hours, nearly an hour after Lown's early arrival, P/O Hugh Wilkie's aircraft touched down on the runway at Mildenhall, with one dead motor. The latter's machine, Stirling bomber, BF470, LS-G, had suffered starboard inner engine failure, leaving 'Wendle' Wilkie no choice but to abort his mission. Flight Lieutenant Megginson brought his aircraft, Stirling bomber, BK818, LS-O, home with flak damage to the tailplane.

Three nights after the loss of Stirling, BK611, a replacement aircraft took to the skies bearing the code LS-U. The new bomber, BF571, piloted by Sgt Allen, participated in a raid against Wuppertal. The attack, carried out on the night of 29th/30th May, was a success, and undertaken without loss to the Squadron.

The first ten days of June 1943 saw minor operations being undertaken by Bomber Command in general, whilst a continuous training program was initiated for No.XV in particular. The first major operation scheduled for this month, occurred on the night of 11th June, when 783 aircraft were detailed for a major assault against Dusseldorf. No.XV Squadron's Battle Order carried the names of eighteen crews and aircraft, all of which took-off, although two returned early. The attack was considered a success, with bombing destroying a large part of the city center. Unfortunately though a Mosquito aircraft is reported to have inadvertently dropped its target indicators some miles to the north east of the City; an error which led to some of the main force aircraft releasing their bomb loads over open countryside. Flight Lieutenant Hugh Wilkie however, was not amongst the former. He held his regular mount, Stirling bomber, BK470, steady on its bombing run and released the deadly cargo from a height of 12,000'. As he was setting course away from the target area, he was warned of the presence of another Stirling flying very close to BK470. The gunners saw the other bomber but paid little attention to it, that is until its guns opened fire at them! The constant speed unit was put out of action as bullets slammed into the port outer engine, causing BK470 to vibrate. Sergeant Frank Diamond, the navigator, thought the aircraft had been hit by flak, and looked out of the port side window to see what was going on. At the same time, John Ledgerwood the mid-upper gunner informed the captain the propeller looked in a bad way. As he imparted this information, the propeller, complete with boss, sheared off and cartwheeled up and over the top of the aircraft, missing the mid-upper turret by inches! Having regained control of the aircraft, Hugh Wilkie nursed it across hostile skies and landed safely back at Mildenhall.

Apart from the two No.XV Squadron machines which returned early, one was listed as missing. It later transpired that Stirling bomber, BF571, LS-U, piloted by P/O Reginald Allen, had been shot down. The aircraft, which was on only its second major operation, crashed into the North Sea 25kms north-west of Den Helder, Holland, at 03.04 hours. The body of Reginald Allen was washed ashore on 1st July, and buried four days later. The rest of the crew, including Pilot Officer Aston 'Tony' Crawford Lake, the navigator, were all killed. 'Tony' Lake, who was born at Baker Street, London (home of the fictional detective Sherlock Holmes), joined the RAF in 1941. On completion of his training at Chatham, New Brunswick, Canada, Sergeant Lake was posted to No.XV Squadron. Although he was a member of the Civil Defence Service, before the war 'Tony' Lake embarked on a modeling career and was known for advertising a well known brand of British cigarettes. He was what one might call a 'model airman'. As so often happened, both Aston Lake and Reginald Allen received their respective commissions just prior to their deaths.

A replacement crew headed by P/O George Judd, reported for duty with No.XV Squadron at RAF Mildenhall, on 15th June. Four days later P/O Judd flew his first 'second dicky' trip with W/C Stephens and his crew, on a raid against the Schneider Aero Works at Le Creusot, France. Another relatively new pilot, Sgt Towse, also flew as a 'second dicky' on this mission. He was teamed with the experienced crew of P/O Curtis.

The thirteen crews from No.XV who attacked the target, all reported having bombed from between 4,000 and 6,500 feet! Although Wilfred Towse had some operational flying experience, it was a real baptism of fire for P/O Judd.

George Judd flew his second 'second dicky' trip two nights later, on 21st June, when he flew with P/O VanEupen on a raid against Krefeld. Another pilot flying in the same capacity as George Judd was Sgt Jack Newport, who flew as second pilot with A/S/L Megginson. Unfortunately, Sergeant Newport would never complete an operation as captain of his own aircraft and crew.

Sergeant Towse was allocated a Stirling of his own for this mission, it carried the serial EH890, and wore the code LS-U. Although this mission was completed without incident to the bomber, the curse which befell aircraft coded LS-U was soon to strike again!!

All Eighteen Stirlings detailed by No.XV Squadron to join the total attacking force of seven hundred and five aircraft, took-off. There were no early returns on this raid, although one machine, whose crew were on their nineteenth mission, was lost. Stirling bomber, BK815, LS-V, flown by Pilot Officer Eric Curtis crashed in Belgium having been shot down. Only the mid-upper gunner, Sgt A. Waugh, managed to bale out.

Flight Lieutenant Wilkie's aircraft was approximately sixteen miles south-west of the target when it was attacked by a Ju.88 nightfighter. Sergeant Palmer, the rear gunner, spotted the enemy aircraft as it crossed from port to starboard, about 1,200 yards astern of the bomber. He immediately ordered the pilot to take evasive action, following which he opened fire with a short burst. Fire spat from the guns of the Ju.88 as the German pilot responded, also, with a short burst. In order that Sgt Palmer could keep the nightfighter in sight, Hugh Wilkie adopted a sweeping banking maneuver to port and starboard. Stirling bomber, BF470, flew in this manner for several minutes, until the attacking fighter was seen to dive down and disappear into cloud. Although Sgt Palmer claimed hits on the Ju.88, no claim was made against the destruction of this aircraft. The Stirling and its crew completed their task and bombed Krefeld at 02.01 hours. They returned to Mildenhall without further incident.

Nightfighters accounted for two of the Squadron's aircraft the following night, when No.XV despatched eighteen Stirlings to attack Mulheim. Sergeant Newport had the misfortune to encounter an enemy nightfighter on the outward route, whilst still flying over the sea. Part of the Stirling's bomb load was jettisoned off the Dutch coast in an effort to gain more height, but this action had little or no effect. Although the bomber pilot

took evasive action, a burst of cannon fire from the enemy aircraft struck the Stirlings's port outer engine causing damage to the fuel system. Sergeant Newport decided to turn back, but there was to be no escape. A second nightfighter appeared and attacked from the starboard quarter low. The port inner motor and fuel tank took the full impact as cannon shells, again, struck the Stirling. As the remainder of the bomb load was jettisoned, the wing caught fire, leaving the crew no option but to abandon the blazing bomber. Five of the crew were successful in leaving the aircraft, but Sgt Wilfred MacAulay, the rear gunner, died in his turret where, it was recorded, he had been trapped by his flying clothing. Stirling bomber, EF348, LS-N, struck the ground at Kessenich, Belgium, at 01.35 hours. Sergeant Jack Newport, the pilot, went down with the aircraft and was also killed in the ensuing crash. The bomber was claimed by Oblt Hans Autenrieth, of VI./NJG1, and credited as the German pilot's second victory.

Of the five crew members who baled out three were captured, but P/O Turner, the navigator, and Sgt Arthur Kellet, the bomb aimer, managed to evade capture. Furthermore, the latter was successful in making a home run, and arrived back in England on 6th December 1943.

Thirty-two minutes after the loss of Stirling, EF348, Stirling bomber, BK656, LS-A, also from No.XV Squadron, was shot down. This machine, piloted by F/O Hawkins, succumbed to the guns of Hauptmann Wilhelm Dormann's aircraft, of III./NJG1, who caught the bomber over Holland. The Stirling crashed in the National Park 'De Hoge Veluwe', near the statue of Christiaan De Wet, 5kms from Otterlo, at 02.07 hours.

Stirling bomber, BK816, LS-X, had a narrow escape on returning from the target, when it encountered a Bf110 nightfighter. The bomber was flying at approximately 11,000' when the rear gunner, Sgt Lamarche, saw the silhouette of the enemy machine reflected on white clouds 3,000' below. The nightfighter had obviously seen the bomber and climbed into the attack. Closing the range, the Bf110 opened fire with both cannon and machine guns. The bomber took evasive action by adopting a 'corkscrew' maneuver, whilst diving for cloud cover. A battle ensued between the enemy aircraft and Sgt Larmache, who claimed hits against the fuselage of the nightfighter. With the bomber now racing across the top of the clouds, the enemy machine was forced to attack from above. As it came in on the starboard quarter high, the rear gunner and the mid-upper gunner both fired off a long burst each. The enemy machine was seen to peel over and dive away, having sustained strikes by the gunners' shells. On returning to base, both Sgt Lamarche and Sgt Gomersall made a claim for an enemy machine damaged. However, the Stirling did not return unscathed. The bomber incurred damage to two petrol tanks in the port wing, damage to the main spar and bullet holes in the port elevators.

An extraordinary example of courage, by two members of the same No.XV Squadron crew, was displayed on the night of 24th/25th June, whilst participating in an attack against Wuppertal. Sergeant Wilfred Towse was at the controls of his Stirling bomber, EH890, LS-U, as one of eighteen aircraft detailed for the raid. He took-off from Mildenhall at 23.25 hours, and set course out over the North Sea. The outward leg of the journey was uneventful until just after the bomber had crossed the Dutch coast, when it was attacked by a nightfighter. The Stirling's gunners, being vigilant and alert, were aware of the enemy aircraft's presence and as it came in to attack, they opened fire and shot it down. The bomber continued on to the target, where its deadly cargo was released at 01.29 hours, from an altitude of 14,000'. As the bombs exploded on the red target indicators below, Sgt Towse turned for home. However, before the aircraft could clear the area, it was struck a violent blow by hot shards of exploding flak, some of which penetrated four of the Stirling's fuel tanks. As the pilot fought to regain control of the juddering bomber, one of the tanks erupted into a ball of flame. Immediately, Sgt Towse plunged the huge bomber into a violently dive, in an effort to extinguish the fire. Fearing the worst, and believing the aircraft to be doomed, Sergeant Martin, the bomb aimer, baled out. Unfortunately for Sgt Martin, who was to become a PoW, Sgt Towse was successful in his endeavor, and managed to quell the flames. The crew, however, still had to faced the daunting prospect of flying across enemy held territory, and then the North Sea, in a crippled bomber. There was also the question of fuel, much of which had leaked out when the tanks were punctured. The answer to the latter question came shortly after the Stirling had crossed the enemy coast, when the engines began to cough and splutter. As the motors died due to fuel starvation, the bomber began gradually to descend towards the sea, whilst the crew were alerted to adopted ditching procedure. The huge propellers turned freely in the rush of wind, as Sgt Towse fought to keep the descending bomber on a relatively even keel. When the aircraft struck the water at 04.00 hours, approximately fifteen miles off the English coast, it sank quickly, taking Wilfred Towse with it. Fortunately, the pilot was able to extricate himself from the sinking machine and, via an open hatch, make good his escape. He broke surface fairly close to a dinghy, into which he was hauled by a much relieved Norman Pawley. As wireless operator/air gunner on the crew, Sgt Pawley had remained at his station throughout, transmitting messages until the very last moment. Those messages rendered valuable information, enabling the crew to be rescued without undue delay. For their courage and resourcefulness, Sergeant Wilfred Towse and Sergeant Norman Pawley were both awarded immediate Distinguished Flying Medals.

A similar set of circumstance befell Sgt Keen during this same mission. His aircraft, Stirling bomber, EF428, LS-N, was attacked by a Ju.88 nightfighter over the Belgium coast. However, when the mid-upper gunner opened fire, the enemy aircraft broke-off the attack and dived away. The aircraft, which was also to suffer from a shortage of fuel, also experienced a defective rear turret. Although the aircraft managed to negotiate the North Sea on the homeward leg, it was forced to land at Martlesham Heath, a fighter airfield near the east coast.

Four nights later, on 29th June, Sergeant Keen's aircraft was again attacked by a nightfighter whilst participating in an attack against Cologne. On this occasion it was to be the German pilot who was the victor. The enemy machine, flown by Hptm Hoffman, of IV./NJG5, caught the British bomber flying over Belgium on the return leg of its mission. In the ensuing combat, Stirling bomber, BK694, LS-C, fell from the sky and crashed at Lommel, to the south of the Dutch border, at approximately 02.20 hours. The crew, with the exception of the bomb aimer, all perished. Flight Sergeant Duckett, RNZAF, was captured and made a prisoner of war.

Six members of another crew were killed this same night, when their aircraft, EH888, LS-Z, was also shot down. The Stirling, flown by P/O Arthur Saunders, crashed into the River Ijssel at Heeswijk, Holland, at 02.40 hours, following combat with a nightfighter. The bomber, from which only Sgt Mallens the navigator escaped, was on its seventh mission.

Squadron Leader Megginson also participated in this mission, but had to return early when the rear turret of his aircraft became defective. The Stirling he was flying, BK805, carried the code LS-U!!

Earlier during the evening, whilst the crews were preparing for the coming attack, there was a lot of activity at the main gate of the camp. Two new pilots, Sgt Gilbert 'Gil' Marsh and Sgt George Barber, together with their respective crews, arrived from No.1651 Conversion Unit at Waterbeach. The area resembled a transit camp, as fourteen men reported for duty and forty-two kitbags (three for each man) were deposited outside the guardhouse. After much form filling the two crews were directed to their respective billets in the old married quarters; one crew per house. As Gil Marsh fell into his bed, the crews of nineteen aircraft were being driven out to their aircraft. The next morning he heard that two of them had failed to return.

On Saturday, 3rd July, Sgt Marsh and Sgt Barber were both detailed to fly their first 'second dicky' trips, when a raid against Cologne was promulgated. It was a raid which, from the start, had all the ingredients for a disaster for the Squadron. One aircraft failed to take-off, two returned early, one was listed as missing and three returned with flak damage, one of which was forced to land away.

Gil Marsh was detailed to fly with F/L Norris, whilst George Barber teamed up with F/L Hunt. Flight Lieutenant Norris took-off at 23.30 hours, in Stirling bomber, BK654, LS-W, and weaved his way across the North Sea. At one stage, the weaving motion adopted by the pilot took the aircraft down to 5,500 above the sea, but in readiness to cross the Dutch coast, F/L Norris gained altitude. Gil Marsh was under no illusions as to the task ahead of him as a pilot, and observed all that was going on around him. As to the chances of survival, he noted that six aircraft went down in flames between the coast and Amsterdam. By the time they reached the target area, it was a blazing inferno with smoke palls rising to approximately 8000'. Flak burst around them as their bomb load rained down onto the target below. Their job done, F/L Norris hauled the Stirling round onto a new course setting and headed for home.

Unfortunately, George Barber would never see home again. The aircraft he was flying in, Stirling bomber, BK648, LS-J, was shot down and crashed at Menden, Germany, killing the entire crew.

Acting Squadron Leader Megginson's aircraft, Stirling bomber BK818, LS-O, had a lucky escape when it was hit by flak as it left the target area. The port inner engine was set ablaze, and had to be feathered. The loss of power from this motor rendered the front and mid-upper turrets u/s, leaving the bomber almost defenseless. However, the bomber returned to England without further mishap, but was forced to land at Manston in Kent. Squadron Leader Martin's aircraft, which led the attack, was also hit by flak, but he managed to get back to Mildenhall.

Flying Officer Waughs took his Stirling, BK719, LS-B, over the target at 01.34 hours. As Sgt Ray Cole was releasing the bomb load, a 4lb incendiary struck the aircraft. It penetrated the starboard side of the fuselage, starting a fire perilously close to some oxygen bottles. Five minutes after the navigator and wireless operator had extinguished the flames, the aircraft was struck by another incendiary in the region of the port engines. Luckily, this aircraft also managed to get home.

Gil Marsh's first trip with his own crew occurred on the night of 5th/6th July, when he was detailed for a 'Gardening' operation. His aircraft, EF351, LS-L, was one of three machines listed on the Battle Order to drop mines off Juist, in the east Frisian Islands. The Stirling took-off at midnight in torrential rain, driven by very strong crosswinds which created less than ideal conditions for a 'Freshman' crew. Sergeant Marsh found the Stirling heavy to handle, and only just cleared the trees at the end of the runway. However, he did have the advantage of being first away. It was not until he returned to Mildenhall, and went into the debriefing room, that he found to his astonishment his was the only No.XV Squadron aircraft to take-off. Following in the wake of Sgt Marsh's aircraft, the second machine made three attempts to get airborne. Having nearly demolished the control office on the third try the crew were 'scrubbed'; and so too were the third crew.

Three nights later, on 8th July, Sgt Marsh was detailed for another 'Gardening' operation when three aircraft were despatched to sow mines in the mouth of the Gironde River. Having taken-off at 22.25 hours, Gil Marsh experienced problems retracting the aircraft's undercarriage, and only succeeded in doing so after expending much effort and energy. Then, to his dismay, the rear turret went unserviceable necessitating a return to base where, after jettisoning 1,450 gallons of fuel, he landed at 23.10 hours. On this occasion the other two crews completed their tasks and returned to base without incident.

On Friday, 9th July, Sgt Marsh took delivery of his own Mk.I Stirling, which he immediately took on a full air test. Being total unaware of the machine's previous history, or the fact it was badly damaged on the night of 22nd/23rd June, he was more than pleased with the aircraft. It did not take long for the crew to think up a pet name for their charge and, due to the aircraft's individual code letter, they unofficially christened the Stirling *Madame X*. The bomber, which bore the serial BK816, was coded LS-X. The 'rookie' pilot had to wait four days before he could take *Madame X* on her first operation with him at the controls. On Tuesday 13th July, eight aircraft were detailed to attack Aachen, LS-X was one of them. The pilot and crew spent the day preparing their machine. The guns were harmonised, the radio equipment checked and, following a thorough inspection, an air test was flown.

The raid led by S/L Martin, was considered a success, although Gil Marsh had reservations about the bright moonlight casting aircraft shadows onto the clouds below over the target area. His fears about nightfighters were unfounded, as it was the flakships which nearly brought down *Madame X*. A German vessel, off the enemy coast, opened-up with a heavy barrage and placed a couple of shells under the tail section of the aircraft. The exploding flak shook the Stirling violently, causing the aircraft to vibrate, and handling problems for the inexperienced pilot. A short while later the bomber was subjected to the blinding glare of a searchlight, which held *Madame X* for about twenty seconds. During that short period, which seemed an eternity to the crew, the aircraft attracted more heavy flak. For-

tunately, both the Stirling and her crew escaped the tenacious beam, completed their task and returned to Mildenhall without further mishap. The following day, Gil Marsh and his crew moved their weary bones from the horizontal to the perpendicular position at 13.00 hours, and went out to inspect their aircraft. To their relief, and that of their groundcrew, the Stirling had incurred no damage. Although Gil Marsh and his crew were not aware of it at the time they, and *Madame X*, had survived their short part in the 'Battle of the Rhur'. Now, they were to support the Squadron, and Bomber Command, in another battle.

In Some Foreign Field

Above: A panoramic view (looking north) from route D946, of the area around St Genevieve (left), 5 kms NE of Montcornet, France, where Stirling bomber, BF475, LS-T, crashed on the night of 10th/11th April 1943. In this foreign field, seven young men from No.XV Squadron died. *Courtesy of David and John Booth*

Left: Sergeant Harry Wakefield, mid-upper gunner to Sgt Eric Trezise and his crew, who was killed on the night of 10th/11th April 1943, during an attack against Frankfurt. *Courtesy of David and John Booth*

The Move To Mildenhall

An aerial view of RAF Mildenhall, clearly showing the operational and domestic areas of the airfield. The hangars to the right of the picture were built during the 1930s, whilst the runway, which can be seen as a black strip to the right of the photograph, was constructed during 1942. *Author's Collection*

'On Finals'. The view from the cockpit of a Stirling bomber, as it crosses the airfield perimeter, prior to landing at RAF Mildenhall. *Courtesy of Oliver Brooks*

One of the Horsa glider aircraft used for transporting ground equipment, from the airfield at Bourne to No.XV Squadron's new base at Mildenhall, on 14th April 1943. *Courtesy of and via Flt/Lt T. N. Harris, XV (R) Squadron*

Mildenhall – Then and Now

Short Stirling bomber, BK654, LS-W, touches down at Mildenhall, whilst five other aircraft overfly the airfield awaiting permission to land. In the background, having landed earlier, is Stirling, BK476, LS-D. *Courtesy of and via Flt/Lt T. N. Harris, XV (R) Squadron*

The Officers' Mess, photographed in May 1988, remains relatively unaltered since the war years. It is currently used (1998), by the United States Air Force Europe, as an Officers' Club. *Author's Collection*

A view from a window of the Officers' Mess looking across the tennis courts to one of the "A" Type hangars, on the airfield at Mildenhall. *Author's Collection*

A "C" Type hangar, built in 1935 as part of the second phase of expansion at RAF Mildenhall, housed No.XV Squadron's aircraft undergoing major overhaul or battle damage repair. *Author's Collection*

Although concealed from view from the Officers' Mess, by post war expansion and vegetation growth, the "A" Type hangar is still in use over sixty years after it was constructed. The hangar, which has been much modified over the years, is in permanent use by the USAFE. *Author's Collection*

The Local 'Watering Hole' – Then and Now

The local 'watering hole' – Then (1944). The Bird-in-Hand Hotel, like much of the airfield to which it is still adjacent, was constructed in 1935. The red brick building replaced a 100 year old inn, which stood on a nearby site. It was one of the first 'landmarks' visited by some of the crews from No.XV Squadron, on their arrival at Mildenhall. Sergeant Mick Cullen, a wireless operator from New Zealand, liked the place so much he married Brenda Jaggard, one of the barmaids. *Author's Collection via the late D. A. 'Pat' Russell*

The local 'watering hole' – Now (1998). Still adjacent to the airfield perimeter, the Bird-in-Hand Motel has change very little in external appearance. A different generation of airmen, from a different air force, now use the facilities. *Author's Collection*

The Local Diner – Then and Now

Members of both aircrew and groundcrew frequented the Rosemary Cafe, which was situated a few hundred yards from the perimeter fence of the airfield. It is said the holes in the wire were created by servicemen endeavoring to get out to the cafe, and not by saboteurs trying to get in to the airfield. *Courtesy of Geoff Hill*

Following the closure of the Rosemary Cafe after the war, the building remained occupied. Its last use, prior to being demolished in the early 1990s, was as a mini-market selling consumables to both local inhabitants and USAFE personnel. *Author's Collection*

The Loss of Sergeant Greenwood

GVI RI

This scroll commemorates

Sergeant J. W. Greenwood
Royal Air Force

held in honour as one who
served King and Country in
the world war of 1939-1945
and gave his life to save
mankind from tyranny. May
his sacrifice help to bring
the peace and freedom for
which he died.

Above left: A pencil portrait, by an unknown artist, of a young John Greenwood. Possibly drawn in the pre-war years, before the latter volunteered for service with the Royal Air Force. *Courtesy of June Watkins*

Above: The Commemorative Scroll, presented by the Royal Air Force to the parents of Sergeant John William Greenwood, recording the sacrifice made by the airman who 'died twice'. *Courtesy of June Watkins*

Left: A simple wooden cross, nailed to a tree, marks the crash site of Stirling bomber, BK691, LS-F, which failed to return from an attack against Mannheim, on the night of 16th/17th April 1943. The aircraft crashed at Hetzerath, Germany with the loss of the entire crew. *Courtesy of and via Steve Smith*

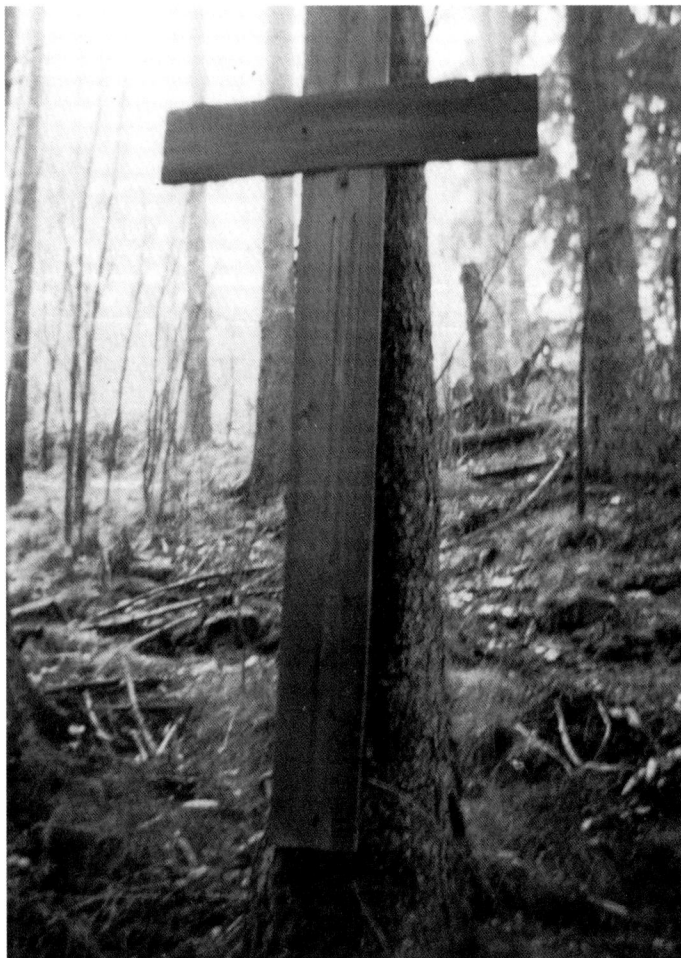

The Loss of Stirling "H" – Harry

Pilot Officer Clive Perring (2nd from right) poses with other members of his crew, not knowing what fate had in store. Two months earlier, the observer/navigator, had sent home to his family, a photograph of Stirling bomber, N3669, with the comment "We fly the new "H" for Harry. That new "H" – Harry crashed in Wiltshire and was immediately replaced by Stirling BF474; also coded LS-H. Unfortunately, BF474 was shot down on the night of 16th/17th April, with the loss of the entire crew. *Courtesy of Steve Smith*

Seven white painted wooden crosses mark the original resting places of the crew of Stirling, BF474, which failed to return from an attack against Mannheim, on 17th April 1943. *Courtesy of and via Steve Smith*

Two of a Kind II

Two almost identical photographs, differing only in the time of day at which they were taken, each showing a Stirling bomber of No.XV Squadron, taking-off on operations. Careful examination of the background of both photographs indicates they were taken from the same spot. *Courtesy of Flt/Lt T. N. Harris, XV (R) Squadron*

Down in Denmark

Five members of the crew of Stirling bomber, BF476, LS-P, which crashed at Krageland, Denmark, having been hit by flak during an attack against Rostock, on the night of 20th/21st April 1943. Included in the photograph are F/L Charles Lyons, pilot (center); P/O Roy Lockwood, B/A (far right); Sgt James Marshall, Nav; Sgt John Shaw, MUG and Sgt Hugh Hipwell, F/E. Not present in the photograph are, Sgt R Howland, W/Op and Sgt M Huggett, R/G. *Courtesy of Don Clarke, MBE, The Mildenhall Register*

Three photographs showing the remains of F/L Lyon's Stirling bomber, BF476, LS-P, which crashed near Krageland, Denmark. The crew attempted to destroy the aircraft by setting fire to it, before effecting an escape. Unfortunately, they were captured whilst hiding in a barn, during the afternoon of 22nd April. *Author's Collection*

As above. *Courtesy of and via Flt/Lt T. N. Harris, XV (R) Squadron*

The Loss of BK657

Wreckage of Stirling, BK657, LS-C, lies strewn across a dyke near Breukelen, Holland, where the aircraft crashed on the night of 26th/27th April 1943. *Author's Collection via Jan Uithol*

Ben Reynders (left), a former escapee listens attentively as Cyril 'Mike' Mora (right), the former wireless operator on BK657, explains how his parachute descent from the blazing bomber, ended in the middle of this greenhouse. *Author's Collection via Jan Uithol*

The Return of BK657

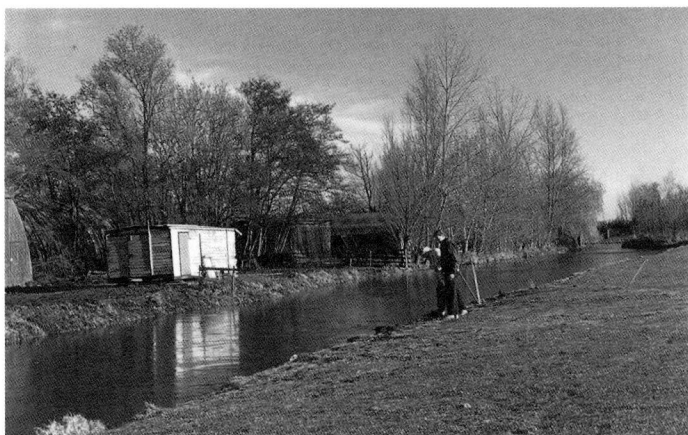

Jan Uithol (left) and Henk Rebel, both members of the Crash Research in Aviation Society Holland (CRASH), inspect the site during the late 1980s and stand on the spot where the Stirling bomber fell to earth. *Author's Collection via Jan Uithol*

After forty-six years in the ground the prop shaft, when cleaned, is found to be in excellent condition. *Author's Collection via Jan Uithol*

The prize, one of the Bristol Hercules engines which powered the Stirling, waits to be cleaned before revealing its condition. *Author's Collection via Jan Uithol*

The day's work at the crash site had finished when the dam holding back the water burst and flooded the excavated area. *Author's Collection via Jan Uithol*

Dortmund Casualties

The crew of Stirling bomber, BK658, LS-K, which was shot down by a nightfighter, on 4th May 1943, with the loss of the rear gunner. The crew consisted of Sgt William McLeod, Pilot (2nd from left, standing); Sgt Archie Eaton, B/A; Sgt Eric Willis, W/Op (front row, left); Sgt Routh, F/E; Sgt Law, Nav; Sgt Harry Flowerday, MUG (front row, center); Sgt Peter McNulty, R/G. *Courtesy of and via Peter Willis*

The mangled remains of Stirling bomber, BK782, LS-X, are inspected by members of a Luftwaffe unit. The aircraft, which was shot down by a nightfighter during the attack on Dortmund, on the night of 4th/5th May, crashed near Utrecht, with the loss of the entire crew. *Author's Collection*

Stirling Bomber, BK611, LS-U, named *"Te Koote"* survived 29 operations before being shot down by flak, at 01.32 hours on the morning of 26th May 1943. *Courtesy of A. W. Edgley*

Five members of the original crew who flew with P/O Jim Stowell are, from left to right: Sgt Jim Banyer, B/A; Sgt William Jennings, MUG; Sgt Ted Kirby, F/E; Sgt Leslie Paterson, W/Op; Sgt "Steve" Stevens, R/G. Jim Banyer, William Jennings and Leslie Paterson were killed, along with three other members of the crew, on the night of 4th/5th May 1943, when their bomber was shot down. 'Steve' Stevens survived, along with Sgt Tom Malcolm the 2nd pilot, to be made a prisoner of war. *Author's Collection*

The Loss of 'Te Kooti'

An unidentified Stirling bomber of No.XV Squadron on final approach to Mildenhall airfield. *Courtesy of A. W. Edgley*

Five members of the crew of BK611 who posed for a photograph are, from left to right: Sgt S Maxted, W/Op; Sgt Patrick Arnott, B/A; Sgt Jack Wilson, pilot; P/O B Cooper, Nav; and Sgt Arthur Edgley, R/G. Patrick Arnott, Jack Wilson and Sgt Ronald Pittard, F/E (not in the photograph), were all killed when the aircraft crashed north west of Venlo, Holland. The rest of the crew were captured and made prisoners of war, although Arthur Edgley and Sgt Maxted evaded capture for six weeks before being betrayed to the Germans. *Courtesy of A. W. Edgley*

Right: A poor quality photograph shows some of the targets attacked by BK611, as painted on the fuselage of the aircraft. Some of the targets are duplicated, but they include from top to bottom: L'Orient, Bay of Biscay, Gironde, Duisburg, Nurnberg, Turin, Hamburg, Dortmund, Berlin, Essen, St Nazaire, Mannheim, Frankfurt, Stuttgart, Rostock and the Baltic. *Courtesy of A.W. Edgley*

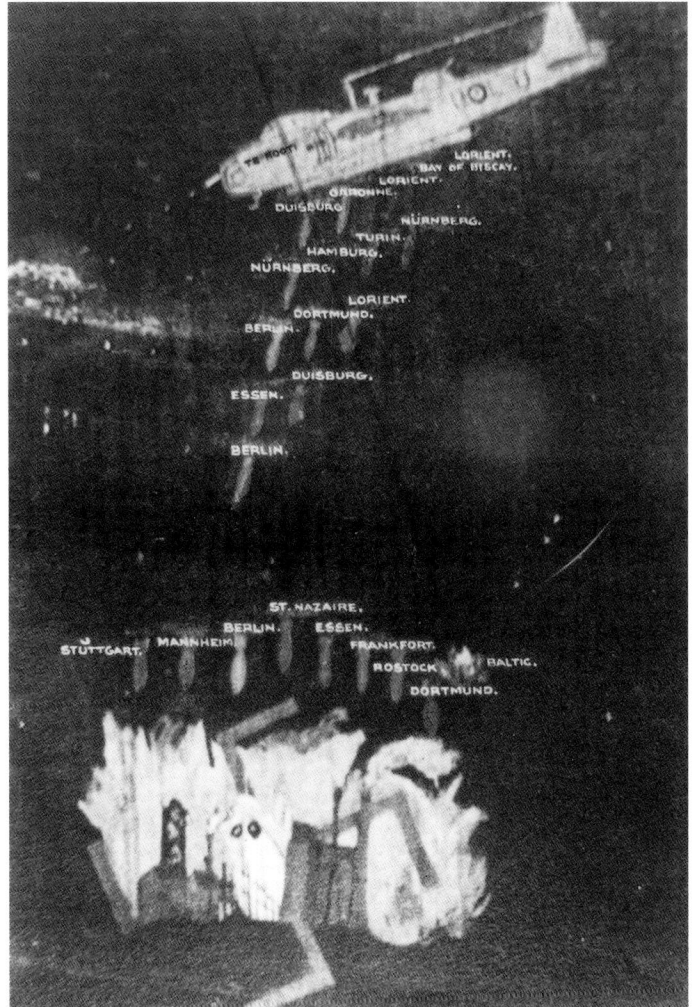

The 'Model' Airman

The training completed, Sergeant Lake relaxes at home with his sister Monica, before being posted to No.XV Squadron, at RAF Mildenhall. *Courtesy of Mrs Ros Dengate, Niece of Aston Crawford Lake*

Left: The model airman, Aston 'Tony' Crawford Lake (3rd from left) messing about in front of the camera at Chatham, New Brunswick, Canada, during his aircrew training period. *Courtesy of Mrs Ros Dengate, Niece of Aston Crawford Lake*

'Wendle' Wilkie and his navigator

The Point of Entry

Looking through the original main entrance gate, on to the camp at RAF Mildenhall during the early 1930s, shows the aircraft hangars still under construction. The building to the right of the picture is the Guard House, which was constructed in 1931. *Courtesy of and via Brenda Cullen*

Left: Pilot Officer Hugh Wilkie, RNZAF, who joined No.XV Squadron from No.1657 Conversion Unit, at the beginning of March 1943. On completion of a 'tour of ops', for which he was awarded a D.F.C, Hugh Wilkie returned to No.1657 Conversion Unit as instructor, with the rank of Flight Lieutenant. *Author's Collection via Frank Diamond* Right: Frank Diamond, navigator to P/O Hugh Wilkie, was Commissioned 'in the field', whilst serving with No.XV Squadron. Having completed a 'tour of ops' on four engined bombers, Frank Diamond volunteered for a second tour on twin-engined Mosquitos. The medal ribbon seen under his navigator's brevet is the 1939/43 Star. *Courtesy of Frank Diamond*

The original Guardhouse, built in 1931, adjacent to the main gate at RAF Mildenhall, where Sgt Gil Marsh and many other young airmen reported for duty. The modified building is still in use (1998), as part of the Support Group facilities, of the USAFE. *Author's Collection*

The Men who went to Mildenhall

A group photograph taken at No.1651 Conversion Unit, RAF Waterbeach, during June 1943, to signify the end of their particular course. These members of aircrew, having satisfactorily completed their training, were posted to operational squadrons. At least two crews from this photograph, those of Sgt Gil Marsh and Sgt George Barber, were sent to No.XV Squadron, at RAF Mildenhall. Identified are: (4th Row) Sgt Jimmy Meaburn, F/E (1st from left); Sgt Gil Marsh, Pil; Sgt Jack Bailey, B/A, RCAF; P/O Richards, Nav; Sgt "Art" Hynam, R/G; Sgt W. Smith, MUG; Sgt George Wright, W/Op (8th, 9th, 10th, 11th, 12th, 13th respectively). (2nd Row) Sgt Jimmy Portsmouth, R/G (6th from left). (1st Row) Sgt Reg Rose, MUG; Sgt George Barber, Pil (6th and 7th from left respectively). Also known to be in the photograph are Sgt Todd, Sgt Scarisbrick and Sgt Abbott. *Courtesy of Reg Rose*

The original main entrance gate to the camp at RAF Mildenhall, photographed during 1998, looking out towards the public highway. On the right of the picture is the original Station Headquarters. *Author's Collection*

The Station Headquarters at RAF Mildenhall, built in 1933, retains the elegance of the period in which it was constructed. The large white Portland Stone commemorative tablet, set in the red brickpaving, records the numbers of the Squadrons which were reviewed by H.M. King George V, at R.A.F. Mildenhall, on the occasion of his Silver Jubilee, on 6th July 1935. *Author's Collection*

The Loss of BK648

The tangled wreckage of Stirling bomber, BK648, LS-J, is inspected by villages from Menden, Germany, where the aircraft crashed, having been shot down on the night of 4th/5th July 1943. Sgt George Barber, who was flying his first 2nd pilot trip, was killed along with the seven members of the usual crew. *Author's collection via Steve Smith*

Madame X

Madame X. Stirling bomber, BK816, LS-X, in which Gil Marsh and his crew flew the majority of their missions, depicted in a painting owned by the former pilot. *Courtesy of Gil Marsh*

Sergeant Gil Marsh (right) with 'Art' Hynams, the rear gunner of 'Madame X', photographed at Mildenhall during the summer of 1943. *Courtesy of Gil Marsh*

14

The Battles of Hamburg and Berlin

Exactly one month after Gil Marsh joined the Squadron, the period known as the 'Battle of Hamburg' commenced, with the first attack taking place on the night of 24th/25th July 1943. Bomber Command despatched 791 aircraft to the target, eighteen of which were detailed by No.XV Squadron.

During the day, Gil Marsh and his crew took *Madame X* for air test and really put the aircraft through its paces. Gil was more than satisfied with the rate of climb, the revs and boost and, after fifty-five minutes in the air, landed a happy man. Having handed the Stirling back to the groundcrew, Sgt Marsh remained close to the aircraft and watched as it was prepared for the forthcoming mission. Even to his still relatively new knowledge, when Gil Marsh saw a 2,000lbs bomb, seven canisters of 30lbs incendiaries and six canisters of 4lbs incendiaries being winched into the Stirling's bomb compartments, he knew a 'sticky job' was likely that night. Having seen a number of paper wrapped bundles being placed in the fuselage of the Stirling close to the flare chute, it was also assumed that propaganda leaflets were also to be dropped.

The raid was led away by F/O Lown who took-off at 22.17 hours, followed three minutes later by F/L Norris and then Sgt Marsh. The latter set course for the east coast, which was crossed at 10,000', before climbing to 15,000'. The flak ships off the enemy coast gave the incoming bombers a warm welcome, and fired off heavy barrages of anti-aircraft fire. On approaching the target, Sgt Marsh dived down at +6 boost, with 2,400 revs on the clock, at an air speed of 260 mph. He leveled out and held 'Madame X' steady at 190mph for the bombing run. As he followed the instruction of Jack Bailey, the Canadian bomb aimer, Gil Marsh was in awe of the sight 13,000' below him, as the city shimmered in the bright burning colours of intense and devastating fire. At the same time, he was highly nervous of the situation of flying straight and level during the bombing run. A pilot's nightmare, especially when a multitude of thirty to fifty searchlights were sweeping the skies in search of a victim. He was also aware that German nightfighters were in the vicinity, but so far they had left him alone. Fortunately, the probing beams had not illuminated *Madame X* either, but shrapnel from an exploding shell did create some minor damage to the airframe. On return to base Gil Marsh, in terms of some merit, recorded in his diary *"holed by flak – first time"*. Had he known at the time, he may well have included an entry relating the fact that his aircraft was the first No.XV Squadron machine to bomb the target that night.

All the Squadron's aircraft returned safely to base, from what the participating aircrews reported as *"the best raid so far"*. One factor which helped in the success of this mission was the use of 'Window', the paper wrapped bundles Gil Marsh had seen being loaded into the aircraft earlier in the day. 'Window' was strips of aluminium foil, each cut to half the length of the German radar frequency. When cut open, and dropped over the target via the flare chute of each aircraft, the foil strips acted as a reso-nating dipole aerial and re-radiated more of the radar pulse than the aircraft. This in turn gave a stronger 'echo' on the German radar screens, giving the impression a vast armada of bombers was on its way to Hamburg.

The next night nineteen aircraft and their respective crews from No.XV Squadron, were despatched to Essen as part of a total attacking force of 705 bombers.

Of the nineteen aircraft which took-off from Mildenhall, one returned early due to problems with the starboard inner engine. Flight Lieutenant Norris, piloting Stirling bomber, BK654, LS-W, jettisoned the bomb load safely into the sea at 00.25 hours, before turning for home.

Pilot Officer Gabel, RCAF, had a lucky escape after the starboard inner engine of his aircraft caught fire, following two separate combats. Having been forced to lose altitude, the bomber, BK774, LS-K, jettisoned its bomb load over Essen at 00.45 hours. Apart from the engine fire, the aircraft also sustained damage to the rear turret, the oxygen supply and the intercom system. From his position on the flight deck, P/O Gabel was aware of something out on the wing gradually bending, and thought the mainplane was warping from the heat of the fire. As he was about to give the order to bale out, the propeller and reduction gear sheared off and cartwheeled away into the night. Following the loss of the propeller, the fire died down and was eventually extinguished, allowing the Stirling to fly home on three engines. However, Sgt Towse, DFM, who had ditched his bomber in the North Sea a month earlier, was not so fortunate. His Stirling on this mission, BK805, LS-U, was attacked and shot down by a nightfighter piloted by Hauptmann Wilhelm Dormann, of III./NJG1. The crew, including Sgt Pawley, DFM, managed to bale out and were taken prisoners of war. Unfortunately, Sgt Wilfred Towse, DFM, was killed when the bomber crashed at Osterwick near Ahaus, Germany, at 00.40 hours.

In general terms, the raid was successful and, apart from inflicting severe damage on the industrial areas of the city, major damage was also inflicted on the Krupps armament works.

After a good night's sleep, Sgt Marsh woke up at 11am on the morning of 26th July feeling refreshed and happy. On this day he, and the rest of the crew were going to put the war behind them; for one week at least. The day was spent arranging various matters like pay, travel warrants, passes and handing in their parachutes. For this one day only, the whole Squadron was going to forget the war in favor of the Station's sports day.

The following morning Gil Marsh was up early, shaved, washed and ready to go, even though he and the crew had to wait for publication of the Battle Order. When it arrived, they sighed with relief to see their names were not included for the night's forthcoming operation, which meant they wasted no further time in getting away from Mildenhall. With regard to how he spent his own leave, Gil Marsh wrote in his diary, *"Sorry! But the next six days are secret!!!"*

The Battle of Hamburg continued that night, when eighteen Stirlings from the Squadron joined 769 other aircraft from Bomber Command for the attack.

Unbeknown to the returning crews, who reported a successful mission, was the fact that concentrated bombing in a densely built-up area had contributed, together with climatic conditions, to producing a firestorm which devastated the city. During their debriefing, the aircrews all reported palls of smoke rising many thousands of feet into the air, and fire glows which could be seen from one hundred and fifty miles away on the return journey.

Unlike the previous attack on Hamburg, the Squadron did not get away without casualties. On this occasion one aircraft, piloted by F/L Childs, failed to return, whilst another crashed on the perimeter of the airfield at RAF Mildenhall during its landing approach. The former Stirling had the misfortune to be hit by flak over the target area, then attacked by an enemy nightfighter. It was the German pilot, Uffz. Luschner, of III./NJG3, who inflicted the coup de grace. The Stirling, EH893, LS-J, crashed at Hamburg-Ochsenwerder. Seven of the eight man crew were killed, including Sergeant Alfred Holden, who was flying a 'second dicky' trip. The only survivor was Sgt Hurley, the wireless operator/air gunner.

The crew of the second Stirling were more fortunate and survived, albeit with minor injuries. Due to flak damage to their aircraft sustained over the target, both inner engines failed during the return flight. The Stirling, EF437, LS-Z, was unable to maintain altitude and crashed into two trees. The wings were sheared off and the bomber crashed at West Row on the perimeter of Mildenhall airfield, at 03.50 hours.

Flying Officer Waugh had taken-off from Mildenhall at 22.25 hours with a complete crew. However, when he landed back at base on return from the raid, only he, Sgt Watson the flight engineer and Doug Boards the second pilot, clambered out of the aircraft. The members of the crew, including Sgt Ray Cole, the bomb aimer, thought this was going to be an easy target, a well earned bonus to go towards their tour after all the Ruhr operations. The only thing out of the ordinary on this mission, compared to other raids, was the inclusion of a second pilot and the bundles of *'Window'*, which the luckless newcomer had been elected to push out every couple of minutes once they had reached the target area. As Sgt Cole had also anticipated, the outward leg of the mission was, so far, uneventful. From his position in the nose of the Stirling, Ray Cole looked down and identified the T-shape island of Sylt before it disappeared from view under the aircraft, as the bomber crossed the enemy coast. He had scarcely reported this pin-point to the navigator when Sgt Slatter, the wireless operator, advised the flight engineer that the oil pressure on the starboard inner engine had dropped. The latter responded by saying it was only the gauge, implying the instrument was at fault. As Alex Slatter made comment about the temperature, the engine in question burst into flames. The fire gained control and started to spread along the mainplane, towards the fuselage and wing bomb-bays. Bob Waugh threw the Stirling into a dive in an effort to extinguish the flames, whilst Ray Cole attempted to jettison the bomb load. Unfortunately, due to the ever increasing "G" force, the bomb aimer found this a difficult task. When both port engines stated to give trouble, and fearing little more could be done to save the aircraft, the pilot gave the order to bale out. There was no panic as the crew clipped on their parachutes, released the escape hatch lever and jumped out into the night. Sergeant Cole thought it was just like an exercise at Mildenhall, that is until he looked up from under his parachute and saw a tongue of flame flying through the darkness above him. Ray Cole's descent took half an hour and, as he floated down to internment, he could see the diminishing tongue of fire receding into the distance. Unbeknown to him and the four other members of the crew who baled out, Sgt Doug Boards and Sgt Watson had remained on the aircraft with the pilot. The fire was eventually extinguished and, whilst F/O Waugh concentrated on keeping the bomber in the air, Sgt Board and Sgt Watson jettisoned all loose equipment. Stirling bomber, BK719, LS-B, returned to Mildenhall where it landed at 03.45 hours, damaged but in one piece.

Sergeant Midgley may well have had reservations about this operation when he saw his name and that of his crew on the Battle Order. It was not the target which gave cause for concern, but the aircraft allocated to him; it was Stirling bomber, EE912, coded LS-U!!. Although the aircraft returned safely from the raid, it did come home with an incendiary canister hung-up in the bomb bay. However, this did not deter Sgt Midgley who adopted the aircraft as his own, and put his personal mark on it.

He had the aircraft adorned just below the cockpit, on the forward fuselage, with a painting of a forlorn looking clown. Above the head and shoulders of the figure was the inscription *Midgley's Flying Circus*, whilst between the clown's legs, below its 'knocked-knees' was the name of the artist, Laura Knight.

Bearing this unorthodox, but highly valued insignia, Sgt Midgley was to fly the aircraft on a number of missions, before it too succumbed to the LS-U curse.

Laura Knight, who was renowned for her pre-war circus pictures, held the honor of Dame Commander of the British Empire, and was a member of the Royal Academy of Art. Now, in her capacity as an official war artist, she was on an official visit to RAF Mildenhall to gain knowledge and atmosphere for a painting she was to undertake depicting Bomber Command.

Having been operational on three of the last four missions, P/O Judd and his crew were hoping to have a night off on 30th July, in order to go out on the town. Therefore, it came as quite a surprise when Sgt Doug Fry, the mid-upper gunner, informed the crew their names were on the Battle Order for that night, for a raid against Remscheid. The list having been posted in the crew room in "A" Flight hanger.

Another surprise awaited the crew when they attended the briefing at 20.00 hours. An RAF film unit had set up a camera and arc-lamps, and were preparing to film the briefing, for inclusion in a film about Bomber Command entitled *'The Biter Bit'*. Later, whilst the crew were standing chatting, waiting for transport out to their aircraft, Doug Fry pointed out to George Judd that they were being filmed. The cameraman was standing only a few feet from them. It was to be the last recorded picture of George Judd, and was subsequently to produce a shock for Doug Fry's mother.

Pilot Officer Gabel led the Squadron away, when he took-off at 22.35 hours. P/O George Judd took-off seven minutes later, when Stirling bomber, EF427, LS-A, lumbered into the air. The outward journey was without incident, apart from a few searchlights and some desultory flak encountered as they crossed the enemy coast. From his vantage point, in the mid-upper turret, high on the back of the Stirling, Sgt Fry identified the target. Looking straight ahead he saw the glow of fires on the ground, whilst bursting shells and photo-flashes momentarily lit up the night. Stiff white fingers of brilliant light swayed gently back and forth, as searchlight beams probed the sky, occasionally crossing like two giant sword blades in unseen hands.

Flying towards the 'cauldron', the crew prepared for the bombing run. Sergeant Long, the bomb aimer, was already in position. By the time Syd Long called, *"Bomb doors open"* the Stirling had already reached the outer defenses. Suddenly, whilst Doug Fry was scanning the skies to the starboard side of the aircraft, a blue master beam swept across the tip of the mainplane. Before he was able to report its presence to the pilot, the beam had swung back and locked onto the bomber. As the other beams endeavored to form a cone with the master beam, the Stirling continued with its bombing run. Flak was bursting closer and closer to the aircraft, as the ground gunners began to get the range. Following his announcement that the bombs had gone, Sgt Long requested his pilot to waste no time getting out of the area. Pilot Officer Judd reminded the bomb aimer that they still had to get their target photograph first, which meant another ten or more seconds flying straight and level. Those ten seconds were to seal the fate of the Stirling and its crew. The gunners had found the range, and flak was bursting around the bomber, peppering it with shrapnel holes as each shell exploded. Above the drone of the engines, Doug Fry heard an unfamiliar noise within the aircraft which caught his attention. Glancing down from his turret, he saw flames flickering and licking their way along the port side of the aircraft. With all the flak activity around LS-A, Doug Fry knew there would not be any enemy fighters in the area. He therefore decided to evacuate the turret, in order to see what he could do to tackle the fire. As he climbed down, with one foot still on the last step of the ladder, a shell burst towards the rear of the aircraft. At the same time Sgt Fry felt a hefty blow to his stomach, which caused him to double up in pain and fall to the floor. An almighty explosion in the nose of the aircraft caused the Stirling to plunge into a near vertical dive, as the pilot lost control of the bomber. The mid-upper gunner found himself floating in the fuselage of the stricken aircraft, but even in his semi-conscious state he knew it was not a dream. Having resigned himself to the fact he was about to die, Doug Fry was suddenly aware the bomber was slowly beginning to pull out of the near vertical dive. Realising he may not get another chance of survival, Sgt Fry staggered to his feet but collapsed again almost immediately. As he thought about trying to locate his parachute, he saw the item slide across the floor of the aircraft towards the flames. Fortunately, the 'chute came to rest against a bulkhead, thus enabling him to crawl over and retrieve it. Having clipped it on to the harness he was wearing, the air gunner removed his flying helmet and, with superhuman effort, made his way to the rear escape hatch. Tired and exhausted, Doug Fry sat on the edge of the open hatch, his feet dangling down into the slipstream. Without effort, he rolled over and fell into the darkness. The aircraft, with long tongues of flame streaking out in its wake, continued to fall in a westerly direction, towards the village of Manheim, north of Buir. With seconds to spare the bomber turned to starboard, away from the village, and crashed into the surrounding fields. Local inhabitants of the village who heard the bomber screaming down to destruction, felt a major catastrophe was about to occur. Many believed P/O George Judd, in his dying moments, averted that disaster.

Sergeant Douglas Fry, who was wounded in the stomach by shrapnel, floated down and landed in the garden of a house owned by Frau Margarete Franzen. He was immediately apprehended and helped into the house, where his wounds were attended to, prior to him being taken to hospital. His condition gave cause for concern on many occasions, as the wounded airman drifted in and out of consciousness. Eventually, Doug Fry made a full recovery but spent the rest of the war as a PoW, where like so many others, he endured numerous hardships. Sergeant Ken Banks, the rear gunner, and Sgt Dick Richards, the flight engineer, both of whom baled out, were also captured and made prisoners of war. The names of the four crew members who were killed in the crash, were recorded on the No.XV Squadron Roll of Honour.

During the period that Doug Fry was listed as missing, before confirmation had been received stating he was a prisoner of war, his mother visited the cinema, where she in turn was to receive a shock. At this stage Winifred Fry did not know whether her son was dead or alive but, in the hope of getting away from the trauma that war brings, she settled back in her seat to watch the movies. The house lights dimmed and the projector flickered into life, casting its beam through the darkness onto the silver screen. Patriotic music filled the auditorium, as the title of the film *'The Biter Bit'* appeared to view. The movie was a ten minute propaganda film showing the role played by aircraft and crews of Bomber Command. As she watched in silence, lost in her own thoughts, Winifred Fry suddenly saw the image of her son Douglas projected onto the screen. It was no illusion, there he was standing looking straight at the camera, whilst talking to his pilot George Judd. An involuntary scream pierced the air, as Mrs Fry fainted from the shock. The cinema staff, assisted by Doug's sister Winnie who had accompanied her mother to the picture house, quickly removed the distressed mother from her seat. In the quiet confines of the manager's office, the reason for Winifred Fry's anguish was quickly explained. Being sympathetic to her plight the cinema manager arranged, with the necessary authorities, for Doug Fry's mother to have a private viewing of the film. This in turn was followed, a while later, by the news that although wounded, her son had survived and was a prisoner of war.

Further good news was received on the Squadron when F/S Cullen, the wireless operator on S/L Megginson's crew announced he was getting married. Graham 'Mick' Cullen had arrived in England, from his native

New Zealand, earlier in the year. He had been sent to No.11 Operational Training Unit, at Westcott, where he met and joined the new crew being formed by (the then) F/L Megginson. On completion of their training, this crew were posted to No.XV Squadron at Mildenhall. Getting their priorities right, one of their first missions was to search out the best local 'watering hole', which proved to be the Bird In Hand Hotel. One of its main attractions was the fact it was situated across the road from the airfield, next door to the Officers' Mess and fairly close to the main gate of the camp. However, for Mick Cullen, there was another attraction in the shape of a pretty young barmaid. Brenda Jaggard, the barmaid in question, was a local girl who lived just along the road. As the crew spent a lot of spare time at *The Bird*, Mick and Brenda quickly struck up a friendship. Eventually, loved blossomed and they subsequently announced their engagement.

Another announcement was made at Mildenhall on Tuesday, 10th August, a few days after Gil Marsh and his crew had returned from leave. The communiqué informed all members of the Squadron that, as from that date, "C" Flight of No.XV was to form the nucleus of a new unit, which would be known as No.622 Squadron. Being part of that flight, Sgt Gil Marsh, his crew and *Madame X* all became part of the new Squadron. Although the Stirling retained its identity, the new unit code of GI was applied on the fuselage, in place of the LS code used by No.XV. Fortunately, as the new squadron was to operate alongside its 'mother' squadron at Mildenhall, the crews were not concerned about having to relocate to a new operational base. Unfortunately, as all the crews were battle experienced, it was decreed they would operate on the night of their formation.

No.XV Squadron detailed thirteen aircraft for an attack against Nuremberg, on the night of 10th August. Although 10/10th cloud was recorded over the target area, all crews reported good concentrated fires, the glow of which could be seen permeating through the clouds from one hundred miles away.

The Squadron lost one aircraft during this raid, Stirling bomber, BF460, piloted by F/S Lewis. The aircraft, coded LS-F, had completed its bombing run and was on its way home. Although its task had been completed, the crew were still at their respective stations. Sergeant John Sparrow sat in the rear turret of the Stirling scanning the night sky, searching for hostile aircraft, but all was quiet. At least, it was in his sector of the sky. Suddenly and without warning, the nose of the Stirling erupted into a mass of shattered, flying plexi-glass and twisted metal. A Bf110 nightfighter had crept out of the darkness, making a head-on attack, and raked the bomber with cannon fire. As flames engulfed the nose section and starboard wing of the Stirling, the enemy pilot, Hpt Johannes Hager, of II./NJG1, turned in towards the bomber for another attack. Albeit too late to save the stricken aircraft, both the rear and mid-upper gunners turned their respective turrets to meet the incoming attack. Simultaneously they opened fire and saw their shells strike the target. The Bf110 is reported to have taken the full impact of the avenging shells, keeled over and dived away. By this time the fire ravaging the bomber had taken full hold, and the aircraft was losing height rapidly, leaving the crew no option but to bale out. Unfortunately, only three of them would make it, Sgt Cave, the wireless operator, Sgt Bryne, the navigator and John Sparrow. Sergeant Frank Poole, the bomb aimer, is thought to have perished after jumping from the blazing bomber without his parachute. The remaining members of the crew died when the aircraft crashed at Doische, Belgium.

Acting Squadron Leader Megginson's ability and experience as a pilot, was put to the test on the night of 16th/17th August, during a raid against Turin. The details were somewhat reminiscent of those endured by S/L Wyatt nearly a year earlier. Bomber Command had despatched a total of 154 aircraft to attack the target, for which No.XV Squadron detailed fourteen Stirlings. One of the latter being Stirling, BK818, LS-O, piloted by A/S/L Megginson. Apart from his usual crew, Megginson was also to carry an important passenger on board his aircraft; Air Commodore Herbert Kirkpatrick, D.F.C, from No.3 Group Headquarters. The latter's function on the mission was merely to observe what proved to be the last attack against an Italian city.

As Stirling BK818 was overflying the South of France, the aircraft began to experience trouble with the engines, which had started to overheat. Climbing to gain altitude in order to cross the Alps only added to the problem, and one of the motors seized up completely. Megginson elected to continue with his mission, but was forced to fly through the mountain range rather than over it. Having safely negotiated the Alps, the Stirling arrived over the allocated target area only to find it shrouded in low cloud. A decision was made to bomb the main target which, owing to the engine problems endured by the Stirling, could have had dire consequences. Whereas the main stream was bombing from an altitude of approximately 16,000', A/S/L Megginson was forced to carry out his bombing run at 11,000'. The task safely completed, the pilot was now forced to decide whether to fly on to North Africa, where he could land and initiate repairs to the aircraft, or attempt a return to Mildenhall. Following a short debate with the crew, it was decided they should endeavor to get their passenger back to Mildenhall.

Once again they successfully negotiated the valleys through the Alps, feeling somewhat small and vulnerable as they gazed at the mountain peaks which towered above them. However, the strain on the aircraft was beginning to show and, as they crossed out over France, a second engine began to overheat. Knowing he was in an emergency situation, Robert Megginson feathered the defective engine and decided to land at the first available friendly base. Praying the two remaining engines would continue to function a while longer, Megginson brought the bomber over the south coast of England, and landed at the fighter airfield at Tangmere. As he touched down, with a sigh of relief from the whole crew, a third engine seized completely. An inspection the following morning revealed the heat endured by two of the engines had completely melted the pistons to the engine blocks. The crew were given a few days unexpected leave whilst new power plants, together with members of the groundcrew who would fit them, were flown down to Tangmere.

Mick Cullen, the wireless operator, was looking forward to getting back to Mildenhall, but on his return was given the cold shoulder. In the excitement of the occasion, he had omitted to contact his fiancee, Brenda Jaggard, to let her know he was all right.

Bomber Command, having received orders to carry out an attack against the German research establishment at Peenemunde on the Baltic coast, detailed the mission for the night of 17th/18th August. A total of 596 aircraft participated, including five from No.XV Squadron.

Being a raid of special significance, several new ideas were introduced by Bomber Command for the success of this mission. However, the German Air Force also had a new idea of their own, which they used for the first time this same night.

The new idea was a weapon, fitted to the cockpits of some of their nightfighters, which consisted of twin upward-firing cannons. The device which was known as *'Schrage Musik'*, was to become very successful and accounted for the loss of at least six of the bombers which failed to return from the attack. One of those bombers was a Stirling from No.XV Squadron. The aircraft, EE908, LS-V, which was piloted by Sgt Grundy, crashed into the Baltic sea. Robert Grundy, who was flying his 12th mission, was killed along with Sgt Ernest Honeybill, the flight engineer, and Sgt Colin Hudson, the navigator. The four remaining members of the crew, who survived the ordeal, were captured and made prisoners of war.

On discovering that his name was not on the battle order for an operation the night of 22nd August, Mick Cullen decided to spend the evening at the Bird in Hand Hotel with his fiancee. As it was Brenda's birthday, the rest of the crew decided to join the couple and make an evening of it. They all enjoyed themselves so much that, when the hotel closed, they decided to continue the party at the home of Brenda's parents. There, the piano pounded into life under the dexterous fingers of 'Mitch' Mitchell, whilst the rest of the group heartily sang popular songs of the day. The atmosphere in the Jaggard household was one of laughter and happiness, with no thought of the war. By contrast, it was to be a totally different picture the following evening, when Robert Megginson and his crew were detailed for a mission against Berlin. Their aircraft was one of thirteen Stirlings

detailed by the Squadron, to join a total force of seven hundred and twenty-seven Bomber Command aircraft despatched to attack the city. With the days beginning to draw-in, and the periods of darkness slowly getting longer, this raid signaled the start of a renewed action in the 'Battle of Berlin'.

Megginson's Stirling, BK818, crossed the Norfolk coast at Cromer and headed east. With the excellent weather conditions which prevailed, it was obvious that enemy nightfighters would be very active and he warned the crew accordingly.

At a point thirty miles from Berlin, A/S/L Megginson received a course change from P/O Burrows, the navigator. The pilot banked the Stirling onto the new heading, aiming now straight for the Reich Capital. The crew carried out their bomb run without incident and, following Andy Haydon's call *"Bombs gone"*, Megginson again banked the Stirling over for a course change away from the target. As he did so cannon fire ripped into the bomber from astern, following an attack from an unseen nightfighter. Having made one passing attack, the enemy aircraft disappeared back into the cloak of darkness. The pilot called for a damage report and, one by one, each member of the crew responded with the exception of the rear gunner, P/O Mitchell. Mick Cullen was instructed to make his way aft, check out the situation and report back. Feeling very apprehensive about what he would find, the wireless operator did as he was bade. His fears grew when, on reaching the rear turret, he found the mechanism jammed and the access doors stuck fast. It was only with the aid of an axe, that Sgt Cullen was able to break open the latter and gain entry to the turret. To his horror, he found the nightfighter's cannon shells had turned the turret into a twisted mass of tangled metal and shattered plexi-glass; the unmoving form of P/O Mitchell was slumped over the gun butts. Sergeant Cullen gently eased 'Mitch' back, only to find the rear gunner had been killed by a solitary bullet through his left eye. Realising he would need help to remove Mitchell's body from the turret, Mick Cullen called on the intercom for Andy Haydon to assist him. Together, they extricated the lifeless form of the rear gunner from the wreckage. Although he would have been unable to take defensive action against an attacker, Sgt Cullen was ordered by the pilot to occupy the shattered turret and warn of further incoming attacks. For the wireless operator turned acting look-out, the trip home was cold, uncomfortable and unenviable.

Pilot Officer Desmond Mitchell, RAFVR, was buried at Dunchurch (St Peter's) Churchyard, in his home town of Rugby. Among the mourners who attended the funeral, apart from his parents, were the members of his crew who had all been given a few days leave.

Apart from the loss of P/O Mitchell, seven other Squadron members were lost during the attack on Berlin when their Stirling bomber, EH875, LS-S, was shot down. The aircraft, which was flown by P/O Eric Cornell, carried the name *Thornaby-on-Tees*, having been purchased and presented by the town of the same name.

It is not known whether any pre-flight ritual was adopted relating to Dame Laura Knight's artwork on *Midgley's Flying Circus*, but Stirling, EE912, LS-U, had two lucky escapes during the Berlin raid.

During the outward journey, Sergeant Vincent Gowland, the rear gunner of LS-U, was looking out of his turret, musing on the subject of enemy nightfighter tactics. The bomber had just crossed the enemy coast, when he saw a Bf110 high on the port quarter rear. In receiving a warning to 'corkscrew', Sgt Midgley gave the aircraft left rudder and pushed the control column forward, causing the bomber to fall in a spiraling dive. The enemy aircraft followed with guns blazing, the trajectory of its fire seen to streak over the top of the turret. Sgt Gowland kept his nerve, and waited for the nightfighter to get a little closer before returning fire. When the gap was down to one hundred and fifty yards Vincent Gowland squeezed the triggers, and the turret's four guns burst into life. The darkness around the aircraft became illuminated as Sgt McCool, RCAF, joined the fray and opened fire from the mid-upper turret. Suddenly a ball of flame emanated from the starboard motor of the twin-engined fighter. The enemy aircraft was last reported going down in a dive. Reporting the outcome of the fight, Sgt Gowland advised the pilot it was safe to resume course for Berlin. The rest of the journey passed without incident, as did the bombing run. However, as the bomb doors were closing, the beam of a searchlight cut through the night sky, sweeping towards the Stirling. In an effort to avoid the tenacious beam, Sgt Midgley, again, threw the aircraft a corkscrew but, unfortunately, was too late with the maneuver, and the searchlight lit up the bomber. For five minutes the beam followed the twisting, turning, diving aircraft, as the pilot tried desperately to evade the blinding ray. In an attempt to fool the German searchlight crew into thinking they were under attack, what was reported as a four and a half pound incendiary device, was dropped from the Stirling. The ruse appeared to work as the searchlight was doused, but not before the bomber lurched under the impact of a flak burst. Calling for a damage report, Sgt Midgley was informed by Sgt Bill Rhodes, the bomber aimer, that *"there's a hole the size of a kitchen sink down here in the belly, and it's bl**dy draughty"*.

Apart from some anxiety with regard to navigation, the return journey was made without incident, although it was accompanied by apprehension. *Midgley's Flying Circus* landed safely at Mildenhall, 04.25 hours.

Whilst casually perusing the Battle Order on 25th August, it came as somewhat of a surprise to Reg Rose to see his name listed for a mine-laying operation that night. Since the loss of their pilot on 3rd July, Sgt Reg Rose and the rest of Sgt Barber's crew had been relegated to the position of 'spare' aircrew. Not having been informed he had been crew-up, Reg Rose made some inquiries and found he had been seconded to join the crew of Sergeant Joseph Russell. Russell, a Canadian from Saskatchewan, had joined the Squadron only a week or two earlier and required a replacement mid-upper gunner; hence Reg Rose joining the crew. Feeling somewhat apprehensive, Sgt Rose made himself known to his new crew. One of the first to greet the air gunner was Sgt Bentley, the bomb aimer and, although neither of them knew at the time, Leslie Bentley's first acknowledgment of the newcomer was to be the start of a life long friendship. Having overcome the initial awkwardness of getting to know each other, the crew took their aircraft on an air test. After flying over the Suffolk countryside for an hour, during which time various checks and tests were carried out, Russell landed back at Mildenhall.

The excitement and trepidation of the forthcoming mission, which had been building-up within Reg Rose during the day, was forgotten as he carried out various checks in his turret prior to take-off. At 20.10 hours, and on receipt of the green Aldis lamp signal to take off, Sgt Russell increased power to the four Bristol Hercules engines and released the brakes. Stirling bomber, BF569, LS-V, began to increase speed and slowly lifted into the darkening sky. Having reached the 'Gardening' area, the mines were dropped without incident and the aircraft and crew returned safely to base.

However, fate decreed the name of Joseph Russell would soon be recorded in the Squadron's annals, along with those of his crew.

Acting Flight Lieutenants Norris and Wilkie, together with their respective crews, were declared 'tour expired' during August and posted to No.1657 Conversion Unit on 26th of the month. Unfortunately F/L Hugh Wilkie, D.F.C., was to lose his life in a flying training accident on 18th April 1944.

Another flying accident was to claim the life of the Squadron's Commander on 30th August. Wing Commander John Stephens, D.F.C., had been undertaking some local flying in a Boulton Paul Defiant, when the machine crashed killing him and a member of groundcrew flying with him. The Defiant, which had been airborne for forty-five minutes, developed a coolant leak which penetrated into the cockpit. Wing Commander Stephens, who was not wearing flying goggles, succumbed to the fumes and was unable to regain control of his aircraft when the engine failed and the machine went into a stall. The Defiant crashed at Beck Row, on the edge of the airfield, at 15.45 hours.

That night whilst participating in an attack against Munchen Gladbach, Sgt Derek Brock was killed in similar circumstances to those of P/O Desmond Mitchell. The former was flying as flight engineer to F/S Thomas's crew, on Stirling bomber, EE954, LS-J. They had bombed the target, and were on the homeward journey, when their aircraft was attacked by a German nightfighter. The enemy machine made a firing pass at the bomber, which resulted in the death of the flight engineer. No other injuries were reported with regard to the rest of the crew.

The following night the curse of LS-U resurrected itself, when another crew took *Midgley's Flying Circus* on a raid against Berlin. Eleven aircraft were detailed by No.XV and, as Sgt Midgley had just gone on leave, Sgt Joseph Milner, RCAF, was detailed to fly the former's Stirling.

The searchlights and flak over the target area were recorded as ineffective, but the nightfighters had found another method by which to illuminate their quarry. The first recorded use of flares dropped from the nightfighter aircraft, enabled the Luftwaffe aircrews to identify the routes used by the bomber stream, to and from the target. It is not known whether *Midgley's Flying Circus* was caught in the glare of one of these flares, but it is known that the aircraft crashed at Roskow, near Brandenburg, with the loss of the entire crew.

Five aircraft returned early from the attack against Berlin including Stirling, BK816, LS-R, piloted by Sgt Russell. A combination of a defective compass and an overheating starboard inner engine, caused the Canadian pilot to abandon the mission and return to base.

Wing Commander A J Elliot officially assumed command of No.XV Squadron, on 3rd September, following the untimely demise of W/C Stephens four days earlier.

The unprecedented award of a Distinguished Flying Medal was made at the end of August to a pilot who, apart from not flying operationally with his own crew, had not even captained his own aircraft. The award was promulgated for the part played by Sgt Doug Boards in assisting F/O Waugh to bring a crippled bomber home, on the night of 27th/28th July. The pilot was awarded a D.F.C., whilst Sergeant Watson, the flight engineer, who also remained with the aircraft, was awarded a D.F.M. Douglas Boards was initially promoted to Flight Sergeant, but was later commissioned to the rank of Pilot Officer, with effect from 13th August 1943. Flying Officer Waugh completed his tour of operations a couple of weeks later, and was posted to No.1657 Conversion Unit, at Stradishall, on 15th September.

'Midge' Midgley, who four days earlier had been promoted to the rank of Pilot Officer, returned to operations on the night of 15th September, having taken receipt of a new aircraft. The replacement for his *Flying Circus* was Stirling bomber, EH940, also coded LS-U !! Thirteen nights later, this aircraft was to be damaged by an incendiary bomb dropped from a 'friendly' aircraft over Hannover. This in turn meant another new machine for P/O Midgley.

There was cause for celebration towards the end of the month when, on 25th September, Air Vice Marshal Richard Harrison, C.B.E., D.F.C., A.F.C., Officer Commanding No.3 Group, signed a recommendation for the immediate award of a Distinguished Flying Cross to A/S/L Megginson. Robert Megginson, who was half way through his second tour of duty, had successfully completed fifty-one operational missions. Two days later Joseph Russell and his crew had cause to celebrate, when the latter was promoted to the rank of Flight Sergeant.

An attack against Hannover, on the night of 27th/28th September, nearly brought about the loss of a No.XV Squadron aircraft, but not due to enemy action. Stirling bomber, BK719, LS-B, had taken-off from Mildenhall at 19.50 hours, to join a main force of 678 aircraft. The outbound journey was without incident, but once over the target area the Stirling's fortunes changed. Pilot Officer Woodley held BK719 steady, and guided the aircraft up to the target as directed by the bomb aimer. As the Stirling crossed the aiming point a Lancaster bomber, flying at a higher altitude, released its load. One of the bombs went through the mainplane and fuel tank of the Stirling, causing it to enter a spin. With needles flickering backwards and forwards across dials and instruments, P/O Woodley fought to regain control of his aircraft. Successful in his endeavors, the pilot hauled the Stirling back onto an even keel and called for a damage report. Ascertaining there were no injuries, and with Hannover just 3,000' beneath his aircraft, P/O Woodley slowly regained altitude and headed for home. On his return to Mildenhall, Frank Watson, the wireless operator, recorded the mission in his Log-Book, adding the comment *"Quiet Trip"*!

A pilot flying his fourteenth mission failed to return, along with his aircraft and crew, following an attack against Kassel on the night of 3rd/

4th October. The pilot was P/O Alex Wood (who had been commissioned on 11th September) and the aircraft was Stirling bomber, BF470, LS-G. The Stirling, which had completed a number of operations under the command of F/L Hugh Wilkie, RNZAF, was one of twelve aircraft detailed by the Squadron for the mission. It was believed to have crashed at Haste, on the north-eastern fringes of Osnabruck, having been shot down by Oblt Fritz Lau, of III./NJG1. Three members of the crew were killed, including Sergeant Rex Blanchard, the mid-upper gunner, who was unable to locate his parachute; he died in the ensuing crash. Those who survived were captured and made prisoners of war.

Another casualty on this raid was Stirling Bomber, EF459, LS-S, which was also attacked by a nightfighter ten minutes after leaving the target area. The aircraft, having bombed the target at 21.27 hours from a height of 16,000, was on its way home when a 'Boozer' warning alerted the crew to the presence of an enemy aircraft in their vicinity. 'Boozer' being a radar tuned to the German nightfighter Lichtenstein frequencies, the pilot, F/S Russell, took immediate evasive action. As he threw the Stirling into a corkscrew maneuver, he alerted the two gunners to the situation and advised them to be vigilant. He then ordered Sgt Marsh, the flight engineer, to climb up into the astrodome and act as another pair of eyes in support of the gunners. The bomber was down to 12,000' when the warning light went out. However, approximately one minute later, as the signal came on again, gun fire struck the side of the bomber. Such was the ferocity of the attack, that the force of the cannon shells slammed into the fuselage of the Stirling, seriously wounding F/Sgt Burns, the navigator and Sgt Rose, the mid-upper gunner. The former had been sitting at his plotting table when a shell exploded against the back of the pilot's armor-plated seat. The resulting blast encompassed the navigator, causing him to suffer a bad head wound. He was later found by Sgt Marsh in an unconscious state, slumped across the charts and maps. On receiving the pilot's warning of an enemy nightfighter, Sgt Rose had turned his turret to face aft in anticipation of an attack. When it came, on the port quarter high, the impact velocity of the enemy cannon shells spun the turret round. On regaining his senses, having been momentarily knocked unconscious by the violent motion of the spinning cupola, it took a few seconds for the occupant to understand how he came to be facing the opposite direction. He was however aware of the exceedingly cold wind blasting in through the broken plexi-glass of his shattered turret. Due to a severe arm wound, Reg Rose was helped out of the tangled wreckage by Sgt Hayles, the wireless operator. It was only after he had been laid on the bed, and his Irving leather flying jacket had been cut open, it was found the air gunner's right arm had been severed at the shoulder. The white heat of the cannon shell which caused the injury had also cauterized the wound, thus preventing the loss of too much blood. A slither of flying metal had also caused a wound to the back of the pilot's neck but this, fortunately, was only a minor injury.

Fortunately the rear gunner had remained unharmed and, whilst the drama was unfolding in the fuselage behind him, got on with the job in hand. Although he was unable to see the attacking fighter clearly, Sgt Forrest aimed his guns at the center of the apex of the incoming tracer and opened fire. True to the Squadron's motto, his aim was sure and flames lit up the night sky on the port quarter as the enemy aircraft peeled over and entered a dive. William Forrest reported the outcome to his pilot, and watched as the nightfighter plunged earthward to explode on impact with the ground.

Damage to the Stirling bomber was extensive, with both elevator and rudder trims cut. Both the port engine controls were damaged, as was the port outer rev counter and the air speed indicator. The Gee set was totally destroyed, and the port inner engine mixture lever was shot away. External damage had been inflicted on the elevators, the rudder and the No.7 gas tank, which was holed and leaking fuel into the fuselage.

As F/S Russell nursed the crippled bomber through the hostile sky, the still unconscious navigator's head wound was attended to. Although morphine was administered to both of the seriously injured crew members, the mid-upper gunner remained awake and in reasonable spirits throughout the homeward journey. He was not, at this stage, aware of the gravity of his wound. His over-riding priority was to quench his insatiable thirst, no doubt brought about by the effects of the morphine. In endeavoring to achieve his goal, Reg Rose drank the crew's complete ration of orange juice and coffee.

Having ascertained the aircraft was going to remain airworthy, F/S Russell instructed the bomb aimer to obtain the navigator's maps and report the courses for home indicated thereon. Unfortunately the charts had soaked up the blood from the navigator's head wound, thereby leaving Leslie Bentley no option but to report back to his pilot that the maps were virtually unreadable. An S.O.S was transmitted by the wireless operator, who was able to provide a fix for their position. He also issued a request for a flarepath landing on the nearest available airfield. In an effort to get an accurate bearing, "darkie", the emergency radio system which helped lost and battle damaged bombers get home was also called. A very faint signal was picked up from RAF Detling, a Fighter Command airfield, 3.5 miles north-east of Maidstone in Kent. However, a stronger signal was received from West Malling, another Fighter Command Station, five miles west of Maidstone. Russell decided to head for the latter.

The journey home had not been without its moments of anguish, as the aircraft had developed a tendency to enter a stall at the slightest relaxation on the controls by the pilot.

Russell found the only way to keep the bomber in level flight was to haul back on the control column, a feat which required a great deal of exertion and strength. To overcome the problem he took the strain through his legs and knees, but tiredness soon entered the equation and the task became a two man job; requiring the combined strength of both F/S Russell and Sgt Bentley. It did not help matters when the flight engineer found the No.5 tank control, and the "S" gear supercharger controls for the port inner engine, had been shot away.

As the English coastline passed underneath the stricken bomber another worry entered the thoughts of the pilot, the condition of the landing gear. He did not know whether it would lower and remain stable. Although the undercarriage mechanism appeared to respond to the controls when operated, the pilot had no way of knowing if the undercarriage had locked into position. Fearing the gear might collapse on contact with the ground, Russell warned his crew to take-up crash positions. Unable to adjust the revs on the port engines, and with only 1/3 use of flaps, he brought the aircraft in for a safe landing.

The moment the aircraft touched-down, emergency service vehicles swung into action and chased the bomber along the runway. Both the wounded crew members received immediate medical attention before being removed from the bomber, prior to being taken to hospital. Obviously, surgeons could do nothing to save the air gunner's upper limb, and four days were to pass before Reg Rose realised he had lost his right arm.

Prior to being posted to No.XV Squadron, Reg Rose had acquired a rabbits foot on a chain. He wore this amulet during his service, in the hope it would bring him luck. Although it could be said luck deserted him on that particular night, it could equally be argued the loss of his arm ultimately saved his life.

For his action in bringing the stricken bomber home, F/S Joseph Vincent Russell, RCAF, was awarded the Conspicuous Gallantry Medal (Flying). For their respective parts in the operation, Sgt Leslie Bentley and Sgt Raymond Hayles were both awarded Distinguished Flying Medals.

Pilot Officer Midgley participated in this raid, piloting his replacement aircraft for the first time. The new Stirling wore the serial EF133, and was again coded LS-U!

Apart from an attack against Bremen on the night of 8th/9th October, only a few minor mine-laying operations were detailed for No.XV Squadron during October. It was during one of these mine-laying missions that F/S Frank Watson, wireless operator on Stirling bomber, BK818, LS-O, encountered St Elmo's fire for the first time. The aircraft, under the command of P/O Woodley, had taken-off from Mildenhall at 17.35 hours on the night of 2nd October, for a 'Gardening' sortie in the Skagerrak. During the flight, which lasted over seven hours, BK818 was caught in an electrical storm and St Elmo's fire danced all over the aircraft. Blue 'flames' six feet high, engulfed the bomber, as lightning struck the trailing aerial. An electrical currant ran up the aerial and burned out the winding handle. On his arrival back at Mildenhall, Frank Watson informed the groundcrew chief that the aerial had been lost. The chief immediately adopted the procedure of fining the wireless operator for the loss of an aerial, the property of His Majesty, King George VI. Indignantly, F/S Watson explained the full story to the crew chief, and invited the latter to inspect the damage. Although the disgruntled crew chief stated he would still have to report the episode, Frank Watson never heard another word about the incident; neither did he pay the fine.

Because the Squadron was soon to convert to Lancaster bombers, the pattern of undertaking only minor mine-laying operations followed into the first half of the following month. On 15th November, as part of the preparations for receiving the new aircraft, F/L R Lown, D.F.C., and the recently commissioned P/O Russell, C.G.M., were selected to attend the Lancaster conversion course at R.A.F. Swinderby, Lincolnshire.

The Squadron detailed twelve Stirlings for an attack against Mannheim, on the night of 18th/19th November. One aircraft and its crew were lost when Stirling bomber, BK707, LS-G, failed to return. The aircraft, piloted by F/S James Calder, RNZAF, crashed at Souain, France.

Two aircraft returned early, one due to the mid-upper gunner being taken ill, the other due to failure of the starboard outer engine. The latter aircraft, Stirling bomber, BK818, which now wore the code LS-R, was piloted by Sgt Oliver Brooks. This 'rookie' pilot, who had joined the Squadron on 1st October, was later to be decorated in the field for his courage.

The following night nine Stirlings took-off for an attack against Leverkusen, north of Cologne. One of the bombers, EF460, LS-N, was piloted by S/L Megginson who was flying the last mission of his second tour. When he landed at RAF Mildenhall, at 21.46 hours, S/L Robert Megginson, D.F.C., D.F.M., was declared 'tour expired'. He was posted to No.1653 Conversion Unit, Chedburgh, ten days later. His crew were disbanded and posted to various training units where they would impart their knowledge and experience to others. Mick Cullen, who had been commissioned on 12th October, went to No.17 Operational Training Unit. However, the crew were re-united on 7th August 1944, when Mick and Brenda married at St John's Church, Beck Row. The R.A.F. Chaplain officiated at the wedding, which was attended by over 100 guests.

The final attack of the month, against a major German target, occurred on the night of 22nd/23rd November, when Bomber Command carried out a raid on Berlin. A total of seven hundred and sixty-four aircraft participated, including one from No.XV Squadron. Stirling bomber, EF177, LS-S, piloted by Sgt Oliver Brooks, was the sole aircraft detailed by the Squadron for the attack. The mission was carried out without incident, and Sgt Brooks returned safely to Mildenhall.

During the last few days of November, six crews undertook mine-laying operations in the Frisian Islands, off the Cherbourg Coast and in the Gironde River. It was on a mission to the latter mining area that Pilot Officer Joseph Russell and his crew took McDonald Hastings, a British war correspondent, on 30th November. Also flying with Russell's crew on the operation, as wireless operator and bomb aimer respectively, were Mick Cullen and Andy Haydon from Megginson's crew. These missions heralded the end of the Stirling era for No.XV Squadron.

Unfortunately, December was to herald the end of another era when Pilot Officer Bill Prune, the Squadron's bulldog mascot, was killed whilst participating in one of his favourite pastimes. Prune, who had been with the Squadron since May 1942, enjoyed nothing more than to chase motorcycles and sink his teeth into the rear tyre of the machines. Usually the rider in trying to take evasive action to avoid hurting the bulldog, would come off worse when he inevitably fell from his machine. Unfortunately, one day in December 1943, this emblem of British spirit, tried this trick with a British Army truck which did not take evasive action, and Bill Prune paid the supreme sacrifice. He was buried on the airfield at Mildenhall, in front of one of the hangers, where his friends could visit him before going on 'ops'.

Madame X's Gentlemen

Sergeant Jack Bailey, RCAF (right) and Sgt George Wright, were bomb aimer and wireless operator respectively to Gil Marsh's crew on Stirling BK816, named 'Madame X'. In August 1943, they became part of No.622 Squadron, when the new unit was formed from "C" Flight of No.XV Squadron. Sergeant Bailey was granted the immediate award of a Conspicuous Gallantry Medal, having piloted 'Madame X' back to Mildenhall, from an attack on Berlin, on 23 August 43, during which the pilot was severely wounded in action. *Courtesy of Gil Marsh*

The men who pampered 'Madame X' and kept the aircraft in good order – the groundcrew – pose for a picture by the aircrew entry door of their Stirling. *Courtesy of Gil Marsh*

Guest of the Reich

Right: Prisoner of War No.222454 F/S Arthur 'Ray' Cole, photographed by his captors, prior to his internment at Stalag 4B, in 1943. Ray Cole was promoted to the temporary rank of Warrant Officer, on 09 August 44, whilst still in captivity. *Author's Collection via Mary Evelyn McCarten*

Midgley's Flying Circus

The crew of Stirling, EE912, LS-U, named 'Midgeley's Flying Circus' watch intently, as war artist Dame Laura Knight adorns the fuselage with the painting of a clown surmounted by the aircraft's name. *Courtesy of and via Vincent Gowland*

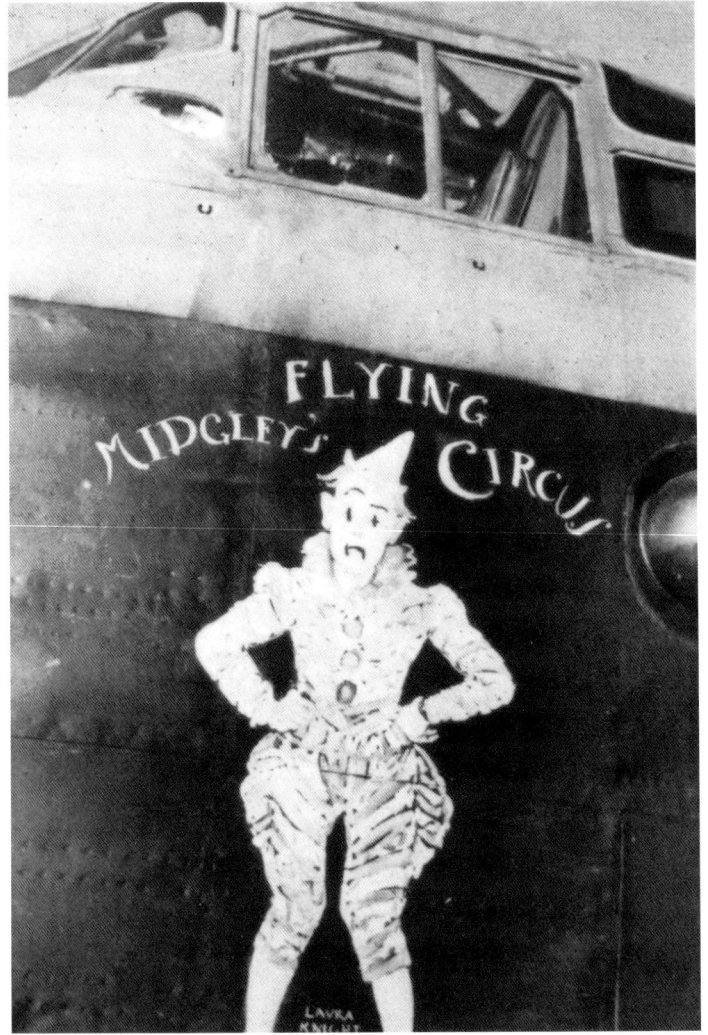

War artist Dame Laura Knight takes time out from her painting to inspect an 8,000lb bomb, which was about to be loaded into the bomb bay of Stirling, EE912, LS-U. The original photograph carried the legend *"One of a clutch for Adolf – Laura Knight, 15.10.43"*. *Courtesy of and via Vincent Gowland*

Close up detail of the artwork painted by Dame Laura Knight, on Stirling bomber, EE912, LS-U. The aircraft, which was delivered to No.XV Squadron on 29 June 1943, failed to return from an attack against Berlin approximately two months later, on 31 August. The Stirling was flying its 13th operational mission. *Courtesy of and via Vincent Gowland*

The crew of 'Midgeley's Flying Circus' included: Sgt Midgeley, pilot; Sgt Ellis, nav; Sgt Rhodes, B/A; Sgt Fenner, W/Op; Sgt McCool, MuG; Sgt Vincent Gowland, R/G; Sgt Knight, F/E. *Courtesy of and via Vincent Gowland*

The Biter and the Biten

Stirling bomber Mk.III, EF427, photographed at Rochester, Kent, before being delivered to No.XV Squadron at Mildenhall, where it became LS-A 'Apple'. The aircraft was shot down on the night of 30/31 July 1943, during an attack against Remscheid. Four members of the crew, including the pilot P/O George Judd, were killed in the ensuing crash. *Courtesy of and via Doug Fry*

The sting in the tail. The rear turret of Stirling bomber, EF427, LS-A. Unfortunately, the guns operated by P/O Ken Banks, RCAF, were of little use the night the aircraft was shot down, as it was a victim of flak. *Courtesy of and via Doug Fry*

Photographed in full flying kit, at Air Gunnery School, Dalcross, Invernesss, Scotland, whilst still under training, 19 year old Doug Fry exudes an air of experience. He later became mid-upper gunner to P/O George Judd's crew. Although wounded, he was to survive being shot down on the night of 30/31 July 1943. *Courtesy of and via Doug Fry*

Whilst Sgt Doug Fry (center) was talking to P/O George Judd (left) on the airfield at Mildenhall, prior to the operation against Remscheid, a film unit meandered amongst the participating aircrews with cameras rolling. Within hours of the occasion being recorded, George Judd had been killed in action and Doug Fry had been made a prisoner of war. The latter's mother was not sure at the time whether her son was dead or alive, as he had been posted as 'missing'. She was to see this image of her son on a British cinema screen, before learning officially of his survival. *Courtesy of and via Doug Fry*

Sparrow with 'clipped wings'

Sergeant John Sparrow was shot down by a Me 110, during his third operational mission with No.XV Squadron, whilst returning from an attack against Nuremburg on 11 August 1943. He parachuted to safety and landed at Heer in Belgium, where he was eventually assisted by of the local resistance unit. He was later captured along with other members of Allied aircrew, having been betrayed to the Germans. Following internment at Fresnes Prison in Paris, he was sent to Frankfurt for interrogation, prior to being transported to Stalag Luft V1 at Heydekrug. During April 1945, John Sparrow managed to escape from a column of PoWs being marched away from the advancing Allies. He met a British tank crew who assisted him and, with whom he stayed for two days. Eventually he was repatriated to England. *Author's Collection via the late John Sparrow*

Off-Duty

Members of No.XV Squadron groundcrew unwittingly emphasis the size of the mighty Stirling bomber, whilst using the aircraft for an unconventional group photograph. The aircraft used for the occasion was, BK818, LS-0, the usual mount of S/L Megginson. *Courtesy of 'Mick' Cullen*

One of the many pastimes when off-duty was a drink at a local 'watering hole'. Squadron Leader Megginson and his crew enjoy some liquid refreshment 'al-fresco' style at an unidentified English pub. The crew comprises, from left to right: P/O D. Mitchell, (R/G); F/O Andy Haydon, (B/A); Sgt E. 'Geordie Soulsby, (F/E); P/O H. 'Bunny' Burrows, (Nav); Sgt G. Mick Cullen (W/Op). Kneeling at the front are S/L Robert Megginson, (pil); and Sgt P. 'Joe' Woollard (MUG). Desmond Mitchell was killed in action on the night of 23rd/24th August 1943, following an attack by an enemy nightfighter, during a raid against Berlin. *Courtesy of G. 'Mick' Cullen*

The Antipodean Air Gunner

A fine study of Sgt Mick Cullen, in full flying kit, at his station on the Stirling bomber. The photograph is reputed to have been taken during a mine-laying operation. *Courtesy and via G. 'Mick' Cullen*

The Event of the Year

The local event of the year at Mildenhall, occurred on 7th August 1944, when P/O Graham Cullen, RNZAF, married Brenda Jaggard, at St. John's Church, Beck Row. The wedding, which was attended by over one hundred guests, was officiated by the Air Force Padre. Flying Officer Andy 'Butch' Haydon, an ex-New Zealand All Black rugby player, who endeavored to embrace the bridegroom in an armlock during the taking of the wedding photographs, acted as Best Man. *Courtesy of Brenda Cullen*

A formal portrait photograph of Graham 'Mick' Cullen who, on completion of his 'tour of duty' with No.XV Squadron, was commissioned to the rank of Pilot Officer. Note the New Zealand shoulder flash at the top of his tunic sleeve. *Courtesy of G. 'Mick' Cullen*

Flying Officer Andy Haydon (holding the ball), photographed with the Station Rugby Team, at RAF Mildenhall. Standing to the extreme left of the photograph is S/L Archie Marsdon, the Senior Intelligence Officer at RAF Mildenhall. *Courtesy of and via Ken Lewis, DFC*

'Spiritual Guidance'

The Very Rev. H A Jones, B.Sc., Provost of Leicester Cathedral, chats informally to members of aircrew, at RAF Mildenhall, during September 1943. *Courtesy of and via Flt/Lt T. N. Harris, XV (R) Squadron*

Having chatted to the aircrews, the Very Rev. H A Jones accepted an invitation to inspect the interior of Stirling bomber, BK654, LS-W. This aircraft, a mark III variant, was at this time the usual mount of Sgt Anderton and his crew. *Courtesy of and via Flt/Lt T. N. Harris XV (R) Squadron*

Four from the Commonwealth

Four members of the same crew unknowingly represent the allegiance of the Commonwealth countries. They are, from left to right: F/S Gordon Wright, RCAF, B/A; F/S Jack Curtis, RNZAF, Nav: F/S William Highland, A/G, RAAF, and F/S Alex Wood, RAFVR, Pilot. Their aircraft was shot down on the night of 2nd/3rd October 1943, during an attack against Kassel. F/S Highland was killed when the Stirling crashed at Haste, near Osnabruck. F/S Wright and F/S Curtis were captured and sent to Stalag 4B whilst F/S Wood, who was also captured, was sent to Stalag L3. *Courtesy of and via Steve Smith*

F/L Woodley and Crew

A crew of mixed fortunes, and one which never completed a 'tour of ops' together, was that headed F/O Charles Woodley. When the latter was posted away the crew split up, leaving each individual to join another crew. Left to right in the photograph are: Sgt Frank Watson, W/Op; F/O Charles Woodley, pilot; F/S Stanley Nystrom, A/G; F/O "Art" Cantrell, RCAF, nav; Sgt Alf Beazley-Long, F/E; F/S Millard, B/A; and Sgt Frank Tyler, A/G. Frank Watson and 'Art' Cantrell both joined F/L Dengate's crew and completed a tour. Charles Woodley returned to No.XV, but was K.I.A. on 8th June 1944. Alf Beazley-Long joined F/L Miller's crew and was shot down on 28th April 1944. He was one of three crew members to survive the action. Frank Tyler was K.I.A on 8th May 1944. *Courtesy of Frank Watson*

The Mid-Upper and the Bomb Aimer

Left: A poor quality photograph of Sergeant Reg Rose, who lost his right arm, as a result of action with an enemy fighter on the night of 3rd/4th October 1943, whilst flying as mid-upper gunner with F/S Russell and his crew. *Courtesy of Reg Rose*
Right: A poor quality photograph of Sgt Leslie Bentley, who was slightly wounded during the altercation with an enemy nightfighter, during which Sgt Reg Rose lost his right arm. Whilst other members of the crew attended to Sgt Rose, Leslie Bentley assisted the pilot in keeping control of the badly damaged Stirling. For his courage and devotion to duty, on the night of 3rd/4th October 1943, Sgt Bentley was awarded a Distinguished Flying Medal. *Courtesy of Leslie Bentley, DFM*

A Squadron of One

Left: On the night of 22nd/23rd November 1943, Sergeant Oliver Brooks and his crew were the only members of No.XV Squadron to attack Berlin. Their aircraft, Stirling bomber, EF177, LS-S, joined 764 other Bomber Command aircraft for the raid. They were in effect, 'a squadron of one'. *Courtesy of Oliver Brooks, DFC*
Right: Study of a Stirling bomber at rest. The aircraft, flown by Sergeant Oliver Brooks, was photographed at RAF Mildenhall, during the later part of 1943. *Courtesy of Oliver Brooks, DFC*

Reporting on the War

McDonald Hasting, the British war correspondent, climbs aboard an unidentified Stirling bomber of No.XV Squadron prior to taking-off on an operation mission. *Author's Collection via G. 'Mick' Cullen*

Having returned safely from the mission on which they were accompanied by McDonald Hastings, the crew of P/O Russell attend a debriefing. From left to right (clockwise round the table, with nearest the camera) are: Unidentified; P/O 'Mick' Cullen; Section Officer Grace 'Archie' Archer, WAAF; McDonald Hastings; Unknown (possibly F/O A Haydon, RNZAF); P/O Joseph Russell, RCAF; Unidentified. The picture is thought to have been taken during the early hours of the morning of 1st December 1943, on return from a mine-laying operation in the Gironde. 'Mick' Cullen and Andy Haydon, both originally part of S/L Megginson's crew, flew only this one mission with P/O Russell. *Courtesy of and via Leslie Bentlet, DFM*

Pilot Officer William 'Bill' Prune

Pilot Officer William 'Bill' Prune, the Bulldog mascot of No.XV Squadron. 'Bill', who was born on 23rd May 1938, is rumored to have flown on operations with various members of aircrew. However, nobody has admitted to engaging in this practice. *Courtesy of Don Clarke, MBE, The Mildenhall Register*

'Bill' Prune's favorite pastime was to chase motorcycles and attempt to bite the tyres of the selected machine. This practice usually resulted in the dislodgment of the rider who, in an effort to avoid hitting the dog, fell from his machine. Unfortunately, one day in December 1943, 'Bill' attempted to up-grade to an army lorry. There was no contest and Bill paid the supreme sacrifice. He was laid to rest on the airfield, near the dispersal areas, where members of the Squadron could visit his grave. *Courtesy of Don Clarke, MBE, The Mildenhall Register*

15

The Lancaster Era

No.XV was stood-down, as an operational squadron, during December 1943 and January 1944 whilst it undertook conversion and training flights with its new aircraft. Unfortunately, however, these flights were not without risk as was illustrated on 13th January when No.XV recorded the first loss of one of its Lancaster bombers. The aircraft, Lancaster Mk.III, ED826, LS-W, was detailed for a loaded climb and cross-country training flight, from which it failed to return. The weather was far from ideal for the exercise, with 10/10th cloud at 1,500'. Carrying a crew of six, the Lancaster, piloted by F/S Walter Houston, RCAF, was last seen going down in a steep dive, with thick smoke trailing from it. The aircraft crashed into the Wash, off the north Norfolk coast at 14.50 hours, killing all those on board. A Court of Enquiry later concluded the pilot lost control of the machine, following an engine failure or fire.

During this same period, a number of escape and survival exercises for members of aircrew were also carried out. Sergeant Dye, an air gunner who had joined the Squadron in December 1943, found himself on one of the former exercises. The procedure was for the 'escapees' to be taken by lorry to a location some ten miles from the base. The participants then had to make their way back to the airfield, without being caught by the police. Bernard Dye teamed-up with Joe Hayes, another sergeant air gunner from the same crew, with whom he would conspire to get back to Mildenhall. As they were walking down a lonely country lane, they became aware of a figure in uniform on a bicycle riding towards them. Without a word to each other the two 'escapees' dived in opposite directions into the hedgerows on either side of the lane. The figure, who turned out to be a mailman, stopped his bicycle near the two men, dismounted, and started to answer a call of nature. As the mailman was in mid-flow, Joe Hayes indicated to Bernard Dye that they should light a couple of the 'thunderflashes' they had in their pockets. On an agreed signal, the two airmen tossed the explosives into the road. They landed fairly close to the unsuspecting figure, and went off with a resounding bang, frightening the life out of the poor mailman. He did not wait to finish his business, but grabbed his mailbag, jumped astride his bicycle and made off down the lane as fast as he could peddle. The two pranksters emerged from their respective hiding places and, full of merriment, continued with their mission.

Although No.XV had taken delivery of its new aircraft during December 1943, the Lancaster era did not commence operationally until the night of 14th/15th January, when the Squadron detailed twelve aircraft to attack Brunswick. However, in the event, only nine of the twelve Lancasters attacked the target, due to the fact one failed to take-off and two returned early.

On their return to base after the attack, some of the pilots, including Len Miller who was flying his first mission since being commissioned, reported a very quiet trip. However, it was not the same for all the crews. Doug Boards, who had returned from an eventful trip over to Hamburg on 27th July 1943, whilst flying as second 'dicky' to F/O R Waughs, was one of the latter.

Pilot Officer Boards, D.F.M., was holding Lancaster, W4355, LS-A, steady on its bombing run when it was attacked by a Bf110 nightfighter. The enemy aircraft inflicted damage upon the Lancaster, but F/S John Angus the Canadian rear gunner, replied with an accurate burst of machine-gun fire which drove off the attacker. Free from further interruption the crew of LS-A resumed their task and bombed the target through cloud, using the marker flares as an aiming point. His mission completed Doug Boards turned the Lancaster onto a heading for home, but another enemy nightfighter pilot had other ideas. For the second time within fifteen minutes, F/S Angus found himself defending the bomber. Coolly and calmly he watched as the enemy aircraft, which he had identified as a Ju.88, crept ever closer. Suddenly the night sky behind the Lancaster lit up in a multi-coloured blaze of light, as fire spat from its rear turret. The light increased in intensity as the German aircraft burst into flames, peeled over and fell to earth. The streaming tongue of flame trailing out behind the stricken aircraft, indicated its path to destruction. Its impact with the ground was marked by an explosion. The Lancaster returned safely to Mildenhall, where it landed at 20.55 hours.

The courage displayed by F/S John Angus, which ultimately saved both the aircraft and its crew, was recognised by the award of a Distinguished Flying Medal gazetted a month after the action.

Six days later, on 20th January, after further training had been undertaken, the Squadron detailed twelve aircraft for an attack against Berlin. Although all the aircraft took-off, four were forced to return early due to various instrument and mechanical defects. The remaining eight machines continued with the mission, completed their primary task, and returned safely to Mildenhall.

Flight Sergeant Robert Butler and his crew were lost without trace, having failed to return from the first major attack against Magdeburg, on 21st January. Ten No.XV Squadron aircraft had been detailed to join a main force of 648 bombers, but four of the ten allotted machines were withdrawn from the battle order during the course of the day.

The five crews who returned to Mildenhall, following the attack, all reported good results with little opposition. The crew led by F/L Moore, RAAF, also reported that their aircraft, Lancaster, LM441, LS-T, had sustained flak holes to the bomb bay. The crew could not recall having had any close encounters with a flak barrage and, as the damage had only been discovered after the aircraft had landed, they were unable to state where or when it had been inflicted.

An attack against Berlin on the night of 27th/28th January was an inauspicious one for No.XV Squadron. Initially, fourteen aircraft and crews were listed on the battle order for the operation, but during the course of the day one aircraft and its crew were removed from the list. Six of the participating bombers returned early due, five of them with mechanical defects and one with instrument failure. To add to the Squadron's misery, another of its aircraft failed to return, with the loss of P/O George Clarke and W/O Godfrey White, RCAF, the pilot and bomb aimer respectively. The rest of the crew of Lancaster bomber, ED323, LS-D, survived and were made prisoners of war.

Berlin received further attention from Bomber Command the following night, when 677 bombers attacked the City. No.XV Squadron detailed eleven aircraft for the operation but three were later canceled. A taxying accident, prior to take-off, prevented at least one of the latter from getting any further than a few yards from its dispersal area. Lancaster, W4355, LS-A, was preparing to taxi out with A/F/L Doug Boards at the controls. On receipt of the appropriate signal, flashed by torches held by members of the groundcrew, the pilot released the brakes and W4355 rolled forward into the darkness. A few minutes after midnight the Lancaster was hit by a similar aircraft carrying out the same maneuver. The incident invoked a review of taxying procedures, to ensure a similar accident did not re-occur.

Lancaster bomber, R5904, LS-G, which was usually flown by an Australian pilot, had been giving cause for concern. Various inspections could find no reason for the numerous malfunctions incurred by the aircraft. During the day, prior to the attack on Berlin, the groundcrews worked feverishly on the bomber to ensure it was fit for the forthcoming mission. During the course of the overhaul the armoured glazed screen, which protected the flight engineer, was removed and not replaced. The consequences of this action were dire.

The aircraft was allocated to P/O Miller who, knowing of the aircraft's problems, took-off at 00.20 hours. Shortly after crossing the east coast of England, the majority of instruments failed. As he still had use of the compass, altimeter and turn and bank indicator, Len Miller decided to continue with the mission.

Although Berlin was shrouded in a thin layer of cloud, P/O Miller had no difficulty seeing the target indicators marking the target ahead. He was holding R5904 steady at 20,000', ready for the bombing run, when he spotted a Ju.88 nightfighter turning in for a low frontal attack. As he yelled a warning to the gunners, the enemy aircraft opened up with bursts of machine gun and cannon fire. The enemy's fire overshot, as P/O Miller threw the bomber into a side-slipping bank. Intent on a kill, the Ju.88 broke away to starboard prior to turning in for a second attack. When it came, from the starboard front quarter, cannon fire raked the mainplane and engines, but fortunately did not put the latter out of action. As the Ju.88 continued on its chosen path, cannon shells and machine gun bullets slammed into the flight deck and peppered the entire length of the fuselage. Alf Pybus, the flight engineer, unprotected due to the removal of the armoured glazed screen, collapsed to the floor having been struck in the head by a bullet. The nightfighter's flightpath took it to the tail end of the Lancaster, where the rear gunner was waiting with four Browning machine guns. As the enemy aircraft emerged into view Sgt Peter Slater opened fire with a long sweeping burst, striking the Ju.88 in the cockpit canopy and engines. As the nightfighter reared up, peeled over and fell away, Wilbert Cully yelled a warning of another attack from a second Ju.88. From his vantage point in the mid-upper turret, Sgt Cully had spotted the new aggressor about to make a head-on attack. Tracer streaked across the night sky, as the German pilot and the mid-upper gunner opened fire simultaneously. Much to the astonishment of the bomber's crew, the nightfighter broke-off the attack and dived away to port side of the Lancaster. However, it was not long before Peter Slater warned his pilot of the presence of a single -engined aircraft, which the former had seen approaching from astern. Having identified it as a FW.190, the rear gunner fired-off a long warning burst. Realising his quarry was alert to his presence, the enemy pilot turned his aircraft away and dived into the darkness without having fired a shot.

The gunners remained in their respective turrets whilst other members of the crew attended to the flight engineer. Although Alf Pybus had shown no signs of life since the initial attack, he was put on oxygen and covered with blankets to keep him warm.

Mindful of the condition of his aircraft, and the plight of his flight engineer, Len Miller made the smoothest landing he had ever made. Unfortunately Alf Pybus would never know of his pilot's achievement, he was already dead.

The aircraft continued to roll down the runway until, without warning, the port undercarriage leg collapsed shredding the tyre in the process. A shower of sparks flew through the air as the port wing dropped and the propeller blades buckled, having made contact with the concrete runway. The Lancaster pivoted off the runway and came to rest on the grass, leaving a great furrow in the earth.

Sergeant Alfred Pybus, RAFVR, was buried at Newcastle-Upon-Tyne (All Saints) Cemetery. His funeral was attended by P/O Miller and the rest of the crew, all of whom had been granted special leave. It was a sad occasion for them all and, as they stood at the graveside, many memories came flooding back.

They had first formed together as a crew at No.26 Operational Training Unit at Wing, in Bedfordshire, during early 1943. They soon formed a bond and became a close knit team who lived, worked and played together. Len Miller remembered how the seven of them would, on warm summer evenings, ride to suitable pubs on two bicycles tied together. One bike carried four people, whilst the other carried three. This was achieved by one person sitting on the handlebars, one standing on the peddles providing the means of motion, one sitting on the saddle and one balancing on the rear wheel wingnuts. In this manner the crew, Alf Pybus included, would hurtle around the countryside in search of a few beers.

Upon their return to Mildenhall, Len Miller's crew found a replacement aircraft waiting for them. It was a Mark.I Lancaster, built by Armstrong Whitworth, powered by four Rolls Royce Merlin 24 engines. It wore the serial LL801, and was coded LS-J 'Jig'. The aircraft had no H-2S, but had the advantage of speed over the Squadron's other aircraft.

Although they had experience as a Lancaster crew, Len Miller dictated that they all got to know the new aircraft intimately, and frequent exercises and air tests were undertaken to accomplish this. However, the groundcrew of LS-J would often wait in trepidation for the aircraft to return from the latter, especially if F/O Billy Goat had gone along for the ride. The above named was not an officer of the Squadron but a goat or kid, which Len Miller had acquired after a party. Although he did not fly on operations, Billy Goat often flew on air tests, and it was during one of these flights he discovered his liking for *'Empire'* tape which was used to bind cables etc. If he had flown on an air test, the groundcrew knew they were in for additional repairs after the aircraft had landed.

On one occasion F/O Billy Goat munched his way through a Lancaster maintenance manual, which some people said made him the most 'genned-up' goat on the Lancaster in the Royal Air Force!

Berlin was attacked again on the night of 30th/31 January. No.XV Squadron despatched eight aircraft, all of which returned to base without incident.

No operations were undertaken by No.XV during the first two weeks of February. Instead, the Squadron utilised the time with a number of flying exercises. These ranged from local flying, air tests, night training exercises, loaded climbs and cross-country exercises. During this period, F/O Woodley was rested and posted to No.1483 "B" Flight Supernumary.

The first operation, following the two week stand-down, occurred on the night of 15th/16th February when Berlin was attacked by a major force of eight hundred and ninety-one aircraft. Eleven of the five hundred and sixty-one Lancasters which participated in the raid, were from No.XV Squadron. Fourteen machines had originally been detailed, but one was canceled during the day, one caught fire on take-off and one returned early. The machine which caught fire was Lancaster, ED310, LS-M, piloted by S/L Lown. The aircraft, which was still in the aerodrome circuit, was climbing away from the airfield when the port inner engine burst into flames. Squadron Leader Lown, having no opportunity to jettison his cargo, landed immediately with the bomb load still in the bomb bay. The aircraft which returned early had managed to take-off without incident, but during the outbound flight, the rear turret on Lancaster, LM456, LS-C, went unserviceable. Not wishing to leave his crew without the protection of any form of firepower at the rear of the aircraft, the pilot, P/O Alan Amies, turned his aircraft for home.

Unfortunately the Squadron lost an aircraft and crew on this raid, when Lancaster, ED628, LS-O, failed to return. The machine crashed into the Baltic, near the island of Hiddensee, off the coast of Rugen. The entire crew perished, including the pilot, F/L William Harris, RNZAF, whose body was found on the island seventy-two days later. Although a total of forty-six aircraft were lost on this raid, the remaining No.XV crews all completed their tasks and returned safely to base.

Amongst the crews who returned safely to Mildenhall, following the Berlin raid, was the one headed by P/O Fisher. Their safe return to Mildenhall saw the successful completion of their first operational mission, as a Lancaster crew with No.XV Squadron. However, for the majority of the crew, their initiation into night bombing operations could so easily have been their first, and last, operational flight. Their aircraft was heading for the target at an altitude of 20,000', when the rear gunner gave warning of an incoming enemy fighter. The twin-engined machine, thought to have been a Do.217, swung into the attack from the starboard quarter low. It was less than 400 yards away when the rear gunner fired a short burst, which caused the enemy fighter to break-away without firing a shot. Repositioning itself for a second attack, the Do.217 again came in from the starboard quarter. As the range between the fighter and the bomber decreased, both the mid-upper and rear gunners of the Lancaster opened fire. Both saw their fire strike the hostile aircraft, which peeled over and dived away. The fact it was last seen going down through cloud, enabled the gunners to later claim the nightfighter as damaged.

Although this crew had been formed at O.T.U, and posted to No.XV at the end of 1943, the man who guarded their tail, Sgt George Allen, was a very experienced member of aircrew and no stranger to Mildenhall. Furthermore, the quiet, unassuming nature of this rear gunner, belied his courage and devotion to duty. George Allan had volunteered for aircrew training at Mill Hill, North London, during the summer of 1940. Following training, as a wireless operator/air gunner at Blackpool, he was posted in late 1941 to No.149 Squadron who were, at that time, based at Mildenhall. Against all the odds he survived a tour of approximately thirty missions, all of which were carried out on twin-engined Vickers Wellington bombers. When he became 'tour-expired', George Allen was posted to the Middle East where he joined No.40 Squadron who also flew Wellington bombers. Compared to No.149 Squadron the new unit was a complete contrast, being part of the Desert Air Force which flew in support of the British Eighth Army in North Africa. It was during this period of his operational career, that George Allen's devotion to duty was recognised by a Mention-in-Despatches. This award was signified to others by the wearing of a small bronze oak leaf, sewn on to a tunic or 'battledress' blouse under the aircrew brevet, on an appropriate medal ribbon. On completion of his second tour, Sergeant Allen returned to England to take up a posting at an O.T.U, where he spent six months as an instructor. Another move occurred during the summer of 1943 which, due to a flying accident during training, was very nearly his last posting. The incident occurred at Ossington, Nottinghamshire, where George Allen had crewed up with Stan Fisher in preparation for a third tour of operations. Taking-off one night in a Wellington bomber with his new crew, the aircraft's port outer engine cut out. The pilot was unable to retain control of the machine, and it crashed into a field sown with potatoes. Upon impact with the ground the airframe broke in two, propelling the fuselage across the field. Large amounts of the crop were scooped up as it skidded along and, by the time the fuselage came to rest, George Allen's turret was well and truly covered with earth and mashed potatoes. Initially, the air gunner experienced problems extricating himself from the wreckage but, he had survived two tours and an air crash, so he was not going to be beaten by a crop of 'spuds'. Fortunately, the crew survived to undertake and complete their conversion to the four-engined Stirling bomber. On completion of this course Sgt Fisher and his crew were sent to No.XV Squadron which, for George Allen, meant a return to his former base. However, on arrival at Mildenhall, the crew were informed the Squad-

ron was converting to Lancaster bombers, and they would be required to undertake further training. In early February, following a successful completion of the Lancaster course, Sgt Fisher and his crew were declared combat ready. Their first mission, from which they successfully returned, was the one against Berlin on the night of 15th/16th February. For this attack they were allocated Lancaster, L7566, LS-E, an aircraft which Stan Fisher was to fly on a number of occasions. The bomber, one of the first two such aircraft to be blooded on a bombing operation, first saw service with No.44 Squadron. Because of its age, this particular machine was slower than the later Lancasters and rarely managed to attain the designated operational altitude. Initially, the crew found this to be an advantage as, during a mission, they would be flying well below the main bomber stream and were therefore able to evade the German flak and nightfighters. With the exception of their first mission that is. They soon discovered, however, there was a major disadvantage when flying over the target area. On more than one occasion they experienced the unpleasant feeling of bombs raining down on them from above, a worrying moment for George Allen as he watched them tumble past his rear turret. Fortunately, as if by divine intervention, the cascading projectiles somehow missed the Lancaster.

Four nights after the Berlin raid, on 19th/20th February, the Squadron lost an aircraft, the crew of which included members of three Commonwealth Air Forces. There were three Australians, P/O Max Hurley, the pilot, F/S Alan Woodford, the wireless operator, and F/S Gordon McMasters, the rear gunner. Flying Officer Frederick Chalmers, a Canadian, was the bomb aimer, whilst P/O Charles Benson from New Zealand, flew as second pilot. Sergeant Hubert Moroni, the mid-upper gunner, and F/O John Fenley, the navigator, were both members of the Royal Air Force Volunteer Reserve. There was one survivor from the crew, Sgt D Frame, who was captured and made a prisoner of war. Their aircraft, R5739, LS-K, which was one of twelve machines detailed for an attack against Leipzig, crashed near Jena, approximately fifty miles south west of the target.

Pilot Officer Miller participated in the attack against Leipzig, but never reached the target. It was the first mission for his new aircraft, LS-J, and his first operation since the loss of Sgt Alf Pybus. Unfortunately, due to the failure of both turrets, Len Miller was forced to abort the mission and return to base. It was an ominous start also for Sergeant Alf Beazley-Long who, having formerly flown with F/O Woodley, joined Len Miller's crew as the replacement flight engineer.

The following night Bernard Dye's crew were detailed for their first operational mission with No.XV Squadron. Although they had joined the Squadron during the previous December, the conversion period to Lancaster bombers had delayed their operational initiation. Now their respective names were listed, as a crew, on the Battle Order for an attack against Stuttgart.

The day had progressed much the same as any other when 'ops' were scheduled. There had been the usual mission briefings, the pre-operation ritual of eggs and bacon, and the last minute checks over the aircraft. Approximately one hour before take-off, Sgt Dye and the rest of the crew were on board their aircraft checking on equipment and making last minute adjustments. When the allotted time came the pilot, Sgt Wheeler, started the four motors, finalised the cockpit checks with the flight engineer, and waited for the signal to taxi out. When it came, he increased the power to the engines, released the brakes and the bomber slowly started to move forward. However, as the Lancaster snaked its way around the taxi-track, an incident occurred amongst some members of the crew. Sitting high on the back of the Lancaster in the mid-upper turret, Bernard Dye was uncertain of what was occurring below him. He became increasingly concerned when it became apparent the aircraft was not lining-up for take-off. As the incident could not be resolved, Sgt Wheeler contacted the control tower for advice and was instructed to abort his take-off procedure. On return to dispersal, the aircraft was met by an official deputation and certain members of the crew were escorted away for questioning. Due to the severity if the incident, the crew was disbanded, a result of which Sgt Bernard Dye was posted to No.622 Squadron, the 'daughter' of No.XV, which also operated from Mildenhall. Having joined another crew, headed by F/O Arthur Horton, Sgt Dye went on to complete a tour of operations.

The attack on Stuttgart was not without cost to the Squadron as Flight Lieutenant Russell and his crew failed to return from the operation. Their aircraft, Lancaster bomber, LM456, LS-C, had completed its task and was on the homeward leg of the mission, when the flight engineer informed the pilot of a fire in the fuselage. The verbal report indicated the metal skin of the fuselage was burning and, although extinguishers were immediately pressed into service, the flames could not be controlled. As the scenario unfolded, Sergeant Bentley wondered what could have caused the fire. There had been no indication of flak activity since leaving the target area, neither had there been any shuddering impact from a strike by a bomb from above. As he monitored the situation, the bomb aimer concluded the fire was being fed by phosphorus from a target indicator. He reasoned the T.I. could have fallen from another aircraft flying above them, having previously failed to release over the target area. The fire, which could not be contained, was beginning to spread and endanger both the aircraft and its crew. From his position in the nose of the Lancaster, Sgt Bentley heard the pilot give the order to bale out. Pausing only long enough to acknowledge the command verbally, he opened the forward emergency escape hatch and tumbled out into the darkness. As he floated down from an altitude of 16,000', Leslie Bentley looked up. The fully deployed parachute canopy obscured his view, but he assumed the rest of the crew had followed him out of the stricken aircraft. In fact, only Sgt 'Darkie' Wharam, the navigator, had responded to the order to jump.

Looking up as he hung suspended in the night sky, Sgt Bentley became aware of a flickering light circling slightly above him and off to one side. He suddenly realised the object was a Lancaster bomber on fire; the same aircraft he had recently vacated. In the prevailing darkness, he was unable to see whether there were any other parachutes in the vicinity. Not being able to do anything else, the redundant bomb aimer watched as the 'flaming torch' continued to fly round in circles, losing height with each completed orbit. Without warning the aircraft erupted into a ball of fire, as it flew into the side of a hill. Lancaster bomber, LM456, LS-C, crashed in the forest at Cleebronn, approximately 30 kilometers north-north-west of Stuttgart, at 04.10 hours. The noise of the explosion woke the inhabitants of Cleebronn, who immediately attended the scene to see what assistance they could offer. Initial reports stated the pilot of LM456 was still breathing when extricated from the wreckage, but died whilst being transported to hospital.

Unaware the aircraft was still occupied, Leslie Bentley felt a tinge of sadness as he witnessed the end of his aircraft. It was not until he was locked away in the confines of a German PoW camp, that he was informed of the true facts. Flight Lieutenant Russell, C.G.M., together with four members of his crew, had been killed.

It was mooted by Leslie Bentley, that Russell would not have baled out of the aircraft had one of his crew been injured, or unable to jump through the lack of having a serviceable parachute. In such a case, it is thought the pilot would have attempted to crash-land his aircraft in order to save his crew. Unfortunately, in possibly attempting to carry out this action, the aircraft struck the ground.

Joseph Vincent Russell was a courageous and experienced pilot who had received an immediate award of a Conspicuous Gallantry Medal, together with a commission, following his exploits during a mission the previous October. Flight Sergeant Raymond Hayles, the wireless operator, one of the four to perish in the crash had been awarded a D.F.M. for the same incident, the details of which had been gazetted only six days before he died.

Another tragic accident was to account for the deaths of a further five Squadron members on the morning of 26th February as Lancaster bomber, ED383, LS-C, was attempting to land. The aircraft, returning from an operation against Augsburg, was diverted to Lakenheath due to poor weather conditions at Mildenhall. Unfortunately, conditions at the former were not much better. As the bomber approached the airfield perimeter on finals, Sgt Haydock, the flight engineer, warned the pilot the flaps would not fully extend. The response was immediate as Tony Davis, the pilot, hauled back on the control column and overshot the runway. Sgt Davis went round

again and lined-up for a second approach, but the problem re-occurred. However, before the pilot could take further action, Sgt Haydock warned there was insufficient fuel to go round again. Realising the aircraft was committed to land, Sgt Bysouth, the mid-upper gunner, decided to vacate his station and take up a crash position in the fuselage of the Lancaster; it was a decision which saved his life. The aircraft struck the runway hard and fast, slewed onto the grass and cartwheeled over onto its back. The noise was horrendous, as plexi-glass shattered and wrenching metal twisted and buckled. Flames danced around parts of the wreckage, as the mangled remains of ED383 came to rest. Warrant Officer Franklin, the wireless operator, was thrown clear of the aircraft which gyrated across the grass, leaving Harry Bysouth and 'Mick' Harbidge, the rear gunner, both trapped in the wreckage. The small dancing flames were the last thing Harry Bysouth saw before he lost consciousness. When he came to, he recognized the figure of George Franklin bending over him, endeavoring to extricate him from the wreckage. Sergeant Harbidge was less fortunate and was trapped under the rear turret. Franklin could not assist the injured rear gunner, who had to await the arrival of the emergency crews before being released. Unfortunately, Sgt William 'Mick' Harbidge, RAFVR, aged 19, died as result of his injuries. His name was added to the Roll of Honour, along with the names of the other five crew members who had been killed instantly.

Harry Bysouth spent some time recovering from the ordeal which had affected him deeply, whilst after a short rest period, George Franklin continued to fly operations with other crews.

At 05.25 hours, fifteen minutes after Sgt Davis's crash at Lakenheath, P/O Miller attempted to land at Mildenhall. Although the fuel tanks of his aircraft were almost dry, he was diverted to Bourn, approximately eight miles away. The conditions at the latter base were no better, with low cloud and torrential rain. Having received an affirmative answer to his request for permission to land, flying blind, Len Miller approached the threshold and put the Lancaster down in a sea of spray. As the bomber rolled along the runway, P/O Miller received another message from the control tower instructing him to turn right at the next exit. He acknowledged the message and made the turn at the same time; all four engines of the Lancaster dying from fuel starvation as he carried out the maneuver!

Two days after his landing at Bourn, Len Miller was promoted to the rank of Acting Flight Lieutenant, along with two of his friends. Pilot Officers' John Hebb and Alan Amies, both pilots, were promoted to the same acting rank.

Apart from an attack on Stuttgart on 1st March, when five of the seventeen aircraft detailed by No.XV abandoned the mission and returned early, no other operations were carried out by the Squadron during the first two weeks of the new month.

Pilot Officer Douglas Boards, DFM, was declared 'tour expired' and posted to RAF Feltwell on the 8th March.

Lancaster bomber, W4355, LS-A, was one of sixteen aircraft detailed by the Squadron to participate in a mission against Stuttgart, on the night of 15th/16th March; it was the only aircraft which failed to return. The Lancaster, piloted by F/L Blott, had taken-off from Mildenhall at 19.15 hours, after which nothing more was heard from its crew. The bomber was shot down on the outbound journey by a Bf110 which is thought to have fired a rocket projectile at the four-engined machine. The port inner motor caught fire and the wing was badly holed. Further damage was inflicted on the hydraulics, creating problems for the gunners in their respective turrets. When the aircraft began to fill with smoke, F/L Walter Blott gave the order to bale out. Later reports implied F/O Cedric Nabbaro, the navigator, sustained a broken leg, whilst W/O Millard, the bomb aimer, was reported to have incurred a fractured spine. Even with the reported injuries, the entire crew evaded capture and were all listed as safe in Switzerland less than two weeks after being shot down.

Flight Lieutenant Len Miller, piloting "J"-Jig, abandoned his attack against Stuttgart due to the failure of the aircraft's rear turret. He jettisoned the bomb load and returned to Mildenhall.

Three nights later, on 18th/19th March, Len Miller and his crew participated in an attack against Frankfurt. Due to the continued unserviceability of their own aircraft, they took Lancaster, ED473, LS-H, the machine usually favored by S/L Lamason. On this occasion, they completed their task without incident.

Another crew change occurred on the night of 22nd/23rd March, when Sgt Watson joined F/L Dengate's crew for an attack against Frankfurt. Frank Watson, who had previously flown with F/O Woodley, replaced Sgt Bill Hopewell as wireless operator. Like Alf Beazley-Long, with whom he had flown, Frank Watson also had cause to remember his first flight with his new crew. Their aircraft, Lancaster bomber, ED395, LS-K, was subjected to a barrage of flak which peppered the fuselage and severed the rear gunner's intercom.

An unusual and nerve-shattering experience befell A/F/L Brooks and his crew, during an operational mission towards the end of the month. Bomber Command had scheduled an attack against Berlin for the night of 24th/25th March, which was later to become known as *'the night of the strong winds'*, and Oliver Brooks had good reason to remember it. His Lancaster bomber, LL827, LS-O, was one of sixteen machines detailed by No.XV for the mission.

The outbound route to the target took LL827 north, out over Denmark. The winds, which were judged to be gusting far in excess of the forecasts given at the briefing, were blowing many of the aircraft off track. Sergeant Ken Pincott, the navigator, was kept busier than usual and constantly passed course corrections to his pilot. The Lancaster droned on through the night sky with no sign of flak or enemy fighters, although the crew remained vigilant. Suddenly, without warning, came the deafening blast of a small explosion. Oliver Brooks instinctively ducked as part of the canopy erupted and pieces of shattered plexi-glass flew into the cockpit, followed by the twisted, bloody, lifeless form of a large bird. The unfortunate creature had flown into the path of the Lancaster and been blown through the cockpit canopy, depositing its feathers and parts of it anatomy around the flight deck. Sergeant Chandler, the flight engineer, ironically known to everybody as 'Chick', had the unenviable task of clearing up the mess and disposing of the remains.

The strong winds continued to create havoc over the target, and blew the target indicators to the south-west of the aiming point. No.XV Squadron's crews, including A/F/L Brooks, bombed on the T.I's seen through the patchy, scurrying clouds. Pilot Officer Thompson, who had received his commission only the day before, courted trouble when his aircraft overshot the target area. The pilot hauled the bomber, LM465, LS-U, back on track and orbited the target, but could not see any sign of the T.I's. Flight Sergeant Lemky, the bomb aimer, therefore released the bomb load on fires already burning. As the Lancaster turned for home, it was attacked by an enemy nightfighter. Although P/O Thompson took evasive action, some of the enemy's gun fire found its target and inflicted damage to the bomber's rear turret ammunition ducts, the elevator trim and severed an oxygen tube. The enemy fighter did not continue the attack, and the Lancaster returned to Mildenhall without further incident.

Of the 811 bombers which participated in the raid, seventy-two aircraft were lost including two from No.XV Squadron. Lancaster bomber, LM441, LS-T, piloted by F/L William Grove, was shot down by flak whilst on the homeward journey, and crashed 3kms north of Bonn. The aircraft's impact with the ground was so great the airframe, which was consumed by fire, was totally destroyed. Although there were no survivors, the bodies of three of the crew were never found. Sgt Arthur Jackson, navigator, Sgt John Johnson, flight engineer, and Pilot Officer James Sills, wireless operator, have no known graves. Jimmy Sills had joined the Royal Air Force as a boy entrant at RAF Cranwell, before the war. By the time he was posted to No.XV Squadron he had already completed a tour of operations and received a Mention in Despatches. Pilot Officer Sills was killed whilst undertaking his 60th mission, the final operation of his second tour. The second machine lost that night, Lancaster, LM490, LS-L, was shot down by a nightfighter; it crashed at Tetlow, south-west of Berlin. The pilot, F/S Leslie Wheeler, whose original crew had been disbanded a month earlier

following an incident whilst taxying out for a raid on Stuttgart, was killed along with his 'new' crew. No.XV Squadron recorded the loss of fourteen members of aircrew that night. Although their aircraft was damaged following the bird strike, Pilot Officer Brooks and his crew returned safely to base.

Two nights later another seven names were added to the Roll of Honor, when P/O Thomas Marsh and his crew were lost during a raid against Essen.

The night of 30th/31st March 1944 was an infamous one in general for Bomber Command, when 795 aircraft were despatched to bomb Nuremberg. It was a bright moonlit night, and the enemy fighters were waiting over the Belgium/German border. Apart from those which returned early due to mechanical or engine defects, eighty-two bombers failed to reach the target. They were shot down by fighters or flak on the outbound journey. Another thirteen aircraft were lost either over the target, or on the return leg. Lancaster bomber, LL827, LS-O, piloted by P/O Brooks had a lucky escape as it approached the target area. The rear gunner spotted and reported the approach of a twin-engined fighter out on the rear port quarter high. The German pilot had obviously seen the bomber, as he turned in towards the four-engined aircraft and dived. The enemy aircraft passed under the Lancaster and re-appeared on the starboard side low, from where 'Chick' Chandler saw it emerge. He quickly warned the gunners, who rotated their turrets to the starboard quarter rear. The nightfighter executed a climbing turn, allowing Harry Marr in the rear turret to identify the aggressor as a Ju.88. The enemy aircraft turned in for the attack, illuminating the sky with dancing stars of deadly fire as its cannons opened-up. Oliver Brooks took immediate evasive action and dropped the starboard wing, but not before some of the illuminated stars 'danced' along the mainplane and struck the number three fuel tank. One of those deadly stars penetrated the tank and exploded but, fortunately for the crew, the tank was dry; it had not been used for this mission. Flight Sergeant Marr saw the fighter coming in and returned the fire, whereas Ron Wilson, in the mid-upper turret, had his vision obscured by the Lancasters' tailplane. As the enemy machine broke away, it was visible to both gunners who opened fire with short bursts. The Ju.88 disappeared from view and was not seen again by P/O Brooks and his crew, although the enemy nightfighter obviously went off in search of other quarry.

The total loss of ninety-five bombers during this operation was the greatest number incurred by Bomber Command during the war. Sergeant Pincott, the navigator, who had been logging each bomber as it was seen to go down advised his pilot, when the tally reached fifty, he was going to abandon the exercise.

Although No.XV Squadron had detailed thirteen Lancasters for the attack, two were later withdrawn. The remaining eleven machines took-off, completed their task and returned safely to base.

The month of April was ten days old before the Squadron recorded another operational mission. The target was railway yards at Laon, France and seventeen machines from No.XV were despatched. In general the raid was not considered particularly successful, but at least all the Squadron's aircraft returned safely.

On the 14th day of the month, John Pratt had the misfortune to witness a tragic accident which claimed the life of one of his colleagues. Corporal George Tyler, an armorer, was directing the bombing-up of a Lancaster when one of the bombs broke free from the winch. As it crashed to the ground, the bomb struck Cpl Tyler across the head, splitting his skull open. The thirty-one year old N.C.O., from Southwark, London, was buried in the churchyard at St John's Church, Beck Row, on the opposite side of the road to the airfield.

The 15th April saw a change of commanding officer for No.XV, when Acting Wing Commander W D Watkins arrived from RAF Feltwell to assumed command of the Squadron from W/C Elliott. The latter, who had occupied the position since the previous September, was posted to No.14 Base.

Inclement weather curtailed some operations during the early part of the month, permitting only training exercises to be undertaken. Sergeant

Harry Bysouth who had recently returned from leave, during which he considered his operational future, flew as a passenger with F/S Thompson and his crew on a cross-country exercise. Although Sgt Bysouth was to have a future, Carlton Thompson was not.

A raid on the night of 18th/19th April enabled No.XV to make another entry in its annals. In conjunction with No.622, its 'daughter' squadron, No.XV was to record the highest tonnage of bombs airlifted out of Mildenhall, on one night, for a single operation. The total weight of the bomb load, which was destined for the railway yards at Rouen, was recorded as 203 tons. The returning crews all reported the raid as a success, with bomb bursts seen on both railway buildings and tracks.

Although all eighteen participating crews from No.XV Squadron returned safely to base following the attack on Rouen, an ex-No.XV Squadron member was killed that same night, in tragic circumstances. Flight Lieutenant Hugh Wilkie, D.F.C., who had completed a tour of operations with the Squadron the previous July, had been posted to No.1657 C.U. His posting to Stradishall, for instructional purposes, had taken effect from 26th August 1943. On the night his old Squadron went to Rouen, Hugh Wilkie was giving flying instruction to a rookie crew. The Stirling in which they were flying, EJ108, was engaged in night circuits and bumps. As the aircraft was taking-off for the fourth time it collided with three American servicemen who were cycling across the active runway. The American soldiers and their cycles were all thrown into the air, the latter causing damage to the underside of the Stirling. A wing panel flew off, causing a dinghy pack to brake free and foul the tailplane. As the aircraft turned back towards the airfield, a loud report was heard from the port mainplane causing the Stirling to enter a steep climb. Hugh Wilkie ordered the crew to abandon the aircraft and all, with the exception of the pilot himself and flight engineer instructor, did so. Unfortunately Sergeant Atkin, the pupil flight engineer, fell out of his incorrectly fitted parachute harness and was killed. The Stirling stalled in its involuntary climb, peeled over and crashed into the ground at Little Glenham, Suffolk, killing the two men still on board. The American servicemen, who were taking a short cut across the airfield following a night out, were all killed. Pilot Officer Rickard, RNZAF, one of the seven rookie crew members who baled out, received injuries of an unspecified nature. This crew, with the exception of P/O Rickard, completed its training and were posted to No.75 Squadron. Unfortunately, they were lost without trace, when their aircraft failed to return from a mission on 30th July 1944.

Another act of courage and determination by one of its crews enabled a further entry to be recorded in the annals of No.XV Squadron, following an operation on the night of 22nd/23rd April. Bomber Command's directive for that night was to attack Dusseldorf with a main force of 596 bombers, of which eighteen were detailed by No.XV. One of the latter aircraft was Lancaster, ND763, LS-W, piloted by Oliver Brooks. The crew compliment consisted of the usual seven men plus F/L John Fabian, the Squadron Navigation Leader who, as 'Y' operator, would operate the H2S 'blind navigation' radar. The bomber took-off at 22.49 hours, and set course eastwards out across the North Sea. Ahead, as they approached enemy occupied territory, Oliver Brooks could see sticks of white light criss-crossing the dark velvet curtain of night, as enemy searchlights swept back and forth.

As the aircraft neared the target area, Allen Gerrard, the bomb aimer, prepared himself and his equipment for the impending task. At the same time Robert Barnes, the wireless operator, left his station in order to take up position by the photo-flashes, further along the fuselage. The Lancaster was now on its bombing run, and Gerrard could be heard over the intercom guiding the pilot to the aiming point. At approximately 01.16 hours, Gerrard pressed the 'tit', and the deadly cargo tumbled from the bomb bay to fall on the target 22,000' below. As Barnes ensured there were no photo-flash hang-ups, and before Gerrard could call *"Bombs gone"*, a hail of cannon shells, fired by a prowling German nightfighter, slammed into the port wing of the bomber. At the same moment, and as the enemy aircraft peeled away, a heavy predicted flak shell exploded directly under the fuselage of the Lancaster. Reverberation from the blast entered the aircraft via the bomb bay, and traversed in both directions along the fuselage. Sergeant Allen

Gerrard, who was due to be married two weeks later, had only minutes to live. Robert Barnes was also a victim of the blast, having one of his legs almost severed by hot shrapnel. Despite his agonising pain Sgt Barnes dragged himself along the floor of the bomber towards the rear turret, crying for help. Flight Sergeant Harry Marr responded to those cries, and clambered out of his turret. The severity of the wireless operator's wounds told Sgt Marr he could not deal with the situation alone, so he called over the intercom for assistance. The call was answered by F/L Fabian, who collected a first aid kit, then made his way aft. Sergeant Ron Wilson, who had received a minor injury to his hand, was called down from his mid-upper turret and instructed to assist John Fabian. This action allowed Harry Marr to resume his station in the rear turret in order to give some defense to the rear of the struggling bomber, in the event of fighter attack.

The severely injured wireless operator lay motionless whilst his two companions endeavored to render first aid, but their efforts were all in vain. Slowly, in the cold friendless confines of the rear fuselage of a battle scarred Lancaster bomber, Robert Barnes' life ebbed away.

Having set the aircraft on a homeward course, Oliver Brooks battled to keep the stricken bomber in the air; he knew there was no other choice. He had called for a damage report and was horrified to learn that three parachutes were unusable and the dinghy had been lost when the flak burst hit the aircraft. The report also disclosed the fact that Ken Pincott, the navigator, had sustained an injury which, unfortunately, resulted in his blood splattering onto the maps and navigational charts. 'Chick' Chandler, the flight engineer, had escaped serious injury by inches. At the instant the flak shell exploded, 'Chick' had leaned forward momentarily, as a lump of shrapnel punctured the fuselage and sliced through the back strap of his parachute harness.

As the aircraft crossed the French coast at an altitude of 4,000', searchlights suddenly swept across the sky and illuminated the bomber in a cone of light. For a moment P/O Brooks thought all his efforts to keep the bomber in the air had been in vain, and that he was now going to lose the fight. Each member of the crew thought *"This is it"*, offered up a small prayer, and waited for the inevitable. They waited, seconds, minutes, nothing! No flak, no fighters. They wondered what was happening. Silent fear and emotions ran high, then suddenly the searchlights were doused and they flew on into the darkness. With aching limbs and bathed in perspiration, and with the aircraft constantly losing altitude, Oliver Brooks nursed the bomber across the North Sea. He was tired, almost exhausted and at first could not believe his own eyes, but there ahead in the distance, he saw the threshold of the emergency runway at Woodbridge. As the Lancaster crossed the English coast at 500' he knew that his troubles were not yet over. He had to land on the first attempt, due to the fact the elevator and rudder trims were both u/s. Furthermore, he had not been able to close the bomb doors, or lower the undercarriage! Unable to take up their normal crash positions, the crew remained at their respective stations during the landing and put their faith in their pilot. The Lancaster crossed the threshold at 120mph IAS, with the throttles jammed open. At 03.55 hours, ND763 touched terra firma and slid across the ground in a shower of debris, dirt and dust. The emergency services swung into action and chased the bouncing, bucking, airframe across the airfield. The screeching, tearing sounds of twisting, buckling metal ceased as the bomber slithered to a halt, and was quickly surrounded by the fire and ambulance crews. Working quickly, but with reverence, the bodies of Robert Barnes and Allen Gerrard were removed from the wreckage, whilst the rest of the crew were taken to the Station Sick Bay.

On the same day that Oliver Brooks went to Dusseldorf, a number of promotions had been announced. Amongst those named were Stan Fisher and Oliver Brooks, both of whom were given the acting rank of Flight Lieutenant. However, following the events of that night, there were no celebrations or revelry for Oliver Brooks and his crew.

Unbelievably, the names of A/F/L Brooks and the surviving members of his crew were recorded on the battle order for operations the next night. The Squadron C.O., W/C Watkins, D.F.C., D.F.M., added his name to the crew list and flew with them in the capacity of bomb aimer. Two other replacements were found, one to cover for the unfortunate Barnes, and one to stand in for Sgt Wilson who had been detained in the sick bay.

Wing Commander Watkins need not have worried about Oliver Brooks ability and confidence as a bomber pilot and crew captain. When his replacement aircraft, LL827, LS-O, was attacked during a mission the next night, A/F/L Brooks threw the Lancaster around the sky and evaded two attacks by an enemy nightfighter and two sets of searchlight beams.

Len Miller, another experienced Squadron member and friend of Oliver Brooks, was a born survivor. Having been brought up in London's East End during the years of depression, he quickly learned how to look after himself. As a bomber pilot, he was fully aware of the possibility of being shot down whilst on operations. With that thought in mind, and in addition to the standard escape equipment carried by all aircrew members, Len always carried extra items sewn into the seams of the denim coveralls he wore when on 'ops'. He was also aware of the chances of survival in the event of being shot down and, in an endeavor to increase those chances, he insisted his crew knew their aircraft blindfolded, literally! Three times a week, when not flying, he and his crew practiced ditching procedures and aircraft evacuation. They used blindfolds as an aid, in order to learn exactly how to remove escape hatch covers in the dark. These exercises ensured that, should the need arise, each crew member knew the aircraft intimately and knew what action to take. That need arose on the night of 27th/28th April, when No.XV Squadron detailed sixteen aircraft to attack Friedrichshafen.

Acting Flight Lieutenant Miller pulled his coveralls on over his battledress blouse, then tucked his uniform cap inside the latter. Finally, he pulled on his 'escape' pattern flying boots, the upper part of which could be cut off to leave a pair of sturdy walking shoes. Gathering together the rest of his flying equipment, Len made his way out to the crew bus.

Out at dispersal, prior to boarding the aircraft, A/F/L Miller tried to conceal his apprehension of the fact that Peter Slater, the rear gunner, would not be guarding the tail of the Lancaster during the mission. Unfortunately, sometime earlier, Sgt Slater had failed to clear a barbed-wire fence whilst attempting to jump over it, and had almost left his testicles partially decorating the wire. Needless to say Peter Slater found it uncomfortable, if not impossible, to sit in the confines of the rear turret for hours on end. Len Miller's apprehension increased when, whilst talking with the replacement air gunner, he found out it was the latter's first operational mission. The experienced pilot was not happy, and decided Bert Cully, the mid-upper gunner, should occupy the rear turret, whilst Robert Watson, the replacement gunner, would 'ride' in the mid-upper turret.

Lancaster bomber, LL801, LS-J, piloted by Len Miller, took-off at 21.44 hours and set course towards the south. The aircraft crossed the English coast over Shoreham, to the west of Brighton, and flew out over the English Channel on a direct course across northern France to the target. The lack of searchlights or flak over the Continent worried Len, as this usually meant the presence of enemy fighters. Also, he was painfully aware his friend and trusted rear gunner, Peter Slater, was not part of the crew.

At a point approximately eighty miles from the target, near Strasbourg, the crew saw the first signs of enemy action. Flak bursts began to pepper the sky, signifying the arrival of the first wave of bombers over the aiming point. Suddenly the darkness which surrounded LL801 turned to daylight as a flare, fired from a marauding nightfighter, illuminated the Lancaster. The enemy aircraft, a Bf110, piloted by Oberleutnant Martin 'Tino' Becker, of I./NJG VI, swung into the attack and opened up with a blast of cannon fire. Instinctively, Len Miller took evasive action and threw the bomber into a corkscrew maneuver, but not before cannon shells raked the length of the fuselage. The port outer engine also sustained hits, setting it ablaze and causing one of the fuel tanks to ignite. A long tongue of flame trailed out behind the bomber as it entered a dive, and Len Miller realised the aircraft was doomed. His final order on "J"-Jig was for the crew to abandon the aircraft but, even after he had given the command, he could still hear the bombers' guns firing. Although he had vacated his own station, the pilot was concerned that some of the crew were still on board the aircraft. Without thought for his own safety, Len Miller clambered back into the pilot seat, re-connected the intercom and screamed a warning for them

to get out. At that moment the port wing folded and there was an almighty explosion, which killed four of the crew, including Sgt Robert Watson. The Lancaster crashed on the French bank of the River Rhine, near the village of Schoenau, with the four unfortunate crew members still inside its fuselage.

Sergeant George Mead, the bomb aimer, and Sgt Alf Beazley-Long, the flight engineer, had both jumped before the aircraft exploded, and floated down to captivity. Rendered unconscious as he was blown out of the cockpit, A/F/L Len Miller, D.F.C, regained his senses as he fell through the night sky. Due to the fact the shoulder harness straps of his parachute had become loose, presumably occurring when blown out of the aircraft, he had to grapple with the pack and release the canopy by hand. Although his descent ended in a tree, from which he managed to extricate himself, the downed pilot was about to start a new fight on the ground. The born survivor evaded capture and, following a number of adventures and close encounters with the Wehrmacht, crossed into Switzerland where he was interned. Internment however was not to Len Miller's liking and, approximately four months after being shot down, he and a friend, 'escaped' from Switzerland, crossed the Alps, and re-entered France. Their expedition was not without its problems and, due to their appearance and general demeanor, they soon came to the attention of a local French Resistance unit. The two grounded R.A.F flyers were passed along a chain, until they were passed to a Free French Group. Len Miller and his friend spent some time with the Frenchmen, during which they assisted the latter in their fight against a common enemy, before being repatriated to England.

The Squadron lost another aircraft that same night, also due to fighter action. Lancaster bomber, LL805, LS-M, is thought to have crashed near Singen, north of the Swiss border, some forty miles east of the target area. The pilot, Flight Lieutenant Sinclair Soper, RCAF, and his crew all perished when the bomber was shot down.

On the penultimate day of April, Harry Bysouth again took to the skies in an effort to regain his confidence. He flew a fighter affiliation exercise with P/O Jones during the late morning, and a Bullseye with the same pilot during the late evening. The two 'missions' boosted his moral, but a week later Thomas Jones was dead. P/O Jones and his entire crew were killed when their aircraft, Lancaster bomber, ED473, LS-D, crashed during an attack against Nantes airfield.

The last day of the month was spent with a number of aircraft and crews undertaking practice bombing exercises at Rushford ranges, whilst others took to the skies on air tests. It was whilst taking-off for one of the former exercises that Lancaster bomber, LL858, swung off the runway and crashed at 22.35 hours. The aircraft, piloted by F/S Thompson, ground looped causing extensive damage to the undercarriage.

The courage and devotion to duty displayed by P/O Brooks, during the recent attack against Dusseldorf was rewarded during the first week of May, when it was announced he was to be awarded an immediate Distinguished Flying Cross. Furthermore, he was to be promoted to the rank of Flight Lieutenant. It was also announced that, in recognition of the respective parts they played in the same operation, F/L John Fabian was to be awarded a Bar to his D.F.C, whilst F/S Ken Pincott was to receive a Distinguished Flying Medal.

On the night of 1st/2nd May, a Lancaster new to No.XV was detailed to carry out its first operational mission with the Squadron. The aircraft, which wore the serial LL806, was one of a batch of 350 Mark.I variants built by Armstrong-Whitworth. Outwardly there was nothing special about the machine but, although it was not known at the time, this particular Lancaster was destined to become one of the Squadron's most well known aircraft.

LL806 had been taken-on-charge by the Squadron as a replacement machine following the loss of F/L Miller's Lancaster, which had been shot down three days earlier. Like Len Miller's aircraft, LL806 was coded LS-J; it was the second "J"-Jig.

That night thirteen aircraft were detailed for the raid, an attack against the railway stores and repair depot at Chambly, in Northern France, amongst them was "J"-Jig.

Flown by P/O Sparks, RNZAF, Lancaster LL806 took-off from Mildenhall at 22.15 hours, followed at regular intervals by the other participating aircraft. The flight out went without incident, but there was a moment of concern for the crew whilst over the target. Bombing on the yellow target indicators as instructed by the *'Master of Ceremonies'*, the crew of LL806 saw their bombs burst either side of the aiming point. However, not all their bombs released; a 1,000lb bomb hung-up in the bomb bay and refused to jettison. Pilot Officer Sparks was forced to return to Mildenhall with the weapon still on board the aircraft. A week was to pass before LL806 undertook its second mission, a night attack against enemy coastal gun positions at Cap Griz Nez. Although returning crews reported seeing their bombs exploding in the target area, later evidence revealed the raid was not as successful as first thought.

As a period of calm prevailed over the Squadron during the first two weeks of May, a number of flying training exercises were undertaken, giving both the aircrews and groundcrews a brief respite from the rigors of attacks against major targets.

It was during this period, on 10th May, that Harry Bysouth flew on an air test with F/L Amies, who was an experienced and much liked pilot. Back on the ground, after the flight, Alan Amies asked Sgt Bysouth to join the crew as mid-upper gunner on the next mission but, still feeling apprehensive, Harry Bysouth declined. It was a decision which saved his life.

Although the Unit did not undertake any major attacks during the first half of the month, No.XV did participate in a small number of minor operations during this period, despatching a total of 120 individual aircraft on ten separate bombing operations. Seven of these attacks were against targets in France, whilst two were against targets in Germany. There was one 'Gardening' operation, which saw four aircraft dropping mines off the coast of Denmark. Unfortunately, these attacks cost the Squadron three aircraft, together with the lives of twenty-two members of aircrew. Named amongst the latter was F/L Amies and his crew, who were lost on the night of 11th/12th May. When Harry Bysouth heard the news of the loss of Alan Amies he was mortified, and reflected on thoughts of what might have been.

Although Oliver Brooks had participated in a number of the attacks mentioned above, since the loss of his bomb aimer and wireless operator, his crew had been made up of spare 'bods'. However, on the night of 19th May, he 'adopted' a crew which became permanent and flew the rest of his tour with him. Ken Pincott was retained as navigator, whilst F/S Watson was taken on as wireless operator. Sergeant Richard 'Basher' Hearne, who was later to be awarded a D.F.M., was flight engineer, and F/O Jones was bomb aimer. Flight Sergeant Vic Reid and W/O Eddie Orchard were taken on as mid-upper and rear gunner respectively. For this raid, an attack against Le Mans, F/L Brooks also had F/L Fabian again acting as 'Y' operator. The mission was carried out without incident, and the aircraft and crew returned safely to Mildenhall.

On the night of 21st/22nd May, "J"-Jig participated in an attack against Duisburg. The aircraft, which was flying its sixth mission, was one of eighteen Lancasters detailed for the raid. On board the aircraft, flying with the usual crew, was F/O Leslie who had been posted to the Squadron that day. Although this officer was flying as second pilot, he would soon fly "J"-Jig on a number of occasions as captain in his own right.

Apart from flying an aircraft which carried the same code as Len Miller's aircraft had carried, F/O Leslie had another affiliation with F/L Miller; their respective crews both looked after Flying Officer Billy Goat. Following the loss of the first "J"-Jig at the end of April, F/O Leslie's crew adopted the animal and endeavored to keep it out of trouble. However, it became John Pratt's duty to ensure F/O Goat was safely locked away each night; the latter being billeted in a shed adjacent to the aircraft's dispersal area, where at least one member of groundcrew was usually on hand.

During the early stages of the outward flight on the aforementioned attack against Duisburg, F/S Pincott, D.F.M., experienced a strange coincidence which he was never to forget. He was engaged in conversation by a young sergeant pilot, who was flying with the crew to gain operational experience. The young pilot, who spoke with an accent, introduced himself as Tom Rudall. Ken Pincott recognised the pilot's accent as being that of the West of England, and asked Tom from which part of the West Country he came. When the young pilot stated he came from Taunton, Somerset, Ken replied with surprise that that was where he also came from. The next question obviously, was to enquire the name of the street in which Sgt Rudall lived. To Ken's astonishment the answer came back Castle Street, to which the navigator replied that that was the same street in which he lived. Not being able to believe the coincidence thus far, F/S Pincott cautiously asked for the house number. When given, it transpired that Sergeant Rudall lived at the opposite end of the same street. For two young men who had lived in the same street and not known each other, the circumstances of their first meeting astounded them both.

The following night, Sgt Rudall took his own crew to Dortmund but, like four of his Squadron colleagues, was forced to abandon the mission due to heavy icing. Although his crew had experienced an operational mission on 19th May when they attacked Le Mans, they had to wait until the last day of the month before their names were again recorded on the Battle order.

Flight Lieutenant Stan Fisher took "J"-Jig to Aachen on the night of 24th/25th May, when fifteen aircraft were detailed by the Squadron to join a main attacking force of four hundred and forty-two bombers. Two crews from No.XV abandoned their missions due to hydraulic and intercom failure on their respective aircraft, but Stan Fisher completed his task and returned without incident. George Allen had not flown with his crew on this mission, but was with them four nights later when they went to Angers. The target, a railway junction and marshaling yard, was attacked by 126 Allied aircraft who inflicted serious damage to tracks and rolling stock.

On the night of 31st May/1st June "J"-Jig participated together with fourteen other aircraft from the Squadron, in an attack against the railway yards at Trappes, near Paris. The Captain of the bomber on this occasion being F/O Leslie. The target, which was bathed in bright moonlight, was located easily enough. However, F/O Leslie later reported having bombed on the red target indicators, as he had not heard any instructions from the Master Bomber.

One aircraft not heard from again after it had taken-off from Mildenhall, was Lancaster bomber, LM121, LS-C, piloted by P/O Peter Dombrain. It was later reported that the aircraft, which had taken-off at 23.57 hours, was shot down over France, having been attacked by a German nightfighter piloted by Hptm Fritz Sothe, of II./NJG4. The bomber crashed in flames at Lormaison, approximately 20kms south of Beauvais, at 01.50 hours. The crew, all of whom were killed, comprised of four members of the RAAF, one member of the RNZAF, and two members of the RAFVR. One of the latter was Sgt Raymond Geoffrey Norris, the wireless operator/air gunner, who came from the Isle of Wight, off the south coast of England. 'Geoff' as he was known to his friends, had lost both parents by the time he was nine years old. He was brought up by an aunt and uncle who ensured his education did not suffer. On leaving school, Geoff Norris obtained employment as a sales agent with Messrs. Fisk and Fisher, a wholesale farmfood and agricultural merchants. He remained with the company until he volunteered for service with the RAF, in December 1941.

During the day prior to the fateful mission, the young wireless operator, using RAF terminology, wrote a letter home. In it, he wrote, *"This evening we're off to get number 8 up. At the moment we're using a brand new kite, C – Charlie, and she's fairly wizard and flies like a bird"*. The letter went on to say how that morning the crew had taken the aircraft on an air test, and how they had taken a member of the groundcrew with them. Less than twelve hours later the young letter writer was dead. Sergeant Raymond Geoffrey Norris had served with No.XV Squadron for just one month, and was participating in his eighth mission, when he was shot down. The aircraft which *flew like a bird*, had only completed two operational missions, both in the hands of P/O Peter Dombrain.

As the month of May closed, speculation began to increase as to when the Allies would mount an invasion of the Continent. The enemy knew it would happen, the Allies knew it would happen, but only those at very high level of authority had any idea as to when it would happen. Although the aircrews of Bomber Command may not have realised it, they had been providing the overture, now the curtain was about to be raised.

Miller's 'Men'

Sergeant Len Miller (later F/L, D.F.C) and his original crew. From left to right are: Back row: Sgt Arthur Mathews, W/Op; Sgt Wilbert Cully, MUG; Sgt Len Miller, Pilot; Sgt Alf Pybus, F/E. Front Row: Sgt George Mead, B/A; Sgt Johnson, Nav; Sgt Peter Slater, R/G. Alf Pybus was killed in action by a nightfighter, on 28th January, 1944 during an attack against Berlin. Exactly four months later, on the night of 27th/28th April, Sgt Matthews and Sgt Culley were killed as a result of fighter attack, during a raid against Friedrichshafen. Len Miller and George Mead survived the attack. *Courtesy of Len Miller, DFC*

An unofficial member of F/L Miller's crew was F/O Billy Goat, who used to fly with him on air tests. The Squadron adopted the goat after Len Miller and his crew were shot down, with care of the animal being the responsibility of F/L Leslie and his crew. *Courtesy of Len Miller, DFC*

Pilot Officer Peter Slater, rear gunner to Len Miller and his crew. The camaraderie between Peter Slater and Len Miller was second to none. Due to an earlier non-combatant injury, the air gunner was not flying with Len Miller the night the latter's aircraft was shot down. *Courtesy of Len Miller, DFC*

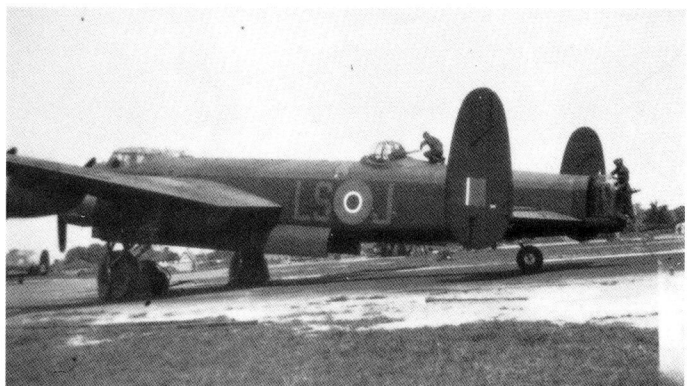

Avro Lancaster Mk.I., LL801, LS-J, built by Armstrong-Whitworth, was the regular aircraft flown by F/L Len Miller, DFC. The bomber was shot down by an Me.110, piloted by Oblt Martin 'Tino' Becker, during an attack against Friedrichshafen, on the night of 27th/28th April 1944. *Courtesy of Len Miller, DFC*

Stan Fisher and Crew

Pilot Officer Stan Fisher and his crew, who undertook their first operation, on a Lancaster bomber, on the night of 15th/16th February 1944. The crew consisted of, left to right;- Sgt Eric Tiplady, MUG; Sgt George Allen, R/G; Sgt Jimmy Crew, W/Op; P/O Stan Fisher, Pil; F/S Ted Grimshaw, Nav; F/O Geoff Wasteneys, B/A, RCAF; and Sgt Norman Berryman, F/E. *Courtesy of George Allen, DFM*

Reunion at Elvington. Four members of Stan Fisher's former crew, meet at the Yorkshire Air Museum, at Elvington, near York. Included in the picture are, from left to right: Jimmy Crew; Stan Fisher; Eric Tiplady; George Allen and Norman Berryman. *Courtesy of George Allen, DFM*

From 'Rookie' to 'Tour Expired'

Three members of F/S Wheeler's original crew pose for a formal photograph. Sergeant Bernard Dye, MUG (left), later joined F/O Arthur Horton's crew on No.622 Squadron, where he successfully completed a tour of operations. Sergeant Keith Hollinrake, F/E (center), completed nineteen operations before being killed in a flying accident on 21st June 1944. Sergeant Joe Hayes, A/G (right), who engaged in pranks with Bernard Dye during training exercises. Their original pilot, F/S Leslie Wheeler, was killed during an attack against Berlin on 24th March 1944. *Courtesy of Bernard Dye*

Flight Sergeant Bernard Dye, having completed a tour of operations, poses for a photograph with his 'adopted' crew. At the time his original crew was split up, No.622 Squadron pilot Arthur Horton was looking for a replacement air gunner. In the photograph, which was taken on 12th August 1944, are from left to right: Sgt B. Grant, R/G; Groundcrew; F/O Arthur Horton, Pilot; Sgt 'Jock' White, W/Op; Sgt Dave Parsons, F/E; Sgt Ken Monether, Nav; Groundcrew; Sgt Brian Gray, B/A; Groundcrew; Sgt Bernard Dye, MUG. The aircraft, Lancaster Mk.I. bomber, LM241, GI-Q, was built by Armstrong Whitworth. It failed to return from an attack against Russelsheim, fourteen days after the photograph was taken. *Courtesy of Arthur Horton*

Loss of a Hero

Sergeant Charles Marsh, flight engineer to Joseph Russell's crew, at his station on an unidentified Lancaster bomber. Marsh was one of the four crew members to die, when their aircraft crashed in flames whilst returning from Stuttgart. *Courtesy of and via Leslie Belton, DFM*

Left: Lost Courage. A photograph, taken at night, of F/L Joseph Vincent Russell, C.G.M, who lost his life on the night of 20th/21st February 1944, during an attack against Stuttgart. *Courtesy of and via Leslie Bentley, DFM*

Air Gunners

Sergeant William 'Mick' Harbidge, friend and fellow crew member to Harry Bysouth. The 19 year old rear gunner lost his life on 26th February 1944, when Lancaster bomber, ED383, LS-C, crashed at Lakenheath following its return from an attack against Augsburg. *Author's Collection via the late Harry Bysouth*

Sergeant Harry Bysouth whose fears of returning to operational flying, following an horrendous crash, were eventually overcome and resulted in the award of a Distinguished Flying Medal. *Author's Collection via the late Harry Bysouth*

Fireworks for Frankfurt

Lancaster bomber, ED473, LS-H, waits to be 'bombed-up'. On the night of 18th/ 19th March 1944, F/L Len Miller piloted this aircraft on an attack against Frankfurt. Note the 4,000 lb 'cookie' on the bomb trolley, to the extreme left of the picture. *Author's Collection*

The same 4,000 lb 'cookie', is used as an unconventional seat by three unidentified members of aircrew, before it is loaded into the bomb bay of Lancaster, ED473, LS-H. *Author's Collection*

Wing Commander W. D. Watkins, D.F.C., D.F.M.

The Dusseldorf Catastrophe

Wing Commander William Watkins, D.S.O., D.F.C., D.F.M., took command of No.XV on 15th April 1944. Seven months and one day later, his aircraft was shot down during an attack against Heinsberg. The crew, with the exception of W/C Watkins, were all killed. The latter was captured and spent the rest of the war as a PoW. *Courtesy of Don Clarke, MBE, The Mildenhall Register*

Sergeant Oliver Brooks original crew consisted of, left to right: Ken Pincott, Nav; Harry Marr, A/G; Oliver Brooks, pilot; 'Chick' Chandler, F/E; Ron Wilson, A/G (concealed); Les Pollard, W/Op; Allen Gerrard, B/A. Les Pollard was replaced as wireless operator by Robert Barnes. The latter was to be killed in action during the attack against Dusseldorf, on the night of 22nd/23rd April 1944. *Courtesy of Oliver Brooks, DFC*

The 'main office' of a XV Squadron Lancaster bomber (possibly ND763, LS-W) showing the pilot's position to the left of the photograph, the flight engineer's panel to the right and the bomb aimer's compartment to the bottom center of the picture. *Author's Collection*

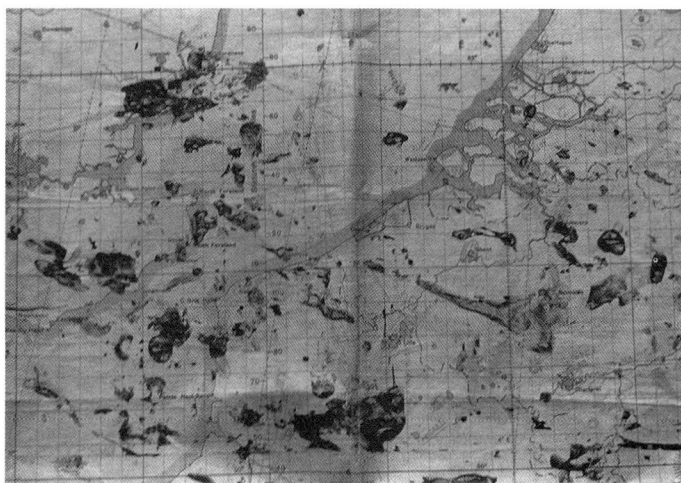

The flying maps and charts used by Sgt Ken Pincott, on the night of the Dusseldorf raid, became blood splattered when he was injured during an attack by an enemy nightfighter. Unfortunately, two of his fellow crew members were killed as a result of the same attack. *Author's Collection*

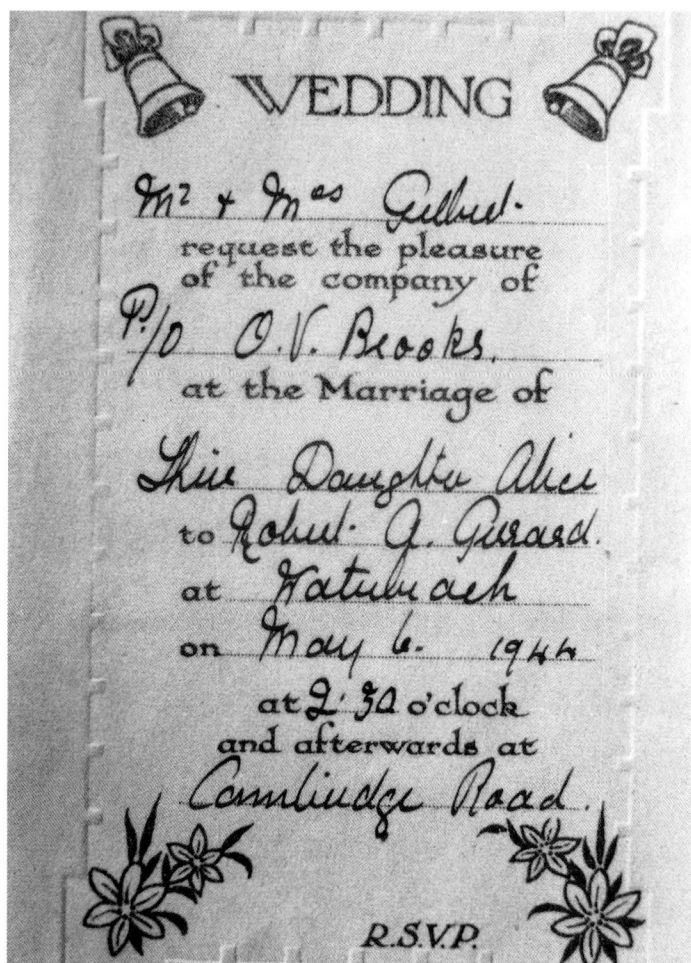

The invitation sent to Oliver Brooks, inviting him to the wedding of his bomb aimer, which was scheduled for 6th May 1944. Unfortunately Allen Gerrard, the bridegroom to be, was killed in action two weeks before the ceremony was to take place. *Author's Collection*

The Loss of the first "J"-Jig

The village of Schoenau, France, where Lancaster bomber, LL801, LS-J, crashed on the bank of the River Rhine, on the night of 27th/28th April 1944, having succumbed to the guns of Oblt 'Tino' Becker. *Courtesy of and via Ken James*

The last resting place, in Schoenau Churchyard, of four gallant airmen of Len Miller's, who died on the night of 27th/28th April 1944. The four Commonwealth War Graves Commission headstones record the names of, from left to right: F/S Arthur Mathews, W/Op; Sgt John Eastman, Nav; Sgt Robert Watson, A/G; F/S Wilbert Cully, A/G. Nineteen year old Sgt Robert Watson was flying his first operational mission with the Squadron. He was acting as a replacement for Peter Slater, the crew's usual rear gunner, who had been grounded due to an injury sustained whilst off-duty. *Courtesy of Ken James*

Old Enemies, New Friends

Oberleutnant Martin 'Tino' Becker (left), the Luftwaffe nightfighter pilot who shot down Len Miller's aircraft, congratulates a member of his own crew on the award of the latter's Knights Cross. *Courtesy of Len Miller, DFC*

Former enemies become new friends. Photographed at Bergheim Cemetery, during 1994, are from left to right: Karl-Ludwig Johanssen (Luftwaffe); Unknown (Luftwaffe); Herr Thomas; Martin Becker (Luftwaffe); Len Miller (RAF). *Courtesy of Ken James*

The enemy at play. Martin Becker takes time out from his nightfighting duties to spend time with a canine friend. The antenna of the Lichtenstein radar, on the nose of his aircraft, enables the pilot to maintain his balance. *Courtesy of Peter Hinscliffe, OBE (Mil)*

A restaurant in a French hotel provides the venue for a gathering of eagles fifty years on. Enjoying good food, good wine and good company are, from left to right: Len Miller (RAF), Wolfgang Falk (Luftwaffe); Frau Falk; Frau Rumpelhardt (wife of Fritz Rumpelhardt). Sitting with their backs to the camera are from left to right: Martin Becker (Luftwaffe) and Karl-Ludwig Johanssen (Luftwaffe). *Courtesy of Ken James*

The Second "J"-Jig

Avro Lancaster Mk.I., LL806, LS-J, the second aircraft of its type to carry the "J"-Jig code, was taken-on-charge by No.XV Squadron, three days after the loss of the original "Jig". *Author's Collection*

The second "J"-Jig waits to be bombed-up, in preparation for a mission later in the day. Two other No.XV Squadron Lancaster aircraft can be seen in the background. *Courtesy of Don Clarke, MBE, The Mildenhall Register*

Alan Amies and Crew

Left: Flight Sergeant Victor Reid, (left) and F/S Eddie Orchard, (right) both flew regularly as air gunners with Alan Amies. They were not with Amies on the night of 11th/12th May 1944, when the latter failed to return from an operation during which he had captained a completely new crew. Both men, who subsequently joined the crew of Oliver Brooks, were later promoted to the rank of Warrant Officer. On 17th November 1944, they were both awarded the Distinguished Flying Cross. *Courtesy of Oliver Brooks, DFC*

Left: Flight Lieutenant Alan Amies, a popular pilot and friend of both Len Miller and Oliver Brooks, was killed on the night of 11th/12th May 1944, whilst participating in an attack against railway yards, at Louvain, Belgium. *Courtesy of Don Clarke, MBE, The Mildenhall Register* Center: Sergeant Stan Watson, Wireless Operator/Air Gunner, was another of Alan Amies's crew who flew the rest of his operational tour with Oliver Brooks. *Courtesy of Oliver Brooks, DFC* Right: F/Sgt Richard 'Basher' Hearne, F/E, who was awarded a Distinguished Flying Medal on 14th November 1944, not only flew with Oliver Brooks but was to remain a life long friend. *Courtesy of Oliver Brooks, DFC*

The Taunton Flyer

A poor quality photograph showing Sgt 'Tom' Rudall and four members of his original crew. They are: Sgt Long; Sgt Tom Rudall (Pil); Sgt R. Bell (B/A), RAAF; Sgt W. Summerton (Nav), RAAF; Sgt P. Smith (W/Op), RAAF; *Courtesy of Don Clarke, MBE, The Mildenhall Register*

Right: Henry Thomas Rudall was posted to No.XV Squadron, at RAF Mildenhall, during May 1944. On being declared 'tour expired' on 27th August 1944, he was posted to No.17 Operational Training Unit, at Silverstone, as an instructor. 'Tom' as he was known, was awarded the French Croix de Guerre. *Courtesy of Tom Rudall, C.du.G (F)*

The Lad from the Isle of Wight

Raymond 'Geoff' Norris (2nd from left) photographed in full flying kit, possibly at No.11 Operational Training Unit, at Bassingbourne, Cambridgeshire. The friends pictured with him are only identified as W Butler, W Fry and W J Henderson. A Sgt William Henderson, who served on No.XV Squadron at the same time as Geoff, was killed in action on 5th November 1944. *Courtesy of Geoff Reynolds, nephew of 'Geoff' Norris*

The crew of Lancaster, LM121, LS-C, which was shot down at 01-50 hours, on the morning of 1st June 1944. From left to right, back row: Sgt Robert Grant (?), A/G; Sgt Leonard Gearing, F/E; P/O Peter Dombrain, Pil, RAAF; Sgt Arthur Long, Nav, RAAF; front row: F/S Frank Reid, A/G, RAAF; Sgt Geoff Norris, W/Op; F/S Lawrence Jamieson, B/A, RNZAF. *Courtesy of Geoff Reynolds, nephew of 'Geoff' Norris*

16

Return to Fortress Europe

At its narrowest point, between Cap Gris-Nez and Dover, the Strait of Dover is approximately twenty-one miles wide. On a clear summer's day, from the cliffs overlooking the Kent Channel port, the coastline of France is visible. However, due to the inclement weather during the first week of June 1944, visibility was very restricted. The gray concrete fortified defenses of the Atlantic wall, like the coastline it defended, blended in with the mist and rain.

The low cloud and squally weather, which also prevailed over Britain, had not been allowed to interfere with Bomber Command's bombing policy. Many targets of varying nature had been attacked over the last few weeks, with the main priorities being coastal defenses, railway yards and minelaying operations off French and Dutch Coasts.

No.XV participated in an attack against coastal batteries at Wissant, north-east of Cap Gris-Nez, on the second day of the month, but the weather conditions prevented accurate bombing.

A similar operation was carried out on the night of 3rd/4th June, when three aircraft from the Squadron attacked a gun battery at Calais.

Many members of aircrew, including those on No.XV Squadron, realised there had been a change in Bomber Command's tactics. For approximately the last six weeks, most of their missions had been against targets in France, Belgium and Holland. Many may have guessed this was the prelude to an invasion, but they did not know when or where it would occur. The major intention was to keep the enemy guessing but, at this period of the war, it was common knowledge there would soon be a return to 'Fortress Europe'.

In the early hours of the morning of 6th June, twenty Lancaster bombers from No.XV Squadron, took-off from Mildenhall and headed south. They formed part of a total force of over 1,000 aircraft detailed for the operation. Their particular target was defenses at Ouisterham, on the Normandy coast. Although the No.XV Squadron crews reported varying amounts of cloud over the aiming point, they all later reported bombing on the target indicators. However, it was not where their respective bomb loads had fallen the crews wanted to report on, they had seen something far more interesting. As the Lancasters headed home and the cold light of dawn broke the darkness, there beneath them on the 'choppy' surface of the sea, the aircrews saw an armada of ships and boats. Momentarily taking their eyes off the skies around them, each man paused to view the scene below. Never before had any of them seen such a sight; a massive fleet of ships heading out across the English Channel. Excitement grew as the airmen looked down and realised the armada was on a course heading for France, or more particularly Normandy. The view from the air prompted F/S Frank Watson, on his return to base, to record in his Logbook, " "D" Day Morning. Wonderful Sight".

The invasion had started and No.XV, along with other Squadrons of Bomber Command had, unknowingly, played a major role in the preparations for the commencement of Operation 'Overlord'. Each man returned to Mildenhall with his own thoughts and memories of the momentous event but, F/O Musgrove, had more reason to remember it than most. The early morning attack on the coastal defenses on "D"-Day, marked the end of his tour of operational duty.

Gerry Musgrave had initially reported for duty with No.XV Squadron during June 1942, when he was bomb aimer to P/O Meredith and his crew. However a few weeks later, on the night of 28th July, P/O Meredith was forced to crash-land at Coltishall following an altercation with both enemy fighters and flak during a raid against Hamburg. As a result of the crash, Sgt Musgrove sustained a broken right arm and slight concussion. The rest of the crew being basically fit returned to operations, but the young bomb aimer was transported by ambulance to Norwich General Hospital, where he was admitted for four days observation. Following his stay at N.G.H, the wounded airman was transferred to the R.A.F Hospital at Littleport, where his arm was reset. This in turn was followed by a period of convalescence.

By the time Sgt Musgrove returned to the Squadron, towards the end of January 1943, No.XV had relocated to Bourn. Upon his arrival at the new station, he was informed that his pilot and crew had been reported 'missing in action' following an attack against Lubeck on the night of 1st/2nd October 1942. Flying Officer James Lloyd Meredith, RNZAF, had only been promoted to his new rank a week before he was reported missing.

Following a number of operational flights with other pilots and crews, and a period on the Base Staff at Mildenhall as Assistant Bombing Leader, Gerry Musgrove finally joined the crew headed by S/L Lamason. However, although F/O Musgrove had become 'tour expired' following the attack on the Normandy beach-head, Phil Lamason and his crew still had five more operations to do to complete their second tour. It was with this thought in mind that the Canadian bomb aimer sought permission from the Royal Canadian Air Force Headquarters, to do five more trips. It was a decision which he would never forget, and he would always associate with "D"-Day – 6th June 1944.

The euphoria which had prevailed on the Squadron two nights earlier, dissipated on the night of 7th/8th June, after No.XV had despatched seventeen aircraft for an attack against railways installations at Massey-Palaiseau. A total of 337 Bomber Command aircraft were involved in the operation, of which twenty-eight failed to return. Four of those aircraft were No.XV Squadron machines. Three of them were lost over France, whilst one succumbed to battle damage and crashed in England. Flying Officer Musgrove was a crew member of one of the Lancaster bombers shot down over France. His request to fly a further five operations with S/L Lamason was granted by the RCAF, thereby enabling the 'tour expired' bomb aimer to stay with his crew. It was a decision which nearly cost Gerry Musgrove his life.

Lancaster bomber, LM575, LS-H, took-off from Mildenhall at 00.42 hours, and climbed towards the moonlit sky. The crews had been informed at the briefing earlier, that a thin layer of cloud would prevail at between 6,000' -7,000' over the target area, and bomb aimers had been instructed to make their bombing runs below the cloud in order to be certain of their aiming point. As S/L Lamason held the Lancaster steady on course, Gerry Musgrove became aware that the cloud, with the full moon behind it, gave a perfect back-drop for any would-be aggressor below in the shadows. As the Lancaster made its final turn in for the run up to the target, a twin-engined nightfighter attacked from below and astern. However, the first burst from its cannons gave warning to the Lancaster crew of its presence, as the gun fire passed ahead of the bombers nose. In a moment of instant reaction, and together as if one voice, the crew all gave warning of the attack. The pilot had also instinctively thrown the Lancaster into an eva-sive maneuver but as he did so, a second burst of cannon fire hit the air-craft. It swept across the mainplane, from port wingtip to starboard wingtip, setting the machine aflame in the process. The pilot warned the crew they should prepare to abandon the aircraft, which they all acknowledged. This warning was followed seconds later by the order to jump from the stricken aircraft. For a heart-stopping moment, Gerry Musgrove found the escape hatch cover would not move from its mounting, even though the release handle had activated properly. Brute force and self-preservation overcame the obstacle, and with a little pressure the defiant cover came away. With one last look back over his shoulder, F/O Musgrove rolled out into the darkness and began counting. Having reached the number ten he pulled the ripcord of his parachute, and seemed to stop in mid-air as the canopy snapped open. With his decent now on a more even kneel, he began to look around and was horrified to see a stream of aircraft apparently heading straight towards him. In a moment of apprehension, Gerry Musgrove gave way to human instinct and tried to draw himself up into the confines of his harness for greater protection, but the aircraft passed harmlessly above him. The rest of his descent was without incident, only his pride was hurt when he landed on French soil.

Although it is thought enemy vehicles were in the vicinity where he landed, Gerry Musgrove managed to evade capture. He was later helped by brave French citizens, who initially gave him shelter and food, before passing him on to members of the resistance who gave him a new identity and formulated plans for his escape. Flight Lieutenant Marpole, the flight engineer, and F/O George, D.F.C, also managed to evade capture, but Phil Lamason and Ken Chapman, the pilot and navigator respectively, were made prisoners of war. There were two fatalities on the aircraft, W.O. Robertson Aitken, RAFVR, the mid-upper gunner and F/O Thomas Dunk, RAFVR, the rear gunner. The latter, who was born in Southern Rhodesia, was found dead in the wreckage of his turret, after the aircraft crashed on the south-west suburbs of Plaisir. The unfortunate rear gunner, who had joined No.XV Squadron on 4th December 1943, was buried in the town's communal cemetery. Around 1950, in memory of the fallen warrior, the inhabitants of Plaisir paid for and erected a fitting memorial on Thomas Dunk's grave. In June 1996, F/O Dunk was further remembered by the inhabitants of Plaisir when a street in their town was named Rue du Lieu-tenant Dunk, in his honor.

The two other aircraft lost over France during this raid were Lancaster, LL945, LS-M, piloted by F/L William Palmer, and Lancaster, LM534, LS-A, flown by F/O Woodley. The latter pilot, Charles Woodley, RCAF, had been reposted to the Squadron, for the start of another 'tour of ops', only fourteen days earlier. He was participating in the fifth mission of his sec-ond tour, when his aircraft crashed at Bonnelles. The whole crew perished.

Only one crew member was to perish from the aircraft which crashed on English soil. The Lancaster, LL781, LS-L, had the misfortune to en-counter two Me 410 nightfighters approximately five minutes after leav-ing the target area. The first enemy aircraft attacked from the port side, but was deterred by the two gunners. With the latter's attention diverted away from him, the second nightfighter took advantage of the situation. His open-ing burst of cannon fire killed Sgt C W Kirk, the navigator, instantly when a shell exploded inside the bomber. The Lancaster sustained damage to the center section of the fuselage, the wings and the starboard aileron which broke away. Sergeant Brennan, the mid-upper gunner, and Sgt Brookfield,

the rear gunner, both responded immediately to the challenge and returned fire. The enemy aircraft was later claimed as destroyed due to the fact it spiraled down and exploded in mid-air. Flight Lieutenant Bell, who had been wounded in the attack, fought to keep the Lancaster in the air. The damage report by the crew indicated the loss of certain flying instruments, the front turret out of action and unserviceability of both brakes and flaps. Ignoring the pain of his wound, F/L Bell put the crippled bomber down at Friston, a Fighter Command airfield in Sussex where, due to the lack of the flaps, it overshot and burst into flames. The six surviving members of the crew all suffered superficial burns. The name of Walter John Bell was added to the annals of No.XV Squadron, when he was awarded a Distinguished Flying Cross, for his skill, courage and determination to fly his damaged bomber home.

Three different crews took "J"-Jig on operations during June, but only on one occasion was trouble encountered. During an attack against French railway installations at Dreux, on the night of 10th/11th June, LS-J was intercepted by a Messerschmitt Bf109. Immediately, at the first sign of trouble, F/L Payne took evasive action, whilst the two gunners opened fire at their attacker. They were rewarded with the sight of the enemy aircraft peeling over and diving earthwards.

The curse of a No.XV Squadron machine bearing the LS-U code re-turned on 12th June when Lancaster, LM465, was lost during an attack against Gelsenkirchen. The aircraft, piloted by P/O Carlton 'Carl' Thomp-son, RCAF, was attacked and shot down by a German nightfighter. The bomber crashed at Meerlo, between Venray and the Dutch/German border, killing six of the seven crew members on board. The sole survivor of the crew was Sergeant Jack Trend, the wireless operator. Having baled out of the doomed aircraft, he landed safely by parachute and managed to evade capture. The pilot of Lancaster LM465, P/O Carl Thompson, was an Ameri-can from Detroit, Michigan who, because of a leg disability, failed his medical to join the USAAF. Undeterred, he made his way to Canada, where he was accepted for pilot training by the Royal Canadian Air Force.

Also lost on this raid was Lancaster, LM156, LS-R, piloted by F/O Simon Phillips, RNZAF. The seven members of the crew were killed when the aircraft crashed in the confines of a gas works, having been shot down by flak.

The following night only one aircraft was detailed for war operations by No.XV Squadron. The lone aircraft, Lancaster bomber, LL923, LS-O, piloted by F/S Dan Capel, took-off from Mildenhall at 22.59 hours and headed south. This Australian pilot, who had flown his first 'second dicky' trip the night before, was taking his novice crew on their first mission; a mining sortie off the coast of Brest. The trip, which was uneventful, lasted four and three-quarter hours. At the post 'op' debriefing, on return to base, the crew reported favorable weather conditions with good visibility over the mining area. Any anxiety they may have experienced before or during the operation dissipated as the aircraft landed back at Base, at 03.43 hours on the morning of 14th June.

Hydraulics failure to Lancaster, PB115, LS-W, gave cause for con-cern to both its crew and the Squadron during an attack on the night of 14th June. The aircraft piloted by F/S Tom Rudall, was one of fourteen ma-chines detailed to attack the port installations at Le Havre. The first indica-tion of anything being amiss occurred over the target area, when the bomb doors failed to open. Although attempts were made to rectify the situation, it became obvious to the pilot that he had no option but to abort his mis-sion. Using the emergency system, nine 1,000lbs H.E. bombs were jetti-soned over the Rushford ranges prior to landing at the emergency airfield at Woodbridge, Suffolk. Although F/S Rudall had taken-off with the rest of the Squadron at approximately 23.30 hours, his late return to Mildenhall had the Squadron fearing the worst.

Shortly after breakfast on the morning of 21st June, W/O George Franklin went to visit his friend and fellow survivor, Harry Bysouth. Al-though it was almost exactly four months to the day since their crash at Lakenheath, the latter was still having trouble coming to terms with the thought of operational flying again. However, by comparison, George Franklin had recovered his nerve very quickly, and had flown a number of operations with various crews. He was in fact, on that particular morning, about to fly on an air test with F/L Arthur Jarvis. On hearing this news Harry Bysouth felt very apprehensive, but said nothing as his friend bade him farewell and walked out through the door. At 11.20 hours, an explosion was heard and a pall of black smoke rose into the air. Less than one hour after leaving his friend, W.O. George Franklin, RNZAF, was dead. The Lancaster in which he was flying, LM576, LS-D, had been airborne for less than five minutes. The bomber was seen to emerge from low stratus cloud, on fire and out of control. It crashed at West Row, close to the airfield at Mildenhall, with the loss of all seven crew members.

Harry Bysouth was devastated by the news of the crash, and in particular by the loss of George Franklin. As the sole survivor of his crew, he began to wonder if he was a 'jinx' to other members of aircrew. The list, which seemed almost endless, raced through his mind. First Tony Davis and four other members of the crew killed in the crash at Lakenheath. These were followed by P/O Jones, F/L Amies and P/O Thompson and their respective crews. Now, after the loss of twenty-six young men, his friend and companion George Franklin. Trying not to sink into an abyss of despair, Harry Bysouth's only comfort and support came from his young wife Renee, whom he had married in November 1941.

The bomber had been one of a number of Squadron aircraft carrying out air tests that morning, prior to an operation later in the day. The mission, when it was announced, was an attack against Domleger, but it did not go according to plan. The raid, which signified No.XV's return to daylight operations, was hampered by ten/tenth cloud preventing the target from being identified. On instructions from the Master Bomber, all crews returned home with their bomb loads.

One of the pilots flying on this operation, who reluctantly complied with the order, was F/L Oliver Brooks, DFC. He was flying the final mission of his operational tour, and had not envisaged it finishing in this manner. Having completed thirty sorties, he was granted leave prior to being posted to No.1651 Conversion Unit. This experienced and competent pilot was to take up his new post as an instructor, at Wratting Common, on 16th July 1944. Accompanying him to No.1651 C.U. was Ken Pincott, his navigator, who had been commissioned to the rank of Pilot Officer, on 27th May.

With his failure to return to aircrew duties, Sgt Harry Bysouth knew that time was running out with regard to his operational flying career. He knew he would either have to re-crew or re-muster. However, tragedy was to strike again, in circumstances beyond his control, which would dictate his future. Inclement weather conditions hindered operations during the final days of June, but improved sufficiently to allow an attack to take place on 30th of the month. The target was a road junction which two German Panzer Divisions, assembling near the village of Villers Bocage, would have to use in order to mount a counter-attack against British and American forces on the invasion front. The seventeen aircraft despatched by No.XV Squadron, were detailed to fly in three 'V' formations, one behind the other, with six aircraft in each of the first two 'vics' and five in the third. The first formation was led by W/C Watkins, the second by F/L Sparks whilst the third was led by F/L Stan Fisher.

The formations were passing over Midhurst, West Sussex, when tragedy struck. Lancaster bomber, ME695, LS-Z, piloted by F/S Johnny French, was holding station as 'wingman' on the starboard of the third 'vic', when it was struck from behind by another aircraft. As a result of the collision, F/S Pawlyk, RCAF, the rear gunner of Johnny French's crew was killed, when his turret took the full force of the impact. Damage was also inflicted on the starboard tail assembly and rudder. The latter machine, a Lancaster Mk.III bomber, PB178, from No.514 Squadron, disintegrated in mid-air and fell to earth. Although still intact, the No.XV Squadron machine began to lose altitude, prompting Johnny French to give the order to bale out. The aircraft was at 11,500' when the bomb aimer and flight engineer jumped. By the time the wireless operator and mid-upper gunner complied with the command, the Lancaster was down to 6,000'. At this relatively low level, F/S French, a tall, powerfully built man, managed to regain control of his aircraft. Fortunately, F/S Geoff Stubbing, the navigator, had remained with

the aircraft and was able to assist the pilot in jettisoning the bomb load. Having happily completed this task, Johnny French crash-landed the aircraft just north of Ford, on the south coast of England. Two of those who baled out, Sgt J Hope and Sgt H Slingsby, the wireless operator and mid-upper gunner respectively, received slight injuries.

The attack on the road junction was considered a great success, with all crews reporting excellent weather conditions and good concentrated bombing. In fact, their reports were confirmed the following day when a message of congratulations and thanks was received from General Sir Bernard Montgomery. The British Commander also commented on the accuracy of the bombing and the results achieved by the participating aircrews. The German armoured columns were prevented from using the road junction, and thereby did not proceed with their planned attack.

Further messages of congratulation were in evidence on 5th July, when His Majesty King George VI paid an official visit to Mildenhall. Flight Lieutenant Oliver Brooks had been on leave just over a week when he received a telegram, ordering him to return to the Suffolk airfield immediately. His presence was required for an investiture, at which he would be officially presented with the award of a Distinguished Flying Cross.

All Station personnel and groundcrew were on parade, together with members of both No.XV and No.622 Squadrons. Also present were members of No.1 Defense Squadron, who provided the guard of honor to His Majesty. The King arrived at Mildenhall accompanied by Queen Elizabeth and their eldest daughter Princess Elizabeth (later to become Queen Elizabeth II). The Royal Party were escorted into No.1 Hanger where the recipients of awards were standing proud and erect. As their names were read out, each man marched forward and stood before the King. The first to step forward was F/L Fabian, D.F.C., who was to receive a Bar to his existing award. The New Zealand signaler was followed by F/L Oliver Brooks, who stood taller than his Monarch but whose eyes stared straight ahead. Other recipients of the D.F.C included F/L Robert Meggison, who had completed two tours of operations with No.XV Squadron and F/O Midgley, the pilot of the Stirling bomber known as *Midgley's Flying Circus*. In all, twelve D.F.C's were awarded to members of No.XV Squadron. The two final honors, Distinguished Flying Medals, were presented to F/O Angus and Ken Pincott. The latter, who had stood motionless throughout the proceedings, was the last man to be called. When his name was read out, he followed the ritual of those who had proceeded him, and marched forward with dignity and pride. With the D.F.M. pinned above the left breast of his tunic, Ken Pincott saluted, half turned and marched proudly back to his allotted place.

Following the investiture, the Royal Party inspected the parade and then joined members of both Squadrons for an informal gathering. As both No.XV and No.622 Squadrons were detailed for operations that night, and as the participating members of aircrew had arrived direct from their pre-flight briefing, they were permitted to greet the King wearing their battle-dress clothing.

The target they attacked that night was a flying bomb storage site under construction at Wizernes. The night was clear and the target was bathed in bright moonlight, which enabled the attacking crews to bomb accurately. The prevailing conditions also meant enemy nightfighters were active although, on this occasion, the Luftwaffe may have been the hunted instead of the hunters. The gunners on Tom Rudall's crew claimed a FW190 shot down, whilst the gunners on F/S Ferguson's crew claimed a Bf110 probably destroyed. Of the fourteen aircraft which were despatched by No.XV Squadron, only one failed to return. Lancaster bomber, LL890, LS-T, piloted by F/O Michael Golub, RCAF, was lost without trace along with its crew.

The majority of operations carried out during the second week of July were hampered by adverse weather conditions, which prevented the crews from seeing the results of their work. On more than one occasion the Master Bomber canceled the operation, and instructed the crews to return home with their respective bomb loads. On one such attack, during the early evening of 12th July, the crews were actually over the target when the mission was aborted. The target on this particular evening was the railway marshaling yards at Vaires, on the outskirts of Paris. A total attacking force of one hundred and fifty-nine Bomber Command aircraft were despatched on this raid, including eleven machines from No.XV Squadron.

Flight Sergeant Dan Capel, piloting Lancaster bomber, PB137, LS-U, took-off from Mildenhall at 17.31 hours. This raid constituted the fifth mission of his tour, but the first he had undertaken during the hours of daylight. Climbing up to clear the overcast sky, he was astounded at the sight which awaited his gaze as the aircraft broke cloud cover. There, stretched out across the sky ahead of him, were scores of Lancasters bombers all heading for France. Above the Lancasters, although F/S Capel could not initially see them, was a large contingent of escorting Spitfires.

Having crossed the enemy coast, the aerial armada flew on over the French countryside. All seemed quiet with no hint of enemy activity until, up ahead, Dan Capel saw the angry black bursts of exploding flak peppering the sky. Bombers started to weave from side to side, as those being fired at took evasive action, when suddenly black smoke started to issue from one of them. Lancaster bomber, ND958, LS-H, pilot by P/O Cowell, had been hit and received flak damage to the port mainplane and one engine. The crippled bomber was seen to drop out of the formation, jettison its bomb load, and turn for home. Although the bombers continued to take evasive action, the flak soon found another victim. Lancaster, PB112, LS-K, piloted by F/L Frank Dengate, sustained damage to the starboard inner engine, which created further problems when the aircraft landed at Mildenhall. Due to the lack of brake pressure, the bomber overshot the runway on landing, crashed through the perimeter fence and came to rest in a cornfield. One member of the crew of this aircraft had already experienced a crash-landing in a Lancaster, but in more dramatic circumstances. Flight Lieutenant John Fabian, D.F.C. and Bar, who on this occasion was flying as navigator, had previously endured the trauma of a crash-landing at Woodbridge, with F/L Oliver Brooks three months earlier. Likewise the wireless operator, F/S Frank Watson, was another experienced crew member, having flown a number of operations with Flying Officer Woodley. The other members of the crew, all of whom were uninjured, were F/L Cox, bomb aimer, Sgt Fred Coney, mid-upper gunner, Sgt Douglas Davies, rear gunner, and Sgt Bob Kitchen, flight engineer.

Whilst F/L Dengate was heading for home, F/S Capel was weaving a path through the maze of flak, which was bursting all around his aircraft. As he swung the Lancaster into a clear patch of sky, a flak shell burst close by. A fragment of hot metal punctured the plexi-glass windscreen, flew into the cockpit, and grazed the pilot's right cheek. Having run the gauntlet of the flak, the crews were instructed by the Master Bomber not to attack the cloud covered target. Although they could have bombed the small aiming point on instruments, it was felt the risk to local French people living near the target was too great.

All eleven Squadron aircraft returned safely to base, but three, each flown by an Australian pilot, had sustained battle damage.

A major attack against Caen, detailed for 16th July, had to be postponed due to adverse weather conditions. An attempt was made to carry out the raid the following day but, having got airborne, the aircraft were immediately recalled. It was not until the early hours of the morning, on 18th July that the fifteen aircraft detailed for the mission finally got away. The target was a steelworks to the south of the town, in an area where an element of enemy forces was holding up the progress of the British Second Army. The attack, made in daylight by a total of 942 Bomber Command aircraft, was considered a great success. Although no enemy fighters appeared over the target area, a total of six bombers were shot down by flak, including one Lancaster. Shortly after the British machines had turned for home, approximately 700 Liberator and Fortress bombers of the USAAF flew over to continue the attack. These in turn were followed by light and medium twin-engined bombers of both Air Forces, who carried out the same task. A conservative estimate recorded that a total of over two thousand Allied aircraft dropped up to 8,000 tons of bombs in a space of one hundred and eighty minutes.

Some of the crews who operated on the early morning raid against Caen found their names on the Battle Order for operations that night, when the Squadron again detailed fifteen aircraft for an attack against a railway

junction at Aulnoye. Although smoke and haze hampered visual identification of the target, all crews followed the instructions of the Master Bomber and bombed on the T.I's. The raid, which was later considered successful, inflicted severe damage to the tracks and deprived the enemy of rail transport access to the battle front.

Flight Lieutenant Bell piloting "J"-Jig, was forced to return early from the latter raid, when the aircraft developed a problem with the starboard outer engine. Although the bomber had participated in the earlier mission against Caen, there had been no sign of malfunction from the now defective motor.

Two nights later, on 20th/21st July, a total of 158 aircraft were detailed to attack the synthetic oil plant at Homberg. Fifteen Lancasters from No.XV Squadron were detailed to participate in the raid which, although reported as successful, cost Bomber Command twenty aircraft, including one from No.XV. The latter machine was Lancaster, R5904, LS-L, piloted by F/L Walter Bell, D.F.C. The aircraft crashed in Belgium at 01.40 hours, with the loss of six members of the crew.

Another crew participating in this raid, who were constantly under attack from Luftwaffe nightfighters, managed to avenge the loss of F/L Bell and shoot down one of their adversaries. During the course of the homeward leg of the mission, F/S Dan Capel's aircraft, Lancaster bomber, PB137, LS-U, was the subject of no less than twelve separate fighter attacks. Due to the alertness of the mid-upper and rear gunners, one German fighter was prevented from returning to its base that night. Attacking out of the darkness, an enemy nightfighter opened fired at the Lancaster before it was in range. Warned of the fighters presence by the incoming cannon fire, Sgt Glyn Thomas rotated his mid-upper turret towards the attacker and prepared to open fire. Sergeant Arty Barrett, the rear gunner, quickly adopted the same procedure, and together they waited. When the fighter, now identified as a Bf109, was approximately 250 yards away both gunners activated their weapons. From his turret, high on the back of the bomber, 'Taffy' Thomas saw their combined fire strike the single-engined fighter, causing it to break-off the attack and dive to one side of the Lancaster. Suspicious of its intentions, and wishing to keep the hostile aircraft in view, Sgt Thomas continued to observe it. Suddenly, the starboard wing of the Bf109 folded and snapped off, causing the machine to burst into flames as it fell from the sky. Glyn Thomas was credited with a confirmed 'kill', to which he later added a further three enemy aircraft damaged. His skill as an air gunner was to earn him promotion to Flight Sergeant and, on completion of his tour of duty, the award a Distinguished Flying Medal.

The luck which had precluded "J"-Jig from being attacked did not extend to Lancaster, LM142, LS-A, which failed to return from a raid on Stuttgart on the night of 24th/25th July. The bomber, which was piloted by F/L Reynolds, crashed in France with the loss of the entire crew. Thomas Reynolds, who had joined No.XV during the previous February as a sergeant, was an experienced pilot and captain. A month after his arrival on the Squadron, he was granted a commission and promoted to the rank of Pilot Officer. On 22nd May, approximately two months before his demise, he had been appointed to the rank of Acting Flight Lieutenant.

On the night of 28th/29th July, Bomber Command despatched 449 bombers to Stuttgart for the last in a series of raids on the city. No.XV Squadron detailed eighteen aircraft for the attack, four of which returned early. One of the latter was Lancaster, LL923, LS-O, piloted by F/S French. The aircraft had only been airborne for thirty-five minutes when a fuel tank, in the starboard wing, developed a leak. This did not create a problem for Johnny French, who jettisoned part of the bomb load in the Wash, off the Norfolk coast, before landing safely back at base. However, the situation invoked a high degree of nervous tension in the crew's rear gunner.

Earlier that day, in idle curiosity, Sgt Harry Bysouth cast a glance at the battle order for the forthcoming operation. Looking down the list of aircraft and crews, he was startled to see his own name listed as rear gunner to F/S French. Apprehension, in the form of cold sweat, began to invade his body when he reached the name of the target; Stuttgart!!

Having taken-off from Mildenhall, Sgt Bysouth watched the airfield recede into the distance, as the aircraft climbed away. He was still very nervous, but made every effort to conceal his emotions. Suddenly, a feeling of doom came over him as he heard the pilot announce the aircraft had developed a fuel leak. His mind raced, he felt physically sick and memories came flooding back. His first mission to Stuttgart, with Sgt Davis, had been abandoned due to the overpowering smell of petrol. The crew were given a verbal 'rocket' for their action. Their second encounter with Stuttgart concluded with five of his friends being killed. However, on this occasion, by the time they were back on the ground Sgt Bysouth had regained his composure; he realised Johnny French was an experienced and competent pilot, who he felt sure would get him through.

Harry Bysouth was by no means alone in his state of nervous tension on the mission against Stuttgart. Although he was not aware of it, another rear gunner was experiencing similar symptoms, but for a different reason. Occupying the rear turret of Lancaster, PB115, LS-W, was Sgt Harry Jackson who, along with his crew, was over an enemy target for the first time. His training had not, and could not, prepare this young man for the sights and sounds he encountered. The searchlights scanning back and forth across the night sky, the angry black balls of exploding flak, the marauding nightfighters with their illuminating flares and the sight of burning aircraft falling from the sky shook the 'sprog' air gunner. He perspired so much with fright, his electrically heated flying suit short-circuited and scorched his shirt!

One of those marauding nightfighters seen by Harry Jackson, encountered Lancaster bomber, PB137, LS-U, piloted by F/O Sanders. In the ensuing combat, the pilot of the bomber temporarily lost control of his machine, which had sustained damage to the port aileron. Fortunately, the curse of LS-U did not intervene, and F/O Sanders was able to regain control of the aircraft.

The final raid of the month, on 30th July, saw fourteen Lancasters from No.XV Squadron join a main force of nearly seven hundred aircraft for an attack against enemy positions in the Normandy battle area. The raid, which was carried out in the early morning, went without incident and all the Squadron's aircraft returned safely to base. This operation proved of great significance to Harry Bysouth, as it was the first mission he completed without mishap or incident. It was a great boost to his confidence and personal morale. He had cleared the first hurdle.

An Allied Airmen Remembered

A group photograph taken at No.1657 Conversion Unit reveals both Thomas Dunk and Phil Lamason with a number of members of aircrew who had previously served with No.XV Squadron, or were destined to do so. They include, left to right: Back Row: Rusty Norman; Tommy Dunk, XV, 1944; Phil Lamason, XV, 1944; Art Johnson; ?, Ginge Howarth; Leslie Smith; Wilbur Crebbin; Tommy Thompson. Center Row: Hank Tilson, XV, 1944; Bob Hooke; Hugh 'Wendle' Wilkie; XV, 1943 ?, Arthur 'Dicky' Craddock, XV, 1942/43; Fred Ely. ? Front: Eric Whitney; Jerry Jerromes. ex.149 Sqdn ? *Author's Collection via Jock Whitehouse*

Phil Lamason and Tommy Dunk with their crew at Mildenhall, spring 1944. From left to right: F/O K Chapman, Nav; F/O Gerry Musgrove, B/A, RCAF; F/O L George, DFC, W/Op; F/L J Marpole, F/E; S/L Phil Lamason, DFC, pilot, RNZAF; F/O Thomas Dunk, R/G; W/O Robertson Aitken, MUG. Thomas Dunk, from South Africa, and Robertson Aitken, from Scotland, were both killed when their aircraft was shot down on 8th June 1944. Dunk was buried at Plaiser Communal Cemetery, whilst Aitken buried in the Jouars-Pontchartrain Churchyard, France. Although the rest of the crew survived the action, S/L Lamason and F/O Chapman were taken prisoners of war, whilst Musgrove, George and Marpole all evaded capture. *Courtesy of Don Clarke, MBE, The Mildenhall Register*

Avro Lancaster, LM575, LS-H, the usual machine flown by S/L Lamason, photographed at Mildenhall being prepared for an operation. This aircraft was lost on the night of 7th/8th June, during an attack against railway installations at Massey-Palaisseau, France. *Courtesy of Don Clarke, MBE, The Mildenhall Register*

In June 1996, the Ville De Palaisir honored the memory of F/O Thomas Dunk by naming a street in the village after him. The upper part of the street sign carries the name of the village, whilst the lower portion bears the name of the honored airman. *Courtesy of Mme Josselyne Lejeune-Pichon*

RAF Friston

Where bombers once landed, sheep now graze. The site of the former Friston airfield on the Sussex Downs, looking towards the coastal area of Cuckmere Haven. Lancaster, LL781, LS-L, piloted by F/L Walter Bell, flew in over the sea on 8th June to make a forced-landing on the airfield. The bomber had been badly damaged following enemy nightfighter attacks over France. *Author's Collection*

Another view of the former Friston airfield (now reverted to farmland) showing the area where Lancaster, LL781, came to rest after overshooting the runway. The skill and determination shown by F/L Bell in bringing his crippled aircraft home, was recognised by the award of a D.F.C. *Author's Collection*

The Curse of LS-U

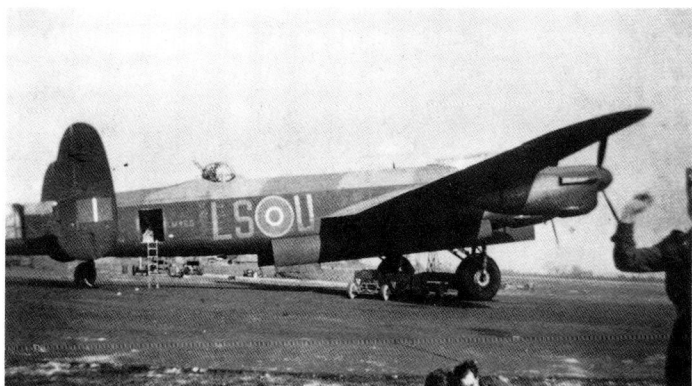

Lancaster, LM465, LS-U, photographed at Mildenhall prior to its loss on the night of 12th/13th June 1944. The American born pilot, Carlton 'Carl' Thompson, can be seen lying on the ground facing the camera. *Courtesy of Jack Trend*

The incendiary canisters are parked under Lancaster, LM465, LS-U, prior to being loaded into the aircraft's bomb bay. *Courtesy of Jack Trend*

The crew of LM465, LS-U 'Uncle', seen at Mildenhall before the 'curse' struck again, are from left to right, back row: Sgt Carlton 'Carl' Thompson (Pil), RCAF; Sgt Thomas Stubbs (A/G); F/S Ronald 'Spike' Lemky (B/A), RCAF; Sgt Richard Mobbs (A/G); front row: Sgt Maurice Pelham (F/E); Sgt Jack Trend (W/Op); F/S Roderick McMillan (Nav), RCAF; The only member of the crew to survive the action on the night of 12th/13th June, was Sgt Jack Trend, who evaded capture. The rest of the crew perished, including Carlton Thompson who was buried with the rank of P/O, and Lemky, Mobbs and McMillan, who were all buried with the rank of W/O.II. *Courtesy of Jack Trend*

Oliver Brook's Last Mission

Two photographs taken from the cockpit of Lancaster, PB115, LS-W, during the outward journey for a raid against V.1. 'Flying Bomb' sites at Domleger, on 21st June 1944. For F/L Oliver Brooks, who was piloting PB115, this mission signified both his first daylight attack and the final trip of his operational tour. *Courtesy of Oliver Brooks, DFC*

Johnny French and His Lancaster

F/S Johnny French seated in the pilot's 'office' of an unidentified Lancaster of No.XV Squadron. On 30th June 1944, the rear gunner of French's crew was killed, when his aircraft was struck by a similar machine from No.514 Squadron. *Author's Collection via the late Harry Bysouth*

Avro Lancaster, NG364, LS-P, was flown by a number of pilots on No.XV Squadron, including Johnny French and his crew. *Author's Collection via the late Harry Bysouth*

Royal Visit to Mildenhall

On 5th July 1944, during a ceremony at Mildenhall, F/L Oliver Brooks stood proudly before his Monarch to receive the award of a Distinguished Flying Cross. In the photograph, from left to right are: H.R.H. Princess Elizabeth; Air Commodore 'Square' McKee; H.M. Queen Elizabeth; The King's aide-de-camp; Group Captain R Young, D.S.O, A.F.C; H.M. King George VI and Flight Lieutenant Oliver Brooks. *Courtesy of Oliver Brooks, DFC*

Following the presentation of awards and decorations at the Field Investiture ceremony, the Royal party were introduced to members of aircrew. As the latter were standing by in readiness for an operation later in the day, they were permitted to parade in their battle dress clothing. The main party includes, from left to right: H.R.H Princess Elizabeth; G/C R Young, D.S.O, A.F.C, (Station Commander); H.M. Queen Elizabeth; H.M King George VI. The King is seen talking to F/S Smith, RAAF (3rd from right) and F/S Bell, RAAF (2nd from right). *Courtesy of and via Arthur Horton*

H.M. Queen Elizabeth chats informally with Bill Farrow whilst, a few paces behind her mother, H.R.H. Princess Elizabeth listens attentively to the conversation. *Courtesy of and via Arthur Horton*

On Sunday, 8th May 1988, nearly forty-four years after the Field Investiture at Mildenhall, Oliver Brooks (center) and two former members of his crew meet for a service of remembrance, at the Commonwealth War Grave Cemetery at Reichswald, West Germany. Although the pilot had remained friends with Ken Pincott (left) after the war, contact with the former flight engineer had been lost. However, various avenues of research re-established communications with 'Chick' Chandler a year before the meeting took place. *Author's Collection*

The Target at Wizernes

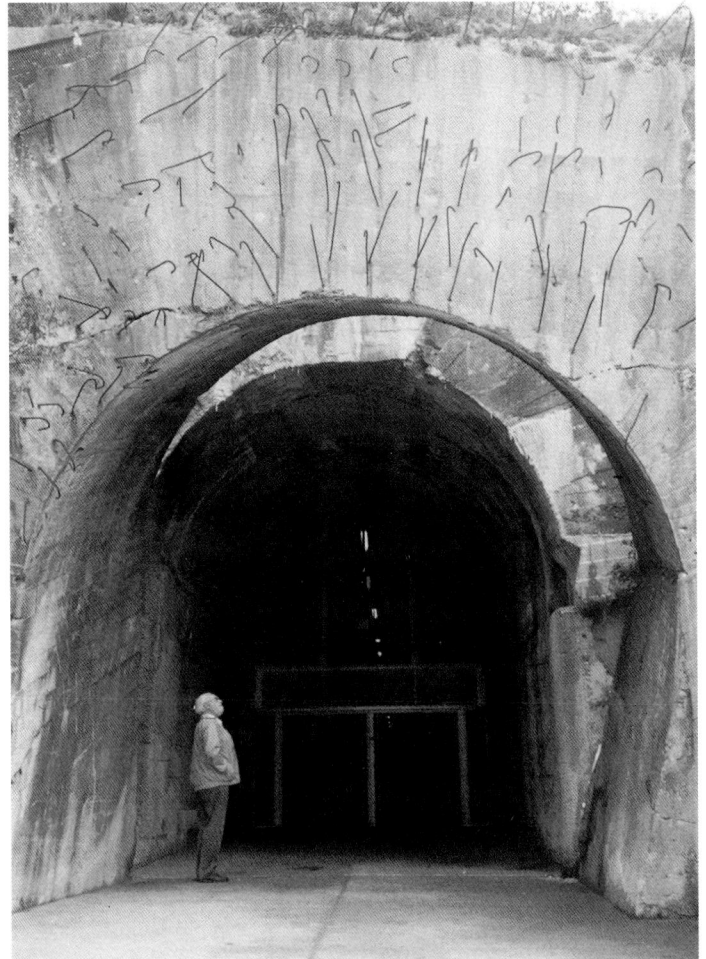

The target for night of 5th July 1944, following the Royal visit to Mildenhall, was the flying bomb storage site under construction at Wizernes, France. Attacks against the massive concrete dome and underground structures proved successful, insomuch as work on the project was stopped. The site remained untouched for many years until the late 1980s, when it was decided to convert the structure into a museum. Following restoration work to some of the tunnels and the interior of the dome, La Coupole as it is now known, opened to the public in May 1997. *Author's Collection*

Right: The concrete structure forming part of the entrance to the main tunnel is inspected by Robb Jones, the author's father and an ex-Royal Engineer with the British Army during the Second World War. *Author's Collection*

Frank Dengate's Two Crews

Frank Dengate original crew consisted of (left to right): Sgt Doug Davis (R/G); Ted (Nav); Sgt Fred Coney (MUG); F/L Frank Dengate (Pil), RAAF; F/O Joe Ell (B/A), RCAF; Sgt Bill Hopewell (W/Op); Sgt Bob Kitchen (F/E); *Author's Collection via Bill Hopewell*

F/L Frank Dengate, DFC (Pil), RAAF; Sgt Fred Coney (MUG); F/O 'Art' Cantrell, DFC (Nav), RCAF; F/S Frank Watson (W/Op); Sgt Douglas Davis (R/G); Sgt Bob Kitchen (F/E); F/O Joe Ell, DFC (B/A), RCAF; *Courtesy of Don Clarke, MBE, The Mildenhall Register*

What Curse ?

The crew who were subjected to no less than twelve separate fighter attacks in one night are, from left to right: F/S Glyn 'Taffy' Thomas (MUG); F/S Dan Capel (Pil) RAAF; F/S Edward 'Ted' McLachlan (B/A); F/S William 'Bill' Kendall (W/Op); Sgt Clive Ball (Nav); Sgt A. 'Bob' Barrett (R/G); Sgt Derek Henderson (F/E). *Courtesy of Glyn 'Taffy' Thomas, DFM*

Sergeant (later Flying Officer, D.F.C) Dan Capel, RAAF, pilot of Avro Lancaster, PB137, LS-U. *Courtesy of Glyn 'Taffy' Thomas, DFM*

Two Faces of an Air Gunner

Left: 'Taffy' Thomas, an air gunner under training, photographed in May 1943, when the uncertainty of what lay ahead showed in his face. *Courtesy of Glyn 'Taffy' Thomas, DFM* Right: Flight Sergeant Glyn Thomas, D.F.M., an experienced air gunner with one confirmed 'kill', a Bf.109 shot down, and three confirmed damaged to his credit. Glyn Thomas was recommended for the award of a D.F.M. on 10th December 1944. *Courtesy of Glyn 'Taffy' Thomas, DFM*

17

Towards the Final Chapter

Although many German cities and towns were still receiving the attention of Bomber Command, the majority of targets during early August 1944 were on French soil. Apart from assisting ground forces in the Normandy battle areas, the Royal Air Force attacked many other targets including flying bomb supply and storage sites, railway installations and oil storage depots; all productive in taking the war towards the final chapter.

Following a briefing at 11.30 hours on the morning of 4th August, fifteen crews from No.XV Squadron prepared for an immediate daylight mission. The target, an oil refinery/storage depot at Bec-d'Ambes, north of Bordeaux, was to be attacked by a total force of 288 Lancasters from Bomber Command.

The first aircraft away from Mildenhall was Lancaster, ME848, LS-N, piloted by F/O Tom Rudall, who took-off at 13.07 hours. Thirty-two minutes later the last machine climbed into the air, leveled out at 1,000', and formed-up on the other Squadron aircraft. At this altitude they flew out over the English Channel and headed towards the enemy coast, where they then increased height to 9,000' in preparation for the attack. The clear weather conditions gave unrestricted views of the target, which initially allowed the crews to identify the aiming point and bomb accordingly. However, as more and more bombs tumbled down and exploded amongst the oil tanks, great plumes of orange and black smoke billowed into the air. The clear sky, which had caused concern for the aerial gunners, darkened in places as the smoke encountered the prevailing wind currents. The gunners on the attacking aircraft however need not have worried, as no enemy fighters were reported or confronted. By the time the Squadron returned to Mildenhall most of the aircraft had been airborne for over eight hours, making this operation the longest daylight mission No.XV had undertaken to date.

Another raid, using the same tactics against a similar target at Bassens, also near Bordeaux, was undertaken the following day. Whilst the aircraft were flying over southern England prior to crossing the south coast, a breakaway barrage balloon was sighted floating through the air. Permission was sought and given for the air gunners to fire off a few rounds, as and when the opportunity arose. Those gunners within range opened fire but nothing appeared to happened, apart from the flight commander telling the gunners what a useless bunch they were. Suddenly, when the aircraft had passed, the balloon exploded in a ball of flame and fell to earth. The gunners, needless to say, felt vindicated.

Flight Sergeant Owen Sylvestre, who was born in Trinidad, loved low-level flying, and took the opportunity to get as low as possible during the flight out to the target. Having crossed the coast, he took his Lancaster, ME848, LS-N, down to wave top height. The bomb aimer, Sgt George Allom watched with youthful exuberance from his station, as water splashed up onto the plexi-glass dome in front of him. However, the fun came to an end when F/S Sylvestre hauled back on the control column and gained altitude in preparation for the bombing run.

Again, over the target, reasonably good weather enabled the participating crews to carry out a successful mission. Although no fighter opposition was encountered, the port outer propeller hub, fuselage and rear turret of Lancaster, PB139, LS-B, piloted by F/S Sleeman, was damaged by flak.

One pilot who had not flown on this operation was Johnny French, who apparently had the day off, which was just as well as he was probably out with his crew celebrating his commission and promotion to the rank of Pilot Officer.

The Squadron returned to night operations on the evening of 7th August, when fourteen aircraft were detailed by No.XV to attack enemy concentrations at Rocque Court, in Normandy. The Luftwaffe made an unwelcome return, and were responsible for a number of attacks on the bombers. Flight Lieutenant John Ball and his crew were killed when their Lancaster, NN700, LS-Q, crashed into the English Channel. The body of F/O Edmund Leah was recovered from the sea, having been washed ashore on the Sussex coast. Flying Officer Leah was buried in his home town of Todmorden, Yorkshire.

On the night of 8th/9th August, F/S Marshall, flying his second mission, took the controls of "J"-Jig for an attack against an oil depot in the Foret De Lucheaux. The rear turret was occupied by a less nervous Harry Jackson who recorded the operation as *"a nice quite trip, with flak only on the French coast"*. Again, the weather was favourable to the crews, who reported clear skies and good bombing. The German nightfighters were conspicuous by their absence, thus allowing LS-J and the twelve other participating aircraft to return home safely.

The next night "Jig" was in the hands of another crew, when F/O Moran and his crew took the aircraft to attack a fuel dump in Fort D'Englos, north-west of Lille. The Squadron had detailed only eight Lancasters for this raid, but a further six machines were detailed for a mine-laying operation in the Gironde river. Enemy fighters and flak made their presence known to the participants engaged on the former mission, but made no impact against the attackers. One German aircraft was, however, claimed as damaged by F/O Kelly and his crew.

Flying Officer Cato took "J"-Jig to Lens on 11th August, for a daylight attack against the marshaling yards. The mission was carried out without incident and the aircraft returned safely to Mildenhall. The following night W/O MacDougall took command of the aircraft, when he was named as captain of one of two crews detailed to carry out an attack in support of ground operations in the Falaise area. Unfortunately, a malfunction in the bomb switch release circuit caused a 1,000lb bomb to hang-up. The offending ordnance could not be persuaded to vacate the bomb bay, thus necessitating a return to base with the bomb.

The Squadron continued to support the Allied Armies in Normandy during the middle of August, by attacking enemy troop concentrations and the German nightfighter airfield at St. Trond. The latter operation was carried out in daylight, with the bombers crossing the English coast at Orford Ness. It was the first daylight mission carried out by Harry Jackson, who had never seen so many aircraft in the air as on that day. He almost yearned for the cover of darkness.

Fourteen aircraft from No.XV Squadron attacked the latter target, which was one of nine airfields in Holland and Belgium selected for attack by a total force of over 1,000 Bomber Command aircraft. Although in total three Lancasters were lost, the fourteen machines from No.XV, including "J"-Jig piloted by F/O Jennings, returned safely to base.

The night of 16th/17th August saw a change of tactics, when Bomber Command resumed operations against Germany. The target was Stettin, against which a total force 461 Lancasters was directed. No.XV Squadron detailed fifteen aircraft for the raid, but only fourteen got airborne. Whilst taxiing out in preparation for take-off, F/L Oliver Brook's former aircraft Lancaster, PB115, LS-W, now piloted by F/L Parke, burst a tyre. This incident did not hinder or delay the take-off period, as all participating aircraft were away within twenty minutes.

The bombing at Stettin was later reported by some crews as scattered, whilst others mentioned that due to interference, they could not hear the instructions of the Master Bomber. One of the pilots who reported the latter problem was F/O Conn, who was flying his first operation with the Squadron, since being posted to No.XV eight days earlier. Flying Officer Conn also reported the build-up of cloud over the target, which obscured visual confirmation of where his bomb load had exploded. However, some target photographs showing large areas of devastation, smoke and fires, were brought back by the returning crews.

At mid-day on 18th August, W/C Watkins led thirteen aircraft on a formation bombing practice. The Squadron C.O. acted as Master Bomber during the session, which lasted three hours. Upon their return to Mildenhall the same crews found their respective names on the battle order for a raid that night. Two additional aircraft and crews were added to the list, which named Bremen as the target. The total number of aircraft in the main force against Bremen was small by comparison with some raids, being only 216 Lancasters, 65 Halifaxes and 7 Mosquitos.

The Lancasters detailed by No.XV Squadron snaked around the perimeter track in single file, and lined up for take-off. A crescendo of noise rent the air as, one by one, the power of their Rolls Royce Merlin engines was increased in an effort to lift the heavily laden bombers into the night sky. With less than a one minute interval between them, the Lancasters roared down the runway at Mildenhall and climbed into the darkening canopy above.

A few German nightfighters were in evidence to 'greet' the attacking formations, but were decreed ineffective by the aircrews. The German flak also responded to the attackers, but this too died away before the conclusion of the raid. However, one or the other could have been responsible for the fire in the port outer engine suffered by Lancaster, PB112, LS-K. The pilot of this aircraft, W/O MacDougall, who had also lost the use of the machines rear turret, was forced to abandon the primary target and return home. Flying Officer Cage experienced similar problems, when power was lost on the starboard outer engine of his aircraft. Flying on the outward leg of the mission, Lancaster, LM473, LS-P, was crossing the enemy coast when the defective motor had to be shut down. Vernon Cage continued on to the target, which he bombed from an altitude of 7,500'. His aircraft arrived back at Mildenhall at 03.30 hours on the morning of 19th, having flown most of the return trip at an altitude of approximately 500'!!

Johnny French and his crew had further cause to celebrate on 22nd of the month, when promotions of rank were advised. These included elevation to Acting Flying Officer for French himself, and Flight Sergeant status for Harry Bysouth.

Due to the lack of operational directives being promulgated by Bomber Command, the crews of No.XV Squadron had an unexpected respite from the rigors of battle during the third week of August. They returned to the fray on the night of 25th/26th, when fifteen crews were detailed to join a main force of 412 Lancaster aircraft, for an attack against the Opel motor factory at Russelsheim. The raid, and the bravery of one particular pilot, was later to be recorded in the annals of No.XV Squadron.

During the outward leg of the mission, whilst flying over the Mannheim area, Lancaster bomber, ME844, LS-C, was hit three times by flak. The aircraft sustained damage to the engines, one of which had to be feathered. The hydraulic system was severed, thereby curtailing use of the rear turret, and a petrol tank was holed. The loss of hydraulics also meant the bomb bay doors would not operate mechanically, and therefore had to be opened manually. Although his aircraft received further damage on the run up to the target, F/L Geoff Stokes elected to continue with the mission. He made a direct approach to the target, where the 4,000lb bomb was released manually. After nearly nine hours in the air, F/L Stokes arrived back at Mildenhall where he executed a safe landing. Geoff Stokes' courage, and determination to complete his mission, was recognised two months later with the award of a D.F.C.

The following night the Squadron divided its resources and carried out two operations in the same area. Eight aircraft took-off for a mining operation off Kiel, whilst a second formation of seven Lancasters headed for the city itself. Those on the mining operation completed their task and returned without incident, except for Lancaster, ND958, LS-H, which had returned early due to equipment failure. However, the attack against the city itself was undertaken at a price. The aircraft flown by F/O William Moran, Lancaster bomber, NF952, LS-Q, which was on its first mission with the Squadron, failed to return. Another seven names were added to the Roll of Honor.

Flying Officer Alan Fleming and his crew had a lucky escape from enemy nightfighters when Stettin was again selected as the target by Bomber Command, for the night of 29th/30th August. The Australian pilot, flying Lancaster, NF953, LS-A, took-off from Mildenhall at 20.54 hours, and followed the fourteen other bombers detailed by the Squadron for the raid.

During the outward leg of the mission, NF953 was attacked by an Me410 twin-engined nightfighter. However, the crew of the bomber were vigilant and gave warning of its approach. Flying Officer Fleming threw the Lancaster into a violent evasive maneuver, which succeeded in losing the aggressor. Unfortunately, in taking this action, they had flown into the path of a prowling Bf109. Instinctively the German pilot opened-up with a burst of cannon fire which, without taking proper aim, struck the bomber in the starboard aileron. As the single-engined fighter made a passing turn aft of the Lancaster, Sergeant Rarp, the rear gunner, fired a four second burst which is thought to have struck the nightfighter. Discretion being the better part of valor, the enemy pilot broke off the attack, and unaware of the extent of the damage to his aircraft, F/O Fleming elected to abort his mission and return to base. However, his troubles were not completely over as, on the return trip, NF953, was attacked by the third enemy nightfighter. Against the odds, the crew of LS-A fought off another attack and returned to Mildenhall without injury to themselves.

Flight Lieutenant Leslie and his crew also participated in the attack against Stettin, and took their regular aircraft, "J"-Jig, on its 50th mission. They dropped their 4,000lb bomb from an altitude of 17,500', on markers in the center and to the northwest of the city. The pyrotechnics of the exploding bomb mingling with the already burning fires and illuminating flares cascading down from above. Having released the bomb load, F/L Leslie held "Jig" steady for the aiming point photograph, which was successfully obtained and brought back to Mildenhall.

A combined operation was carried out on the last night of August when ten aircraft from No.XV Squadron, led by Stirling bombers from No.218 Squadron, carried out an attack on Pont Remy, in Northern France. The bombing was carried out in formation and was considered very successful, with all crews reporting clear visibility and visual identification of the target.

Just after 13.15 hours on the afternoon of 3rd September, ten Lancasters from No.XV Squadron took-off from Mildenhall for a raid on Holland. Their target, on this particular day, the fifth anniversary of the outbreak of the war, was a Luftwaffe airfield near Eindhoven. In order to reach the target, the aircraft and crews had to deal with the vagaries of the weather. Climatic conditions on the outbound leg were less than ideal, with cloud up to approximately 20,000'. Furthermore, the bombers were subjected to severe icing. It was due to this latter problem that Lancaster, NF958, LS-M, piloted by F/S Ron Hastings, was forced to return early. The aircraft could not be coaxed to gain altitude, or maintain speed to remain with the formation, thus endangering the aircraft and its crew.

The weather conditions changed in the aircrew's favor two days later, when they carried out an attack against German positions at Le Havre. No.XV Squadron detailed fifteen Lancasters to join a main force of 348 Bomber Command aircraft, in the first of a series of raids targeted at enemy garrisons circumvented by the advancing Allied troops.

The first aircraft away from Mildenhall that afternoon, at 16.35 hours, was Lancaster bomber, LL923, LS-O, piloted by F/L Bob Cameron, RNZAF. On board the aircraft, apart from his usual crew, was Brigadier General Schlatter of the American 9th Air Force. Flying as an official observer, Schlatter could not have failed to be impressed by the visual marking of the target, and accuracy of the bombing; all carried out in excellent visibility.

A feeling of deja vous overcame the Squadron the following day, when the battle order was posted. Although some of the crews had been changed, the target was the same, the aiming point was the same and take-off time was almost the same. Furthermore, a similar number of aircraft from both Bomber Command in general, and the Squadron in particular, would be operating. One aircraft, Lancaster bomber, PB115, LS-W, would even be carrying a high ranking officer as a passenger; albeit one from the RAF! The only thing that did change was the weather, which became very cloudy with reduced visibility. However, the crews overcame the latter problem by reducing altitude and bombing from a height of 7,000'. This also enabled Group Captain Reynolds, the passenger flying with F/L Parke, to observe the action in more detail.

Le Havre was again visited on the morning of 8th September, when fourteen Lancasters from the Squadron joined a total force of 333 aircraft. The deterioration in the weather conditions continued, forcing the bombers down to a bombing altitude of between 2,500' and 5,000'. Compared to the previous raids on this target, the results were considered poor due to the overshooting or forward creep of the exploding bomb loads.

The last No.XV Squadron aircraft to have taken-off from Mildenhall for the attack against Le Havre, was Lancaster, PB137, LS-U. It was being flown by P/O Bob Jones who, along with his crew, was flying his first operational mission with No.XV. Many of his new colleagues thought it was also his last mission, due to the fact the LS-U landed twenty-three minutes after all the other aircraft had arrived safely back at Mildenhall.

Although Le Havre continued to be the subject of Bomber Command's attention again on 10th and 11th September, No.XV only flew against this target on the former date. On the latter date the Squadron was engaged in an attack against the synthetic-oil plant at Kamen.

Sergeant Jackson experienced 'flak' of a different kind during the Kamen raid, when he got a rebuke from his pilot during the flight out. The rear gunner was concerned about the proximity of another bomber, which he reported to his pilot as being on the port beam. Having checked the location of the other bomber the pilot, F/S Marshall, replied that the other machine was in fact on the port quarter, not the port beam as reported. A debate then ensued over the intercom, with harsh words being said by both parties. When Sgt Jackson was reminded of what it had cost to train him, both in time and effort, he responded by telling his pilot to concentrate on flying the aircraft. The young air gunner, who still found operational flying somewhat of a tense experience, felt close to tears and stayed silent for the remainder of the mission.

The following night the Squadron detailed thirteen aircraft for an attack against Frankfurt. Although all the bombers took-off safely, two of their number failed to return. Lancaster, LM110, LS-G, piloted by Warrant Officer Allan MacDougall, RAAF, crashed at Frankfurt at 22.40 hours, with the loss of the entire crew. Twenty minutes later, Lancaster, NF958,

LS-M, flown by Flying Officer Norman Overend, RNZAF, fell from the sky and impacted with the ground near Neckar, in the suburbs of Heidelburg. Flight Sergeant Harry Beverton, the mid-upper gunner, perished with his pilot, whilst the rest of the crew who baled out, were captured and made prisoners of war.

Despite the losses, the attack was considered successful with large areas of the city left in flames.

Bomber Command scaled down the number of operations during the middle of September giving the squadrons, including No.XV, a respite from operations. However, the rest was only temporary, and No.XV returned to the fray on the night of 17th.

A large raid, involving 762 aircraft, was directed by Bomber Command against Boulogne, where German troops were still ensconced. No.XV despatched sixteen aircraft to the French coastal town, led by the C.O. W/C Watkins. The attack, similar to the recent operations undertaken against Le Harve, saw over 3,000 tons of bombs dropped on the enemy garrison.

Further raids were made on French and German targets during the third week of September, with Calais being attacked on 20th, Neuss on 23rd and Calais again on 24th. No.XV provided ten aircraft for the latter raid on Calais, which was carried out at low-level during the late afternoon. The cloud base over the target area was only 2,500', causing most crews to bomb from that same altitude. Two aircraft were lost during the attack, including one from No.XV Squadron. Lancaster bomber, LM109, LS-E, was on its bombing run when the port outer motor erupted into flames, as flak burst around the aircraft. The pilot, F/L Smith continued with his mission whilst the flight engineer, Sgt Mayer, feathered the engine and extinguished the fire. Following release of the bomb load, LM109 was again struck by flak and received further damage. The pilot who had been wounded by this second salvo, ordered the crew to bale out. As no reports or comment had been heard from Sgt Shearer, the rear gunner, during recent events, Sgt Paterson, before obeying the pilot's order, took it upon himself to ensure his colleague was unharmed. Believing the crew to have complied with his order, the pilot attempted to leave the aircraft through the upper escape hatch, but evidence suggests his parachute caught in the hatch and refused to break free. Unfortunately, both F/L Philip Smith and Sgt John Paterson, the pilot and mid-upper gunner respectively, died in the crash. Sergeant William Shearer is thought to have been killed by the first bursts of flak. The four crew members who survived, had baled out of the stricken aircraft from an altitude of only 1,800'. They were all fortunate enough to have landed in Allied held territory where they were 'picked up' by the Canadian Army. By means of being passed down the line by various army units, the four airmen were eventually flown back to England, where they were debriefed before being returned to the Squadron on 28th September.

Owing to a deterioration in the weather over England, a number of the returning bombers were diverted to other airfields. Mildenhall being closed, some of No.XV's aircraft were instructed to land elsewhere, which precluded the Squadron from detailing a full compliment for a battle order the following day.

The Calais area of the Channel coast became the main target for No.XV Squadron during the final week of September. On each occasion the attack was carried out in clear conditions, from an altitude of approximately 5,000', with bombing being considered concentrated and accurate.

On 3rd October eleven crews from No.XV Squadron were detailed for a daylight attack on Walcheran, in an effort to breach the dyke at Westcapelle and flood the island. It was hoped this action would render the German gun batteries in the area non-effective, and assist the Allied armies in their campaign. The crews were briefed to bomb from 5,000', with strict instructions to ensure they were right on the aiming point. Conditions over the target were less than favorable, with 7/10ths cloud, but this did not deter the bombers. Sergeant George Allom, complying with the instruction given at the briefing, placed his bombs accurately and watched with delight as sea water poured through the breaches in the wall. As the dyke succumbed to the full force of the explosives dropped by a total attacking force of 252 bombers, the water level rose. In the main, results were effective but some buildings and gun emplacements remained above the tide level. Although they were surrounded by water, some of the latter continued firing at the attacking bombers, but their efforts were in vain. All Bomber Command aircraft returned safely to their respective bases.

Apart from a mining operation off Anholt island, Denmark, on the night of 4th/5th October, the early part of October saw No.XV Squadron return to attacking targets in Germany. Saarbrucken, Dortmund and Kleve all felt the weight of attacks by the Squadrons of Bomber Command, including No.XV. However, on the 14th day of the month, Duisberg was to receive an unprecedented two raids in one day. Code-named *'Operation Hurricane'*, the intention was to carry out a maximum effort attack in the shortest possible time. The inclusion of the USAAF VIIIth Bomber Force demonstrated the overwhelming air superiority held by the Allies. For the first attack No.XV despatched a total of twenty Lancasters, all of which took-off between 06.41 and 07.02 hours. First to take-off from Mildenhall was S/L Payne, on whom the rest of the Squadron formated for the flight to the target. As Duisburg was approached, the Squadron broke formation and attacked individually. Although 4/10th cloud was encountered over the target, the crews bombed visually on instructions from the Master Bomber, and later reported satisfactory results. The last aircraft to return to Mildenhall was Lancaster, HK620, LS-V, piloted by F/O Kelly, who landed at 11.45 hours.

Whilst the returning crews were being debriefed, having their post 'ops' meal and getting some sleep, the groundcrews were working feverishly, in order to prepare the aircraft for a second attack later that day. On this occasion, the Squadron despatched twenty-two Lancasters, all of which took-off between 22.06 and 22.31 hours. Flying Officer Coulter, who had been the last but one crew home on the earlier raid, was first to take-off for the second mission. He, like the rest of the crews, found hazy conditions over the target, but this did not hamper the attack. By the time the No.XV Squadron aircraft were turning for home, large fires were taking hold which in turn were emitting columns of thick black smoke.

Apart from being the day of the double raid, Saturday, 14th October also held a special significance for Australian pilot Ron Hastings. It was the day his commission in the rank of Pilot Officer was granted. Unfortunately, due to the job in hand, any celebrations the crew had in mind were put on hold.

The 18th October 1944, heralded a new operating policy for No.3 Group, when, at the instigation of its commander, Air Vice-Marshal R Harrison, Bonn was attacked by G-H blind-bombing apparatus equipped aircraft. Although other Groups in Bomber Command had been using the apparatus for some time, a number of No.3 group's aircraft had recently been equipped with the instruments, which would enable crews to undertake missions on days when cloud cover obscured ground details. Those aircraft so equipped were detailed to lead the attack, whilst the remaining aircraft would follow them in to the target area and bomb accordingly. As a means of identification, each G-H equipped aircraft had two horizontal yellow bands painted on the tailfins, immediately above the national marking.

No.XV Squadron detailed fourteen aircraft for the mission, all of which took-off safely between 08.34 and 08.57 hours. Prior to the actual attack the Squadron's aircraft divided into pairs, each of which was led by a G-H equipped aircraft from No.218 Squadron. Although some of the equipment failed, forcing a number of crews to bomb visually, the raid was considered a great success. Many fires started, causing much of the old town to be burnt out. One of those pilots who bombed visually, F/L Fleming, had the satisfaction of seeing his bombs explode in the town center. He also had the satisfaction of knowing when he touched-down at Mildenhall, where he landed at 12.59 hours, he was 'tour expired'.

The battle order for the night of 19th/20th October, detailing a total force of 565 Lancasters to attack Stuttgart, included the names of Harry Bysouth and his crew. However, although he still experienced pre-op' nerves, the thought of attacking the prescribed target no longer held the same fear for him. His aircraft, HK619, LS-Y, piloted by Johnny French, was one of nineteen aircraft detailed by the Squadron for the attack. The

crews found the target obscured by 10/10ths cloud and, although some serious damage was caused, the raid was considered to be scattered.

Two battle orders were posted by No.XV Squadron on the night of 22nd October. The first listed five aircraft for a mine-laying operation in the Baltic, whilst the second detailed twelve aircraft for an attack against Neuss. Amongst the crews listed for the former mission was that headed by F/O Marshall, for whom it was very nearly their final operation.

It had been decided earlier during the day, to assess how quickly the five Lancasters on the mining operation could be got into the air. This was to be achieved by having one aircraft mid-way along the runway whilst the bomber ahead was taking-off and another following, just starting its take-off run. At 16.15 hours the first aircraft away, Lancaster, MF849, LS-F, piloted by F/O Morrissey started its take-off run. One minute later Lancaster, PB115, LS-W, piloted by F/O Marshall, was given the green Aldis light signal from the airfield controller and commenced its take-off run. All seemed in order as the latter aircraft rolled along the runway, gathered speed and left the ground. Suddenly, at an altitude of approximately 100 feet, PB115 caught the slipstream of the preceding aircraft. The ground below took on a strange aspect, as a wing dropped and the Lancaster alarmingly veered round to starboard. At his station in the radio compartment, 'Rockie' Knight, the wireless operator, heard the pilot grunting and straining in an effort to keep the aircraft on an even keel. Eventually, much to the relief of the entire crew, F/O Marshall succeeded in his endeavors and slowly brought the Lancaster back to level flight. The seven men aboard the bomber were all shaken by their unexpected close encounter with the ground, and it took a while for them to settle down. In fact the crew of the aircraft were not the only ones who were shaken. The whole scenario had been witnessed by members of the groundcrew, who saw the Lancaster on the horizon dip below the level of the airfield perimeter. They were fully aware of the outcome had the six mines on board the aircraft exploded. Fortunately, those mines were later 'sown' in the sea off Bornholm, an island in the Baltic, south of Sweden. During the return trip, the five Lancasters were diverted to Lossiemouth in Scotland, where they landed at approximately 23.30 hours. They all returned safely to Mildenhall the following day. Needless to say the 'rapid' take off experiment was quietly forgotten, but the wireless operator still refers to the occasion as the 'Knight of shivers'.

On 23rd October, Flight Lieutenant Alan Fleming, RAAF, said goodbye to his friends and departed from Mildenhall. He had been posted, along with his crew, to No.186 (Lancaster) Squadron where, exactly five months later, he would be awarded a D.F.C.

Later that same day, Lancaster, "J"-Jig, undertook its 70th mission. Piloted by F/S Hastings, the bomber participated in an attack against Essen. This raid, the heaviest to date against the city, was to be the only occasion on which the Australian pilot and his crew flew "Jig". The Lancaster was one of fifteen aircraft detailed by the Squadron to join a total force of 1,055 Bomber Command aircraft. The attack went well, with the glow of fires being reported as reflecting through the clouds. Flak and enemy fighters were in evidence and although five bombers in total failed to return, "J" was not one of them.

Essen was again the target two nights later, when nearly 800 aircraft attacked the Krupps steelworks and other war industrial areas. On this occasion the Squadron detailed eighteen aircraft, all of which took-off. The reception over the target was unwelcome as moderate to heavy flak 'greeted' the attacking crews. As the raid progressed, the intensity of the flak decreased, and with no enemy fighters to worry about, the bomber crews could concentrate on their allotted tasks.

Consternation grew rapidly when P/O Hastings and his crew looked at the Battle Order on 28th October, and saw that the Squadron C.O. was detailed to fly on operations with them that morning. The target was listed as a daylight attack on Flushing, and the aircraft they had been allocated was Lancaster, HK620, LS-V. However, neither of these facts did anything to help them overcome their trepidation of having to fly with a Wing Commander on board the aircraft. As it turned out the C.O. himself had feelings about the matter, and deemed F/S Hasting's crew too inexperienced to act as Master Bomber crew. The former amended the battle order so as to enable him to fly with F/L Hall, who was detailed to be the first to take-off. Lancaster, ME884, LS-C, with F/L Hall at the controls, left Mildenhall at 08.33 hours. Within fifteen minutes, the thirteen aircraft detailed for the mission were airborne and heading towards the east coast of England. The attack was carried out in clear conditions and with good visibility, which assisted the bomb aimers greatly. As if to show the C.O. whether they were an experienced crew or not, P/O 'Bobbie' Burns, bomb aimer on Ron Hasting's crew, placed the bomb load on the dockside, adjacent to some oil storage tanks.

October concluded with a daylight attack against an oil plant at Bottrop, during which No.XV sustained casualties. The Squadron had detailed twelve Lancasters to join a main force attack of 101 bombers. The flight out had gone without incident, but circumstances changed over the target. With a number of crews all bombing at the same time, thick black smoke rose into the air and mingled with the cloud. Enduring the prevailing conditions, and taking instruction from his bomb aimer, F/O Robert Ostler started his bombing run. As they were nearing the aiming point, a barrage of flak exploded in the path of Lancaster, PB115, LS-W, killing Sergeant B Holt, the bomb aimer, instantly. Although the aircraft sustained some damage it remained airworthy, thus allowing the bomb load to be released by the flight engineer.

Flying Officer Sleeman had completed his task and was on the homeward leg of the journey, when the port outer engine of Lancaster, ND958, LS-H, caught fire. The flames were quickly extinguished, and the engine shut down. However, even though it proved necessary to feather the port inner engine for a short time, the aircraft returned safely to Mildenhall.

Fate took a strange twist for F/O Marshall and his crew, when tragedy struck the Squadron on 2nd November. No.XV had detailed fourteen Lancasters to join a main force of 184 bombers for a G-H attack against the synthetic oil plant at Homberg. The raid culminated with the loss of two Squadron aircraft and the deaths of twelve men. Following the operational briefing, Laurie Marshall and his crew were out at dispersal carrying out final checks on the aircraft they were detailed to fly. Due to the fact their usual machine, PB115, LS-W, was supposedly undergoing a major overhaul, they had been allocated Lancaster, LL923, LS-O. The dispersal area for the latter aircraft was opposite that of LS-W, whose engines were being 'run-up'. The thought of another crew flying the aircraft annoyed F/O Marshall, but all protestations to the Wing Commander were unheeded. The agitated pilot, whose pride was hurt, was advised the battle order had been issued and had to be adhered to. To add insult to injury, F/O Marshall had the indignity of having to line up behind LS-W, and follow his regular aircraft into the air on take-off. The raid was carried out in cloudy conditions, but most crews were able to bomb visually. Although he had not taken-off in the best of moods, F/O Marshall's aircraft was one of the first from No.XV Squadron to bomb the target. The crew later reported their bombs had slightly undershot the aiming point, but bombing later became more concentrated. It was during the return leg of the operation that fate intervened and Harry Jackson, rear gunner to F/O Marshall, witnessed the loss of their beloved *Wondering Willie*. At approximately 14.15 hours, whilst flying over Holland, LS-W collided with Lancaster, HK612, LS-L, also of No.XV Squadron. The former aircraft, flown by F/O John Hoggard an experienced pilot on his twenty-first mission, was attempting to formate on the latter machine when the two made contact. Harry Jackson watched in horror from the confines of his rear turret, as both Lancasters went down. Lancaster, PB115, LS-W, crashed approximately one kilometer south of Veghel, whilst HK612, LS-L, piloted by F/L Earley, hit the ground at Keldonk, to the south east of Veghel. Bernard Earley and his crew all perished in the crash, whilst only two men from the crew of LS-W survived. Warrant Officer Donald Garber, bomb aimer, was seriously injured but eventually recovered at an unknown hospital, whilst F/S Jeffrey, rear gunner, was only slightly hurt and was transferred to Wroughton Hospital, near Swindon, three days after the incident.

Having returned safely from this raid, F/O Capel and his crew were declared 'tour expired' and sent on leave. All but eight of the crew's 32

missions had been carried out whilst operating on Lancaster, LS-U. They had overcome the curse of the "U"-uncle code and survived to tell the tale. They were also to become the most highly decorated crew on the Squadron, earning a total of four Distinguished Flying Crosses and two Distinguished Flying Medals.

On 4th November the Squadron participated in a daylight attack against Solingen, when 176 Lancasters bombed the town center. No.XV Squadron detailed thirteen aircraft for the operation, all of which returned safely.

The following day another raid was scheduled against Solingen, with the steel works being the target. Apart from a couple of additions, more or less the same crews as before were listed for the operation. Having completed the briefing, the aircrews went about their business, passing the time until take-off. Some went out to their aircraft, whilst others engaged in other activities. Harry Jackson was lying on the grass on the airfield, indulging in 'horse-play' with F/L Cato, when the latter asked him about the previous days raid. Remembering the minimal opposition from the flak Sgt Jackson intimated, with an air of false bravado, it had been *"a piece of cake"*. This answer proved satisfactory to Hughie Cato, who announced that he had volunteered for the mission. The fifteen aircraft detailed by the Squadron took-off between 09.51 and 10.05 hours and headed towards the Rhur. Apart from the G-H apparatus on one of the aircraft failing, the raid went according to plan until a massive explosion rent the air. Again the rear turret of a Lancaster bomber was to provide the viewpoint for the drama, as Harry Jackson again watched in horror as the fireball fell from the sky. He immediately called his captain on the intercom and reported the destruction of an aircraft, but was told not to be bloody stupid. When advised it was probably a 'scarecrow' or artillery shell blowing up, the rear gunner responded by tersely replying there was no flak and asked the navigator to record the episode in his log.

Back at Mildenhall, Harry Jackson was standing outside the Squadron office when he was approached by his skipper who was obviously in some distress. Flying Officer Marshall advised Harry that Lancaster, NF916, LS-Z, had failed to return; its pilot was F/L Cato. The missing Lancaster, which was acting as a G-H Leader, was seen by other crews to have been struck by a bomb released from an aircraft flying above. The bomb detonated on impact, killing the whole crew. Apart from being fellow Australians, Laurie Marshall and Hughie Cato shared a billet together, thus adding to the former's deep sorrow and sense of loss.

Sergeant Harry Jackson, who by this time was an experienced flyer with many operational missions to his credit, had still not fully come to terms with operational flying. He felt a nervous wreck and the continuing loss of aircrews, some of whom were friends, convinced him that he would not survive. His last chat with Hughie Cato, when he made light of the attack, also played on his mind. The rear gunner had had enough and wanted to say so, but one thing stopped him; the air gunners' brevet he wore over the left breast pocket on his uniform. Like all the other airmen who wore an aircrew brevet, he had worked and trained hard to win his half-wing. The thought of the brevet being stripped from his uniform filled him with dread, and gave him the courage to go on.

Two days after the loss of Lancaster, PB115, LS-W, a replacement aircraft was taken-on-charge by No.XV Squadron and given the code LS-W. It was in this aircraft, HK626, that Air Gunner Jackson continued his war, on 6th November, when he participated with sixteen other crews in an attack against Coblenz. Although enemy fighters were active, and flak was reported as intense, all the Squadron's aircraft returned safely to base.

A bizarre incident occurred on 8th November during a daylight attack against Homberg. Lancaster, PB112, LS-K, piloted by F/O Morrissey, was one of sixteen aircraft despatched by the Squadron to join a main force of 136 Bombers. As the aircraft was being flown on a straight and level heading towards the target, F/S Schofield was in the nose of the Lancaster guiding his pilot onto the aiming point. The absence of enemy nightfighters over the target enabled the anti-aircraft gunners to pepper the sky with an intense barrage of flak, most of which appeared to be aimed at PB112. The plexi-glass dome on the nose of the aircraft, fronting the bomb aimer's compartment, reflected the pattern of the exploding flak. Endeavoring to ignore the 'fireworks', F/S Schofield prepared to release the bomb load. Approximately fourteen seconds before the Lancaster reached the aiming point, the plexi-glass dome shattered with an explosive crack as shrapnel from an exploding shell penetrated the nose of the bomber. The bomb aimer flinched in shock as he was struck on the head by flying debris and, as he instinctively endeavored to protect himself, he accidentally pressed the 'tit' on the bomb release. Fortunately F/S Schofield was not seriously hurt and, although the load had been dropped early, the bombs were seen to explode on a railway line. Both the aircraft and crew returned safely to Mildenhall, where they landed at 12.32 hours.

The next raid undertaken by Lancaster, PB112, LS-K, was an attack against the synthetic-oil refinery at Castrop-Rauxel, on 11th November. The aircraft, which was flown by P/O Alan Sellwood, was one of sixteen machines detailed by the Squadron to join a main force of 122 bombers. However, the pilot was forced to abort the mission before reaching the target, due to the aircraft's starboard inner engine developing a problem. Again, the aircraft and crew returned safely to Mildenhall.

The American First and Ninth Armies, who were about to launch an offensive on German lines between Aachen and the Rhine, issued a request for help from the Royal Air Force. In response to the request, aircraft from Bomber Command carried out attacks on the towns of Duren, Julich and Heinsberg. Aircraft from 1, 5 and 8 Groups were detailed to attack the first target, whilst bombers from 4, 6 and 8 Groups raided the second target. The last target was attacked by aircraft from No.3 Group. The raids, which were intended to cut communications behind German lines, took place in daylight, on 16th November.

No.XV Squadron, who detailed eighteen aircraft for the attack, was led by W/C Watkins who also acted as Master Bomber for the No.3 Group aircraft. William Watkins was a very experienced bomb aimer, and a highly decorated officer. As a sergeant, in 1940, he was awarded a Distinguished Flying Medal following an attack on a convoy of enemy transport, during which five runs were made though fierce anti-aircraft fire in order to ensure maximum damage with his bomb load. At the end of 1942, whilst an acting squadron leader, he was awarded a Distinguished Flying Cross in recognition of the numerous sorties he had undertaken, including a raid on Genoa where four runs were made over the target before the bomb load was released. His continued bravery, determination and leadership also led to promotion to Wing Commander, and the award of a Distinguished Service Order.

To enable him to undertake his duties as Master Bomber, W/C Watkins elected to fly with F/L Sanders and his crew on Lancaster, PB137, LS-U. Flight Lieutenant Cox, who was named as the Deputy Master Bomber, flew on Lancaster, ME844, LS-C, with S/L Payne and his crew.

The participating bombers from No.XV Squadron commenced take-off from Mildenhall at 13.02 hours, and were all airborne within twenty-five minutes. Setting course towards the North Sea, the bombers headed east and flew out over Orford Ness, on the Suffolk coast. The trip out was quiet, but it was a different story as they approached the target area. Flak was moderate to intense, and was giving some crews cause for concern. As Lancaster, HK626, LS-W, piloted by F/O Marshall, arrived over the target, the crew could hear the Deputy Master Bomber calling his leader. Several times F/L Cox was heard endeavoring to establish contact with W/C Watkins on the radio, but on each occasion there was no response from the latter's call-sign. Eventually, the D.M.B. announced to the participating crews that he was taking over command of the raid. His first action was to cancel the indicators which he observed had been misplaced, and re-mark the aiming point. The bombing, which had become scattered due to the lack of instruction from the Master Bomber, now became concentrated. Unbeknown to F/L Cox, W/C Watkins was in no position to issue instructions, as the aircraft in which the latter was flying had been destroyed. Lancaster, PB137, LS-U, which crashed in an area to the north of Heinsberg, was the only No.3 Group aircraft lost during the operation. Wing Commander William Watkins, who is thought to have been blown out of the bomber as he was trying to effect an escape from the stricken Lancaster, was the only survivor. He was later captured and spent the remainder of the war as a PoW.

Having returned from leave, in order to obtain their clearance chits and posting details, Dan Capel and his crew learned to their dismay of the loss of their beloved LS-U. The groundcrew, who had looked after the aircraft, were also fairly downhearted. The former therefore decide to entertain the members of groundcrew to an end of tour party. All went well until 'Taffy' Thomas invited 'other ranks' into the Sergeants' Mess at the end of the evening, for a few more beers. Unfortunately, due to the noise they were making, the 'illegal' party was discovered by the Station Warrant Officer who demanded an explanation. By this time Taffy had succumbed to the effects of alcohol, so it was left to Bob Barrett, the rear gunner, to give an explanation which, unfortunately, led to the groundcrew being put on a charge for 'being out of bounds'. Sergeant Barrett himself was charged with 'Inciting airmen to do wrong'. The charge against the 'other ranks' was eventually dropped on the grounds they had been invited into the Mess by the aircrew sergeants. Bob Barrett was let off with a reprimand from the flight commander. Taffy Thomas, who had instigated the whole thing and had fallen asleep in the armchair, got away with it.

The Squadron returned to operations on 20th November, when the synthetic-oil plant at Homberg was again selected as the target. However, due to a combination of inclement weather and inaccurate bombing, the attack was not considered successful.

Bomber Command, not being satisfied with the previous day's attack, sent its crews back to Homberg to complete the job. A total of 160 Lancasters were detailed for the task, of which fifteen were despatched by No.XV Squadron.

A vast improvement was found in the weather conditions over the target area, allowing the crews to see their objective. Initially the bombing was scattered, but as the attack progressed the situation improved. Large 'mushrooms' of black smoke and orange flame billowed up into the air, rising to an estimated 6,000', as bombs exploded on the factory complex. Bomber Command was pleased with the results of this attack.

Whilst the aircrews of No.XV had been attending to their duties over Homberg, Wing Commander Nigel MacFarlane, their new commanding officer, arrived at Mildenhall. Due to pressure of work, inclement weather and a lack of operations, eight days were to pass before he was able to lead the Squadron on a mission.

The early hours of the morning of Wednesday, 29th November, were clear but frosty. In the darkness, broken only by the pin-points of light from hooded vehicle lamps, the sound of voices floated across the airfield as fifteen crews stood by their respective aircraft. Some were engaged in forced chatter, brought on by pre-ops nervous tension, whilst others acted out a small good luck ritual around the tail wheel of their bomber. As the time drew nearer, each pilot motioned his crew into the aircraft to prepare for take-off. A short while later the darkness was shattered by a crescendo of noise, as sixty Rolls Royce Merlin engines roared into life. At 02.47 hours, the first Lancaster lifted off the runway at Mildenhall and climbed into the darkness. Within fourteen minutes the remaining aircraft were all airborne including Lancaster, HK695, LS-V, piloted by the new commanding officer. In the bomb bay of the aircraft was a 12,000lb high capacity 'Tallboy' bomb. The destination for this bomb, and all the others carried by the Squadron's aircraft, was Neuss where they were to attack the railway marshaling yards. Unfortunately, when they arrived over the target, the crews found the clear skies had given way to 10/10th cloud, the markers scattered and the attack running late. Wing Commander MacFarlane was not impressed. However, enemy opposition being light, the crews were able to complete their task and return to base without incident.

The last day of November saw an attack against Bottrop, when a total of sixty Lancasters from No.3 Group attacked a coking plant. No.XV Squadron detailed sixteen aircraft for the raid, all of which returned safely. One Lancaster, NF953, LS-A, piloted by F/O Mason, was forced to land at Woodbridge, Suffolk, due to a lack of brake pressure on the aircraft.

Flying Officer Hastings landed back at Mildenhall at 14.45 hours. He taxied Lancaster, PA170, LS-N, to the dispersal area where he opened the bomb doors, switched off the booster pumps and turned off the master engine cocks. Once the engines had stopped, the pilot switched off the ignition system before placing all electrical switches in the off position. It was a procedure he had carried out many times before, but on this occasion it was for the last time. Flying Officer Ron Hastings and his crew were 'tour expired'. Ron Hasting and his navigator, F/O Robert 'Bob' Smith, both returned to Australia; their native land. The rear gunner, Sgt Don 'Ike' McFadden, returned to his North London home, but could not settle down. As a youngster, 'Ike' had always been fascinated by the Wild West. It was an interest which stayed with him through the war years and thereafter. His restlessness continued until 1948, when he finally emigrated to America, acquired work on ranches in the west, and ultimately became a cowboy. Fame caught up with him in later years, due to his work in films, T.V. commercials and personal appearances.

On 2nd December, No.3 Group issued a directive for an attack to take place against the Hansa benzyl plant, at Dortmund. A total of 93 Lancasters participated in the mission, of which twelve were detailed by No.XV Squadron. Amongst the twelve bombers from Mildenhall was Lancaster "J"-Jig, which was flying its 80th operation. At the controls of the aircraft was F/O Marriott, who was piloting "J" for the first time. The Squadron flew in formation across the enemy coast unhindered, and on to the target area where they encountered moderate heavy flak. The attack, which was carried out in 10/10th cloud with the use of G-H equipment, was considered a success. All 93 aircraft returned to their respective bases, including "Jig".

It was a different story two days later, on 4th December, when the Squadron lost two aircraft and seven members of aircrew. No.XV, who had been selected to lead the raid, detailed thirteen aircraft to participate in an attack on Oberhausen, against which a total of 160 Lancasters from No.3 Group had been directed. Take-off from Mildenhall commenced at 11.45 hours, with the Squadron getting all its aircraft airborne within twenty minutes. Lancaster bomber, HK626, LS-W, piloted by F/O Davies, took-off at 11.51 hours, followed one minute later by Lancaster, PA170, LS-N. The latter aircraft, which had been taken on charge during the middle of October, was flown by F/O Ostler.

The weather conditions and circumstances of the attack were almost identical to the raid undertaken two days before, when 10/10th cloud prevailed. There had been no intervention by enemy fighters on the outbound journey, but as the bombers approached the target area moderate heavy flak began to pepper the sky. The deadly black balls of exploding metal found a victim when Lancaster, PD419, LS-P, flew into the path of the flak. The machine, which was flying in the No.2 position of the leading 'Box' formation of four aircraft, sustained extensive damage and dropped out of formation. The pilot, F/O Bob Jones, was incapacitated having received blast wounds to his face, which induced temporary blindness. Although in a state of severe discomfort, he remained composed enough to regain control of his aircraft and pull out of the dive. With the help of F/S Gordon Sims, the wireless operator. the pilot slowly regained his position in the formation, from where he followed through with his attack as briefed.

As the lead formation from No.XV Squadron flew over the target, a huge explosion rocked the air. A brilliant flash lit up the clouds, as a multicolored fireball filled the sky. The fireball, which a few seconds earlier had been Lancaster, HK626, LS-W, took the lives of seven young men. As the aircraft, which had been flying No.3 in the 'Box' disintegrated, Derek Lord, the flight engineer on PD419, watched in horror as debris flew into the path of Lancaster, PA170, LS-N. As the bomber, which was flying No.4 in the formation, dropped away and disappeared from sight, the fireball turned to a pall of dense black smoke. It hung momentarily in the air before dissipating and drifting away on the wind.

Derek Lord was to learn later that Flying Officer Ostler, the pilot of LS-N, managed to make a belly-landing behind Allied lines without injury to himself or his crew.

Amongst the crew members who perished on Lancaster, HK626, were F/O Vincent Davies, W/O Jack Lowe and F/S Robert Skilbeck all from Australia. The latter airman, who was wireless operator, had completed twenty-five operational missions. He enlisted for service in his native Victoria and having been accepted for aircrew, was sent to Britain via America. Although he had visited many famous places during his training

and travels, he treasured one outstanding memory which occurred whilst on a visit to Scotland, during the latter half of 1944. Having decided to spend some time in Edinburgh whilst on leave, Robert Skilbeck was privileged to meet the King and Queen of England. Details of his travels, and who he met, were recorded in his letters home; although he had never forwarded details of his crew to his family. This omission was rectified on 23rd November in a letter to his younger brother Ralph. The letter, which contained names and details of each member of the crew, was the last communication Robert Skilbeck had with any member of his family. Twelve days later he, along with the rest of the crew, paid the supreme sacrifice.

However, F/O Bob Jones was still in trouble and relying on directions and instructions from his wireless operator. He was determined to get both his aircraft and crew home safely. It was not until they were flying over the sea, on the homeward leg of the mission, that the pilot regained the sight in one eye. Eventually , he too, like F/O Ostler, made a safe landing in a damaged bomber. For his courage, determination and fortitude, F/O Jones was granted the immediate award of a Distinguished Service Order.

The Oberhausen raid was the last mission Lancaster bomber, HK647, flew wearing the LS-L code. On its return to Mildenhall the aircraft was taken out of service, overhauled, and given a change of identity. It returned to operations just over a month later, adorned with the code letter "K". The 'rejuvenated' bomber flew its first mission as LS-K on 7th January 1945, when F/L Clayton took the aircraft to Munich.

During the first two weeks of December, a number of targets in Germany were attacked by No.XV, including the Schwammenauel Dam in the Roer Valley, on the 5th, when F/O David Bone undertook his first mission with the Squadron. David Bone was a very experienced pilot, who had already completed a tour of operations flying Blenheim bombers with No.18 Squadron. On his tunic he wore the ribbon of the Distinguished Flying Medal, which he had been awarded just over a year earlier. He arrived at Mildenhall late in the day on 16th November, and was escorted to the Officers' Mess, where he met two members of his new crew. Like himself, Andy 'Butch' Hayden, who had previously flown with F/L Meggison, and 'Scotty' Dunlop were both second tour men.

The raid on the Schwammenauel Dam (across the River Roer), a daylight mission carried out in brilliant sunshine, went without incident. No enemy fighters were seen, and there was minimal flak over the target area. An escort of Spitfire and Mustang fighters wheeled around the sky above the bombers, taking advantage of the conditions in which they found themselves. It was the ideal sort of raid for a 'new' crew to get to know each other under operational circumstances.

Other targets attacked during December included Merseberg (6th), Duisburg (8th) and Osterfeld (11th). The raid against Duisburg witnessed the ability of another new crew who had recently joined the Squadron. Flying Officer Ken Lewis and his crew were detailed for a daylight attack on the railway yards. Piloting Lancaster, LM238, LS-T, Ken Lewis took-off at 08.37 hours, the fifth of thirteen aircraft detailed to do so. The mission went without incident, apart from the fact the new pilot had to face the prospect of landing with a 1,000lb bomb still hung-up in the bomb bay. However, the aircraft landed safely at Mildenhall, and both aircraft and crew survived to fight another day.

Unfortunately, an attack against Witten on 12th December had similar overtones to the mission against Oberhausen, eight days earlier. As with the latter raid, the Squadron detailed thirteen aircraft for the attack, which was against the Ruhrstahl steelworks. It too was carried out in poor weather conditions, with low cloud and reduced visibility. All aircraft bombed as directed, but as F/L Marsh flew over the target his aircraft took a direct hit from flak. The Lancaster, HK627, LS-F, exploded and fell from the sky. It crashed in flames at Stockum, north east of the target, killing the entire crew.

A few minutes after 13.00 hours on Thursday, 14th December, a solitary aircraft lined-up for take-off on the runway at Mildenhall. The Lancaster, NG357, LS-K, piloted by F/L Percy, was the sole representative from No.XV Squadron, detailed for a mine-laying mission in the Baltic. At six minutes past the hour, F/L Percy increased the throttle power, released the brakes and LS-K surged forward. The aircraft lifted off the runway and climbed into the grey overcast sky. The weather conditions did not relent and, having dropped six mines into the sea using H2S, F/L Percy was forced to land at Lossiemouth, Scotland, on the return journey.

In the early hours of Saturday, 16th December, the German Wehrmacht launched its last great offensive in the Ardennes area of Belgium. Under the command of Field Marshal von Rundstedt, Army Group B was intent on striking westward and dividing the advancing Allied armies. The operation, which was also devised with the intention of capturing Brussels and Antwerp, took the Allied Armies completely by surprise. The inclement weather was found favorable by the German troops, who used the low cloud and mist cover to gain ground. Unfortunately, the Allies did not share the same sentiment as the enemy, as it was to be a full week before either the RAF or the USAAF were able to assist their colleagues on the ground.

The weather over England that same day consisted of low cloud and poor visibility, but this did not prevent 108 aircraft from Bomber Command from attacking the railway yards at Siegen. No.XV Squadron despatched twelve aircraft, all of which left Mildenhall between 11.28 and 11.36 hours. Lancaster, NF957, LS-X, piloted by F/S Williams, was forced to return early due to severe icing. The remaining eleven crews pressed on and found the now customary 10/10th cloud blanketing the target. Being experienced at blind-bombing techniques, they completed their task and turned for home. Although there had been no evidence of flak over the target area, the crews encountered both hostile fire and enemy fighters on the return journey. One particular burst of flak, which peppered the sky around the bombers, sent shards of hot metal into both the port and starboard inner engines of Lancaster, NG357, LS-K. Fortunately the pilot, F/O Slaughter, managed to make an emergency landing on a friendly airfield at Louvain, Belgium. The remaining aircraft all returned safely to Mildenhall.

The railway yards in the ancient town of Trier, Germany, approximately ten kilometers from the German/Luxembourg border, were the subject of daylight attacks on 19th, 21st and 23rd December respectively. On each occasion the flak was light to moderate in intensity and, although the cloud thinned during the last attack, no enemy fighters were evident. However, Bomber Command did lose one of its aircraft during the latter attack, which was later reported as the worst raid the town had had to endure during the war.

One Squadron member who was feeling somewhat fed up at this period of time was F/O Bone, who had been grounded by the Medical Officer on 15th December due to an attack of impetigo. Another visit to the M.O., on the 19th, saw the pilot grounded for a further two days. Although two days later the impetigo had improved, David Bone was frustrated to learn he was showing the signs of a stye to his right eye. The outcome was the M.O. grounded him for another two days!

An attack, scheduled for Christmas Eve, was canceled early during the day, when the continuing bad weather cast fog over East Anglia. However, not wishing to deprive the enemy of a 'present', when the weather improved No.3 Group re-briefed its crews. The revised attack, using night bombing tactics, was directed against Hangelar airfield, to the north east of Bonn. The fourteen aircraft detailed by No.XV Squadron for the mission commenced take-off at 15.13 hours, but all did not go according to plan.

As Lancaster, NF957, LS-X, climbed into the darkening sky, the port inner engine developed a fault. The pilot, F/L Rosenhain, had no option but to gain altitude before feathering the defective motor. Having jettisoned the bomb load in the North Sea, F/L Rosenhain turned back over the coast and landed the aircraft safely at Woodbridge, the emergency airfield north east of Ipswich.

As far as No.XV was concerned, the attack against the enemy airfield was virtually unchallenged. Flak was not in evidence over the target and, although combats with enemy fighters were seen and reported, no opposition was directed at "Oxford's 'Own'". One Lancaster from No.622 Squadron, No.XV's daughter Squadron, was lost along with four members of its crew. For the crews of No.XV however, the only agitation appertaining to the mission came when they were diverted to Shipdham, home of the 44th Bomb Group, 8th USAAF, due to fog closing down their home base. Such were the weather conditions that the aircraft were left on the American airfield, whilst the crews were returned to Mildenhall by RAF transport.

Persistent fog shrouded the dawn on the morning of 25th December and brought with it a heavy hoar-frost, which covered everything in white crystals. Aircraft which had not participated in the last mission stood on the airfield, their shapes penetrating through the fog in ghostly fashion. The Christmas festivities, and the weather, provided a welcome respite from the rigors of operations for all crews, and gave many the opportunity to observe some of the rituals associated with the occasion. In time honored tradition, the Messes were decorated accordingly and, the Officers served Christmas lunch to the other ranks.

The twelve aircraft which were left in the care of the 44th Bomb group at Shipdham, were still there two days after Christmas. On the morning of 27th December, arrangements were made for crews to bring eight of the Lancaster back to Mildenhall during the course of that day. The four remaining machines, all of which were fitted with G-H equipment, were flown to Woodbridge from where they would operate.

On 29th December, F/L Johnny French and his crew became 'tour expired', following a raid against Koblenz. Having landed back at Mildenhall at 17.25 hours, after an uneventful raid, he shut down the engines of Lancaster, NG364, LS-P, for the last time. As the motors cooled down, and the airframe creaked and groaned, Harry Bysouth prepared to leave the rear turret. His task was complete, and all thoughts of the 'jinx' he thought he carried were erased from his mind.

The last day of 1944 saw an attack against the railway yards at Vohwinkle, with two of the aircraft detailed to operate from Woodbridge joining the raid. The usual conditions prevailed, with respect to both the weather conditions and limited enemy resistance. The fact no losses were again incurred by the Squadron, may well have led some to believe the war was all but over. However, although weakening rapidly the German Forces still had some fight left in them, as the New Year would show.

Although still suffering from less than A.1. health, F/O David Bone would show he too still had some fight left in him. He was sent on leave over the Christmas and New Year period, but reported for a medical in London, on 2nd January 1945. The news he received there was music to his ears. He was passed fit and recommended for a return to normal duties.

Low-Flying Bomb Aimer

Harry Bysouth's New Crew

Having overcome his difficulties in returning to operations, Harry Bysouth completed a 'tour of ops' with Johnny French and his crew. Seen grouped around the crew entry door of Lancaster, NG364, LS-P, are: F/S Harry Bysouth, R/G (standing in doorway); F/S Hope, W/Op (sitting in doorway); F/L Johnny French, Pilot (standing 2nd from left); W/O Stubbings, Nav; P/O Smith, B/A; P/O Slingsby, MUG; W/O Raine, F/E. *Author's Collection, via the late Harry Bysouth*

Sergeant George Allom, bomb aimer to F/S Owen Sylvestre's crew. *Author's Collection*

George Allom (center) chats to his ex-pilot Owen Sylvestre (right) outside the Bird in Hand Hotel, whilst 'Ernie' Fitch looks on. The latter flew as mid-upper gunner to the crew. *Author's Collection*

A portrait of Harry Bysouth taken from the photograph sewn into the lining of his uniform, as was the practice of RAF aircrew. In the event being of shot down, the original photograph (20mm square) could be used on any false papers issued by a local resistance unit rendering assistance to the airman. Fortunately Harry Bysouth never needed to make use of this photograph. *Author's Collection, via the late Harry Bysouth*

The Raid on Stettin 1

F/L L Conn (5th from left) and his crew at Mildenhall during 1944. This pilot flew his first mission on the night of 16th/17th August 1944, when he took Lancaster, LM238, LS-T, to Stettin. *Courtesy of and Via Ken Lewis, DFC*

F/L John Sleeman, D.F.C., RAAF, (Center) and his crew at Mildenhall, also participated in the attack against Stettin. *Courtesy of Ken Lewis, DFC*

Bill Parkes and Geoff Stokes

F/O Geoff Stokes (right) with F/O Bill Parkes photographed at Mildenhall, in front of Lancaster, LM238, LS-T. Geoff Stokes was to be awarded a D.F.C. in October 1944, following an action the previous August when he continued with his mission in a severely crippled aircraft which he subsequently brought home. *Author's Collection via the late 'Gary' Cooper*

Avro Lancaster, ME844, LS-C, was severely damaged during an attack against the Opel motor factory, at Russelsheim, on the night of 25th/26th August 1944. Following major repairs, the aircraft flew further operations with No.XV Squadron as LS-W, before being taken-on-charge as KM-D by No.44 Squadron. *Author's Collection via the late 'Gary' Cooper*

The Raid on Kiel

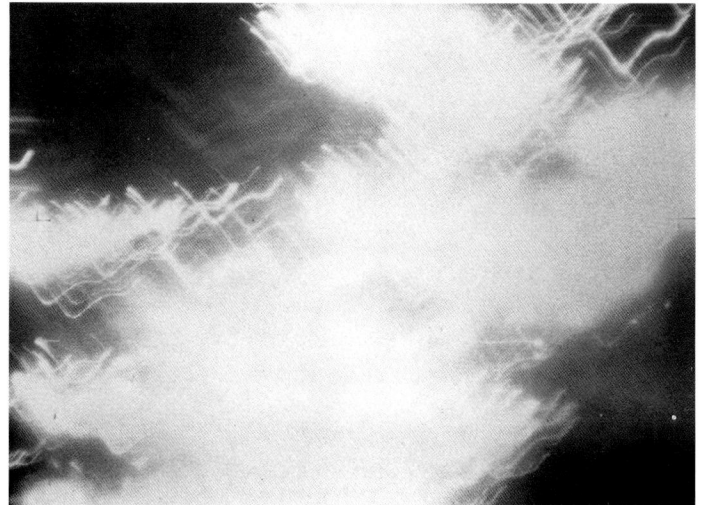

On the night of 26th/27th August 1944, F/L Leslie and his crew took the second "J"-Jig to bomb Kiel. The Squadron divided its resources into two sections, with eight aircraft laying mines off Kiel, whilst the remaining seven aircraft bombed the city. The crew consisted of, from left to right: F/L William Leslie (Pil); F/O Frank Frudd (Nav); Sgt Bill Gundry (F/E); F/S 'Rosie' Rozier W/Op); Sgt 'Red' North (MUG); F/S 'Fen' Fendley (R/G); F/O 'Maxie' McNiece (A/B). *Courtesy of Bill Gundry*

Bombing photograph taken over Kiel, from Lancaster, LL806, LS-J, on the night of 26th/27th August 1944. The aircraft was flying at an altitude of 17,000'. *Courtesy of and via Bill Gundry*

The Raid on Stettin 2

A painting, by aviation artist Keith Aspinall, depicts the events of the night of 29th/30th August 1944, when F/O Alan Fleming evaded the attention of two enemy nightfighters during an attack on Stettin. *Author's Collection*

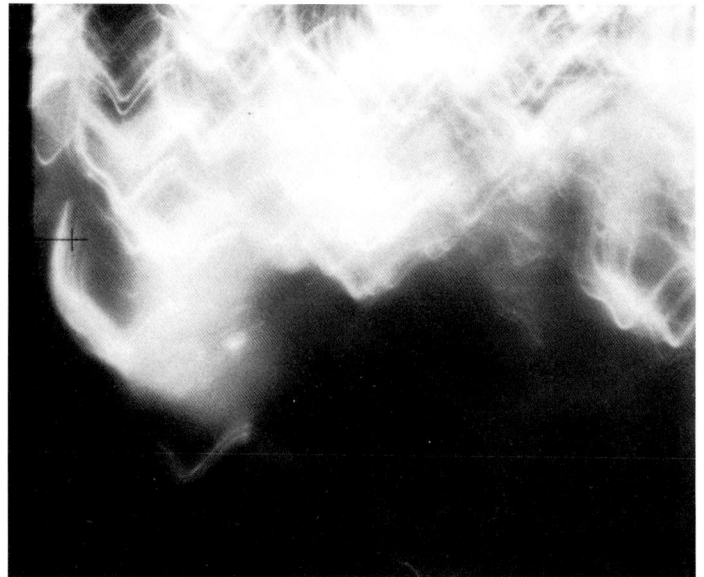

Flares over Stettin, as seen from Lancaster, LL806, LS-J, on the night of 29th/30th August. F/L Leslie and his crew notched-up the aircraft's 50th mission on this operation. *Courtesy of and via Bill Gundry*

The 'Aussie' and the New Zealander

Australian pilot F/O Ron Hastings (crouching) and his crew. From left to right are: F/O Bob Smith (Nav), RAAF; Sgt Mike Malyon (MUG); Sgt Don 'Ike' McFadden (R/G); F/S Vic Pearce (W/Op); P/O 'Bobbie' Burns (B/A), RAAF; Missing from the photograph is Sgt Jock Munro, the flight engineer. *Courtesy of Ken Lewis, DFC*

New Zealand pilot S/L Bob Cameron (5th from left) with members of his crew, which included F/S Fletcher, W/Op (1st from left); Sgt Frank Mckenna, F/E (4th from left); P/O Jim Glasspool, Nav (6th from left); Sgt Bert Mead, MUG (7th from left), F/O Williamson, B/A, and Sgt Hamlett, R/G. A prewar policeman, Frank McKenna eventually rose to the rank of Squadron Leader. Towards the end of the war he obtained a posting to the Special Investigations Branch of the RAF, where he was responsible for tracking down the majority of the Gestapo officers involved with the slaughter of the fifty British PoWs who escaped from Stalag Luft III, in 1944. *Courtesy of Ken Lewis, DFC*

The Loss of Lancaster, LM110

Avro Lancaster, LM110, LS-G, was taken-on-charge by No.XV Squadron during the first week of June 1944. This aircraft was flown by a number of pilots, including F/O Geoff Stokes, F/O William Leslie, F/L Johnston and W/O MacDougall. The latter pilot was killed, along with his crew, when the bomber crashed at 22.40 hours, on the night of 12th September 1944, whilst participating in an attack against Frankfurt. *Author's Collection*

Morrissey's Crew and a Lanc Coded "F"

F/O Morrissey's crew, having disembarked from a train, pose for a photograph in the station yard, before continuing their journey. The crew consisted of F/O Mayne, Nav, RAAF (left, back row); Sgt Feit, F/E (center, back row); F/S Scholfield, B/A (right); Sgt Murphy, W/Op (left, front row). The two gunners, both kneeling in the front row are Sgt Walker and Sgt Clisby. Note the railway porter in the background holding a parcel. The photograph was taken by F/O Morrissey. *Author's Collection*

On the night of 22nd October 1944, F/O Morrissey took Lancaster, MF849, LS-F, on a mine-laying operation in the Baltic. This aircraft, coded "F" is obviously undergoing engine tests. Note the exposed starboard inner engine and the covered tyre of the undercarriage. *Author's Collection*

Most Decorated Crew

Beating the curse of an aircraft coded "U"-Uncle. Having completed a 'tour of operations' with No.XV Squadron the aircrew, along with members of their groundcrew, pose for a photograph with Lancaster, PB137, LS-U. From left to right, standing: Sgt 'Bob' Barrett, D.F.M. (R/G); P/O Clive Ball, D.F.C. (Nav); Groundcrew; F/O Dan Capel, D.F.C. (Pil); Groundcrew; F/S Glyn 'Taffy' Thomas, D.F.M. (MUG); Sgt Derek Henderson (F/E). Left to right, kneeling: Groundcrew; P/O Edward 'Ted' McLachlan, D.F.C. (B/A); Groundcrew; P/O William 'Bill' Kendall, D.F.C. (W/Op). *Courtesy of Glyn 'Taffy' Thomas, DFM*

The feeling of joy that a No.XV Squadron aircraft coded LS-U, had completed a 'tour of operations', can be seen on the faces of the groundcrew who serviced the aircraft. Their charge, Lancaster, PB137, LS-U, can be seen in the background. *Author's Collection*

The Air Gunner who Persevered

Left: Nervousness shows on the face of Leading Aircraftman Harry Jackson, was photographed in full flying kit, at Air Gunnery School, Morpeth, on 23rd November 1943. *Author's Collection via Harry Jackson*

Above: A more confident Harry Jackson, photographed in full operational flying clothing, with Sergeant Jack Kay, RAF Mildenhall on 23rd November 1944. This particular date signified the end of the two air gunner's association with Lancaster, LM238, LS-T, seen in the background. The fact they still had three more missions to undertake before completion of their 'tour of operations', may account for the worried expression on Jack Kay's face. *Author's Collection via Harry Jackson*

F/O Alan Sellwood

Three members of F/O Alan Sellwood's crew were, from left to right: Sgt William 'Bill' Blencowe, R/G; F/S Rodney Pope, A/B and Sgt Danny Boon, MUG. After the war, Rodney Pope took was ordained as a Minister in the Church. *Courtesy of Don Clarke, MBE, The Mildenhall Register*

Left: Flying Officer Alan Sellwood, RAAF, who was forced to abandon his attack on the synthetic-oil refinery at Castrop-Rauxel, on 11th November 1944, due to problems with the starboard inner engine of Lancaster, PB112, LS-K. *Authors Collection via Len Miller, DFC*

Tour Expired

Left and center: Sergeant (later Warrant Officer) Don McFadden, air gunner to F/O Ron Hastings and his crew. On completion of his service as an air gunner with No.XV Squadron, and his de-mob from the Royal Air Force, Don McFadden made a decision which changed his life forever. As a young boy, born and bred in North London, he had always had an interest in cowboys and the American West. In 1948 he emigrated to the United States, where he eventually became an actor in both western films and commercials. His interest expanded into Rodeos and similar western festivals. Known as 'Ike', the ex-air gunner who swapped his four .303 Browning machine guns for two Colt 45s, is a well-known and much loved cowboy. *Author's Collection* Right: Flying Officer Bob Smith, Nav, RAAF (left) was another member of F/O Ron Hasting's crew. He is seen here, in Trafalgar Square, London, with P/O Jim Glasspool who flew as navigator to S/L Cameron's crew. The lion in the background of the photograph is one of four such statues at the base of Nelson's Column. Unbeknown to the two tour-expired airmen, the Author's great-great grandfather was a member of the team which assisted in the construction of the lions during the 1860s. *Courtesy of Bob Smith*

Lost Over Oberhausen

Sgt Pat Wilson (left) and F/S Robert Skilbeck, photographed in July 1944 in Cambridge, both flew as part of F/O Vincent Davis's crew. They were killed on 4th December 1944, when their aircraft exploded in mid-air following a direct hit from enemy flak. *Courtesy of Ralph Skilbeck*

Left: Robert Skilbeck, photographed wearing flying clothing, whilst undertraining as a wireless operator with the Royal Australian Air Force, during 1943. *Courtesy of Ralph Skilbeck*

Lancaster, PA170, LS-N, photographed after its forced landing on 4th December 1944. The aircraft was struck by flying debris following the mid-air explosion on board Lancaster, HK626, LS-W. Flying Officer Ostler, the pilot of LS-N, managed to get his aircraft down without injury to himself or his crew. *Courtesy of and via Flt/Lt T. N. Harris, XV (R) Squadron*

Above and two below: Three photographs of Lancaster, PA170, LS-N, which force-landed behind Allied Lines, near Esch, on 4th December 1944. *Courtesy of and via Ad van Zantvoort*

Confused Identity

Lancaster, HK647, LS-L, being prepared for a forthcoming operation. This aircraft flew its last mission as "L" on 4th December 1944, when it participated in an attack against Oberhausen. On its return to Mildenhall, the bomber was recoded LS-K. *Courtesy of and via Flt/Lt T. N. Harris, XV (R) Squadron*

'Jonah' Jones

When Lancaster, HK647, returned to service as LS-K on 7th January 1945, F/L Clayton (standing 4th from left) flew the 'rejuvenated' bomber on a mission to Munich. Although Henry Clayton and his crew flew LS-K on a number of occasions, they eventually took over Lancaster, NG358, LS-H. *Courtesy of Don Clarke, MBE, The Mildenhall Register*

Flying Officer Bob 'Jonah' Jones was granted an immediate award of the Distinguished Service Order, following an action over Oberhausen on 4th December 1944. Although temporarily blinded by flak F/O Jones, through the assistance of his crew, managed to fly his severely damaged aircraft back to Mildenhall. *Courtesy of Ken Lewis, DFC*

'Aussie', 'Tommy' and Crew

Lancaster, LM238, LS-T, was the first aircraft F/O Ken Lewis, RAAF, flew on joining No.XV Squadron. His first mission with 'Tommy', on 8th December 1944, concluded with a 1,000 lbs bomb hung-up in the bomb bay. Seen with Ken Lewis (left, back row) are four members of the aircraft's groundcrew. Written in chalk across the RAF roundel are the words, *"BARKER OF BOMBER COMMAND"*. The significance of this statement is unknown, but it could relate to the surname of one of the groundcrew. *Courtesy of Ken Lewis*

Right: Whilst Sgt 'Paddy' Langan sits in his 'office' at the rear of Lancaster, NG444, LS-Y, the rest of the crew assemble under the guns for an informal photograph. From left to right are: F/S Roy MacDonald (W/Op) RAAF; F/L Ken Lewis (Pil) RAAF; F/S Bellis (Nav); Sgt Watts (MUG); Kneeling at the front are: Sgt Parry (F/E) and F/S Cowie (B/A). *Courtesy of Ken Lewis, DFC*

Men Who Flew "J-Jig"

Flight Lieutenant Hopper-Cuthbert and crew. *Courtesy of Ken Lewis, DFC*

Pilot Officer Geoff Stokes, D.F.C., and his crew. From left to right, back row: Sgt Eric Bradbury (W/Op); Sgt Bill Scannell (F/E); Sgt Paddy Carson (MUG); Sgt 'Gary' Cooper (R/G). Seated, front row: P/O Wally Airey (Nav); P/O Geoff Stokes (Pil); P/O Arch Rattray (B/A) RCAF. *Author's Collection via the late 'Gary' Cooper*

Left: Flying Officer Marshall and his crew, photographed in December 1944. From left to right: Sgt Tim Bates (Nav); F/O Les Ford (B/A); Sgt Harry Jackson (R/G); F/S Ron 'Rocky' Knight (WAG); Sgt Jack Kay (MUG); F/O Laurie Marshall (Pil); Sgt Ken Kenny. *Courtesy of Ken Lewis*

Flight Lieutenant D Kelly (Center, front row) and his crew. *Courtesy of Ken Lewis DFC*

Squadron Leader Patrick Percy, D.F.C., from Argentina, joined the RAF and flew with No.XV Squadron. *Courtesy of Ken Lewis, DFC*

F/L Bruce Giles, D.F.C., RAAF (extreme left) and his crew. Two members of the groundcrew occupy the flight deck of the aircraft. *Courtesy of Ken Lewis, DFC*

Flight Lieutenant James Slaughter, D.F.C., RAAF (center, front row) with both his aircrew and groundcrew utilises the wingroot of Lancaster, NG365, LS-N, for a group photograph. *Courtesy of Ken Lewis, DFC*

Pilot Officer Bill Mason, RAAF, together with four members of his crew. From left to right are: F/S P Elger (W/Op); P/O Bill Mason (Pil); F/S D Brown (B/A); Sgt H Leigh (F/E); Sgt Norman Bibbing (MUG). *Author's Collection*

Flying Officer Burn's crew consisted of, from left to right: Sgt E Doble (B/A); Sgt Stan Franks (F/E); Sgt M Giddings (R/G); F/O Norman Burns, D.F.C. (Pil): F/S J Nicholson (W/Op); F/S R Dukes (Nav); F/S E Deakins (MUG). *Author's Collection*

The men who kept "J"-Jig flying – the members of the aircraft's groundcrew included, from left to right: 'Cookie' Cook; Unidentified; John Pratt (kneeling); Norman Pratt; Bill Foy. *Author's Collection via John Pratt*

18

"Manna" and "Exodus"

During the early hours of Monday, 1st January 1945, the Luftwaffe High Command launched Operation *'Bodenplatte'* (Baseplate), an aerial offensive directed against Allied airfields in north-west Europe. The objective of the attack, which was to be led by Major General Peltz, was to destroy Allied fighter aircraft based on Continental airfields before they could leave the ground.

Although the offensive had deliberately been contrived to occur during poor weather conditions, fortunately for the Allies, due to inadequate planning, the attack was a disastrous failure. The Luftwaffe lost approximately 100 of its fighters to its own anti-aircraft batteries, who had not been made aware of the operation. The German ground gunners, detecting a mass formation of fighters heading in their general direction, immediately went on the defensive and opened-fire. Including aircraft lost to Allied causes in the short space of just under four hours, the already weakened Luftwaffe lost just over 230 pilots. However, when 143 Lancasters from No.3 Group, Bomber Command attacked Vohwinkel later that same day, the Luftwaffe responded with both fighters and flak. As on previous occasions the target at Vohwinkel was the railway marshaling yards where, apart from the flak barrage which greeted their run up to the aiming point, the attacking crews found clear skies. No.XV Squadron had detailed twelve Lancasters for the raid, but during the day two of the allotted aircraft went unserviceable. The ten remaining machines commenced take-off at 16.00 hours, led by F/L Cyril Hagues. The moderate but accurate flak which greeted them was assisted by searchlights sweeping across the night sky. As shells burst around the bombers sending hot splinters of metal in all directions, damage was inflicted on the rear turrets of two No.XV Squadron aircraft. Lancaster bomber, NF957, LS-X, piloted by F/S Blaxendale, remained airworthy and managed to fly back to England, although it was forced to land at Tuddenham. The second aircraft, Lancaster, NG365, LS-N, flown by F/L Marriot, was less fortunate and crash-landed in Belgium with a wounded rear gunner.

The following night, F/O Hopper-Cuthbert piloted Lancaster "J"-Jig on its 95th mission, when the Squadron detailed thirteen bombers to join a main force of over 500 aircraft for an attack against Nuremberg. The raid was carried out in clear conditions, with good visibility and a rising full moon. The fact there was very light flak over the target assisted in producing a first rate attack. All the Squadron's aircraft returned safely to base, including LS-J.

On rising from his bed at 07.30 hours on the morning of 5th January, David Bone found the day had dawned with grey overcast skies and light snow showers. He also found he was nursing a headache, which was the result of a binge in the Bird in Hand Hotel and the Officers' Mess the night before, to celebrate his promotion to Flight Lieutenant. This did not however prevent him from participating in an attack against Ludwigshaven later that day.

Another cause for celebration that same morning, although no drink was imbibed, was the imminent mission to be undertaken by Lancaster, "J"-Jig. The raid against Ludwigshaven was to be the aircraft's 100th operation and, in honor of the occasion, a number of photographs were taken before the machine's departure.

As the morning progressed the skies showed signs of clearing, bringing the promise of a bright, sunny, winter's day. By the time the formation was west of Mainz, the cloud had cleared altogether, leaving good visibility for the crews.

Although the formation was escorted by both Spitfires and Mustangs, flak and enemy fighters were in evidence, although neither bothered F/L Bone. Looking out the starboard side of the cockpit of Lancaster, HK648, LS-F, David Bone noticed red Verey flares emanating from an aircraft in trouble. Unfortunately, the crippled Lancaster, LL923, LS-O, piloted by F/S David Williams, was seen to fall from the sky shortly afterwards, having been hit by flak over the target area. The only crew member to escape from the stricken aircraft was Sgt D T Darby, the rear gunner, who was subsequently captured and made a prisoner of war.

Two nights later, on 7th January, a new crew carried out their first operational mission with the Squadron, since their arrival at Mildenhall on the penultimate day of 1944. The 'rookie' crew were led by F/O Doug Hunt, an Australian pilot from Mount Barber, South Australia. Their initiation into combat flying was to be overseen by W/C MacFarlane, who elected to fly with them as captain. Their aircraft, Lancaster bomber, HK648, LS-F, was one of twelve aircraft despatched by the Squadron.

Unfortunately, the main force of 645 bombers arrived over the target area early, due to incorrect wind forecasts issued by the Meteorological Officer, and had to await the arrival of the Pathfinder aircraft. However, once target markers were dropped the attack commenced, and the bomb aimer's were able to get on with their allotted tasks. In the bomb aimers compartment on Lancaster, LS-F, was Pat Russell, who was conscious that the Squadron C.O. was leading the crew. The 'rookie' bomb aimer therefore wanted to make a good impression, and do things "by the book". Not being able to see the target indicators on which they had been briefed to bomb, Sgt Russell canceled the first run and instructed the pilot to 'go round' again. Although nothing was said by his fellow crew members, the young bomb aimer had the distinct impression the rest of the crew were not best pleased with him. The task completed, the aircraft and crew returned to Mildenhall, where they landed at 03.05 hours.

Flying Officer Doug Hunt and his crew were safely home having completed their first operational mission whilst F/O Bob Jones, D.S.O., who landed nearly an hour later, was declared 'tour expired'.

A daylight attack against the railway marshaling yards at Krefeld was carried out in very wintry conditions on 11th January. No.XV Squadron detailed twelve aircraft to join a main force of 152 bombers from No.3 Group.

Although runway 110 at Mildenhall had been swept clear, take-off from the base was hazardous, due to five or six inches of snow having fallen overnight. Once airborne, the pilots found that the weather conditions made it extremely difficult for the heavily laden bombers to fly in formation. By the time they reached the target area, where they had to endure severe crosswinds, the formations were almost non-existent. A number of crews unleashed their bomb loads on the Leader's signal, others bombed on E.T.A and some just took pot luck. On return to Mildenhall, the crews of No.XV Squadron had little to report.

The weather and workload were completely forgotten during the evening of 12th January, when the Squadron held a drinking party for the members of groundcrew. The reason, as if any were needed, was the completion of S/L Bob Cameron's first tour. With beer being plentiful (it had been rationed to provide plenty for the party) the 'erks' enjoyed themselves, as did the tour expired Squadron Leader who insisted on playing the bagpipes!

A change in the weather conditions during the second week of the month, enabled bomber crews to carry out a very successful daylight attack against Saarbrucken on 13th January. Bomber Command detailed 158 Lancasters for the mission, of which fourteen were supplied by No.XV Squadron.

As Pat Russell looked out through the plexi-glass dome on the nose of Lancaster, LL854, LS-S, he watched the aircraft ahead 'snaking' round the peri-track, preparing to line-up for take-off. Flying in his own right as captain of the aircraft, F/O Hunt lifted LS-S into the air at 12.05 hours; the last aircraft to take-off. Cloud base over Mildenhall airfield was approximately 1,000', and the Lancaster was lost to view as it gained altitude. Struggling onwards and upwards, it emerged from the gloom at 3,000' into a blue sky illuminated by wintry sunshine. Doug Hunt quickly caught up with the rest of the Squadron, and flew in formation with them to the target. Feeling more at ease without the C.O. on board the aircraft, Pat Russell was able to watch the bomb load whistle down to explode on the target. All crews later reported a very concentrated attack.

Being misled into thinking their third mission was *"a piece of cake"*, Doug Hunt and his crew relaxed a little too much on the homeward journey and nearly paid the ultimate price for their naiveté. Enemy opposition to the raid being almost non-existent, F/O Hunt engaged *'George'*, the automatic pilot, and went forward into the bomb aimer's compartment. George Pitkin, the flight engineer, remained at his station in order to keep an eye on things in the cockpit. Suddenly, whilst the pilot and the bomb aimer were chatting, a muffled splutter was heard and the aircraft went into a dive. Clambering back to his rightful position in the cockpit, Doug Hunt regained control of LS-S, brought the Lancaster onto an even kneel, and endeavored to assess the situation. It transpired the aircraft had inadvertently flown over the French coast, between Calais and Dunkirk, where pockets of German resistance were still continuing the fight. The 'spluttering' noise, heard above the roar of the Merlin engines, was in fact exploding flak, some of which penetrated the aircraft causing minor damage. To add to their problems, F/S Phil Smeeton, the wireless operator, informed his pilot that a diversionary landing instruction had been received. Due to East Anglia being 'cloud bound', all No.XV Squadron aircraft were to divert to, and land at, Predannack in Cornwall. The navigator, Sgt John Sheppard, was not too pleased about the order as he did not have the necessary navigation charts for that area of the country. Late afternoon brought its own problems when, in the fading light of dusk, the Lancaster crossed a stream of Handley Page Halifax bombers heading out for the 'night shift'. Anxiety began to set in when, dropping down through cloud, F/O Hunt realised he was still flying over the sea. Some rocket flares, fired upwards from below were seen but not acted upon as the crew felt they had troubles of their own; they did however log the time and position the flares were seen. Fearing they had overflown the west coast of Britain, Doug Hunt decided to alter course. It was a decision which ultimately saved their lives. Thirty-five minutes after changing course, with the fuel gauges almost reading zero, the crew spotted some searchlights sweeping across the night sky. Staring down into the darkness of the Cornish cliff tops, they suddenly realised there was an airfield situated amongst the searchlights. It was only as they joined the circuit in preparation for landing, they were informed by Phil Smeeton that the airfield was in fact Predannack, where other aircraft from the Squadron had also landed!

Having landed a further shock awaited the crew, when they were informed the rockets they and seen and logged earlier, were in fact part of the *'Granite'* emergency system. The projectiles had been fired from the Scilly Isles, 25 miles past Land's End, to warn the crew their aircraft was heading towards New York and a watery grave!

All the Squadron's aircraft, with the exception of F/O Hunt's Lancaster, returned to Mildenhall during the course of the following day. Doug Hunt and his crew were forced to remain in the West Country for a further three days, whilst their aircraft was repaired and made airworthy.

The coking plant at Enkerschwick, in the Rhur, received attention from Bomber Command on 15th January. No.XV Squadron detailed twelve aircraft for the attack; all of which took-off safely. First away, at 11.50 hours, was Lancaster. NG444, LS-Y, flown by F/L Dollison, RAAF. Flight Lieutenant Rosenhain, RAAF, took-off two minutes later piloting Lancaster, NG338, LS-M, whilst F/L Whittingham who was flying his usual aircraft, Lancaster, HK648, LS-F followed four minutes later. All pilots later reported an effective attack, with minimal opposition from the enemy forces.

Tragedy stuck on 17th January when an accident claimed the lives of seven young men from the Squadron. The crew, who had only recently been posted to No.XV, had not even commenced operational flying. Their aircraft, Lancaster bomber, PB802, LS-F, piloted by F/O John Crone, RAAF, was returning from a cross-country exercise, when it dived into the ground. The aircraft impacted near Roudham Hall, Thetford, Norfolk, at 21.30 hours, killing everyone on board.

The wintry weather took a turn for the worse during the middle of January, with fog, frost and snow causing many operations to be canceled. Apart from an attack against a coking plant at Sterkrade, north-east of Duisberg, on 22nd January, no other raids were carried out by the Squadron until the end of the month.

On Sunday, 28th January, the Squadron despatched ten aircraft for a daylight attack against railway marshaling yards to the south-east of Cologne. A total of 153 No.3 Group Lancasters participated in the raid, which was carried out in reasonable weather conditions. Leaving Mildenhall mid-morning, the aircraft from No.XV flew in open formation to the target, where intense and accurate flak was encountered. Four aircraft in total were lost during the mission, including one from No.XV Squadron. Lancaster bomber, HK618, LS-G, piloted by F/O Bignell, did not achieve its objective, having been hit by flak approximately fifteen miles short of the target. Although the aircraft caught fire, the bomb load did not explode thus allowing the crew to effect an escape. Other crews watched as the bomber went down with an engine on fire, followed by the more genteel descent of seven parachutes.

For No.XV Squadron, the final raid of the month occurred on 29th January when eleven aircraft were detailed to attack the Uerdingen railway yards at Krefeld. These aircraft joined a total force of 148 bombers detailed for the mission by No.3 Group. All participating aircraft returned safely.

The new month commenced with a daylight raid on the railway marshaling yards at Munchen-Gladbach on 1st day of the month, followed by a night attack against a military barracks at Weisbaden on the 2nd/3rd February. Light to moderate flak was encountered on both missions, but no aircraft were reported lost or missing. Lancaster bomber, LM240, LS-R, piloted by F/L Ostler, was forced to land at Juvincourt airfield, France. Although the crew were uninjured and were back at Mildenhall within six days, their aircraft was to remain on the Continent for some weeks.

The Squadron lost an aircraft during an early evening attack on 3rd February, when twelve machines from No.XV Squadron were despatched to bomb the Hansa benzyl oil plant, north-west of Dortmund. A total force of 149 No.3 Group aircraft were detailed for the raid, of which a total of four failed to return. Incorrect meteorological forecasts, combined with rapid changes in the weather conditions, caused the bombers to arrive late over the aiming point, which had been marked with target indicators. The intensity of opposition from the enemy was on a level not recently experienced by the bomber crews during similar operations. To their chagrin, a higher degree of searchlight activity than had been seen or experienced for sometime was operating, as were a small number of enemy nightfighters. Furthermore, a moderate but accurate barrage of heavy flak was peppering the sky. One of the machines lost to the defenses was Lancaster bomber, PD419, LS-V, piloted by F/O Morris. Five of the crew managed to bale out safely, but P/O Harold Slingsby, and F/S Ever Temperton, RNZAF, the mid-upper and rear gunners respectively, paid the supreme sacrifice. The average age of aircrew being approximately twenty-one years, at thirty-eight years of age Harold Slingsby was considered an old man. Evan Temperton, who was only twenty-two years old, was buried with the commissioned rank of Pilot Officer.

Of the one hundred aircraft despatched by No.3 Group to attack the oil plant at Wanne-Eickel, north of Bochum, on 7th February, only seventy-five Lancasters actually bombed the target. Due to the prevailing wintry weather conditions, many of the attacking force abandoned their missions because of heavy icing. Of the thirteen aircraft detailed by No.XV Squadron for this attack, three succumbed to the weather. The pilot of Lancaster, ME434, LS-D, F/L Hammond, decided to turn for home; unfortunately he was destined not to make it. Rather than waste the trip, F/L Hammond, decided to unleash his bomb load on Krefeld during the return trip. As the aircraft was completing its bombing run it was hit by flak. The bomber rolled over onto its back and went into a spin. Fearing recovery was impossible, the order was given to abandoned the aircraft. Fortunately the pilot, and his all officer crew, survived the ordeal but were taken prisoners of war. Unbeknown to the crew their unmanned aircraft recovered from its spin, regained normal flying altitude, and flew a further 150 miles, before crash-landing on Juvincourt airfield!!

The crew of Lancaster, HK620, LS-W, were not so fortunate when their aircraft was shot down during the early hours of the morning two days later. The aircraft, piloted by F/O James Cowie, was one of twelve machines detailed by the Squadron to attack the Hohenbudberg railway yards at Krefeld. The crew of eight, which included F/S Maurice Hathaway, the mid-under gunner, were all killed.

During the second half of January 1945, the eastern frontiers of the German Reich had collapsed and given way to the rapid advance of the Russian forces. On 29th of the month the Red Army was approximately 95 miles from Berlin; two days later they had halved that distance. The Germans, fighting on two fronts inside their own territory, were hard pressed. Both refugees and military personnel were amassing at, amongst other places, Dresden, just inside the German lines. Not being of military significance earlier in the war, the medieval city had never been the subject of attack by Bomber Command. However, that was to change on the night of 13th/14th February, when a total of nearly 800 British bombers attacked the city in two separate raids.

The first assault was to be a lone effort by No.5 Group followed, three hours later, by a collective strike by No's 1, 3, 6 and 8 Groups. The thirteen aircraft despatched by No.XV Squadron, which took-off from Mildenhall between 21.37 and 21.54 hours, were detailed to form part of the second attack. The following day the USAAF kept up the pressure of the attacks by despatching 311 B-17 Flying Fortresses to Dresden.

Such was the accuracy of the bombing that a firestorm evolved and, before the night was over, an estimated 50,000 people had perished in the burning city. Upon their return to Mildenhall, the crews of the No.XV Squadron aircraft reported the fires were visible from over one hundred miles away. One crew had to defer their report due to having landed in France. Lancaster bomber, LL854, LS-S, piloted by F/L Bithell, developed mechanical problems on the homeward journey and required urgent repairs.

The members of groundcrew were kept very busy following a daylight attacks against Wesel, on 16th, 18th and 19th of the month. The Squadron despatched ten aircraft on the first mission, many of which returned bearing varying degrees of battle damage. Flying Officer Bone's Lancaster, HK648, LS-S, returned with flak damage to the tailplane assembly. Thir-

teen aircraft were sent on the second mission, whilst fourteen aircraft undertook the third attack.

The third week of February saw attacks against Chemnitz, Wesel, Dortmund and Gelsenkirchen. The latter raid, which occurred on 22nd of the month, was a daylight attack against the coking plant at Buer. Fourteen aircraft were detailed by the Squadron, all of which took-off safely. The weather was clear with excellent visibility, allowing the Squadron to fly to the target in formation. The clarity of the weather, which provided a panoramic view of events for the aircrew, was also of some benefit to the enemy flak units. With the sky being devoid of cloud cover, the German ground gunners were able to aim visually at the bombers. As they approached the target area, F/S Pat Russell lay in the bomb aimer's compartment in the nose of Lancaster, NF953, LS-A, and watched as the box barrage of flak came up to meet them. As they ran in to the target, there was an explosion just ahead of the Lancaster, but fortunately no damage was incurred. Momentarily shaken, the bomb aimer thought the blast must have been caused by a 'scarecrow' shell, which simulated an aircraft blowing up when it exploded. However, Sgt Russell did not have time to reflect on what might have been, as the Lancaster was by this time over the target. Guiding the pilot to the aiming point, by use of the bomb sight, the bomb aimer unleashed the bomb load onto the target. As he did so, a terrific bang resounded throughout the aircraft, and the Lancaster shuddered under the impact of an exploding shell. Flight Sergeant Phil Smeeton, the wireless operator, yelled a warning that the starboard outer engine was on fire. The flight engineer, Sgt George Pitkin, immediately hit the propeller feathering switch, and activated the Graviner fire extinguisher. Clambering out from his 'office' in the nose of the aircraft, Sgt Russell prepared for a hasty exit. However, thanks to the quick reaction of the flight engineer, the fire was doused and the Lancaster was able to fly home on three engines. By the time the attackers left the target, large columns of smoke rose high into the air and dissipated like a dark mist in the breeze.

Although only one Lancaster (NG450, from No.218 Squadron) was lost during the attack, many more returned to their respective bases showing the signs of battle. The groundcrews on a number of No.3 Group airfields, including No.XV at Mildenhall, spent that night repairing flak holes on various parts of their aircraft.

The following day, Gelsenkirchen received another visit from the Lancasters of No.3 Group, when they again bombed a coking plant with no losses to themselves. The weather could not have been more of a contrast to the previous day, with heavy cloud, showery conditions and poor visibility.

Having safely completed its 113th mission during this operation, another bomb insignia was added to the mission tally on the port side of "J"-Jig's forward fuselage. To mark the occasion, F/L David Bone was asked to arrange for the photographic section to take a number of photographs. Arrangements were made for later that same afternoon, when a corporal photographer took pictures of the 'oldtimer'. Amongst those present, formally arranged in front of the aircraft, were G/C Ken Batchelor, the Base Commander, W/C Nigel MacFarlane, the O.C. No.XV Squadron, F/L Bone and F/L Whittingham. The Squadron Leader Engineering Officer and members of "Jig's" groundcrew were also invited to attend the celebrations.

Following an attack on Kamen, on 25th February, one crew returned to Mildenhall having left their bomb aimer in France. The Squadron had detailed fourteen aircraft for a daylight raid on an oil refinery north of the town near Dortmund. Having taken-off from their Suffolk base between 09.28 and 09.38 hours, the Squadron formed up and headed out over the sea. As the main attacking force of 340 aircraft headed towards Kamen, the flak grew heavier in intensity. However, once over the target area the heavy flak died down, making it easier for the crews to carry out their tasks. Unfortunately the reprieve from the anti-aircraft fire was only temporary as, once the bombers had set course for home, the heavy flak resumed. One shell exploded very close to Lancaster bomber, NG364, LS-P, causing shards of metal to penetrate the fuselage and injure F/O Bender, the bomb aimer. Fearing for the well-being of the wounded crew member, F/L Noble, the Australian pilot landed his aircraft at Liege, in France, where

F/O Bender was removed from the bomber and transferred to hospital for treatment. At 16,02 hours, twenty-eight minutes after he had landed, F/L Noble took-off from the French airfield and returned to Mildenhall.

A further attack was made against Dortmund on 26th February, whilst Gelsenkirchen was the target on the last two days of the month.

A raid on Cologne, during the afternoon of 2nd March, was a failure due to the breakdown of the G-H station in England. No.XV Squadron had detailed thirteen aircraft to join a main force of 155 Lancasters for the attack but, due to the aforementioned circumstances, the crews released their bombs over unrelated areas. Many are known to have fallen on Bonn, whilst others fell as far south of the target as an area south-west of Koblenz.

During the course of the day, on Saturday, 3rd March, a Lancaster from No.XV Squadron took-off from Mildenhall and headed for France. The target was Juvincourt airfield, and the mission was to deliver a crew, who were detailed to collect and fly home Lancaster, LM240, LS-R, which had force-landed there exactly one month earlier. Having returned to England, the aircraft was taken-on-charge by No.149 Squadron with whom it served until it was struck-off-charge on 25th March 1948.

Even at this late stage in the war, new crews were being posted to No.XV Squadron. They were greeted by W/C Nigel MacFarlane, the Unit's C.O. who, as was his rule, always flew with the 'rookie' crews on their first operation. On 4th March he acted as captain of Lancaster, NF953, LS-A, when he took F/S McDonald and his crew to Wanne Eickel. Two weeks later, on 17th March, whilst piloting Lancaster, ME844, LS-C, he initiated F/S Sievers and his crew into the mysteries of operational flying. The target on the latter occasion was Dortmund, from which they returned safely. It was whilst returning from this same raid that F/L Lacey feared for the safety of his aircraft. He was piloting Lancaster, HK799, LS-D, low over the sea when another aircraft formated on his port wingtip. The other bomber, Lancaster, NG357, LS-G, flown by F/O Doug Hunt, RAAF, had moved in close so that the latter's bomb aimer, Pat Russell, could obtain some good air-to air photographs. It was later reported that F/L Lacey was none to pleased about the impromptu photo session.

Other targets attacked by the Squadron during March included Gelsenkirchen, Wesel, Dessau, Datteln, Essen and Dortmund. The majority of aiming points being synthetic oil or Benzene plants. Although no aircraft from No.XV were lost during these attacks, many returned to Mildenhall bearing battle scars and flak damage of varying degrees.

On 21st of the month, the Squadron detailed fourteen aircraft to participate in a raid against marshaling yards and a railway viaduct at Munster. During the outward journey the port inner engine of Lancaster, NG358, LS-H, developed a problem and was feathered by Sgt Clear, the flight engineer. Although flying on three engines, the Australian pilot, F/S McLennan, elected to continue with his mission. Over the target area he, and the other crews, found visibility good with only 2/10th cloud. The run-in to the aiming point was hampered by moderate to intense flak, but all crews completed their tasks and returned safely.

The following day, 100 aircraft from No.3 Group were detailed to attack the residential and commercial areas of Bocholt. No.XV Squadron despatched fourteen Lancasters, which commenced take-off at 10.46 hours, led by S/L Percy. As the twelve preceding aircraft climbed away from Mildenhall, the thirteenth machine lined-up on the runway in preparation for take-off. At 10.55 hours, Lancaster Bomber, HK773, LS-W, with W/O Frederick Newton, RAAF, at the controls, rolled forward. Gaining flying speed, the aircraft took-off and climbed away. As it did so, F/Sgt 'Swifty' Swallow, RCAF, of No.XV Squadron who was standing at the end of the runway, gave them a wave in farewell. Within minutes of leaving the ground disaster struck when a valve inlet spring in the port outer engine dropped, causing the motor to catch fire. At an altitude of only 150', the aircraft was seen to bank round in a turn with flames streaming out of the feathered engine. In a possible attempt to force-land the burning machine the escape hatch was jettisoned, but this was to be of little use to the crew. The bomber fell to earth south of Mundford, in the densely tree populated area of Brandon Forest. An almighty explosion, followed by a huge pall of smoke rising high into the air from between the trees, left the local inhabitants who

had been watching the drama unfold, in no doubt as to the tragic outcome. The seven young men who formed the crew were killed instantly when the 4,000 lb bomb the aircraft was carrying exploded. They were about to undertake their third operational sortie with the Squadron, when they died.

The last aircraft away from Mildenhall, Lancaster bomber, HK789, LS-R, piloted by F/S Eves, took-off at 10.56 hours. It followed the same path as LS-W, but fortunately climbed away without incident.

Although the sky over the target area was clear, and totally devoid of cloud, bombing was carried out using G-H. Most crews were able to check visually where their bombs had impacted, and later reported excellent results.

During the last nine days of March, only two operations were carried out by No.XV. On 23rd of the month, the Squadron detailed eight aircraft for an attack on Wesel in support of British troops who were about to launch an offensive on the town. Six days later, eleven Lancasters were detailed to join a main force of 130 bombers, for an attack on the Hermann Goering benzyl plant at Hallendorf, near Salzgitter. Both raids were carried out successfully, but flak caused damage to one No.XV Squadron aircraft during the latter raid. Lancaster, ME844, LS-C, piloted by F/O Bone, D.F.M., lost the use of the starboard inner engine, which had to be feathered. Fortunately, David Bone and his crew, who were on the homeward leg of their final mission, returned safely to Mildenhall on three engines.

Flying Officer David Bone, who was awarded a Distinguished Flying Cross in July 1945, was one of the very few members of aircrew who served with No.XV Squadron to be awarded both a D.F.C. and a D.F.M.

Weather conditions during the first three days of April were typical for the month, with clear dry spells alternating with scattered showers. During this period only training exercises were carried out, with air tests, fighter affiliations, bombing details and cross-country exercises being the order of the day. The first operational mission for the new month, was carried out on 4th April, when a night attack was detailed against the Leuna oil refinery at Merseberg, near Leipzig. The Squadron detailed thirteen aircraft to participate in the raid, all of which took-off between 18.24 and 18.36 hours. Squadron Leader Hagues was first away, piloting Lancaster, ME455, LS-O. A total force comprising of 327 Lancasters and 14 Mosquitoes was despatched by Bomber Command for this attack.

Some confusion reigned on the approach to the target when dummy marker flares, which were reportedly laid by a Pathfinder Force unit, were bombed by unsuspecting aircrew, including one from No.XV. The remaining twelve machines despatched by the Squadron continued on to the target location, where they arrived as instructed. They were forced to orbit the aiming point for a few minutes as the target had not been marked. Although heavy flak was encountered during this period, the inaccuracy of the aiming meant all No.XV Squadron aircraft returned to base unscathed. Unfortunately Lancasters, HK555, and RA533, of No.115 Squadron and No.186 Squadron respectively, were lost when the two machines collided over Geish, Germany, whilst returning from the target.

On the night of 9th/10th April, the Squadron despatched a total of seventeen aircraft for a raid against Kiel harbour. Two of the Lancasters were detailed to mine the approaches to the harbour, whilst the remainder attacked German naval targets. The Squadron flew in low level formation over the North sea, gaining altitude to bombing height when approximately half way across. The attack was very successful, with severe damage being inflicted upon the Deutsche Werke U-Boat yard and other local buildings. The German pocket battleship *Admiral Scheer* capsized due to damage sustained during the attack, whilst both the *Emden* and the *Admiral Hipper* were badly damaged. The enemy defenses were very active, with searchlights weaving back and forth across the night sky, whilst the flak was both heavy and fairly accurate. Lancaster bomber, NG444, LS-Y, was damaged during the attack, forcing the Australian pilot, F/O Jim Doogue, to abort his mission. The aircraft, however, remained airworthy and returned safely to Mildenhall. This particular Lancaster was the machine usually flown by F/L Ken Lewis, who also hailed from Australia. The latter, who had applied for aircrew training with the Royal Australian Air Force in May 1942 had, like so many of his countrymen, found himself on active service in

England. Ken Lewis undertook his initial training in his native land, attending Elementary Flying Training School in New South Wales. On completion of the E.F.T.S. course, was he posted to S.F.T.S in Alberta Canada. He completed his training on 17th September 1943, was presented with his 'wings' and promoted to the rank of sergeant. January 1944 saw the young Australian on English soil at Smiths Lawn, a training airfield near Windsor Castle. Further advanced training and aircraft type conversion was endured until 26th November, when Sgt Lewis was posted, along with his recently formed crew, to No.XV Squadron.

Apart from the damage inflicted on Lancaster, NG444, the Squadron's aircraft all returned safely to Mildenhall.

A similar operation against Kiel was undertaken on the night of 13th/14th April, when a total force of 377 bombers attacked the port area. No.XV Squadron detailed fourteen aircraft for the attack, one of which aborted due to failure of the port outer engine. The results of the attack were not considered as good as the previous raid four nights earlier.

Whilst the rest of the Squadron were engaged in the mission against Kiel, F/L Gray and his crew undertook a mine-laying operation in the Kattegat. Their aircraft, Lancaster bomber, NG561, LS-L, was the sole representative from No.XV Squadron, amongst a total of over 100 aircraft engaged on similar duties.

The following night a major step forward was taken when Bomber Command attacked Potsdam. Not since March 1944, had the heavy bombers of the Royal Air Force entered the Berlin defence zone. A total of 500 aircraft were despatched for the raid, of which sixteen were detailed by No.XV Squadron.

Enemy opposition to the attack was strong with searchlights actively scanning the sky over the target area, supported by bursts of moderate to heavy flak. German nightfighters also entered the fray, but only managed to account for the loss of one Lancaster bomber. The attack, which was the last operation of its kind undertaken by a major force from RAF Bomber Command, was very successful with severe damage being caused to the city.

Although the weather remained fine, No.XV did not operate during the course of the following four days. When the Squadron did return to the fray, it was to join a major attack against the Luftwaffe airfield and Kriegsmarine harbour installations on Heligoland. This small island, situated in the North Sea approximately fifty miles off the German coast, was attacked during the day on 18th April by nearly 1,000 bombers. The Squadron detailed eighteen aircraft for the mission, all of which took-off safely. The first Lancasters away from Mildenhall circled the airfield at 4,000', thus allowing the tailenders to climb to altitude and form up before setting course. Their route took them on a north-easterly heading, crossing the English east coast at Cromer. Once out over the North Sea the formation climbed to 19,000', the pre-ordained bombing height. The sky over the target was clear, allowing all crews to bomb visually. Flying Officer Phillips, bomb aimer to F/O Jim Doogue and his crew, turned the blue sky black following detonation of their bomb load. Having instructed his pilot to hold Lancaster, PP762, LS-N, steady on its run-in to the target, F/O Phillips pressed the bomb release 'tit'. The projectiles tumbled from the bomb bay, fell to earth and exploded on an oil dump, creating a billowing plume of black smoke. The thick dense smoke swirled up into the sky, momentarily blocking out all sight of the target.

Flying Officer McGrath had the unenviable task of bombing from an altitude of 4,800' after Lancaster, HK693, LS-B, refused to maintain bombing height following failure of the starboard inner engine. Fortunately, by the time his aircraft reached the target, most of the smoke had cleared and the other bombers were turning for home. All crews later reported a very concentrated attack, with no response or hindrance from the enemy defenses.

The following day the Squadron despatched five aircraft to join a total force of 49 Lancasters, for an attack against Munich. No.XV's primary target was the electric power plant which provided energy to part of the city's transport system. Again, the weather was clear and the crews were able to visually identify the target. However, on this occasion, the enemy defenses were active, and one of No.XV's aircraft returned home with a flak hole in the port wing.

Regensburg was attacked on 20th April, when seven aircraft from No.XV Squadron joined a force of 100 Lancasters in a raid on the oil storage depot. Having commenced a series of similar attacks in June 1944, this operation was the last of its type in the campaign against enemy oil targets.

Bomber Command, in support of the British XXX corps who were poised to attack Bremen, despatched over 700 bombers for a raid on the town on 22nd April. The operation did not go according to plan, and a combination of cloud, dust and smoke caused the Master Bomber to abort the mission after only 195 Lancasters had bombed. All but one of the fourteen aircraft detailed for the attack by No.XV Squadron, had completed their tasks before the abort instruction was given.

The third week of April saw No.XV Squadron involved in a series of training exercises, including fighter affiliation, cross-country navigation and special equipment bombing. Although these exercises had been carried out for real many times during the war, a new form of operation was to be undertaken, where the latter two exercises would be of paramount importance.

Due to the combined effects of the Allied aerial bombing, operations by the British 2nd Tactical Air Force and the work of the Dutch Resistance, the Dutch railway system had ceased to function. With the lack of movement by rail, food and general supplies could not be transported into certain parts of Holland. As a result of negotiations during early April, between the Allies, the Dutch Resistance and the Germans, the latter agreed to allow food to be dropped over Holland by low-flying aircraft from the Royal Air Force. The concept behind the mercy missions was for the Bomber Command aircraft to fly unarmed, at low-level, to pre-designated dropping zones, where panniers of food would be released from the bomb bay of each aircraft. Feeling dubious about flying over enemy held territory without the means to defend their aircraft, most crews chose to fly with their guns primed and ready for action. The foodstuff, which consisted of flour, egg powder, dried milk, biscuits and chocolate, was packed into individual sacks which in turn were contained in the specially designed panniers. The cargo was to be released from the panniers, over the designated dropping zone, by the bomb aimer utilising the bomb release mechanism.

These missions, aptly code-named *'Manna'* operations, commenced for No.XV on 29th April, when 12 aircraft were detailed to fly individually to Rotterdam. Each Lancaster was loaded with five panniers of supplies, all of which were to be released over an airfield one mile south of the city.

The first of the twelve No.XV Squadron aircraft took-off from Mildenhall at 12.33 hours, the remaining eleven machines all being airborne within the following ten minutes. They flew low out over the sea; the crews wondering what sort of reception would be awaiting them. As the aircraft roared over the Dutch countryside, from his vantage point in the bomb aimer's position in the nose of Lancaster, HK693, LS-B, Sgt Pat Russell looked down. Through the plexi-glass blister of his bomb aiming compartment, from an altitude of 200', he could clearly see hordes of people waving and gesticulating excitedly. He was also aware that some figures in the throngs remained motionless and did not join in the celebrations. They were easy to spot; they were wearing green uniforms!

The steady drone of the Lancasters engines brought more and more people onto the streets, many of whom carried anything which could be waved at the low flying aircraft. The elation of the starving Dutch people increased when the bomb doors of each Lancaster opened, and the food sacks came tumbling out. For them, the fifty packs dropped by No.XV Squadron was Manna from Heaven.

Although an undertaking had been given by the Germans that the participating aircraft would be given clear passage over Holland, one or two of the Squadron's machines returned to Mildenhall bearing small caliber bullet holes. Lancaster, PP672, LS-N, piloted by F/O Blaxendale, received a bullet hole in the starboard outer engine whilst flying over the almost aptly named Overflakke!

The following day a similar mission was flown, when fourteen aircraft from the Squadron undertook another *'Manna'* operation over

Rotterdam. On this occasion, the drop zone was a field situated two miles to the north west of the town. The scenario was the same, with the scenes on the ground equally enthusiastic. Members of the local populace were seen waving sheets and other items including, for the first time in many years, British Union flags.

One unfortunate incident seen during this mission, later reported by two returning crews, was the setting alight of a house by stray marker flares. Both F/O Nottage, pilot of Lancaster, PP664, LS-U, and W/O Hackforth, who was flying Lancaster, NG338, LS-M, recorded seeing the accidental fire.

The first day of May arrived, accompanied by heavy cumulus clouds which built up during the day and brought with them scattered showers. These showers, however, did not deter the Dutch from venturing out to a sports field near the Hague, where fourteen aircraft from No.XV Squadron had been ordered to drop their next consignment of food panniers.

As he lay in the bomb aimer's compartment of Lancaster, NG358, LS-H, Sgt Pat Russell could not fail to identify the drop zone which lay ahead. The large sports field was a flat area of open land surrounded by a mass of people who were galvanised into a state of emotional excitement by the sight of the approaching bombers. With his pilot, F/O Doug Hunt, holding the aircraft steady, the bomb aimer pressed the 'tit' and the panniers fell away. Having been trained to drop death from the bomb bay of his aircraft, it gave Pat Russell a deep sense of pride that he was also able to drop the means for sustaining life. It was a feeling which remained with him for the rest of his life.

Areas around the Hague remained as the prescribed drop zone for the next four 'Manna' operations, which occurred on 2nd, 3rd, 5th and 7th May respectively. The inclement weather persisted intermittently during all four missions, but did not dampen the ardor of the recipients of the food panniers. No.XV Squadron's final 'Manna' operation occurred on 8th May 1945, when six aircraft were detailed to drop supplies over the same area. The weather changed for the final mission, when the thundery showers gave way to warm south-westerly winds. The Dutch people obviously felt the wind of change as, on that same day, the end of hostilities in Europe was announced. The German Forces had surrendered. Having completed their task satisfactorily, the Squadron formed up over Holland and flew in formation back to Mildenhall.

As the war entered its final phase, the Allied commanders were faced with a new, but somewhat pleasant, problem. The sweep across Europe by the liberating armies meant freedom for the hundreds of men who had been captured and interned by the German Forces. It soon became a major priority to attend to the needs of these men, some of whom were wounded or sick, and repatriate them to England at the earliest opportunity. Given the code-name 'Exodus', it became apparent the quickest method of getting them home was by air, using both bomber and transport aircraft for the purpose.

No.XV Squadron flew its first 'Exodus' mission on Thursday, 10th May, when it detailed fifteen aircraft to fly to the French airfield at Juvincourt, near Rheims. As the Lancaster could accommodate twenty-four passengers, the former PoWs were formed into groups of that number. Having been processed through the security system, for proof of identity etc, they were driven out to the aircraft's dispersal area for embarkation.

One former PoW and ex-XV Squadron member who was repatriated in this manner was Doug Fry, who had been shot down during the summer of 1943. Having explained to the Lancaster crew who were about to fly him home, that he had been shot down whilst flying as mid-upper gunner on a Stirling bomber, the captain of the Lancaster invited the ex-Kriegie to occupy the vacant seat in the mid-upper turret for the forthcoming journey. Needless to say the Doug Fry did not need to be asked twice.

Over the following two weeks, the Squadron flew a further eight 'Exodus' operations, all from Juvincourt airfield. Due to the influx of returning personnel, reception/receiving centers had been set up at a number of different airfields in Southern England. Amongst the airfields used by No.XV Squadron at which it disembarked its passengers, were Tangmere in Sussex, Dunsfold in Surrey and Westcott in Buckinghamshire.

One No.XV Squadron aircraft was destined never to fly into any of the airfields set up as reception centers. Lancaster, PP672, LS-N, piloted by F/L Bagenal crashed at Juvincourt, on 13th May. The aircraft, with its full compliment of returning passengers, crashed on take-off and burst into flames. Although the Lancaster was totally consumed by fire, everybody on board the aircraft escaped.

An episode of a more humorous nature occurred during a lull on an 'Exodus' operation undertaken during the middle of the month. Flight Lieutenant Ken Lewis, who was 'tour expired', hitched a ride to Juvincourt with F/L Jolly and his crew. Having landed at their destination, they were standing under the wing of Lancaster, NG444, LS-Y, sheltering from the rain, when an American army truck arrived carrying 24 ex-p.o.w's. The ex-Kriegies were a 'rag-tag' looking bunch, but were in good spirits and excited about going home. One of their number, a little English 'Cockney' who stood about 5' 2" in height and was built like a brick, approached the crew and asked who was in charge. On being asked the nature of his problem, the east-Londoner stated he had a large parcel to take home, and wanted to know if he could take it on the aircraft. Flight Lieutenant Lewis took the inquisitor to the bomb bay where the empty 'Manna' food panniers where still in place. It was explained to the latter that all the P.O.W's loose luggage would be stowed in these panniers for the journey home. When asked if they would be large enough to accommodate the parcel, the east-Londoner dubiously affirmed he thought so. He further stated he would go and get the parcel, and asked the pilot not to leave without him. Ken Lewis and the rest of the crew watched, with some amusement, as the 'Cockney' disappeared into some nearby woods. A short while later, the little man reappeared carrying a huge box 6' long and 18" wide. Having staggered back to the waiting aircraft, he was asked if the box was a coffin. With a true 'Cockney' accent he replied, *"Na, me old woman always wan'ed a granfarfer clock so I pinched one for 'er"*.

The last Ken Lewis saw of the east-Londoner was after the aircraft had landed back in England, and the latter was seen walking slowly towards the reporting/reception desk, looking like a snail with its house on its back.

Between 10th May and 21st May, No.XV Squadron despatched a total of seventy-nine aircraft on seven 'Exodus' operations, thereby repatriating nearly 1,900 ex-prisoners-of-war back to Britain.

On 23rd May, No.XV Squadron despatched two aircraft for an operation over Germany. The purpose of the mission, codenamed 'Baedeker', was to ferry members of the groundstaff on a tour of the Rhur area. Each Lancaster carried five passengers, who were able to see first-hand the destruction leveled at the enemy by Bomber Command. The first Lancaster, NG444, piloted by F/L Jolly, took-off just after 09.30 hours, followed nine minutes later Lancaster, PA235, piloted by F/L Gray. Although the first aircraft was forced to return early due to low cloud over the continent, the latter machine was able to complete its task.

During the afternoon of the following day, No.XV detailed ten Lancasters for another 'Exodus' operation. As before the aircraft flew to Juvincourt airfield where twenty-four passengers climbed aboard each machine, before returning with them to Oakley, a Bomber Command airfield in Buckinghamshire.

Flight Lieutenant Ken Fisher, who piloted Lancaster, NG444, during this operation, had an extra passenger on board his aircraft, who had boarded LS-Y at Mildenhall. The person in question, a Miss Russell, was a British Broadcasting Corporation representative who had flown out with the crew for the purposes of a wireless broadcast. Although only one other 'Exodus' operation was carried out before the end of the month, a further nine 'Baedeker' sorties were undertaken during the same period.

On 29th May, "C" Flight, No.XV Squadron was formed at Mildenhall from a nucleus of aircraft from No.617 Squadron. The new flight, which bore the Squadron codes DJ on its aircraft, was placed under the command of W/C Charles Calder, DSO and Bar, DFC. Around this same period of time, shortly after the formation of "C" Flight, Bomber Command decided to despatch a force of bombers to the Far East to assist in bringing the war with Japan to a conclusion. No.XV Squadron was advised that it was one of the units selected to form part of the Tiger Force, which was destined for the Far East campaign. However, as preparations for Tiger Force were still being put in place, No.XV Squadron was to continue to operate from Mildenhall until further notice.

Such was the popularity of the *'Baedeker'* trips, which were made on a voluntary basis, that at least one aircraft was detailed for such a mission every day for the first three weeks of June. Amongst the senior officer structure who took the opportunity to fly over the ruins of Germany, were A/Cmdr Herbert Kirkpatrick and A.V.M. Harrison, CB, CBE, DFC, AFC. The former, who also flew on *'Exodus'* operations, often piloted the aircraft himself, whilst the latter was happy to be 'chauffeured' by W/C Macfarlane.

The month of June also saw the Squadron undertake a number of training exercises. Codenamed *'Post Mortem'*, these exercise consisted of the detailed number of aircraft making mock attacks on a pre-determined target; usually Flensburg on the German/Danish border. The 'attacks', carried out at various times of the day, usually took about six hours to complete.

From an operational point of view, the month of July consisted mainly of training missions. *'Post Mortem'* attacks were extended to encompass Kiel, "C" Flight undertook high level bombing trials, whilst other aircraft carried out cross-country and fighter affiliation exercises. Many other aircraft and crews from the Squadron were kept busy with *'Baedeker'* flights, which showed little sign of losing their attraction.

Towards the end of the month, on 23rd July, whilst the remainder of the Squadron waited for news of its posting to the Far East, Lancaster bomber, NX687, LS-A, headed for Rio de Janeiro. On board the aircraft, piloted by W/C 'Jock' Calder, was the Commander-in-Chief, Bomber Command, Air Chief Marshal Sir Arthur Harris, KCB, OBE, AFC. The latter had been invited to South America to join the celebrations marking the return of the Brazilian Expeditionary Force following the end of the war in Europe.

The following day four aircraft were detailed to undertake a trial run for a series of operations which would take the bombers to Italy. Codenamed *'Dodge'*, the intention of these operations was not to attack the country, but to bring home members of the Central Mediterranean Forces who had served there. For the exercise, the four Lancasters despatched flew a circular route encompassing Mildenhall, Stradishall, Bari on the southern Adriatic coast, Tibenham and back to Base. The trip took two days, and all aircraft returned safely to Mildenhall on 26th July.

The first true *'Dodge'* operation was undertaken on 3rd August, when five aircraft took-off for Italy. Unfortunately, Lancaster, LL806, LS-J, encountered engine problems on the outbound journey. Having landed at its destination on three engines, the aircraft was deemed unserviceable. When the remaining four bombers, each carrying twenty passengers, returned to England two days later, the crew of LL806 flew back as passengers on board one of the other aircraft.

A further eight similar missions were flown during August, with each aircraft taking two days to complete each operation.

Having flown out on sorties during the middle of the month, three aircraft were forced to remain at Bari for a third day due to a lack of passengers for repatriation! However, the passengers designated to fly home on 24th August, on Lancaster, NG340, DJ-U, a "C" Flight aircraft, were delayed due to the pilot reporting sick.

On 15th September 1945, RAF Mildenhall opened its gates to the public in commemoration of the fifth anniversary of the Battle of Britain. Being one of the many operational stations to open for the *'RAF at Home'* day, Mildenhall presented technical and domestic exhibitions laid out in the hangers, as well as a total of twenty-two different types of aircraft lined-up for display. The weather, which over the previous fourteen days

had been dull, cloudy and wet, changed to afford the 8,000 visitors the benefit of a pleasant September day. Although a number of local residents attended the event, figures later suggested that a total of 1,600 cars had been parked on the airfield.

The *'At Home'* day was deemed a great success, generating much excitement. However, just four days later the mood of some of No.XV's aircrews was to change to one of apprehension. Because of a hostile attitude which had developed between Russia and the Western Allies over the control of Berlin, No.XV Squadron had been selected to undertake yet another type of operation; one with more sinister overtones. The aircraft detailed for the operation, code-named *'Spasm'*, were instructed to fly into Gatow airfield, on the north-west perimeters of Berlin. Ostensibly, the reason for the flights was to deliver personnel and freight to the city, but in reality it was to retain the security of British tenancy.

The implied threat from the Soviets however, did not stop the down sizing of the Royal Air Force, in terms of men and machines, which had begun earlier in the year. Many squadrons, from both Bomber and Fighter Commands, were either disbanded or reduced in size. Although No.XV had been fortunate to avoid disbandment, the Squadron was reduced to two flights during the third week of September. The reduction of manpower was a fairly easy task to implement as, inevitably, many of the Dominion and Commonwealth Air Force personnel returned home.

There were many for whom the transition from service to civilian life was a traumatic experience. Some, who had had employment prior to volunteering for aircrew duties, were lucky enough to be invited to resume their pre-war careers. Others, who had enlisted straight from school or college and had not started on a career, found themselves in a very competitive market. A fortunate few were invited to stay on in the post war air force, having been granted a permanent commission.

Even with its numbers reduced, the Squadron continued to function operationally. As September gave way to October further *'Dodge'* operations were carried out all without incident, apart from Lancaster, NG444, going unserviceable in Italy. The aircraft, which had been piloted by F/L Woodcraft, had to be left at Bari whilst the crew returned to England as passengers onboard Lancaster, PP670, piloted by F/L Ginone.

A deterioration in the weather during October hampered a number of training exercises and *'Baedeker'* sorties. However, the Squadron did manage to undertake a number of night cross-country exercises during this period.

Weatherwise, the first week of November commenced much the same way as October had finished; with foggy mornings giving a dull start to each day. Although the conditions often improved, no operations were undertaken until the 7th of the month. On this date six aircraft left Mildenhall and flew to Pomigliano, east of Naples, where they were employed on a *'Dodge'* operation. The participating aircraft all returned safely two days later, with a total of 120 repatriated personnel.

Whilst four aircraft undertook *'Dodge'* sorties during the remainder of November, a further four No.XV Squadron machines flew two separate *'Spasm'* operations to Gatow airfield. These two *'Spasm'* operations, together with the earlier one in September, were the only missions of this type to be undertaken by Bomber Command. All three operations were carried out by No.XV Squadron. Apart from the odd mission in December, these sorties were amongst the last flown by No.XV in 1945.

By the beginning of December all Commonwealth and Dominion air force personnel who were still serving with No.XV Squadron, had been repatriated to their native land. Many of their number, however, would never go home. They now lay where they had fallen in battle; in the cities, towns, villages and countryside of Europe and Great Britain. Their epitaph being they fought for freedom, their only reward being they would never grow old.

Two for Enkerschwick

F/L John Dollisson, D.F.C., RAAF (center) and his crew took Lancaster bomber, NG444, LS-Y, to Enkerschwick, on 15th January 1945. Apart from the pilot, the crew consisted of: P/O Thornhill, Nav; W/O House, RAAF, W/Op; F/S Rose, A/B (3rd from left); Sgt Haynes, MUG; Sgt Jackson, R/G; Sgt Earl-Davies, F/E. Note the aircraft's code letter on the forward fuselage, adjacent to the bomb aimer's compartment. *Courtesy of Ken Lewis, DFC*

F/L Harold Whittingham, D.F.C. (4th from left, standing) poses by the fuselage of Lancaster, HK648, LS-F, with members of his aircrew and groundcrew. The aircrew, who took the aircraft to Enkerschwick on 15th January 1945, consisted of: P/O Merrifield, Nav (standing extreme right); P/O Dodd, RAAF, W/Op (Standing extreme left); P/O Graham, A/B; Sgt Hasler, MUG; Sgt Warnock, R/G; Sgt Sims, Mid-Under; Sgt McNish, F/E. *Courtesy of Ken Lewis, DFC*

A Tragic Loss

An 'Able' Aircraft and Pilot

F/O John Crone, RAAF, and his crew were killed in a flying accident on 17th January 1945. The photograph, presumably taken by John Crone, shows the crew members who perished with him. Sgt George Lake, Nav; P/O Lex Riordan, W/Op, RAAF; F/S Leonard Wilkins, B/A; F/O Harry Freedman, A/G; F/O Fuller, A/G; Sgt Richard Devlin, F/E; *Author's Collection via Bob Collis*

F/O Doug Hunt at the controls of NF953, "A"-Able, photographed whilst en-route to Wanne Eickel. *Author's Collection via the late Pat Russell Collection*

Left: F/O Doug Hunt, RAAF, runs-up the engines of Lancaster bomber, NF953, LS-A, prior to taxiing out in preparation for an attack against Wanne Eickel on 7th February 1945. *Author's Collection via the late Pat Russell Collection*

"Back-off" 1

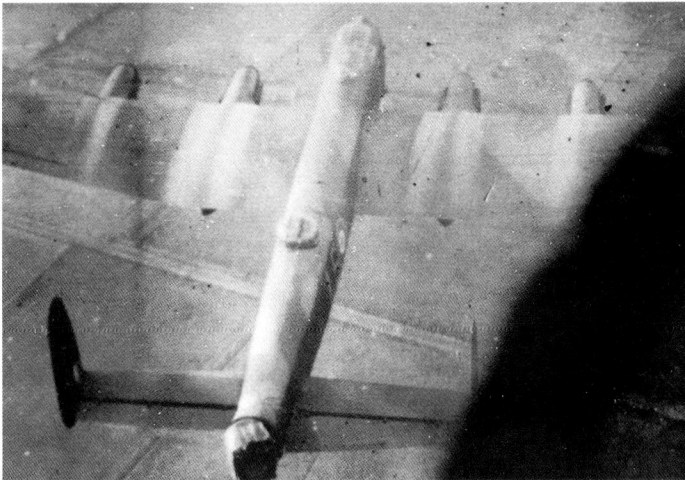

Lancaster bomber, NG357, LS-G, as seen by Pat Russell from the bomb aimer's compartment in the nose of Lancaster "A"-Able. Shortly after the photograph was taken, NG357, "G"-George broke away for a landing at RAF Mildenhall. *Author's Collection via the late Pat Russell Collection*

F/O Doug Hunt, piloting Lancaster, NF953, LS-A, closes up on an unidentified Lancaster of No.XV Squadron flying slightly below him. The photograph was taken on 7th February 1945, during the homeward journey following an attack against Wanne Eickel. *Author's Collection via the late Pat Russell Collection*

MacFarlane's Men

Six of 'MacFarlane's Men' from "B" Flight are, from left to right: F/L Bagnell, RAF; F/O Jim Doogue, RAAF; F/L William Jolly, D.F.C., D.F.M., RAF; F/O Watts, RAAF; F/L Ken Lewis, D.F.C, RAAF; F/O Paddon, RAAF. *Courtesy of Ken Lewis*

Photographed on the steps of the Officers Mess, back row, left to right, are: F/L Nottage, RAAF; F/O Paddon, RAAF; F/O James, RAF; S/L Cyril Hagues, D.F.C., RAF; F/O R Woodcraft, RAF. Front row: F/O Ken Fisher, RAF, F/O Bruce, RAF; W/C Nigel MacFarlane, D.S.O, RAF. *Courtesy of Ken Lewis, DFC*

Lancaster, NG358, LS-H

Two views of Lancaster bomber, NG358, LS-H, wearing G-H Leader markings on the tailfins. *Author's Collection via the late Pat Russell Collection*

Heading for Gelsenkirchen

'Fisher's Ferry'

Above and below: Lancaster bomber, NG444, LS-Y, with F/O Slaughter at the controls, takes-off from Mildenhall and sets course for an attack on Gelsenkirchen, on 22nd February 1945. *Author's Collection via the late Pat Russell Collection*

F/L Peter Rosenhain, DFC, RAAF (center), poses for an end of tour photograph with his crew, at Mildenhall. The crew consisted of: F/S Evans, Nav; F/S Daly, W/Op; W/O Anderson, F/E (3rd from left); F/O Button, B/A (4th from left); F/S Taylor, MUG; F/S Linley, R/G. The aircraft in the background is Lancaster bomber, NG338, LS-M, F/L Rosenhain's regular machine. On completion of his tour, the aircraft was passed to P/O Ken Fisher. *Courtesy of Ken Lewis, DFC*

Having been handed over to Ken Fisher as his usual mount, Lancaster bomber, NG338, LS-M, was immediately christened *'Fisher's Ferry'* by the crew. The aircraft, seen in this photograph with the name just visible under the cockpit, was taken over by P/O Fisher on 2nd March 1945. *Courtesy of Geoff Hill*

The Work Never Ends

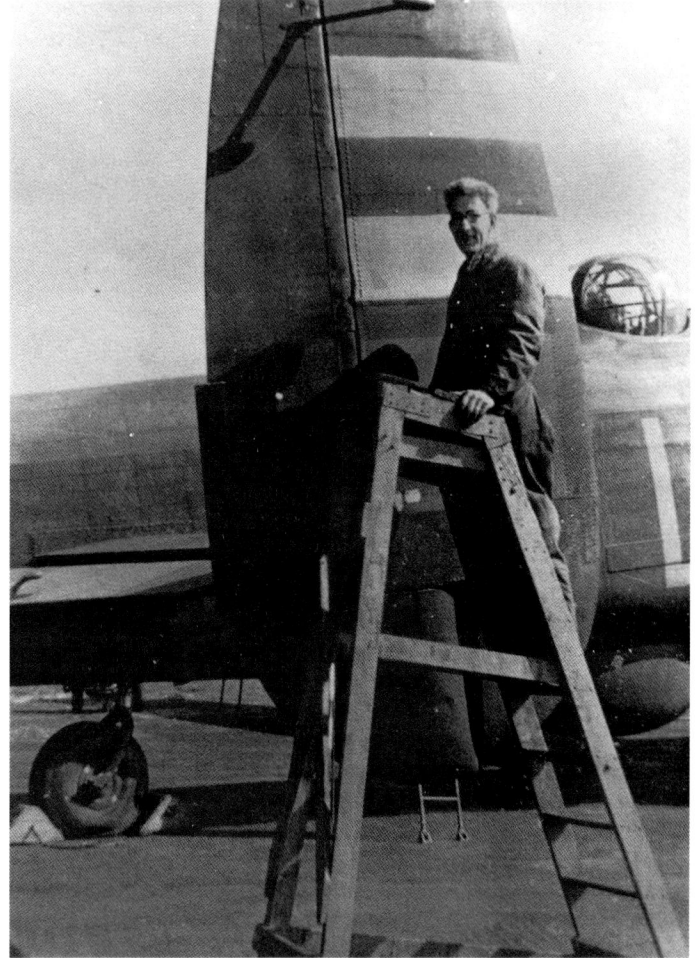

Above and right: An 'Erks' work is never done. Having completed his task at the front of Lancaster, NG338, an unidentified member of the groundcrew repositions the step ladder at the rear of the machine to undertake further tasks. *Courtesy of Geoff Hill*

A Lanc' coded "L"

On 2nd March 1945, F/L David Bone piloted Lancaster bomber, ME849, LS-L, on a daylight mission to Cologne. Unfortunately, due to adverse weather conditions which affected the G-H bombing aids, the participating crews overshot the target. Some bombed Coblenz whilst other, including David Bone, bombed Bonn. This aircraft, which completed its final mission with No.XV Squadron two nights later, was also to see service with No.44 Squadron. *Author's Collection*

"Back-off" 2

As F/L Lacey piloted Lancaster, HK799, LS-D, low over the sea, F/O Hunt decided to practice some close formation flying. The incident, which left the F/L Lacey very unhappy about the antics of the latter crew, occurred on 19th March 1945, whilst returning from an attack against Dortmund. *Author's Collection via the late Pat Russell Collection*

Lancaster bomber, NG357, LS-G, was the aircraft flown by F/O Hunt, on 19th March 1945, when he upset F/L Lacey with his close formation flying. *Author's Collection via the late Pat Russell Collection*

The Men Who Kept Them Flying

Lancaster, NG357, LS-G, is used as a backdrop for a photograph of the men who kept the aircraft flying – the members of "A" Flight groundcrew. Seated 2nd from right is LAC Percy Pluck, an aircraft electrician, who was known to his colleagues as 'Duke'. *Author's Collection*

The members of "B" Flight groundcrew, including Sgt Bobby Gault (4th from right, middle row) pose for a group photograph on the airfield at Mildenhall. *Courtesy of and via Flt/Lt T. N. Harris XV (R) Squadron*

The Admiral Scheer Attack

The box in front of F/L Robert Bithell, DFM (center), bears the legend *Admiral Scheer* and an indecipherable date. On 13th March 1945, No.XV Squadron attacked the Port at Kiel, where two German warships were known to be harboring. F/L Bithell, piloting Lancaster bomber, HK619, LS-V, participated in the attack along with his crew. The latter consisted of: F/O Harris, Nav; F/S Reece, W/Op; F/O Wallace, B/A; F/O Schanschieff, MUG; Sgt Price, R/G; Sgt Draper, F/E. Although he does not appear in the photograph, there was an 8th crew member on this mission. W/O Schanschieff flew as mid-under gunner. *Courtesy of Don Clarke, MBE, The Mildenhall Register*

Australian pilot F/O Jim Doogue (from Kalgoolie) also participated in the attack against Kiel on 13th March 1945. The crew consisted of, from left to right: F/O Johnny Phillips, B/A; Sgt 'Timber' Woods, R/G; F/S Pete Connolly, W/Op; F/O Jim Doogue, Pil, RAAF; P/O Frank Smith, Nav; P/O Alan Palmer (a qualified pilot) F/E; Sgt 'Dusty' Rhodes, MUG. The aircraft in the background is thought to be Lancaster bomber, ME844, LS-W. *Author's Collection*

"All ready Skipper"

F/S Cowie, bomb aimer to Ken Lewis, in the nose of Lancaster, NG444, LS-Y. *Courtesy of Ken Lewis, DFC*

Sgt K Parry, flight engineer, at his station on the starboard side of Lancaster, NG444. *Courtesy of Ken Lewis, DFC*

F/L Ken Lewis, D.F.C., RAAF, at the controls of LS-Y. *Courtesy of Ken Lewis, DFC*

F/S Roy MacDonald, wireless operator, RAAF, in his "office" on Lancaster, "Y"-Yoke. *Courtesy of Ken Lewis, DFC*

F/S Jack Bellis, navigator, inspects the equipment in his compartment on Lancaster, LS-Y. *Courtesy of Ken Lewis, DFC*

Sgt Watts, air gunner, shows of the fire power of the mid-upper turret. *Courtesy of Ken Lewis, DFC*

Right: A poor quality photograph of F/S 'Paddy' Langan, air gunner, in the aircraft's rear turret. *Courtesy of Ken Lewis, DFC*

The Last Crew Lost

Standing left to right: Aus 436058, W/O John Newton, RAAF pilot; 1602108, Sgt. M.F. Matthews, Air Bomber; 1604557, Sgt. P. Cooley, Mid-Upper Gunner; front row left to right; 1893709, Sgt. G.A. Cope, Wireless Operator/Air Gunner; 1801646, Sgt. C.A.J. Church, Navigator; 2219122, Sgt. T.E. Jenkins, Rear-Gunner. *Author's Collection via 'Swifty' Swallow*

Crash Site of HK773

A deep crater, in the forest at Mundford, near Thetford, marks the crash site of Lancaster bomber, HK773, LS-W. *Author's Collection*

Exactly 53 years to the day of its demise, parts of Lancaster, HK773, are recovered by Keith White, Anthony Seeley and Roger Woodcraft. *Author's Collection*

Fragments of metal, recovered from a vast area of the forest floor at Mundford, bears testimony to the force with which Lancaster, HK773, LS-W, exploded when it impacted with the ground on 22nd March 1945. *Author's Collection*

Bob Collis, of the Norfolk and Suffolk Aviation Museum, contemplates on the loss of the seven crew members who died when Lancaster, HK773, crashed five minutes after taking-off from Mildenhall. *Author's Collection*

Operation 'Manna'

F/L Nottage, RAAF, passes the time whilst waiting to participate in the first *'Manna'* operation, on 29th April 1945, by posing for a photograph on the starboard outer engine of Lancaster bomber, PP664, LS-U. *Courtesy of Ken Lewis, DFC*

Bomb trolleys are utilised by the groundcrew to maneuver panniers containing food sacks into place, under the bomb bay of Lancaster bomber, NG444, LS-Y, in preparation for *'Operation Manna'*. *Courtesy of Ken Lewis, DFC*

People wave from the streets below, as Lancaster bomber, NF953, LS-A, flies low over the Dutch countryside on its way to the designated dropping zone, during a *'Manna'* operation. *Author's Collection via the late Pat Russell Collection*

Grateful Dutch people show their appreciation, as sacks of food tumble out of the bomb bay of an unidentified Lancaster. *Author's Collection*

A poor quality photograph, taken from the rear turret of Lancaster, NG444, LS-Y, showing the food sacks falling to earth over Holland. *Courtesy of Ken Lewis, DFC*

Hundreds of food sacks lie scattered on the ground, having been dropped over Holland by aircraft from RAF Bomber Command. *Author's Collection via the late Pat Russell Collection*

Operation 'Exodus'

A line up of at least twelve Lancaster bombers at Juvincourt airfield, on 27th May 1945. The aircraft and crews were awaiting the arrival of a convoy of U.S. Army trucks, which were transporting the ex-prisoners of war who were to be flown home. *Courtesy of Ken Lewis, DFC*

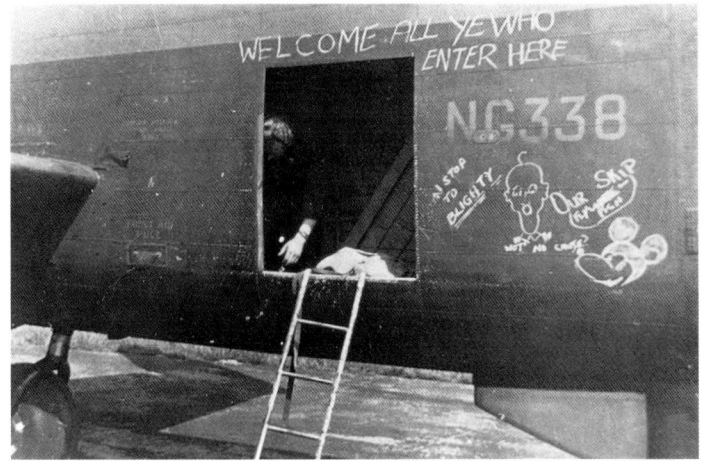

Graffiti adorns the fuselage around the crew entry door of Lancaster bomber, NG338, LS-M, piloted by F/O Fisher. Above the doorway is a sign of welcome for the PoWs returning home from captivity in Germany, which reads, *'Welcome All Ye Who Enter Here'*. Beneath the aircraft serial number is the legend, *'Non Stop To Blighty'* (England). Two captions adjacent to a caricature of the pilot read, *'Our Skip – Mr Fish'* and *'Wot No 'Chute'*. The figure in the doorway is F/S Geoff Hill, bomb aimer to F/O Fisher. *Courtesy of Geoff Hill*

Looking happy at the thought of going home, ex-PoWs make final adjusts to the life vests before boarding Lancaster bomber, ME455, LS-O, for the journey home. *Author's Collection via the late Pat Russell Collection*

'Time to go'. With fourteen of the allotted number already on board the aircraft, the remaining ten ex-PoWs try to look non-chalant and unhurried about boarding the aircraft. *Courtesy of Don Clarke, MBE, The Mildenhall Register*

Two unidentified members of No.XV Squadron aircrew inspect the wreckage of a Focke-Wulf Fw 190. The absence of a number of panels on the aircraft indicates souvenir hunters have been at work. In the background is Lancaster, NG364, LS-P. *Author's Collection, via the late Harry Bysouth*

Three unidentified No.XV Squadron Lancasters taxi out across Juvincourt airfield, in preparation for take-off for the final leg of an *'Exodus'* operation. Each aircraft carried a total of twenty-four ex-PoWs, some of whom could not have failed to notice the wreckage of three ex-Luftwaffe aircraft. *Courtesy of Ken Lewis, DFC*

Having landed at Westcott, Buckinghamshire, an ex-PoW, setting foot on English soil for the first time (possibly in many years) cannot wait to discard his *'Mae West'* life vest. Unfortunately, the unidentified soldier still had to be processed through a reception center before being allowed home. *Author's Collection via the late Pat Russell Collection*

The final stage of the journey. A poor quality photograph showing repatriated PoWs boarding an RAF truck, for transportation to the reception center at Westcott airfield. *Author's Collection via the late Pat Russell Collection*

'Baedeker Trippers'

After the cessation of hostilities, members of RAF Bomber Command's groundstaff were given the opportunity to fly in their Squadron's aircraft, on flights over Germany, to view for themselves the destruction caused by the bombers. Commonly known as *'Cooks Tours'*, No.XV Squadron detailed two such missions on 25th July 1945 under the codename Operation *'Baedeker'*. Although the first group flew on Lancaster, PA235, LS-E, they were photographed on their return to Mildenhall standing by Lancaster, HK799, LS-D. The aircraft's crew consisted of F/O Woodcock, Pil; F/O Hutchinson, Nav: F/S Bates, W/Op; F/S Bumford, R/G; F/S Randall, F/E an F/O Jim Glasspool. The passengers included: F/S Kay Godfrey, WAAF (2nd from left); Section Officer Grace "Archie" Archer, WAAF (3rd from Left) Corporal Glen (Kneeling, left); LAC Barnett; LAC Thompson. *Author's Collection via the late Pat Russell Collection*

Lancaster, HK772, LS-B, piloted by F/S Richardson (extreme right), was also used for *'Baedeker'* operations. This pilot had the enviable task of 'chauffeuring' three young ladies of the Women's Auxiliary Air Force, who worked in the Parachute Section, over Germany to view the damage. The photograph, which was taken on their return to Mildenhall, implies that F/S Winter (center) made a friend during the trip. *Author's Collection via the late Pat Russell Collection*

Lancaster bomber, LL806, LS-J "Jig" notched up 55 missions on 8th September 1944, when F/O Jennings took the aircraft to Le Harve. F/L William Leslie and his crew, who by that time had flown ten missions on "Jig", could not resist the temptation of being photographed with their former machine. In the picture are, from left to right: Sgt, Groundcrew: F/O McNiece, B/A; F/S Fendley, R/G; F/S Rosier, W/Op; F/O Frudd, Nav: Sgt Gundry, F/E; F/L Leslie, Pil; Sgt North, MUG. *Courtesy of Bill Gundry*

The 113th bomb symbol, having been applied to the tally on the aircraft forward fuselage, photographs were taken. In this picture, with seven members of the aircraft's groundcrew are, Group Captain Ken Batchelor, D.F.C., the Mildenhall Station Commander (3rd from left), F/L David Bone, D.F.M., Pilot (4th from left), F/L Harold Whittingham, Pilot (5th from left) and W/C Nigel MacFarlane (extreme right). *Author's Collection via the late Pat Russell Collection.*

On completion of "J"-Jig's 113th mission, flown by F/S Meikle, on 23rd February 1945, F/L David Bone was instructed to organise a photographic session, in order to pictorially record the occasion. One of the pictures shows an unidentified Leading Aircraftman painting the 113th bomb symbol on the aircraft's mission tally. *Author's Collection via the late Pat Russell Collection*

F/L Doug Hunt, with four members of his crew in front of "Jig", No.XV Squadron's top scoring Lancaster. From left to right are: F/S Pat Russell, B/A; Sgt Paddy Kirrane, R/G; F/O Doug Hunt, Pil, RAAF; Sgt George Pitkin, F/E; P/O John Sheppard, Nav. *Author's Collection via the late Pat Russell Collection*

Also photographed with LS-J were Ken Fisher and his crew. They are: F/S Ernie Ferraretti, W/Op (3rd from left); F/L Ken Fisher, Pil (4th from left); F/S Geoff Hill, B/A (6th from left); F/S Doug Able, Nav; Sgt Doug Plumb, MUG; Sgt Ted Parry, R/G; Sgt Pete Burnett, F/E. *Courtesy of Geoff Hill*

The mission tally on the forward fuselage of Lancaster, LL806, LS-J, indicating 134 bombing operations, three *'Manna'* (food dropping) operations and three *'Exodus'* (repatriation of ex-prisoners of war) operations. The aircraft also carries the legend, *'134 Not Out'*. Although not indicated on the tally, LS-J also flew three 'Dodge' operations. The machine, which was built by Armstrong-Whitworth, flew its first operational mission on the night of 1st May 1944. It was struck-off-charge on 5th December 1945, having amassed a total of 765 flying hours. *Author's Collection via the late Pat Russell Collection*

'Tiger Force'

Avro Lancaster B.VII, NX687, LS-A, was the only aircraft of its type used by No.XV Squadron. The aircraft, which was built by the Austin Motor Company, was designed for operations in the Far East. However, it was not used for this purpose but achieved fame as the aircraft in which Air Chief Marshal Sir Arthur Harris, Commander-in-Chief Bomber Command, flew to Rio De Janiero, on 23rd July 1945. *Author's Collection via the late Pat Russell Collection*

Lancaster B.VIII, NX687, LS-A, with W/C Charles Calder at the controls, climbs out of Mildenhall. The color scheme of white painted upper surfaces and black painted lower surfaces are shown to good effect in this picture. *Author's Collection via the late Pat Russell Collection*

19

The Post War Years

Apart from a reduction in manpower, the end of the war also brought with it a need to dismantle an arsenal of surplus weaponry, which included bombs and incendiary devices. During February 1946, No.XV Squadron was detailed to participate in the disposal of the latter weaponry, in an operation code-named 'Sinkum'. As the name suggests the incendiaries were jettisoned into the sea in Heligoland Bight, off the north German coast, and in Cardigan Bay, off the west coast of Wales.

On 12th March, W/C Nigel MacFarlane relinquished command of No.XV Squadron in favor of W/C Dennis Witt, D.S.O., D.F.C., D.F.M. The former had been a very popular leader, held in high esteem by those who served under him. He was particularly noted for putting 'rookie' crews at ease on their first operational mission with the Squadron, by flying with them as captain of the aircraft. By his actions, he was seen as a source of confidence and inspiration by many of those who flew with him.

The incoming C.O. was of a similar ilk. Dennis Theodore Witt joined the RAF as an aircraft apprentice in January 1931. Six years later he was recommended for flying training, and ultimately became a highly experienced bomber pilot. By the time he took command of No.XV Squadron he had flown exactly one hundred operational missions, many of them with the Pathfinder Force.

One of W/C Witt's first duties following his arrival on the Squadron, was to prepare No.XV for bombing trials, which were to commence during that same month. The trials, which entailed dropping 12,000lb 'Tallboy' deep penetration bombs, were to take place against the submarine pens at Farge, a small port on the River Wesser, north of Bremen. Although construction work on the former Reich U-boat pens had not been completed by the time the war ended, they still presented a formidable target. The concrete and steel structure, which boasted 14 foot thick walls and a 24 foot thick roof, was 1,400' long, over 300' wide and stood 80' high.

An incident during the trials at Farge gave one of the Squadron's ex-navigators cause for concern, and an unforgettable experience. Pilot Officer Jim Glasspool who had completed his tour of operations in December 1944, remained at Mildenhall in order to take up duties as the Squadron G-H Officer. Having been one of the first navigators to be trained on G-H, he was the ideal candidate for the job and was responsible for training crews from both No.XV and No.622 Squadrons. More importantly, to him, he was always 'up in the mornings before the other guys' so as to visit the Base Operations Room to receive the G-H broadcasts, which gave target information and bombing co-ordinates. He would then attend the briefings for both Squadrons and impart the relevant information for G-H 'ops'. On completion of their missions, P/O Glasspool would interrogate the returning crews, write-up the de-briefing reports and have the latter teleprinted to Bomber Command H.Q.

The 'Tallboy' trials on the U-Boat pens at Farge, enabled the training officer to acquire some unexpected further flying hours. The first trip occurred on 17th March when he flew as navigator to F/O Bull and his crew on Lancaster, HK765, DJ-Z.

However, the outcome was a bit of a let down for Jim Glasspool who had been looking forward to the operation. On arrival over the 'target', the pilot was instructed not to release his bomb but return with it to base. Any anxieties the navigator may have had about landing with a 12,000lb penetration bomb still on the aircraft were dispelled when the Lancaster landed safely at Mildenhall.

A week later, on 24th March, P/O Glasspool had the opportunity to repeat the exercise when he was again sent to Farge, with the same crew and the same objective. On arrival over the 'target' the pilot carried out four trial runs, dropping a 25lb practice bomb each time, prior to making the run during which the 'Tallboy' would be released. Again, F/O Bull received an R/T message saying enough tests had been carried out for the day and he was instructed to return home with his bomb.

At approximately 15.00 hours, as F/O Bull held the Lancaster steady on its final approach to Mildenhall airfield, Jim Glasspool reached down under the navigator's table to switch off the various radar sets. As he did so, the chart table appeared to 'rise up' and strike him under the chin. In his slightly confused state, he heard a female voice say, *"Will it explode?"* Unbeknown to him, the moment P/O Glasspool bent down, the bomb literally 'fell' from its mounting in the bomb bay. With the aircraft on finals, flying at an altitude of approximately 150' with an approach speed of 130 miles per hour, the bomb was in perfect alignment with the runway. The 'Tallboy' struck the ground in a shower of sparks, whilst the tailfins scored deep marks in the surface of the runway as it rocketed along like an uncontrolled toboggan. As the projectile fell from the aircraft and upset the trim, the nose of the Lancaster rose up and pointed towards the sky. Immediately the bomber stalled and began to side-slip, port wing down, towards the ground. From his precarious position under the chart table, Jim Glasspool heard the power surge through the engines as somebody 'up the front' rammed all four throttles fully open. Members of the groundstaff, as well as those in the control tower, watched aghast as the bomber screamed round in a tight arc, with the port wingtip only inches above the ground. Carving its way through the air, the Lancaster just avoided colliding with the bomb dump, situated on the airfield in the vicinity of Mons Wood. After control of the bomber had been regained, F/O Bull rejoined the circuit and effected a safe landing.

Despite many searching enquiries involving all the specialists, the cause of the incident was never established. Apart from 'Tallboy' weapons being used, as the trials progressed the larger 22,000lb 'Grand Slam' bombs were also used. It was due to the size of the latter bombs, that No.XV Squadron was partially re-equipped with Lancaster B.I.(Specials), which had cut-away bomb bays. Some further adaptations to the aircraft had also been carried out, including the removal of the front and dorsal turrets. Furthermore, they had been fitted with Stabilised Automatic Bomb Sights, which were recorded as being very accurate. The trials had caught the attention of the United States Army Air Force, who showed a keenness to participate. The USAAF detailed four B.29 Super Fortress and three B.17 Flying Fortress aircraft, each equipped with Norden bomb-sights and 20,000lb Amazon bombs. The resulting Anglo-American exercises became known as *'Project Ruby'*.

During this same period No.XV Squadron also participated in *'Operation Front Line'* exercises, which were basically a show of strength by the Royal Air Force. Usually, approximately sixty aircraft from various Bomber Command Squadrons would form up over the east coast of England, prior to heading off in formation across the North Sea toward Germany.

On 19th August, before the completion of the trials, No.XV moved back to Wyton where it had been stationed during the early part of the war. On the same date, W/C Gerald Bell assumed command of the Squadron from W/C Witt. The former officer had, in February 1939, been Personal Assistant to Air Vice-Marshal Babington, C.B., C.B.E., D.S.O., the Air Officer Commanding, Royal Air Force, Far East Command.

The combined trials culminated on 31st August with very satisfactory results for both the RAF and the USAAF

Further reductions regarding aircraft strengths were made to RAF Squadrons towards the latter part of 1946. No.XV Squadron found the number of Lancasters on its charge reduced from ten machines to four B.I. 'Specials' and two B.I's. It was becoming evident that the operational life of the Lancaster, as far as No.XV was concerned, was coming to an end. One of the last major sorties flown by the Squadron, using this type of aircraft, occurred on 5th November. Flying in 'vics' of three, the Lancasters of No.XV Squadron led a *'Front Line'* operation over Cologne, the Rhur area, Bremen and the inland port of Hamburg.

During February 1947, the first three Avro Lincoln B.II bombers arrived at Wyton, destined for service with No.XV Squadron. For a short period Lancasters and Lincolns served side by side, as the crews converted to the latter aircraft. By the end of March, with the conversion period complete, the remaining three Lincoln bombers were taken on charge by the Squadron.

Memories of the war were put out of mind as the crews succumbed to peace time training on their new aircraft. In appearance the Lincoln was very much like the Lancaster, having been designed and built by the same manufacturer. However, the updated machine had more powerful engines, increased fuselage length and modified mainplane.

Shortly after the delivery of the new equipment some alterations were made to the design of the bomber, including an increase in size of the bomb-doors. This modification enabled the Lincoln to carry 'Tallboy' bombs. Two of these modified machines, classified as Mark IIB's, were received by No.XV during April.

Apart from receiving two new aircraft, the Squadron also welcomed a new commanding officer during April. Wing Commander Bell, having been appointed O.C. Flying at Wyton, relinquished command of the Squadron to S/L Leonard Kneil, D.F.M., on 11th of the month. The new C.O. had been awarded his Distinguished Flying Medal on 17th January 1941, whilst serving as a sergeant with No.44 Squadron.

Ironically, No.44 Squadron had been associated with No.XV Squadron for over a year; the former having been posted to Mildenhall on 25th August 1945, whilst the latter was still in residence there. Four days after No.XV had moved to Wyton, No.44 Squadron followed arriving at the 'new' base on 29th August 1946. As part of their respective training programs, the two Squadrons were directed during May 1947 to participate alongside each other in *'Operation Hatbox'*. The nature of the on-going exercise involved the bombing, with 500lb and 1,000lb bombs, of two Royal Navy Ships. These 'trials', against the de-commissioned destroyer *HMS Pathfinder* and cruiser *HMS Hawkins*, continued through the summer until September.

During November, No.XV Squadron despatched a detachment of two aircraft to Shallufa, Egypt, under the aegis of *'Operation Sunray'*. A total

force of six Bomber Command aircraft were sent, with two aircraft from No.90 Squadron and two from No.138 Squadron making up the full complement. The crews, who were detailed for extensive weapons training in the Middle East, found themselves being scheduled for deployment on actual operations. An uprising, in the canal zone of Egypt by a local tribe, had necessitated the use of military force and all six bomber were prepared for action. On 25th November, the six Lincolns were deployed to Khormaksar, where each machine was loaded with 12 x 1,000lb bombs. Two days later Sergeant Balcombe and his crew took off for No.XV Squadron's first offensive mission since the end of the war. The following day, 28th November, the same crew delivered 4 x 1,000lb and 8 x 500lb bombs to the same target area. The latter load, however, was fitted with twelve hour delayed-fuses, a fact the dissident tribe found unhealthy. The uprising was suppressed and the aircraft returned to their respective units.

Aircraft from No.XV Squadron returned to Shallufa on 1st March 1948, when S/L Kneil led a detachment of three machines for inclusion in a continuation of *'Operation Sunray'*. Apart from the scheduled intensive weapons training, desert navigation and fighter affiliation exercises were flown. Exactly one month later, on 1st April, the three machines returned to England.

Following a number of weeks of regular flying training, the Squadron was instructed to detail one aircraft for an unusual photographic sortie. On 21st June, the first trans-Atlantic crossing by jet aircraft was undertaken, when No.54 Squadron flew a formation of De Havilland Vampire aircraft to the United States of America for a goodwill tour. No.XV recorded the historic event on film.

Another change of command occurred on 24th August, when S/L Kneil handed over to S/L John Blount, D.F.C. Like his predecessor, the new C.O. had been awarded his decoration for bravery during 1941. One of his first duties as the new C.O. was to prepare some of the Squadron's aircraft for a return to the airshow scene, as No.XV Squadron had been selected to participate in 'Battle of Britain' displays during the following month.

In November, three aircraft were detailed to participate in *'Operation Sunray'* and were despatched to Shallufa for a one month detachment.

The post war reduction of Royal Air Force Squadrons and personnel continued into the late 1940's, much to the chagrin of many. However, in order to overcome the loss of famous Squadrons or units with distinguished war records, it was decided to 'attach' certain Squadrons by number only. The theory behind this action being that the Squadron number was readily available if and when the Unit was re-activated. Under this dictate No.21 Squadron, on being disbanded, was attached to No.XV Squadron during February 1949. The Unit number thereafter being officially designated as No.XV/21 Bomber Squadron. This remained in effect until September 1953, when No.21 Squadron was re-formed as a Squadron in its own right.

No.XV continued its peacetime training with various exercises being undertaken throughout 1949. Amongst the foremost practice missions carried out during this period were Naval exercises, in March and July, and a U.K. air defence exercise which occurred during June and July. A Norwegian air defence exercise in September was followed two months later by a further participation in *'Operation Sunray'*, when six Lincoln bombers were detached to Shallufa, where they remained until 4th December.

Eleven days after the return of the six aircraft from Egypt, S/L John Denny, M.B.E., D.F.C., assumed command of the Squadron. John Blount, D.F.C., who in later years was to attain the rank of Air Commodore, was to be tragically killed in a helicopter crash on 7th December 1967, whilst acting in his capacity as Captain of the Queen's Flight.

Formation flying had played a major part of No.XV's history. The Unit had, after all, been named the number one light bomber Squadron in the RAF in 1935, a fact recognised when it was chosen to lead the light bomber Wing at the RAF Hendon display in June of that year. The following month, No.XV Squadron was again chosen to lead the light bomber wing during the Jubilee Review flypast, at Mildenhall. In later years, during the closing stages of the Second World War, the Squadron often flew to and from the designated target in formation. It therefore came as no surprise to many when, exactly fifteen years and one day after the Jubilee

Review, five Lincoln bombers from No.XV Squadron led a formation flypast of fifteen Squadrons at the SBAC display at Farnborough in July 1950. Other public appearances made by the Squadron during the latter part of 1950, included numerous Battle of Britain' Open Day displays. These were held in September, at various RAF airfields throughout the Country.

The penultimate month of 1950 saw the Squadron's association with the Lincoln bomber draw to a close when, on 14th November, the Unit relinquished its last aircraft of the type. Two weeks later, on 29th November, the Squadron disbanded and moved to Marham, where it reformed with new aircraft which could fly in excess of 30,000', could achieve a speed of over 360mph and were protected by twelve 5" Browning machine guns.

The escalation of the Cold War during the early 1950's, had made the need for military air power more evident, especially in the United Kingdom. If Britain was to be able to play a major role within the policy of the North Atlantic Treaty Organisation (NATO), signed in April 1949, RAF Bomber Command needed to increase its potential as a striking force. With this view in mind, seventy Boeing B.29A bombers were acquired from America, under the Mutual Aid Defense Program. These aircraft, which would eventually form eight Bomber Command squadrons, were known in the RAF as Washington B.I's.

Having relocated to RAF Marham in preparation for conversion to the new bomber, No.XV Squadron's stay in Norfolk was very short; lasting approximately two months. This was due to the fact the Station, which had been designated the Washington Force Headquarters, was already playing host to four B.29A squadrons. Following completion of the course, and having taken-on-charge four aircraft of its own, No.XV quickly moved to RAF Coningsby where it became part of the second Washington Wing.

In order to get the Squadron up to operational strength as quickly as possible, an advanced party of ground staff arrived at the Lincolnshire base on 22nd January. By the end of February, No.XV was at full compliment regarding its personnel, and had taken-on-charge two additional aircraft.

The aircrews, having got to know their aircraft, flew a number of familiarisation flights. Plotting local landmarks and generally gaining knowledge of the surrounding countryside. The latter being of use when flying cross-country exercises in less than perfect weather. Other exercises carried out during the same period included high-level bombing, radar bombing and gunnery training.

The 16th March 1951 was a poignant day in the Squadron's history; being the last occasion No.XV's aircraft flew displaying the LS code they had adopted in 1939. From this day on, its 'new' aircraft would fly with natural metal finish; only the Royal Air Force roundel would adorn the fuselages.

Two more aircraft were taken on charge by No.XV during June, taking the Squadron's compliment up to eight B.29A bombers.

With the Squadron flying American designed and built aircraft, it necessarily followed that the Americans would show some interest in how the Brits were faring with their new equipment. The opportunity to show the Americans came when F/L Ellis Ware, D.F.C., and his crew were selected to represent Bomber Command in the USAF Strategic Air Command's Bombing and Navigation Competition. Knowing the honor of the Royal Air Force rested on their ability, they left Coningsby on 1st August and headed west across the Atlantic. The crew of the American built RAF bomber returned home at the end of the competition with honor intact, having achieved a final place in the top ten.

Ellis Ware had the opportunity to repeat his recent success, when he was chosen to represent the Squadron in Bomber Command's own Bombing and Navigation Competition, held in December. Participating with F/L Ware was F/L Richard Hardy who, as a young flight sergeant with No.XV Squadron, had been awarded a D.F.M. for his part in the attack against the German warships *Gneisenau* and *Scharnhorst* ten years earlier. Richard Hardy was later posted to No.635 Squadron, where he was awarded a D.F.C., before eventually returning to No.XV Squadron.

The pride of any Army Regiment, apart from its historical traditions, is its Regimental Color. King George VI had announced, as early as 1943,

his intention to present a Banner or 'The Standard' as it was to be known, to any RAF Squadron which merited the Monarch's gratitude. In order to qualify, a Squadron must have either undertaken some outstanding operations or completed a minimum of twenty-five years service to the Crown. No.XV Squadron qualified on the latter, and as a result was notified in January 1952 that it was to be presented with the award of 'The Standard'. In similar manner to Army Regiments, RAF Squadrons who were awarded 'The Standard' were permitted to adorn it with their respective battle honors.

This good news was overshadowed the following month when, on 6th February, His Majesty King George VI died peacefully in his sleep. He was laid to rest in the vault at St. George's Chapel, Windsor Castle, nine days after he passed away. Princess Elizabeth, the late King's eldest daughter succeeded him as Monarch and was crowned Queen Elizabeth II at Westminster Abbey, on 2nd June 1953.

The enthronement of a new Monarch meant the military services, including the Royal Air Force, its various Commands and Squadrons, would need to amend all authorised badges and crests. Under the late King, these emblems and insignia were surmounted by a Tudor crown. However, with the ascension of the Queen to the throne all official emblems now had to be surmounted by an Imperial Crown.

The change of crown was an easy one but, unbeknown to No.XV, the Squadron would have more of a challenge regarding 'The Standard' in the not too distant future.

The intense training which continued throughout the year, again showed results for No.XV in September when S/L Denny and his crew participated in the USAF Strategic Air Command's Bombing Competition at Tucson, Arizona. Upholding the honor of the Squadron, and emulating F/L Ware's efforts, the C.O. and his crew also managed to finish in the top ten. Three months later similar results were achieved, when the Squadron was awarded a fifth place in the Bomber Command Blind Bombing and Navigation Competition held during December.

For No.XV, the celebrations heralding in the New Year were short lived, due to a tragedy which struck on the night of 5th January 1953. In one of the very few accidents involving a B.29A in RAF Service, Washington bomber, WF553, crashed into a field whilst preparing to land. The aircraft, which had previously seen service with No.57 Squadron, was completing a routine exercise and was on final approach to the airfield. The pilot, F/L Rust, and five members of his crew were killed when the bomber struck the ground near Miningsby, north-east of Coningsby. One crew member survived the crash, having been propelled through the nose of the aircraft due to the force of impact with the ground. Flight Lieutenant Read, the navigator/bomb aimer, sustained severe leg injuries incurred as a result of his being partially buried under wreckage.

Although it had no bearing on the decision, three months after the crash the Squadron began to ferry its Washington bombers back to Marham. The exercise, which commenced on 13th April, took seventeen days to complete. It heralded the end of another era in No.XV's history. The American built bombers were the last piston engined aircraft to be flown by the Squadron. No.XV was about to enter the jet age.

During the period the Squadron changed its equipment, a change in leadership also occurred. Squadron Leader Denny, who had been in command of No.XV for the last three and a quarter years, was posted away on 17th April. As there was no formal handover to an incoming C.O., Flight Lieutenant J Vnoucek held temporary Command of the Squadron for a period of approximately seven weeks.

With the advent of the jet age, pilot and aircrew training took on a new dimension. The crews selected to fly the twin-engined jet bombers for No.XV Squadron had commenced training with No.231 Conversion Unit at Bassingbourn, on 7th April. The Cambridgeshire airfield had, during the Second World War, been home to the Flying Fortresses of the 91st Bombardment Group, USAAF. The *'Ragged Irregulars'* as they were known, flew their last mission from the airfield on 25th April 1945. Redeployment of this unit back to the U.S.A. commenced at the end of the following month. Although the Royal Air Force had taken over the base at the end of the war, there was an American presence on the airfield up to, and during, the early 1950's.

The newly trained jet crews arrived at Coningsby on 26th May 1953, followed three days later by the first of the new aircraft. English Electric Canberra B.2., WH724, which flew in on 29th May, arrived from Binbrook. By the end of the following month the Squadron had a full compliment of four machines.

Flight Lieutenant Vnoucek's 'term' as temporary commanding officer came to an end on 5th June, when Squadron Leader John Ayshford, D.F.C., arrived to officially take over the position.

The usual cross-country familiarisation flights and local training missions were all carried out from Coningsby, along with bombing exercises. A major exercise in September, followed by the annual formation flypasts at 'Battle of Britain' air displays, kept the Squadron busy during that particular month. The exercise, known as *'Premraf'*, was set up to test the Country's defenses; particularly the early warning radar system and the response of the Home based fighter units.

Also in September, the link between No.XV and No.21 Squadron, which was inaugurated on 1st February 1949, came to an end. The latter Unit was reformed at RAF Scampton as a Squadron in its own right, on 21st of the month. Like No.XV, it too was equipped with Canberra B.2 bombers.

The end result of any operational bombing mission was, as it had always been, to fly long distances, pin-point the target and destroy it. To enable this action to be carried out successfully required accurate navigation. Many advancements in bomber technology had been made during the Second World War, some to the benefit of navigators. The end of hostilities in 1945 did not, however, mean an end to the research and development programs, which continued during the Cold War. Trials had began in 1951 with the MkIX Bubble Sextant, which enabled navigators to find their way to a target by use of astro-navigation. The trials were extended and in April 1954, No.XV Squadron was given the task of testing the instrument.

After being in residence at Coningsby for three and a quarter years, No.XV Squadron was relocated to RAF Cottesmore, Rutland, on 19th May. Their stay in England's tiniest County was very short, lasting only nine months. However, whilst there, the Squadron adopted the Station badge, which it reproduced on the tailfins of its aircraft. They continued the astro-navigation trials whilst at Cottesmore, and also carried out visual and radar bombing techniques. Amongst the major exercises undertaken whilst based in Rutland, were those named *'Battle Royal'* and *'Lone Ranger'*. The former involved working closely with other units from Bomber Command and Fighter Command, in a combined exercise which also included the Second Allied Tactical Air Force. The *'Lone Ranger'* exercises put to good use the navigational practice missions recently undertaken, when the participating crews flew to overseas locations. Initially, the 'target' was Wunstorf, West Germany, but later Gibraltar and Idris in Libya were added.

Having spent the final two years of the war based in Suffolk, the Squadron returned to the County during February 1955, when it relocated to RAF Honington. The move was made over a three day period, with the first aircraft arriving on the 15th of the month. It did not take the crews long to settle in their new surroundings, and they soon replaced the hunting horn and horseshoe badge of Cottesmore, with the white pheasant badge of the new Station.

As the aircrews flew familiarisation flights in order to gain knowledge of the local landmarks, a new commanding officer settled in to his new office. The latter was no stranger to the Squadron, having previously served with the Unit at Wyton, during 1948. Squadron Leader Allan Scott, D.F.C. and Bar, a pilot who had flown Stirlings, Lancasters and Lincolns, took up his post on 28th March. Later that same month, S/L Scott led the Squadron during the Bomber Command Bombing Competition, in which No.XV achieved third place.

Having originally taken four Canberra B.2 machines on charge the Squadron was, by July 1955, operating a compliment of ten such aircraft. Training on these aircraft progressed in the normal manner, with all the usual exercises being undertaken. The number of 'target' locations used in the *'Lone Ranger'* exercises had increased by this time to include such places as Nicosia and Khartoum. Other locations, such as Egypt and Iraq, were added during the course of the year.

On 3rd February, nearly a year after he took office and three years after the award was approved, S/L Scott collected the Squadron Standard. The initial joy of the occasion was short lived when it was discovered the Squadron number had been embroidered on 'The Standard' in Arabic numerals. During a meeting with Sir William Heaton-Armstrong, the Chester Herald and Inspector of RAF Badges, S/L Scott explained the significance of the Roman numerals and the part they played in the Squadron's history. Eventually, after much deliberation, Sir William gave authorisation for the XV to be amended. However, he issued an edict that no other squadron would be granted the same concession. From that day forward No.XV Squadron's Standard would proudly bear Roman numerals, whilst the numbers on every other squadron Standard in the RAF would be in Arabic.

On 18th April 1956, Marshal Bulganin, Premier of the USSR, and Mr Khrushchev, the Secretary of the Russian Communist Party, arrived in Britain for an eight day official visit. During their stay in the United Kingdom, the two Russian Leaders and their entourage had talks with members of the British Government. A number of official duties and visits were also arranged. One of the latter engagements arranged for the Communist delegation, which was scheduled for 23rd April, was a visit to RAF Honington.

In preparation for the visit all the Canberra B.2 bombers on the Station were drawn up in a line outside the main hangers. Apart from making inspection of the assembled aircraft easier, it also overcame many security problems. There was also the fact all B.2 variant aircraft had been grounded prior to modification. This in turn meant the proposed flypast, in salute to the Russian guests, had to be undertaken be Canberra T.4 aircraft.

The modifications to the B.2 bombers were completed a month later, allowing the Squadron to recommence flying them on 24th May.

The first major task carried out with the up-dated aircraft commenced on 23rd June. The Squadron was detailed to participate in the NATO exercise codenamed *'Operation Thunderhead'*. The assignment entailed flights over Greece and Turkey, where the bombers were 'intercepted' by fighters from the air forces of both those Nations. During the seven day period over which the exercise lasted, No.XV Squadron was based at El Adem, Libya. Whilst a week long detachment to the Mediterranean may have seemed ideal, there were disadvantages; the heat being a major factor. During pre-flight checks, temperatures up to 135 degrees F. were often recorded in the aircraft cockpits. However, although they did not solve the problem, canvas covers fitted over the cockpit glazing helped alleviate part of it.

No.XV Squadron was on show again during July, but not on this occasion to the public. The event, the first Royal Review of Bomber Command by Her Majesty Queen Elizabeth II since her enthronement, was held at Marham, on 23rd July. The Squadron detailed three Canberra aircraft to participate in the flypast, whilst a fourth machine was 'on show' in the static display.

The celebrations at Marham were overshadowed three days later when the British Government found it necessary to prepare for military action. The cause of the problem being the nationalisation of the Anglo-French controlled Suez Canal Company, by the Egyptian President, Colonel Gamel Nasser.

Supported by the French and Israeli Governments, Britain despatched a military force to the Eastern Mediterranean.

Over a nine day period, between 2nd and 11th August, No.XV Squadron assisted in the build-up, and undertook a total of fifty-six flights between Honington and Luqa, Malta. The significance of these sorties, part of *'Operation Accumulate'*, was to deliver a stockpile of bombs to the Mediterranean island for use in any forthcoming action. Each aircraft left Britain carrying six 1,000lb bombs and, on completion of its task, returned to Honington.

With an escalation in the political situation, the Squadron was deployed to Nicosia, Cyprus, as part of *'Operation Reinforced Alacrity'*, on 23rd October. The aircraft, each carrying a load of six 1,000lb bombs, and with instructions to maintain radio silence, staged through Halfar in Malta.

Black and white invasion stripes were applied to the rear fuselage and wings of each aircraft, upon arrival in Cyprus.

The dropping of Israeli paratroopers 30 miles east of Suez, during the afternoon of 29th October, signified the start of the offensive against Egypt.

Hostilities against the Arab nation began at dusk on 31st October, when *'Operation Musketeer'* was commenced. *'Musketeer'* involved attacks against eight Egyptian airfields, by a total force of two hundred and forty British and French aircraft. The Armee de l'Air provided forty F-84 Thunderstreaks, whilst the British contingent consisted of de Havilland Venoms, Vickers Valiants and English Electric Canberras.

Although the attacks on the airfields continued for approximately a week, other targets were also detailed for attention. The Cairo Radio station at Abu Zabac, received 'a visit' from No.XV Squadron on 2nd November.

Three days after Cairo Radio went off the air an airborne assault was made, followed twenty-four hours later by a landing of seaborne forces. Although the objectives were quickly achieved, condemnation against the British invasion was voiced most strongly by the United States of America. Other countries sided with the U.S.A, leaving Britain little choice but to concede to the international demand for a cease fire. This was put into effect at midnight on 6th/7th November.

No.XV Squadron undertook a total of thirty-seven missions during the six day period of fighting, five of which were flown by S/L Scott piloting Canberra B.2 bomber, WD980. His crew for those operations comprised of F/O Bob Weeks, bomb aimer/observer, and F/O Wells, navigator.

Following the execution of the cease fire No.XV, who had dropped a higher bomb load than any other squadron during the conflict, was ordered to return to the United Kingdom with immediate effect. In compliance with this command, Squadron Leader Scott took-off from Nicosia at 02.00 hours on 7th November. After routing via Luqa, and logging a flying time of six and a half hours, he landed at Honington later the same day. The joys of being home based did not last long as, due to the political tension rising in Egypt, the Squadron was re-deployed to Luqa over the forthcoming festive season.

After six weeks at Luqa, where they spent Christmas and New Year practicing their technique on the bombing range, the Squadron was brought home. Again, the joy of being back in the U.K was short lived, when it became known that No.XV was to be disbanded on 15th April 1957.

During the twelve years since the end of World War Two, many changes had been made in aircraft design and weapons technology. The latter had brought about the advent of nuclear bombs, which in turn had an effect on the design of the aircraft designated to carry such weapons. One aircraft to evolve from the many designs put forward, was a very futuristic looking machine classified as the HP80. Built by Handley Page, this aircraft would become universally known as the Victor bomber and, along with two other aircraft of corresponding specifications – the Vulcan and the Valiant, would make up one third of Britain's airborne nuclear capability. This deterrent would, by virtue of the names given to each of the aircraft, become known as the "V" Force.

No.XV Squadron became part of this Force when it was reformed with Victor bombers at Cottesmore, on 1st September 1958. Wing Commander David Green, a wartime bomber pilot with a DSO, OBE and a DFC, was appointed to command the Squadron. It was he who collected the Squadron's first "V" bomber, Victor B.I, XA941, from the manufacturers at Radlett, Hertfordshire, on 7th October. Over the following five months, seven other similar aircraft were acquired from Handley Page in the same manner, all but one of which were collected by W/C Green.

Having taken-on-charge all eight allotted machines, the first major task was for members of both aircrew and groundcrew to get to know their aircraft. The latter in particular found a number of servicing problems which had to be overcome.

A member of Cottesmore's groundcrew team at that time was Jeff Mellor. He had been posted to the Station as a Junior Technician Airframe Fitter (a rigger) the previous February, having completed an overseas tour with the 2nd Tactical Air Force in Germany on a Gloster Meteor Squadron!.

One of the major problems for people like J/T Mellor was getting acquainted with the vast amounts of new ground equipment the "V" bomber required. Equipment such as single and double Giraffes, 25-ton lifting jacks, engine hoists and slings and a most useful piece of equipment – the Platform, Aircraft Servicing, Mobile, Adjustable – which soon became known as the 'Safety Raiser'.

As the "V" bombers arrived at Cottesmore they were given thorough acceptance checks. The servicing, modifications and rectification work which was carried out at that time, provided valuable experience for the groundcrews. Although the Squadron's wartime code of "LS" had gone forever, the groundcrews were permitted to apply No.XV's crest and the numerals XV above the flash on the tailfin of each of its aircraft.

Adorned with these embellishments, Victor bomber, XH594 took-off from Cottesmore on 21st March and set course for Cyprus. The aircraft, which was piloted by the Station Commander, G/C 'Johnnie' Johnson, DSO, DFC, broke the England to Cyprus flight record by completing the journey in three hours, forty-six minutes. The co-pilot on the flight was No.XV's Commanding Officer, W/C Green.

As a rapid response to the threat of hostile action was of prime importance, the aircrew's duty was to get airborne as quickly as possible. Apart from the usual training required for operational duties, one major aim was to be able to scramble four aircraft within four minutes of receiving warning of an impending threat. The drill devised to achieve this aim, and the practice which went into it, paid off. On 1st April 1959, the then Prime Minister of Great Britain, the Right Honorable Harold McMillan, witnessed the first ever four minute scramble undertaken by any of the "V" Force aircraft. The four bombers used for the occasion had previously been parked at an angle on the runway at Cottesmore. On a given signal the crews sprinted across the tarmac, clambered into the cockpit and started up the engines. The fact that the four engines on each aircraft had to be started individually, bears witness to the intensity of the training.

During the summer of 1959 the Squadron participated in Exercise *'Mayflight'*, which placed crews on a high state of alert whilst awaiting orders to scramble and undertake a simulated mission. For the benefit of the exercise, the Squadron utilised the airfield at St Mawgan as a dispersal base. In the event of real hostilities, to ensure their survivability, the Cornish airfield would have become the base from which they operated. They practised long range navigation exercises by flying over Northern Canada, and carried out detachments to the old wartime base at Goose Bay.

One detachment to Canada occurred between 8th and 17th July when W/C Green and F/L Bell flew out to British Columbia to represent the Squadron at a week long airshow in Vancouver. They were accompanied by the Air Officer Commanding No.3 Group, Air Vice-Marshal Dwyer, C.B.E. The event was staged to celebrate the 50th Anniversary of Flight in Canada. The event, which was a great success, drew large crowds of people, many of whom took the opportunity to photograph the visiting Victor bombers.

Members of the British public had the opportunity to photograph the Victors of No.XV during the second week of September, when the Squadron provided a formation flypast of three bombers at the Farnborough Air Show.

Having been one of the two aircraft flown to Canada in July, Victor bomber, XH594 was becoming a world traveler. On 8th October, a month after the Farnborough Air Show W/C Green flew the aircraft to Akrotiri on its second trip to Cyprus. The purpose of the mission being to provide Brigadier Weismann, the Chief of the Israeli Air Force, with a demonstration flight.

The first detachment by a No.XV Squadron aircraft to the United States occurred during October, when an aircraft was despatched to Offut Air Force base. The exercise was code-named *'Western Ranger'*.

The following month, the Squadron participated in Exercise *'Mayflight II'*, which again meant utilising St Mawgan airfield. However, on this occasion, instead of being dispersed for eight days it was decided to extend the period to 7th December. This gave an insight as to the Squadron's capabilities of being dispersed 'in the field' during long periods.

The first day of 1960 gave two causes for celebration. Apart from being New Year's day, 1st January brought with it promotion for David Green, who was promoted to the rank of Group Captain. Although a posting to a new command or unit usually came with promotion, G/C Green remained in command of No.XV Squadron for a further three months. He eventually handed over command of No.XV to W/C John Matthews, A.F.C, on 1st April 1960.

Promotion had also been bestowed on Jeff Mellor who, with the rank of Corporal Technician, was officially posted to No.XV Squadron during May. With eighteen months of servicing Victor bombers to his credit, Jeff Mellor arrived on the Squadron as an experienced Air Frame Fitter. He soon adapted to the two-shift system operated by the groundcrews, which ensured the Squadron's flying program. Workdays operated from 08.00 hours until 17.00 hours, whilst the second shift worked from 17.00 hours until 02.00 hours the following morning; or until the work was completed and the aircraft were serviceable. This changed over each week and also included a full working weekend every four weeks. War or peace, home based or on detachment, little had changed for the hardworking groundcrews, as Exercise *'Mayflight III'* proved. Whilst No.XV was on detachment at St Mawgan, towards the end of the second week of July, it played host to the Chief of the Air Staff, Air Chief Marshal Sir Thomas Pike, G.C.B, C.B.E, D.F.C. Needless to say, the aircraft had to be in tip-top condition.

The workload for some members of groundcrew increased when major modifications, in the form of electronic warfare equipment, was made to some Victor bombers during early 1960. These modifications brought with them a change of designation of the aircraft to B.1A's. The arrival of Victor B.1A, XH613 on the Squadron in July 1960 confirmed No.XV's status as the first unit to operate the type.

With a wealth of air display experience to its credit, No.XV Squadron was again selected to participate in the Farnborough Air Show during early September. It was here, for the first time, the public was permitted to witness the Squadron's ability to 'scramble' four bombers within four minutes.

As well as No.XV's Victor bombers being at Farnborough, No.617 'Dambusters' Squadron flew their Vulcan bombers down from RAF Scampton, whilst No.148 Squadron flew in from RAF Marham with their Valiant bombers. Thus, Britain's nuclear deterrent was represented by all three type of "V" bomber. Whilst both No.XV Squadron and No.148 Squadron 'performed' on two mid-week days, No.617 Squadron displayed on three occasions, including the weekends.

Although No.XV's bombers flew from Cottesmore to Farnborough, the groundcrew party was bussed by RAF coach the previous day. They were billeted at RAF Odiham, where they slept overnight. The following morning the groundcrew party were bussed to Farnborough, where they were briefed by a squadron leader in airshow procedures!! Jeff Mellor, who was there in his capacity as an air frame fitter, listened in amazement as the squadron leader instructed the tug drivers what gears to use whilst towing the aircraft, and what revs to tow at.

Each of the four aircraft used by No.XV during the display, were attended by an aircraft servicing chief (Crew Chief) and five members of groundcrew – all of whom were dressed in white overalls. At a given time, with each aircraft connected by a towing arm to a Tugmaster tractor, the procession set off from the dispersal along the main runway with the supporting groundcrew following behind in Land Rovers. At a predetermined given point, the groundcrews alighted from their vehicles whilst the aircraft were reversed into position along the length of the runway.

For safety reasons, one member from each groundcrew crossed the runway with a CO2 fire extinguisher and waited whilst a Sims-start unit (a bomb trolley filled with aircraft batteries) was positioned in front of each aircraft. The two power cables leading from the trolley were connected to the aircraft, along with the crew chief's intercom lead. The latter was attached to a rope which was passed through the right-hand pilot's DV window and pulled on board the aircraft. Once the appropriate connections were made, the crew chief had two-way communication with the pilot.

Whilst all this was going on, the remaining four members of the groundcrew lined up by the start trolley.

At the firing of a mortar shell the man with the fire extinguisher ran like mad across to the aircraft, the start buttons were pressed and the engines burst into life. On receipt of a signal from the crew chief, the power cables were disconnected, the start panel closed and secured, and the cables wrapped round the wheels to chock the trolley and stop it from being blown away as the aircraft took off. Finally, the crew chief's lead was disconnected and all six members of the groundcrew took shelter behind the start trolley. The thunder of twelve Armstrong Siddeley Sapphire turbo-jet engines pulsating at full throttle, as the aircraft roared along the runway, was an exhilarating experience for the young Jeff Mellor.

The end of the display day for the groundcrews was somewhat of an anti-climax, as they were bussed back to Cottesmore for one day – only to repeated the whole procedure again a day or so later.

On 12th September 1960, four aircraft from No.XV squadron were despatched to Cyprus for a period of five weeks. Whilst there they undertook bombing training exercises at the Episkopi and El Adem ranges, participated in the 'Sunspot' exercises and formed part of a flypast over Nicosia for Archbishop Makarios.

The last quarter of the year saw No.XV Squadron involved with a number of training exercises, some of which included further overseas detachments.

There was no let up in the intensity or quality of the training, which continued into the New Year. In January 1961, a number of 'Lone Ranger' exercises were flown, including one to Nairobi.

During the third week of April, No.XV Squadron participated in the annual Bomber Command Bombing and Navigation Competition, held at RAF Lindholme. Six crews, led by W/C Matthews, were entered against fifteen other Units.

Although prestige and honor were at stake, the outcome of the competition was not as important as some news imparted to No.XV around this same time. The Squadron learned it was to receive its Standard. A big occasion and, as befits the Royal Air Force that meant rehearsing for the big day. The flying program and aircraft servicing still had to continue whilst parade rehearsals were held during the afternoons. After a few sessions it soon became apparent that not everyone was giving his best, possibly the result of a late night on the flightline had dampened some of the enthusiasm. However, the problem was soon sorted out after two drill Sergeants, who brought out the best in everybody, arrived from RAF Uxbridge. With F/O John Laycock, who was to receive the Standard, giving encouragement from the touch-line, things improved further.

During the early 1960's, all airmen were wearing the smooth (T63) best-blue uniform and the light colored poplin shirt with detachable collar. It was decided to send everyone's shirt, complete with a collar, to a special laundry to have the front of the shirt, the collar and cuffs specially starched to give a nice shiny finish. Margaret Mellor, who was making a special journey from home to attend the presentation, never allowed anyone to wash her husband's best shirts and said she would do the job herself. Unfortunately she starched the whole shirt and found afterwards that it could practically stand up on its own! Corporal Technician Jeff Mellor looked splendid in his shirt but, after presenting arms with a .303 rifle a few times, was raw under the armpits.

As the big day approached dress rehearsals were held for the C.O., then the Station Commander, then the Air Officer Commanding the Group and finally, the Commander-in-Chief Bomber Command. On this last rehearsal a strapping Flight Lieutenant paraded around with a placard around his neck bearing the name -HRH Duchess of Kent – (the member of the Royal Family who was to present The Standard). Behind the F/L was the smallest, youngest-looking Pilot Officer in Bomber Command with a placard stating – Lady in Waiting.!! High spirits amongst the troops at seeing this 'vision' brought one or two *"What are you doing tonight Ducky?"* type remarks. The remarks were accompanied by one or two quiet wolf whistles, which were followed by a glare from the C-in-C and a sharp retort from W/C Matthews; the latter being a strict disciplinarian.

On the day itself, 3rd May 1961, everything went according to plan. The RAF Chaplain-in-Chief, the Venerable F W Cocks, BB, QHC, MA, officiated at the dedication ceremony, with Her Royal Highness the Duchess of Kent in attendance.

The sun shone brightly, reflecting off the highly polished medals and decorations worn by the past members of the Squadron who had assembled for the occasion. Amongst their number was Air Marshal Sir Thomas Elmhirst, the former C.O. who first applied the Roman numerals to the Squadron's aircraft.

It was a day for pride, both in the Royal Air Force and No.XV Squadron. Every man on parade that day gave of his best.

Exercise *'Mayflight IV'* was carried out during the same month, with St Mawgan again being used to disperse the Squadron's aircraft. The four Victors despatched from Cottesmore remained at the Cornish airfield for four days, between 10th May and 14th May.

The month of June saw the Squadron deploy to RAAF Butterworth in Malaya, where it participated in Exercise *'Profiteer'*. This deployment was welcomed by many members of the groundcrew, some of whom had never been out East before. Also, at the same time, the C-in-C Bomber Command, Air Chief Marshal Sir Kenneth Cross, K.C.B, C.B.E, D.S.O, D.F.C, was flying out to Australia to welcome a Vulcan bomber of No.617 Squadron from RAF Scampton, which was flying non-stop to RAAF Richmond, near Sydney. No.XV Squadron provided two Victor bombers for the occasion. Sir Kenneth flew in the first machine, piloted by the C.O. of the Operational Conversion Unit at Gaydon, whilst F/L Don Rigby of No.XV Squadron piloted the back-up machine.

A small servicing team of groundcrew, which included Jeff Mellor as rigger NCO, was selected to accompany the C-in-C on a tour around Australia. Much to the envy of the rest of the Squadron, who had to wear the old standard KD uniform, this small band of men had been kitted out with a special smooth tropical uniform, at RAF Innsworth, Gloucestershire, before their departure to RAAF Butterworth.

The main party left Cottesmore by Transport Command Bristol Britannia aircraft, whilst the 'Australia Party' (which included the Vulcan groundcrew) left a couple of days later courtesy of a No.216 Squadron de Havilland Comet. The latter group's first stopover was at RAF Nicosia, Cyprus, followed by a night at the Miniwalla's Grand Hotel, Karachi. Their next stop was at RAF Gan, a tiny atoll in the Maldives, where the runway started in the sea at one end and finished in the sea again at the other end. They took advantage of this island paradise and did little but swim and lounge on the beach. Their sojourn was broken when they joined the rest of the Squadron at RAAF Butterworth, where they serviced their aircraft and enjoyed the hospitality of the Australians. A few days later the selected group continued its journey, stopping over at RAF Changi, where they sampled the delights of Singapore.

Their first stop in Australia was at RAAF Darwin, where they stayed overnight in the luxury of a British Overseas Airways Corporation (BOAC) hotel. Being there on the right night, the group was fortunate to be able to partake of the regular Sunday evening Smorgasbord. The abundance of food the night before did not prevent any of the group from fitting into their respective tropical uniforms, neither did it preclude them from flying on to RAAF Richmond the following day. As the Comet headed south across the continent of Australia, darkness began to descend and had taken hold by the time the aircraft landed at its destination. Looking out of the cabin windows, Jeff Mellor and the others were surprised to see the Australian welcoming contingent wearing, not only their best blue uniforms, but greatcoats as well. On disembarking from the aircraft the RAF party soon found out why – it was winter down under!!

The No.XV Squadron Victor bombers, which were soon adorned with red painted kangaroo symbols, were the first aircraft of their type to visit Australia. The RAF crews soon found themselves being entertained by their counterparts, many of whom had emigrated from Britain. On a visit to the Returned Servicemens League in Sydney, Jeff Mellor met several ex-RAAF men who had served in the United Kingdom during the war,

including one who had flown from Cottesmore. Needless to say quite a few drinks were downed whilst reminiscing.

The next stop on the tour was at RAAF Edinburgh Field, near Adelaide, where the RAF had a permanent staff. A member of that staff turned out to be an old friend whom Jeff Mellor had known at Cottesmore. The two men found it good to meet up again, but Jeff was smitten by the other guy's girlfriend who was the reigning Miss South Australia.

After seven days at Edinburgh Field the tour moved on to RAAF Pierce, near Perth, its final stop before returning to Darwin. Throughout the tour the visitors had been treated like royalty wherever they went. The local people wanted to meet them, entertained them in their homes and showed them every courtesy. No.XV Squadron had made its mark halfway round the world.

The route home was via Singapore, RAAF Butterworth, Colombo in Ceylon, Gan, Aden, Malta and finally Cottesmore. Many adventures befell the tour party on their route home, including giving a ride to a British politician who needed to get back to Britain quickly. There was also the problem of getting caught up in a build up of military equipment at RAF Khormaksah, Aden, due to a crisis in Kuwait.

Back at Squadron dispersal, at Cottesmore, Jeff Mellor was getting his car out of the ground equipment garages, where he had stored it for safety, when he was approached by the Flight Sergeant. The latter stated he had news for the rigger NCO, whose name had apparently been placed on an overseas warning list. That, stated the Flight Sergeant, was the good news – the bad news was the posting was to Aden!!! Two months after receiving the news Jeff Mellor left No.XV Squadron and headed once more for the Middle East.

Although one of their most experienced NCO riggers was no longer with the Squadron, No.XV participated for a third consecutive year in the Farnborough Air Show, during September. However, on this occasion, the program was changed from the 'scramble' routine to a simulated high-level attack over the airfield.

The Royal Australian Air Force played host to aircraft and crews from No.XV in January 1962, when the Squadron again deployed to RAAF Butterworth for another *'Profiteer'* exercise. As on the previous occasion, weapons training and bombing were carried out on the Song Song range whilst No.XV was in Malaya. The detachment concluded on 3rd February, when all the Squadron's aircraft returned home.

Wing Commander Matthews and his crew endured a narrow escape on 14th June when their aircraft crashed five miles north east of Cottesmore. Due to the electrical connectors becoming disconnected from the fuel trays as the aircraft approached the airfield, all four engines powering Victor bomber, XH613, ceased to function. Although only the pilot and co-pilot had the benefit of ejection seats, it is thought the crew evacuated the crippled machine by baling out of the crew entry hatch, within thirty seconds of receiving the Squadron Commander's order to abandon the aircraft. Although there were some slight injuries, the entire crew escaped without loss of life.

In a break with tradition, the 1962 Farnborough Air Show went ahead for the first time in four years, without the participation of No.XV Squadron. Instead, the Unit's aircraft were seen at a number of 'Battle of Britain' air shows, which occurred during the same month.

The last month of the year saw a change of leadership, when W/C Marshall took command of the Squadron on 1st December.

April 1963 saw a dramatic change in the policy of attack regarding the "V" Force. Due to the increase in number of Soviet fighters capable of striking incoming RAF bombers at high altitude, it was decided the "V" bombers should attack their targets at low-level. This decision was based on the knowledge that no Soviet fighter, at that time, possessed a radar capable of looking down and conclusively detecting another aircraft flying beneath it.

Needless to say, once the policy was officially adopted, a concentrated effort in training was required, so as to bring the new tactics into operation as quickly as possible. No.XV Squadron applied themselves to the task over the following few months.

Whilst No.XV Squadron was training for one form of warfare, Malaya was being subjected to another, albeit on a smaller scale. Following the formation of the Federation of Malaysia on 16th September, widespread riots began. As a result of these riots Martial Law was declared in the Country two days later.

Indonesian military forces were also known to have made a series of armed infiltrations, both in the air and on the ground. The Soviet Union had, during 1961, supplied the Indonesian Air Force with TU.16 bombers, and it was these aircraft which kept infringing Malaysian airspace. With the possibility of full scale hostilities breaking out, the British Government decided to send a force of Victor bombers to the area. Under the guise of Exercise *'Chamfron'*, eight aircraft, four from No.XV Squadron and four from No.10 Squadron, were despatched to Tengah, Singapore, during December.

Tension mounted as the situation grew worse. On 7th January 1964, the aircraft of both Squadrons relocated to RAAF Butterworth, where they practiced their new found skills of low level attack procedures. As during their previous visits to Butterworth, bombing practice was carried out by the Squadron's aircraft over the Song Song range. On 22nd January, the range reverberated to the detonation of thirty-five 1,000lb bombs, dropped from the bomb bay of Victor, B.1A, XH648.

The Royal Navy also played their part in this violation, having amassed a force of approximately eighty warships off the coast, in order to prevent a seaborne invasion taking place. The Indonesians responded, during the first week or two of September, by illegally extending the area of their territorial sea boundaries to effectively seal off the straits at either end of the island of Java. To prove the transgression would not be tolerated, the Royal Navy were ordered to sail these waters with a carrier group. Furthermore, the Victor bombers were on standby to attack Indonesian airfields should any attempt be made to interfere with the passage of the warships.

Throughout the period of confrontation, the crews remaining at Cottesmore, who were kept busy with routine exercises and training missions, were periodically exchanged with those in the Far East.

On the last day of September 1964, No.12 Squadron officially took over the mantle of Exercise *'Chamfron'* from No.XV, who flew back to Cottesmore. The Squadron's return to Britain gave little cause for celebration, as they were declared non-operational on 1st October. As that month closed, so did another chapter in No.XV's long and illustrious history. The Squadron was disbanded.

The First Post War C.O.

The B.I. 'Specials'

An unidentified B.I (Special) *'Clapper Lanc'*, climbs out of Mildenhall during mid-1945. The aircraft was so named due to the fact, when the 22,000lbs bomb was released from the bomb bay, the Lancaster *'went like the clappers'*. *Authors Collection, Original Source Unknown*

Avro Lancaster B.I. Special *'Clapper Lanc'*, PD129, of No.617 Squadron, photographed at Mildenhall, prior to being re-designated LS-R of No.XV Squadron. *Author's Collection via the late Pat Russell Collection*

Wing Commander Dennis Witt, D.S.O., D.F.C., D.F.M., was a highly experienced bomber pilot who had risen up through the ranks. He commanded No.XV Squadron between 12th March 1946 and 19th August 1946. *Courtesy of and via Flt/Lt T. N. Harris, XV (R) Squadron*

Surplus to Requirements

The overall size of an 8,000lbs 'Cookie' can be ascertained by comparing it with groundcrew chief Bob Gault, who utilises the bomb as a bench. The bomb was later dropped by F/L Ken Lewis. *Courtesy of Ken Lewis, DFC*

Flight Lieutenant Ken Lewis inspects a 'Ten Ton' bomb (minus its tail assembly), which appears to have been put on display along with a Lancaster bomber, on a bomb site at an unknown location. *Courtesy of Ken Lewis, DFC*

Bombing-up an unidentified Lancaster, in preparation for the trials against the U-Boat pens at Farge, in 1947. *Courtesy of Flt/Lt T. N. Harris, XV (R) Squadron*

When not acting as a transport cradle for the 22,000 lbs bomb, the bomb trolley makes a suitable resting place on a warm summers day. Australian pilot Ken Lewis takes advantage of both the weather and the facility. *Courtesy of Ken Lewis, DFC*

The Lincoln Era

Avro Lincoln B.II, RF532, LS-D, seen undergoing maintenance at Wyton and in flight. *Courtesy of and via Flt/Lt T. N. Harris, XV (R) Squadron*

Two photographs taken during an escape and evasion exercise briefing, at Wyton, during 1948. Unfortunately nobody in either photograph has been identified. *Courtesy of and via Flt/Lt T. N. Harris, XV (R) Squadron*

Poor quality photograph of Lincolns, RF503, LS-F, and RF514, LS-B, at Shallufa in 1948. *Courtesy of and via Flt/Lt T. N. Harris, XV (R) Squadron*

Avro Lincoln B.II, RF532, photographed at Shallufa in 1948. *Courtesy of Flt/Lt T. N. Harris, XV (R) Squadron*

A poor quality photograph of Avro Lincoln, RF370, LS-A, over Liverpool during 1948. The River Mersey can be seen beneath the aircraft. *Courtesy of and via Flt/Lt T. N. Harris, XV (R) Squadron*

Poor quality photograph of five Lincoln bombers from No.XV Squadron, photographed on 24th June 1950, rehearsing their formation flypast in preparation for the Farnborough Airshow. Nearest the camera is Lincoln coded LS-D. *Courtesy of and via Flt/Lt T. N. Harris, XV (R) Squadron*

Two photographs of Avro Lincoln, RF395, LS-E, taken during the aircraft's participation in the RAF Biggin Hill, Battle of Britain Airshow, on 16th September 1950. *Courtesy of and via Flt/Lt T. N. Harris, XV (R) Squadron (19/16) and Author's Collection (19/17)*

The Washington Era

Poor quality photograph of an unidentified No.XV Squadron Washington bomber flying low over the Lincolnshire countryside. *Courtesy of and via Flt/Lt T. N. Harris, XV (R) Squadron*

Washington bomber, WF506, LS-E, photographed at RAF Conningsby during 1951. To the extreme right of the picture is Washington, WF505, LS-D. *Courtesy of and via Flt/Lt T. N. Harris, XV (R) Squadron*

A line-up of Boeing B.29 Washington bombers of No.XV Squadron, photographed sometime after March 1951, following removal of the Squadron codes. *Courtesy of and via Flt/Lt T.N. Harris, XV (R) Squadron*

Flying Officer Robbie Burn, Washington B29A pilot with No.XV Squadron, at RAF Coningsby. *Courtesy of and via Flt/Lt T. N. Harris, XV (R) Squadron*

Flying Officer 'Chic' Fox, navigator/plotter on Washington bomber, VF506, No.XV Squadron, 1951. *Courtesy of and via Flt/Lt T.N. Harris, XV (R) Squadron*

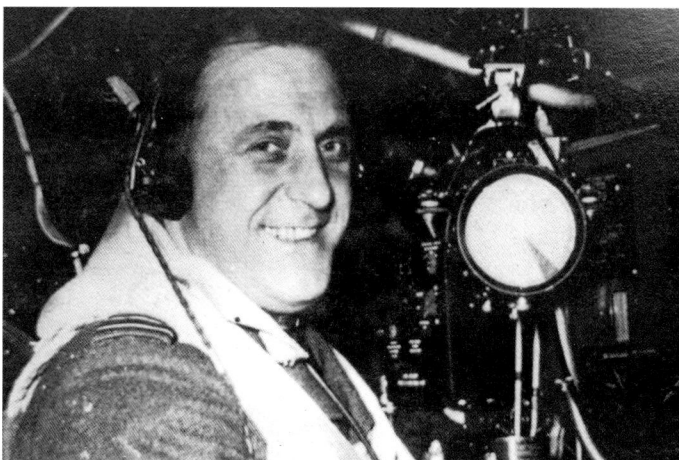

Flight Lieutenant Hank Vnoucel, acted as temporary commander of No.XV Squadron as from 17th April 1952, for a period of approximately seven weeks, following S/L Denny's posting. *Courtesy of and via Flt/Lt T. N. Harris, XV (R) Squadron*

The Canberra Era

English Electric Canberra MK.2, WJ972, of No.XV Squadron photographed at RAF Cottesmore. The Station emblem, a hunting horn across a horseshoe, is depicted on the aircraft's tailfin. *Courtesy of and via Flt/Lt T. N. Harris XV (R) Squadron*

A poor quality photograph of Canberra bomber, WH907, taken at Gibraltar possibly during a 'Lone Ranger' exercise towards the end of 1954. *Courtesy of and via Flt/Lt T. N. Harris, XV (R) Squadron*

A V Roe built Canberra B.2 bomber, WK132, taking-off from RAF Honington during September 1955. *Courtesy of and via Flt/Lt T. N. Harris, XV (R) Squadron*

English Electric Canberra, B.2, XA536, photographed at Luqa whilst returning from a *'Musketeer'* operation. The RAF Station Honington insignia, a Pheasant, is depicted on the tailfin. The aircraft underwent a number of modifications, being converted from a B.2 bomber to a T.11 variant and later to a T.19. *Courtesy of and via Flt/Lt T. N. Harris, XV (R) Squadron*

Squadron Leader Scott, D.F.C., piloting Canberra bomber, WD980. This aircraft, which was used as the backdrop for a number of crew photographs, was one of a production batch of 105 machines. *Courtesy of and via Flt/Lt T. N. Harris, XV (R) Squadron*

The usual crew of Canberra B.2, WD980. From left to right: F/O Bob Weeks (B/A); S/L A Scott, D.F.C. (Pil); F/O Wells (Nav). *Courtesy of and via Flt/Lt T. N. Harris, XV (R) Squadron*

Left: No.XV Squadron. Standing, left to right: F/O Poyser; Unidentified; F/L Harry Bullen; F/L Morgan; F/O Butterworth; F/O Jones; F/O Wheatley; F/O Gordon Swanson; F/L Bromwich; F/L McEwan, RAAF; F/O Hancock; F/L Wickham; Unidentified; Unidentified; F/O Payne. Kneeling, left to right: F/L Thompson; F/O Chris Donne; F/O Bob Weeks; F/O Bentley; F/O Naylor; S/L Scott, DFC; S/L Alan Thompson; Unidentified. *Courtesy of and via W/C Gordon Swanson*

The Victor Era

Photographed following collection of the first Victor bomber (XA941) are, from left to right: Chief Technician Hunt, Crew Chief: Unidentified; W/C David Green, O.C. XV Sqdn; S/L Sherston; F/L Stubbs. *Courtesy of and via Flt/Lt T. N. Harris, XV (R) Squadron*

Handley Page Victor B.1, the first machine of it type taken-on-charge by No.XV Squadron, landing at Cottesmore. *Courtesy of Jeff Mellor, JP*

Victor bomber, XH594, arriving at Akrotiri on 21st March 1959, after a record breaking flight from Cottesmore. The two pilots were G/C 'Johnnie' Johnson, Cottesmore Station Commander, and W/C David Green, O.C. XV Sqdn. Dispersed in the background are two Canberras and a Shackleton. *Courtesy of Flt/Lt T. N. Harris, XV (R) Squadron*

The first Boeing 707-420, in the colours of British Overseas Airways Corporation (BOAC), flies low of the crowds and static aircraft at the Anniversary of Flight Airshow, at Vancouver, British Columbia, Canada, on 11th July 1959. In the foreground is Handley Page Victor Bomber, XH594, of No.XV Squadron. *Courtesy of and via Flt/Lt T. N. Harris, XV (R) Squadron*

No.XV Squadron in 'three-ship' formation flypast at the 1959 Farnborough Airshow. Left to right are XH593, XH588 and XH590. *Courtesy of and via Flt/Lt T. N. Harris, XV (R) Squadron*

Handley Page Victor, XH592, having completed a low slow pass with 'everything' down, commences to gain altitude during a display at the Farnborough Airshow. *Courtesy of and via Flt/Lt T. N. Harris, XV (R) Squadron*

The four Victor bombers of No.XV Squadron, suitably parked at the correct angle on the runway, await the signal to demonstrate the procedure for a four minute 'scramble', at the Farnborough Airshow, during September 1960. *Courtesy of Jeff Mellor, JP*

A poor quality photograph of W/C J. Matthews, AFC, who commander No.XV Squadron between 1st April 1960 and 1st December 1962. *Courtesy of and via Flt/ Lt T. N. Harris, XV (R) Squadron*

Jeff Mellor, who served with No.XV Squadron as an Airframe Fitter (rigger) later rose to the rank of sergeant. *Courtesy of Jeff Mellor, JP*

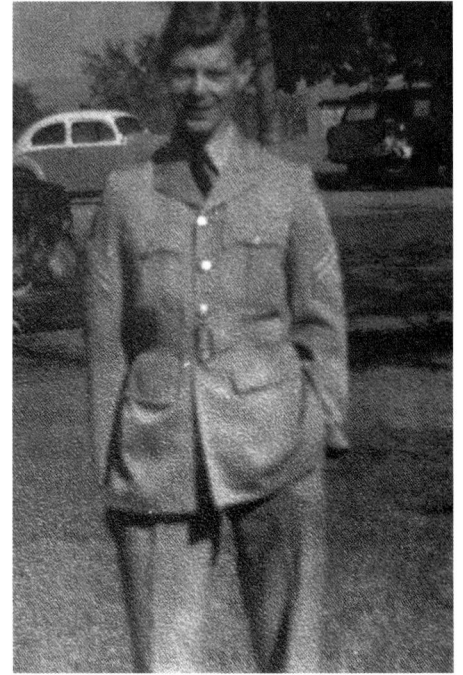

Jeff Mellor, photographed in Darwin, during June 1961, models the smooth fabric tropical uniform which was the envy of his colleagues who went to Malaya. *Courtesy of Jeff Mellor, JP*

The driver of the Tugmaster tractor unit gently eases his vehicle towards the towing bar, which is connected to an unidentified Victor bomber, prior to moving the machine at the Farnborough Airshow. *Courtesy of Jeff Mellor, JP*

The No.XV Squadron groundcrew servicing team, with an unidentified Victor bomber, at Farnborough. Jeff Mellor is in the front row, 8th from left. *Courtesy of Jeff Mellor, JP*

Handley Page Victor, XH589, at the Farnborough Airshow. The Squadron number and crest can clearly be seen on the fin, above the tail flash. *Author's Collection*

20

The Peace Keepers

As a result of a number of military projects being canceled due to political intervention, indecision and incompetence during the 1960's, the Royal Air Force faced the prospect of entering the next decade without a modern aircraft capable of low-level strike penetration. The British Government's eventual answer was to convert a carrier borne aircraft, into a land based bomber capable of the majority of requirements proposed for the role it was to undertake.

The aircraft selected, which was originally conceived as the Blackburn NA.39, Buccaneer, had been in service with the Royal Navy since 1962. Seven years later, following some modifications and redesignated S.Mk2A and S.Mk2B, the Buccaneer entered service with the Royal Air Force. The first unit to acquire the folding-wing aircraft was No.12 Squadron, which reformed at RAF Honington on 1st October 1969. Exactly one year later on Thursday, 1st October 1970, No.XV was reformed, also at Honington, as the second Buccaneer unit. Equipped with the S.2B variant, the Squadron was placed under the command of W/C David Collins.

Initially, the new No.XV possessed only one aircraft, Buccaneer, XW526. However, due to the fact there was no dual trainer version of the Buccaneer, this machine was soon joined by Hawker Hunter, T7A, WV318. The latter aircraft had been fitted with an instrument system which replicated that of the bomber, in order that dual instrument flying training could be given to new crews. The intake of crews who initially formed the new Squadron had all undergone conversion training with No.736 Naval Air Squadron at Royal Naval Air Station, Lossiemouth. However, realising the need for its own training unit the RAF duly set about forming No.237 Operational Conversion Unit, which was to be based at Honington.

No.XV's stay at Honington was destined to be a short one, the Squadron having been detailed for permanent deployment to Germany in the forthcoming New Year. Before their departure to Continental Europe the crews had a busy period ahead of them. Their new role demanded a high standard of ability in high speed, low-level, cross-country flying and, in order to achieve this, they entered a vast program of training exercises. Also scheduled in the training program were low-level navigational sorties, and a number of exercises involving bombing practice and rocket firing.

Away from the flying training the Squadron had two other important duties to undertake; both paramount in its history and tradition. First, as it was to become part of Royal Air Force Germany (RAFG), No.XV sought permission for its aircraft to be adorned with the Squadron numerals. Approval was given to a white "XV" being displayed on the tailfin of each aircraft, where it was to be applied above and behind the fin flash. As it was part of the Squadron's inventory, this approval also included the Hunter aircraft. The second task the Squadron was quick to complete occurred on 15th October, two weeks after No.XV had been reformed. Since its disbandment in 1964, the Squadron's Standard had been laid up in the Royal

Air Force College Cranwell. On 15th October, F/O Barry Chown was authorised to collect the 'Colour' and return it to the Squadron.

In January 1970 No.XV officially became part of RAFG, when the Squadron relocated to RAF Laarbruch, West Germany. To mark the occasion, W/C Collins led a formation flypast of two Buccaneers and the Hunter. On landing, the crews were greeted and welcomed to Germany by the base Commander, Group Captain J Beddoes.

Although they had spent the last few months training, the crews immediately settled in to a further period of learning. Apart from familiarisation flights over a different country, they had to quickly become acquainted with the German low-level flying structure and methods.

Although all contingencies and emergency procedures were instilled into the aircrews, the benefit of training and experience could not always prevent accidents. Unfortunately, on 25th March, an incident occurred which claimed the life of the Squadron C.O. Wing Commander David Collins, who was piloting Buccaneer, XW532, had just taken-off from Laarbruch when the aircraft crashed. Flight Lieutenant Paul Kelly, who was flying as navigator, was also killed.

Wing Commander Roy Watson who had previously been O.C. Operations Wing at RAF Honington, assumed command of No.XV Squadron on 6th April, following the loss of W/C Collins.

The summer months saw No.XV return to the public eye when at the end of May, the Squadron detailed Buccaneer, XW527, to attend the Paris Air Show. The aircraft, which spent a week at the event, was displayed in the static park. Two weeks later, on 12th June, the Squadron paid tribute to Prince Bernhardt of the Netherlands, when it participated in a celebratory birthday flypast over the Dutch Air Force base at Deelen.

In the event of an outbreak of war in Europe, all RAF squadrons based in Germany would have been placed under the control of NATO, as part of the Second Allied Tactical Air Force. To this end, No.XV Squadron adopted a policy of training which would enable it to attain the criteria required by NATO for both attack and strike (nuclear) operations.

It was fortuitous that during September, at the time No.XV implemented this policy, the first batch of new crews began arriving on Squadron from No.237 OCU. Apart from being initiated into the dictates of flying low-level over West Germany, they were also able to participate in the rocket firing and weapons training recently introduced to the Squadron's training program.

The training undertaken during September was put to the test during November when No.XV deployed five aircraft to Decimomannu, Sardinia, for a three week Armament Practice Camp. The Italian Air Force range at Cape della Frasca, was used by the Squadron to undertake all aspects of bombing and weapon delivery. The Squadron excelled itself by achieving the required standards and more, in all categories. The commitment of both the air and groundcrews paid off, when the Squadron was assigned to Nato as part of its attack force the following month.

NATO's acceptance of No.XV as part of the Second Allied Tactical Air Force, did not mean the Squadron could rest on its laurels. Bombing practice continued into the New Year when, during February 1972, over the Vliehors range in Holland, the first 1,000lb bombs were released from No.XV Squadron Buccaneer aircraft.

Amidst all the training, No.XV also found time to entertain guests. Amongst those who visited the Squadron during 1972 were the parents of F/L Paul Kelly. In memory of the son they had lost less than a year earlier, Wing Commander and Mrs Kelly presented the Squadron with a silverware model of a Buccaneer.

Two official visits were made to Laarbruch during May. The first on the 9th of the month when the guest of honor was Chief of the Air Staff, Air Chief Marshal Sir Denis Spotiswood, G.C.B., C.B.E, D.S.O, D.F.C. Sixteen days later, on 25th May, the Squadron played host to His Royal Highness Prince Philip, the Duke of Edinburgh.

Exactly four months after the Royal visit, the British Defense Minister, Lord Carrington, and Herr Leber, the West German Defense Minister, visited the Squadron. Air Marshal Sir Harold Martin, K.C.B, C.B.E, D.S.O., D.F.C, A.F.C, who had gained fame as one of the original *'Dambusters'*, accompanied the two Ministers in his capacity as Commander-in-Chief RAF Germany.

Sardinia was visited again during the Summer, when six aircraft and ten crews deployed to Decimomannu on 3rd July, for an eleven day visit. As with the previous visit, bombing practice was paramount but on this occasion the Squadron adopted a new method of attack.

Apart from the detachment to Sardinia, No.XV also participated in a *'Southern Ranger'* exercise at Akrotiri, during the first week of August. This latest venture was the Squadron's first deployment to the island since July the previous year. A new low-visibility camouflage, introduced at the end of 1972, demanded the removal of all white markings from the Squadron's aircraft. This did not create a problem as far as the national insignia was concerned, but the Squadron was aggrieved at the thought of having to remove the "XV" from the tailfin. To overcome the problem, and in an effort to retain its Roman numerals, the "XV" was painted a dull red.

The white "XV" adorning the Hawker Hunter was also to be removed, but for a different reason. As No.XV and No.16 Squadrons were operating alongside each other at Laarbruch, it was decreed in January 1973 that both units should have use of the Hunter. With this objective in mind, the aircraft was transferred to the Station Flight, which led to the removal of the Roman numerals.

One of the last major engagements undertaken by W/C Watson, prior to relinquishing command of the Squadron, was to lead a flypast of four aircraft from Laarbruch. The formation took to the air on 22nd March, as a farewell gesture to Air Marshal Sir Harold Martin who was retiring from the service. His place as C-in-C RAF Germany was taken by Air Marshal Sir Nigel Maynard, K.C.B, C.B.E, D.F.C, A.F.C. Eighteen days after leading the flypast, on 9th April, W/C Watson officially handed over command of No.XV to W/C Michael Simmons.

One of Michael Simmons' first duties as O.C. No.XV was to welcome the new Commander-in-Chief RAF Germany, when the latter made his first official visit to the Squadron on 27th April.

Another flypast, undertaken on 8th June at Headquarters AFCENT in Brunssum, Netherlands, marked the birthday of Her Majesty Queen Elizabeth II. The Birthday Honors list, when published, revealed the award of a British Empire Medal to No.XV's senior armament specialist, F/S J McClure.

Armament training was carried out during a month long deployment to Decimomannu, between August and September. However, due to commitments at the end of the former month, the Squadron was forced to divide its resources and send them to Sardinia in rotation. Half the Squadron was despatched for the final two weeks of August, whilst the second group went for the first two weeks of September.

Another commitment taken on by the Squadron during 1973, which lasted through to 1976, was the responsibility of maintaining two aircraft at 15 minutes readiness, 24 hours a day in the nuclear strike role.

The Squadron was privileged to receive another VIP guest during September, when Air Chief Marshal Sir John Barraclough, K.C.B, C.B.E, D.F.C, A.F.C visited No.XV. During his visit, this high-ranking officer was taken aloft in Buccaneer, XW534, piloted by F/L Moses. No doubt a memorable occasion for both of them.

The year of 1973 almost had a tragic ending for No.XV Squadron, when one of its aircraft was struck by lightning during a night flying operation. The incident temporarily blinded the pilot, but the quick thinking and subsequent actions of the navigator saved the day. In recognition of his action, F/L Colin Tavner was subsequently awarded the Queen's Commendation for Valuable Service in the Air.

A former member of No.XV who had been posted to No.237 OCU during June 1973, was awarded an Air Force Cross on 1st January 1974. Squadron Leader H Williamson, an ex-Flight Commander, received the honor in recognition of his work with the Squadron.

Even though its daily routine and extensive training program kept the Squadron busy, it always found time to entertain visitors and guests. In early May, members of the Weeze Gemeinde, including the Burgomeister, were shown around the Squadron and invited to inspect a Buccaneer aircraft. During the evening of the same day, the guests responded by inviting the Squadron to a concert held in the Weeze Rathaus. As a token of friendship between the Squadron and the local community, an exchange of gifts was made.

At the end of the same month Meneer Gerrit J Zwanenburg, the head of the Identification and Salvage team, based at the Royal Netherlands Air Force base at Soesterberg, gave a lecture. Again, a presentation was made to the Squadron, when W/C Simmons accepted some pieces of a former Squadron aircraft. Salvaged from an excavation in 1967, the items recovered proved to be from Stirling bomber, BF353, LS-E. This machine, piloted by P/O Leonard O'Hare (a native of the U.S.A.), was shot down on the night of 16th/17th September 1942, during an attack against Essen. Being mindful of its past history, the pieces presented to the Squadron were duly mounted in a display case.

In mid-June, the Squadron returned to Decimomannu for the Armament Practice Camp. As with the previous visit, the Squadron again divided its resources and initially despatched only half the Squadron. Live weapons were dropped during the first week of July, when the Squadron detailed three aircraft and crews to participate in the CENTO fire Power Demonstration at Akrotiri.

The friendship which had developed between the local populace of Weeze and RAF Laarbruch was demonstrated on 13th July when the Freedom of the Town was bestowed upon the RAF base. In appreciation of the honor, which had never before been bestowed on the Royal Air Force in Germany, a parade and flypast was held.

Another cause for celebration arose during October, when the Squadron was given permission to reinstate the white "XV" on the tailfin of its aircraft. The necessary work, which was implemented immediately, was completed by the end of the following month.

For many squadrons, No.XV included, 1975 was a year for celebrations. A number of those which had been formed in 1915 were now, sixty years later, serving together in Germany. To mark the occasion a flypast was arranged for 24th January. Aircraft piloted by the commanding officers of No.XV and No.16 Squadrons, No.17 and No.19 Squadrons and No.20 and No.31 Squadrons all flew over the headquarters of the RAF Germany and the Second Allied Tactical Air Force. The C.O. of No.14 Squadron had the honor of leading the formation.

No.XV Squadron's 60th anniversary was marked by the issue of a First Day Cover. The envelopes, which bore 30pfg Deutsche Bundespost stamps, were illustrated with a BE2c, the Stirling bomber, *MacRobert's Reply*, and a Buccaneer. They were flown on a route covering Laarbruch, Wyton, Brest, Wyton, Laarbruch, in a No.XV Squadron Buccaneer, XX888, coded "C", on 17th February. Flight Lieutenant Taylor, the pilot, and F/L Coop, the navigator, were airborne for two hours fifty-one minutes on a course simulating that taken on 18th December 1941, when nine No.XV Squadron Stirlings attacked German warships in Brest harbor.

No.XV Squadron held its own celebrations at the Royal Air Force Club in London, on 1st March 1975. Amongst the 119 or more guests who attended the function were nine members from the Royal Flying Corps period, nine from the interwar period and thirteen former commanding officers. Named amongst the latter group were two from the interwar period, three from the second war period and eight from the post war period including W/C Michael Simmons.

During the year of celebrations, the Squadron took the opportunity to record its history, a task undertaken by F/L Norman Roberson. After much painstaking work, the end result was a much sought after illustrated edition totaling sixty-four pages. Copies of the book, many of them bearing numerous signatures, are treasured possessions amongst ex-Squadron members.

The celebrations were somewhat marred by an accident which occurred on 16th June, when two No.XV Squadron aircraft were involved in a mid-air collision over the North Sea. Although F/L Haynes and F/L Shaw, the crew of Buccaneer, XW536, were forced to eject, the second machine Buccaneer, XW528, returned safely to its temporary base in Denmark.

Norman Roberson was able to record another piece of the Unit's history first hand on 30th December, when he flew with W/C Simmons on the latter's last sortie with the Squadron. On arriving back at Laarbruch, they found members of the groundstaff waiting for them with a bottle of champagne on ice.

Apart from a posting, the New Year brought with it the award of an Air Force Cross to Michael Simmons. The latter being announced in the New Year Honors List.

Wing Commander Peter Oulton officially took over command of the Squadron on 2nd January 1976. As had become the custom at this time, a number of photographs were taken to mark the occasion.

The Squadron lost one of its aircraft due to an accident, on Wednesday, 3rd March, when Hawker Siddeley Buccaneer, XV166, crashed into Kings Forest near Bury St. Edmunds, Suffolk. The aircraft, which was on a training mission from its home base at Laarbruch, West Germany, was on final approach to RAF Honington when it fell from the sky two and a half miles short of the airfield. The crew, consisting of the pilot, F/L Graham Bowerman, and F/L Chris Davis, the navigator, managed to eject from the bomber seconds before it struck the ground. Although both members of aircrew survived the low-level ejection, the pilot suffered injuries to his back.

Flight Lieutenant Bowerman recovered from his injuries and returned to the Squadron where, on 6th September, he acted as Standard Bearer during a parade at Weeze. During the parade, held to mark the second anniversary of the granting of the Freedom of the Town to RAF Laarbruch, the base personnel exercised the right to march through the town.

Wing Commander Peter Oulton continued to lead the Squadron throughout 1977 when, amongst other commitments, No.XV sent a small detachment to America for a *'Red Flag'* exercise. The exercise, which was carried out over an area of approximately 15,000 square miles of the Nevada Desert, allowed the participating crews to undertake realistic battle scenarios using live weapons. The United States Air Force Aggressor Squadrons, who were experts in Soviet tactics, provided the 'enemy' opposition for aerial combat. Operating under these 'pseudo' battle conditions, the operational efficiency of both the air and groundcrews benefited enormously.

On 18th August the Squadron lost Buccaneer, XX890, when the aircraft crashed whilst on approach to Laarbruch. Fortunately the crew, F/L Mackenzie and F/L Pittaway, were able to eject safely from the machine.

The last Buccaneer aircraft to be taken-on-charge by No.XV Squadron was delivered on 6th October 1977. In a ceremony involving a representative of Hawker Siddeley, F/L Pete Rolfe, F/L Martin and a group of technicians, Peter Oulton accepted the aircraft's Form 700, thereby adding Buccaneer, XZ432. to the Squadron's inventory.

On 4th July 1978, W/C Oulton relinquished command of No.XV Squadron in favor of W/C Trevor Nattrass, AFC. The day to day routine of the Unit remained the same under its new leadership, with all aspects of

flying exercises and weapons delivery being to the forefront of its training.

The Squadron returned to the airshow scene during the summer of 1979, when F/L Williams and F/L Collingworth took Buccaneer, XT275 to the International Air Tattoo. The airshow was held at the USAFE base at Greenham Common, Berkshire, over the weekend 23rd/24th June. The No.XV Squadron aircraft arrived during the day on Friday, 22nd June and remained in the static park until its departure the following Monday. During that period, whilst parked with its wings folded back next to a USAFE F.15 fighter, the machine attracted a great deal of attention from both aircraft enthusiasts and American Servicemen alike.

A deployment to a *'Red Flag'* exercise at Nellis Air Force base in the Nevada Desert, was marred by tragedy when S/L Kenneth Tait and F/L Ruston were killed in an accident on 7th February 1980. Their aircraft, which was maneuvering for a low-level pass near the base, broke-up in the air due to a structural failure of the main wing spar. The two crew members had no chance to escape as Buccaneer, XV345, dived into the ground from an altitude of approximately 100'. A subsequent investigation, during which all Buccaneers were grounded, revealed that a number of this type of aircraft were suffering from similar fatigue. Many of the effected aircraft were repaired and eventually returned to service. However, a great number were beyond economical repair and never flew again.

Shortly after the Squadron's sixty-fifth anniversary, on 8th May 1981, No.XV paraded its Standard in one of the hangers at RAF Laarbruch. The Reviewing Officer, who took the salute at the march-past, was Squadron Leader Peter Boggis, DFC, RAF (Retired), the former pilot of the Stirling *MacRobert's Reply*. The occasion was attended by many former commanding officers, some of whom had achieved higher rank. The guest list included two Air Marshals, six Group Captains, two Wing Commanders and a number of Squadron Leaders. Also invited to attend were Herr and Frau Willems, the Burgomeister of Weeze and his wife. After the Ceremony and service, the latter of which was conducted by the Venerable H. Stuart, Q.H.C., M.A., the Chaplain-in-Chief RAF, the guests retired to the Officers' Mess for luncheon. This was one of the last major occasions over which W/C Trevor Nattrass presided as C.O. of No.XV, before officially handing over leadership of the Squadron to W/C Eddie Cox on 6th July 1981.

Due to the screening of a television series which followed the flying training of a former milkman, No.XV Squadron was brought to the attention of the British public. John McCrea wanted more out of life than delivering milk, and therefore applied to join the Royal Air Force. The series started at the Aircrew Selection board, where only the fortunate few got through; John McCrea being one of them. From that point on, with the television producers and camera crews following his almost every move, the 'rookie' pilot became a household name. On completion of his training, Flying Officer McCrea was posted to his first squadron; No.XV based at Laarbruch, early in 1981. Needless to say, initially, the cameras went with him to record how he settled to Squadron life. Whilst there, the producers of the program also gave an insight to the role played by the Squadron in Germany.

When F/O McCrea represented the Squadron at a gathering of wartime flyers from No.XV at a reunion at Mildenhall, his appearance at the venue generated much interest. He was, in more ways than one, the 'star' attraction. Many cameras worked overtime as people asked if they could be photographed with him.

Cameras also worked overtime at a reunion held at RAF Mildenhall, in April 1982, when S/L Peter Boggis performed an unveiling ceremony. The Squadron was represented by W/C Eddie Cox, who attended the occasion with a small number of aircraft and their crews. Amongst the aircraft which had flown across from Germany to the American operated air base, was Buccaneer, XT287, wearing the tailcode "F". This machine had previously been selected, due to its tailcode, to be the first No.XV Squadron aircraft since the war to bear the name *MacRobert's Reply*. As with the Stirling bombers which had proudly worn the name forty-one years earlier, the Buccaneer was also adorned with the MacRobert Family crest.

Although Saturday, 8th April dawned very blustery, the windy conditions did not prevent a large group of ex-No.XV wartime members and their wives from gathering on the parking ramp in front of the aircraft. The Buccaneer, resplendent in its European camouflage, wore a large red fabric cover on the side of the forward fuselage. After a brief speech from W/C Cox, Peter Boggis was invited to peel back the aforementioned cover to reveal the crest and name of *MacRobert's Reply*. The Squadron's re-association with the MacRobert name was warmly welcomed by the assembled throng who, after the official duties were complete, were invited to inspect the aircraft.

One No.XV Squadron aircraft which aroused a great deal of attention by its presence, apart from Buccaneer, XT287, was a Hawker Hunter T.7. Closer inspection revealed the aircraft to be WV318, the same machine the Squadron had relinquished to the Laarbruch Station Flight many years before. Another interesting feature about the Hunter, which many of those who realised it raised issue on, was the fact the aircraft was wearing the Squadron's white "XV" on the forward fuselage, just below the cockpit.

The occasion awakened many memories for those who served with No.XV Squadron during the war years. It also generated much interest amongst the USAFE personnel who were on duty during the ceremony. Many of the latter showed a genuine interest to learn more about the MacRobert connection and the background leading to the event.

Exactly twelve months after the re-naming ceremony, in response to an invitation issued by W/C Cox, some of those who had attended the reunion at Mildenhall (known as the Mildenhall Register) visited No.XV Squadron at Laarbruch.

On arrival at the main gate to the West German base, the group was met by the familiar figure of F/O John McCrea, who was to act as liaison officer. The weekend stay was well planned, with a number of functions arranged for the guests, including a barbecue held at the Squadron's dispersal and crewroom area on the Saturday afternoon. One of the more solemn duties undertaken on the Sunday morning, was a visit to the Commonwealth War Graves Cemetery, in the Reichswald Forest. Following a short service, during which a wreath was laid, the members paid homage in their own way to those who fell and are buried therein. Following lunch, a trip across the German/Dutch border to Arnhem was made.

The day for departure arrived all too soon, but everyone in the group had enjoyed their visit. Many had learned about the day to day workings of the Squadron at that time, whilst many others had had the opportunity to swap stories and relate their wartime experiences to the younger fliers of No.XV.

Over the following few months some of those younger fliers were to be posted away, due to the fact the Squadron was to re-equip with new aircraft. It had been decided by higher authority that, after nearly thirteen years service with No.XV, the Hawker Siddeley Buccaneer aircraft were to be withdrawn from Squadron service. Although the last No.XV Squadron Buccaneer flight was made in June, the unit spent the remainder of the summer months preparing for its new aircraft.

On Thursday, 1st September 1983, No.XV was officially reformed at Laarbruch as the first Squadron, under the command of Royal Air Force Germany, to be equipped with the Panavia Tornado GR.1, swing-wing bomber. Wing Commander Eddie Cox relinquished command of the Squadron in favor of W/C Barry Dove, who had the distinction of being the first navigator in the RAF to command a Tornado squadron.

Initially, familiarisation flights over West Germany were undertaken by the new crews who had joined the Squadron, but these flights were curtailed following the crash of a similar aircraft from another squadron later that same month. The suspension of flying order was lifted at the end of the first week of October, allowing the Squadron to continue its training program.

An official parade, at which the Commander-in-Chief RAF Germany, Air Marshal Sir Patrick Hine, reviewed the handover of the Squadron's Standard, took place at Laarbruch on the last day of October. In a symbolic gesture of the occasion, a flypast comprising two Buccaneers and two Tornados, drew the proceedings to a close.

The new aircraft had many advantages over its predecessor, which became more apparent as the results of the various exercises and weapons

training sorties were scrutinised. A three week training period on the weapons ranges at Capo della Frasca, Decimomannu, confirmed the results. As the first Tornado squadron in the RAF to visit the Sardinian base, No.XV had little choice but to show the effectiveness of the new machines.

The hazards of training, especially during low-level exercises are all too obvious, but it was an act of nature which was responsible for the Squadron's first loss. Returning from a routine sortie on 6th February 1984, Tornado, ZA451, was struck by lightning. Flames were seen trailing out from under the starboard wing, which left F/L Smith and S/L Ian Travers-Smith, the pilot and navigator respectively, in no doubt that they had to eject, and quickly. The burning aircraft impacted with the ground, on empty farmland, near Jever, north-west of Wilhelmshaven. Two other aircraft from the Squadron, who were flying in the same area that night, were also struck by lightning. However, they were both more fortunate and managed to execute emergency landings without further incident.

The 1984 reunion, held at Mildenhall over the weekend of 12th/13th May, was again honored by the presence of two Tornados from No.XV Squadron. One of the aircraft detailed to fly into Mildenhall was Tornado, ZA446, "F", which had taken on the guise of *MacRobert's Reply*. A poignant moment occurred on the Saturday morning when the Lancaster bomber, operated by the Battle of Britain Memorial Flight, flew over the two Tornados as if in salute to its successors. The camouflaged top surface of the Lancaster's mainplane became clearly visible as the aircraft banked round, prior to lining-up for a landing at the base. Although the two types of aircraft, both of which were operated by the Squadron, were divided by a period of forty years, to see them together on the ramp instilled a sense of pride into the men and women assembled there. The occasion warranted a half-page, illustrated, article in a local newspaper.

The following month, between 18th-20th June, the Squadron participated in its first NATO Tactical Evaluation (TACEVAL). In order to be declared operational to NATO standards, the base and the Squadron had to prove not only its ability to respond immediately to the threat of war, but also to be able to undertake the predetermined procedures as dictated in the event of hostilities. Effectively, the base and the Squadron are at war for the duration of the exercise, with other NATO Countries playing the aggressor. The personnel of No.XV were rewarded for their efforts the following month by being declared operational in the strike (nuclear) role. To add to their pride and esteem, the Squadron received a visit from His Royal Highness The Prince of Wales, who received a briefing from S/L Bruce Monk and a close look at one of the Squadron's aircraft.

A further detachment to Decimomannu was necessitated in order to enable No.XV to achieve the required standards of bombing accuracy for conventional operations. Needless to say, these standards were easily reached and on 1st January 1985, the Squadron was declared fully operational in bothe the strike and attack roles.

In April 1985, the Squadron undertook further training in Canada, first at Goose Bay, Labrador, followed immediately by a deployment to Alberta. During its detachment to Goose Bay (the first time the Squadron had been there), it carried out Operational Low Flying (OLF) as part of Exercise *Western Vortex*. The crews found OLF exercises exhilarating due to the fact, weather permitting, they were allowed to fly at a minimum altitude of 100'.

At Cold Lake, in Alberta, the Squadron participated in a *Maple Flag* exercise. Here they put their bombing techniques into practice in a combat scenario, whilst being opposed by the Canadian and American fighters. Following a successful detachment, No.XV returned to Germany at the end of the month.

On 1st August 1986 W/C Barry Dove, AFC, handed over command of the Squadron to W/C Mike Rudd, thereby returning leadership of the unit to a pilot. One of Barry Dove's final duties, as outgoing C.O., was to fly a sortie in a Tornado piloted by the new commanding officer.

Another first for the Squadron occurred in February 1987, when No.XV deployed to Nellis Air Force base for a *Green Flag* exercise. Although the unit had attended on a previous occasion, it was their first detachment to Nevada since re-equipping with the Tornado. The participating RAF aircrews acquitted themselves very well against their trans-atlantic 'cousins', drawing admiration for the accuracy of their bombing scores. However, the Squadron did not have things all its own way, due to the tactics employed by the aggressor squadrons against whom they were pitched.

Wing Commander Mike Rudd made the acquaintance of some of the Squadron's wartime flyers when he attended the Mildenhall reunion in May 1988. The following month, having previously accepted an invitation issued by the C.O., a number of those former flyers paid a return visit to the Squadron in Germany. Whilst at Laarbruch, the guests were treated to a party in the All Ranks Club, where the assembled throng chatted, danced, drank and generally had a good time. A Families Day was held by the Squadron, where a number of displays and exhibits of general interest were mounted. A hardened aircraft shelter in which *MacRobert's Reply* was displayed surrounded by a number of weapon systems, proved a popular item; as did the barbecue which followed later in the day. The solemn side of the visit occurred when the group again visited the Reichswald Forest War Cemetery for a short service of remembrance.

In another act of remembrance undertaken later that same year, an expedition made up of personnel from both the base and No.XV Squadron, set out with the intention of crossing the Pyrenees following the route used by escapers and evaders during the Second World War. The eight man team of Expedition *Bold Escaper* led by Corporal Craig Bonnington, a Physical Training Instructor, left Laarbruch on 11th October. Armed with knowledge provided by Reg Lewis, a former member of No.XV Squadron aircrew who had carried out the task for real during the war, and all necessary clearances to proceed with the venture, the group headed for Dole, in France. Over the next six days, the members of the expedition made their way though the mountain range, reaching altitudes in excess of 8,000'. Having had the benefit of local guides all along the route, the Spanish frontier was crossed late in the afternoon, on the final day of the walk. The team, who were tired but elated, were aware that their efforts had brought forth many happy memories for those who, many years before, had a more pressing need to guide members of the Royal Air Force across the Pyrenees.

Apart from a New Year, 1989 brought with it another change in the leadership of No.XV. Wing Commander Broadbent who, like Barry Dove was a navigator, took on the mantle of commanding officer on 20th January. Although he may not have realised it at the time, John Broadbent's tenure as C.O. was going to be a particularly busy one, but not necessarily a totally enjoyable one.

The usual routines of training, in its many aspects, were carried out on a daily basis in an effort to continue the extremely high standard of operational ability associated with the Squadron. However, on occasions, things sometimes go wrong.

Exactly six months after the new C.O. took office, the Squadron lost another aircraft. The crew of Panavia Tornado, ZA468, an ex-No.16 machine, were forced to abandon the aircraft when it ran into difficulties shortly after take-off from Laarbruch.

One of the forthcoming events facing John Broadbent was the 75th anniversary of the formation of No.XV Squadron. Although it was not until the following year, the planning of suitable celebrations would take some time and would involve many people.

One young pilot by the name of John Peters, who was serving with No.XV at the time, devised an idea for bringing part of the Squadron's history to the attention of others. To commemorate the anniversary, and to trace the unit's history in France, F/L Peters set about organising an expedition which would cycle between those airfields from which the Squadron operated during two World Wars. Apart from the possible discovery of any memorabilia, the young pilot was also hoping to meet any local inhabitants who may have had an association with No.XV during those times.

The expedition, which comprised John Peters, Rupert Clark, John Orrock, Andy Gorton, Shane Probert and Karl Ballard, set out the following March. Their starting point was the former staging airfield at St Omer. Avoiding the major routes, the team stuck to the country lanes. By the end of the first day they had reached St Pol, where they stayed in the Arras Youth Hostel. The next day the team headed for Marieux, which was situ-

ated on the top of a hill. Much to their delight, the intrepid cyclists were met by the town's Mayor who plied them with copious amounts of alcohol, spirits and wine. As if this were not enough, the team was treated to a steak and pate' lunch. By the time the champagne was served everybody was immersed in the musical strains of the National Anthem.

After the generous libations of lunchtime, the team visited the site of the airfield and a chateau which, they learned, was commandeered as an Officers Mess. They also met an old gentleman by the name of Andre Buchez who, at 85 years of age, said he still remembered scrounging cigarettes from the Squadron members! Continuing their journey, the team cycled to Vignacourt via Vert Galant. Again, the team was met by the local Mayor and town officials, who entertained them to another sumptuous meal. Guided tours of the area, followed by further liquid refreshments, were accorded to the expedition members before they set off on the next part of the venture. The pattern, now set for their arrival in a town, repeated itself at Lealvillers, Courselles-le-Comte, Savy, Lechelle and Longavesnes. Only at Thiepval, where a monument honoring 73,000 dead has been erected, was a solemn and sober attitude prevalent.

On the final day of the trip the team visited Betheniville and Vraux. At Vraux the expedition members were able to see where the Squadron was based during the early part of the war. They were also able to visit the 'Red House' on the edge of the airfield.

Apart from the enjoyment they derived from their trip, John Peters and his team did a great deal to foster good relations, on behalf of the Squadron, with the people they met.

The official celebrations for No.XV Squadron's 75th anniversary took place at Laarbruch over a long weekend, between 17th – 20th May 1990. The format was very much the same as that arranged for the 70th anniversary, with both formal and informal events taking place. Amongst the former was a Parade and March-past, followed by a finger buffet lunch, with a Ladies Dining-in Night in both the Officers' and Sergeants' Messes being held in the evening.

The informal celebrations included a Families Day held during the afternoon on the Friday, with a Rhein cruise scheduled for the Saturday afternoon. The latter was followed by an All Ranks hanger party in the evening.

A number of aircraft from other squadrons were provided for display purposes during the Families Day afternoon. However, none of them generated as much interest as Tornado, ZA549, which had been adorned with a special anniversary color scheme. The aircraft was 'captured' on canvas by aviation artist Keith Aspinall, who presented the finished work to the Squadron.

As with previous anniversary events, a Service of Remembrance at the Reichswald Forest War Cemetery was organised for the Sunday morning. On this occasion, as with the event held in 1975, a serving officer, or N.C.O of the Squadron, stood silently at the head of the grave, behind the headstone, of each known wartime member of the Squadron buried therein. Although not necessarily the same age as the deceased, the serving member paying tribute to the fallen comrade was of equal rank. This simple gesture on the part of the modern day Squadron, was much appreciated by the ex-wartime flyers who attended the Service. On return from the Remembrance service, the guests were treated to a family lunch at the Officers' Mess. This was a very casual affair, taken alfresco style. The sound of music, in the form of Dixieland and Traditional jazz drifted through the warm summer air; that is until the arrival of W/C Bob Munns. Once the dexterous fingers of the ex-wartime flyer started to race across the piano keys, a full-scale 'jam' session occurred. Many a foot was set tapping as people gathered around to listen, whilst laughter and light-hearted chatter in the background all added to the ambiance of the occasion.

Less than a year later the piano had fallen silent, the feet had stopped tapping and the laughter and chatter had given way to fear and concern. No.XV Squadron had gone to war again.

As the result of an Iraqi invasion of Kuwait during August 1990, and Saddam Hussein's failure to respond to the United Nations Security Council demand for the withdrawal of his troops from the oil rich State, the west went to war against the Arab dictator. The aircraft of No.XV Squadron were despatched to the Middle East in two flights, with the second flight leaving on 2nd December. Due to the number of crews and personnel being transported to the 'war zone', a Hercules C.130 transport aircraft was pressed into service. Their ultimate destination, having routed out through Akrotiri and Tabuk, was Muharraq Air Base on Bahrain Island.

Whilst the politicians endeavored to resolve the problems, the Allied forces geared-up for war. Finally, after all else failed, military action commenced against Iraq on the night of 16th/17th January 1991. The attack began with a strike by Tomahawk cruise missiles, backed-up by F.117A Stealth bombers. These were followed by the Tornado bombers of No.XV Squadron, which due to their low flying techniques were known as 'Mudmovers'. The latter attack being undertaken in two waves.

As F/L John Peters and F/L John Nichol walked out towards the crew bus, they met two of their colleagues who had just landed, having participated in the first wave attack. From the look on the latter's faces, the two Johns knew the forthcoming mission would not necessarily be (in RAF jargon) *'a piece of cake'*.

The first problem for the nervous crew came when their fully laden bomber refused to start. After three unsuccessful attempts at igniting the jet engine, the crew had no option but to decamp from the aircraft and repair to a serviceable machine.

With the light of dawn breaking across the sky, John Peters taxied out, fully aware that another crew from his formation would not be getting airborne due to a defective system on the aircraft.

Having completed the task of in-flight refueling from an RAF Victor tanker en-route to the target, the pilot and navigator busied themselves with the necessary checks before going into combat. With the crew working together, almost as one man, the Tornado raced low across the desert towards the target. John Peters counted the seconds down to the pull up point, where the aircraft would initiate a climb and the bombs would be lobbed at the target. With groundfire fire bursting all around the aircraft, F/L Peters pressed the commit button in order to release the bomb load. Unfortunately, nothing happened. Tension in the two cockpits rose as John Nicholl endeavored to check the cause of the problem, hoping it was not due to a mistake on his part, whilst John Peters battled with the aircraft. As the Tornado continued to climb, more and more Iraqi guns were turning their attention to the bomber.

At an altitude of approximately 3,400', the pilot realised he had to get his aircraft back on the 'deck' without further delay. The weight of the bombs on the underside of the aircraft meant it was very slow to respond to his command as he banked the bomber over into a dive. After what seemed an eternity, the Tornado leveled-out and, once in level flight, the navigator jettisoned the bombs. With the aircraft relieved of its 8,000lb weight the Tornado was more agile, but unfortunately that agility was not to save it from destruction. A surface-to-air-missile (SAM) struck the bomber, with a shuddering impact, as they tried to bank away from an Iraqi communications site. Although John Peters attempted to regain control of his machine, with many of the avionic systems knocked out the crew had little choice but to eject from the stricken aircraft. Panavia Tornado GR1, ZD791, plummeted into the ground, where it exploded in a ball of flame.

For F/L Peters and F/L Nicholl the nightmare had only just begun. They were captured almost immediately and held by Iraqi soldiers for forty-seven days, during which time they underwent many traumatic experiences. The full story of their war in the Gulf was later recorded in their book, *'Tornado Down'.*

Initially, it was not known if the two downed airman had survived their ordeal, which invoked a feeling of despair on the Squadron. However, there was a job to do and, in line with the other units in the Gulf theater, the members of No.XV got on with it.

On Thursday, 14th February 1991, No.XV was detailed to attack hardened aircraft shelters on the Al Taquaddam airfield, forty miles west of Baghdad. Amongst the aircraft despatched by the Squadron was Tornado, ZD717, piloted by F/L Rupert Clark. In the back seat, flying as navigator, was F/L Stephen Hicks. Their aircraft formed part of a formation of eight

similar machines which, apart from the American electronic jammer and defense suppression aircraft, were supported by four Buccaneers. The latter aircraft would direct the laser-guided bombs, fired by the Tornadoes, down onto their objectives.

The aerial armada, which took-off at approximately 06.00 hours was led by S/L Pablo Mason, a flamboyant character who sported a large 'handle-bar' mustache. '

Flying through thick cloud cover the formation headed north-west across Saudi Arabia, before turning north toward Iraq. Although the cloud created some difficulties during the in-flight refueling procedures, the attacking force reached the target area without incident. However, the situation was to change rather rapidly. The cloudless sky over the enemy airfield soon became peppered with areas of drifting thick black smoke, as the Iraqi defenses opened fire with heavy Triple A gunfire.

One by one the Tornados flew through the curtain of exploding hot metal. With a mixture of fear, trepidation and excitement, the crews threw themselves into their attack routine. The laser-guided bombs were released and, with the assistance of the Buccaneers, streaked earthwards to locate their targets. Several direct hits were recorded. Unfortunately, S/L Nigel Risdale and his navigator W/C John Broadbent, experienced problems and were unable to release their bomb load. Although they intended to 'go round again', they decided to abort the mission and return to base.

As they set off for home, Pablo Mason and the other members of the formation saw two large black puffs of smoke followed by an explosion. This was followed almost immediately by a second SAM explosion on the opposite side of the Tornado. Having previously heard a warning of surface-to-air missile attacks, they instinctively knew one of their aircraft had been hit. Glancing out from his cockpit, S/L Mason recognised the stricken aircraft as being that flown by Rupert Clark. All prayers and thoughts, willing the crew to eject, remained unanswered. The crippled bomber remained airworthy, but adopted a gliding attitude as it slowly sunk towards the sandy desert floor. Although the Tornado was severely damaged, Rupert Clark managed to retain some control of aircraft which crashed behind enemy lines.

Flight Lieutenant Clark and F/L Hicks were engaged on their 15th war operation when they were shot down. Tragically for the latter, it was to be his last mission. Stephen Michael Hicks, aged 29, had been killed instantly when the missile exploded on impacted with the aircraft. Flight Lieutenant Hicks, who was the only member of No.XV Squadron to lose his life during the Gulf conflict, had been one of the officers who stood behind a headstone at Reichswald Cemetery the previous May.

Rupert Clark survived the ordeal, but was captured by Iraqi soldiers and made a prisoner of war.

Two weeks after the death of Stephen Hicks and the capture of Rupert Clark the war ended. The Iraqi forces had surrendered. A week later the captured Allied airmen, although battered and bruised, were released and repatriated. The other members of No.XV Squadron returned to Laarbruch towards the end of March, where they received a tumultuous welcome.

By comparison, the homecoming of F/L Hicks was a sad and somber affair. On 27th March he was laid to rest with full military honors in the cemetery adjoining St Eval Parish Church, near St Mawgan, on the north Cornish coast. The location, chosen by his widow, is fairly close to where Lyn and their two sons, Philip and Graham, still live.

During the months following the Gulf war, a number of awards were announced including three to members of No.XV Squadron. The Commanding Officer, John Broadbent was awarded a Distinguished Service Order, whilst S/L Gordon Buckley and S/L Nigel Risdale each received a D.F.C.

On 23rd August 1991 W/C John Broadbent, D.S.O, handed over command of the Squadron to W/C Andy White. However, due to the disbandment of No.XV early the following year, the latter's tenure as C.O. lasted only nine months.

Although many thought No.XV had disbanded for the last time, such was the feeling of resentment from those who had served with the Squadron, the unit was not totally abolished. Bowing to pressure, No.XV was reformed as a reserve squadron at Honington, in the Weapons Training role, under the command of W/C Alan Hudson. At a ceremony held at the Suffolk base on 1st April 1992, the latter unit officially accepted the No.XV Standard, and was thereby officially re-titled the Tactical Weapons Conversion Unit/No.XV (R) Squadron.

Apart from the Squadron Standard another important link with No.XV was retained, an aircraft named *MacRobert's Reply*. There was however one noticeable difference with the artwork on the machine. Apart from being adorned with the Family crest and legend, the white Roman numerals and the tailcode "F", the aircraft also bore the coronet and sword emblem of the TWCU on the tailfin.

No.XV operated at Honington as part of the TWCU for the next two years, during which time it welcomed a number of ex-Squadron members to the base. On 9th/10th July 1993, the XV Squadron Association held its annual reunion at Honington. A formal dinner was held on the Friday evening, followed by an informal gathering on the Saturday morning. A display of the aircraft, weapons and exhibits were arranged for inspection in the hangers, whilst the Squadron's photograph albums, flight charts and memorabilia were displayed in the crewroom.

Another occasion when ex-members of the Squadron gathered at the Suffolk base occurred during May 1994, when the TWCU/XV moved from Honington to Lossiemouth. Following a parade, Wing Commander Alan Hudson, who was later to be awarded an O.B.E, led the Squadron in a final flypast salute before heading north.

The Tactical Weapons Conversion Unit/XV (R) Squadron immediately commenced its training program upon arrival at RAF Lossiemouth in Morayshire, Scotland. Wing Commander Graham Bowerman, who had served with the Squadron as a flight lieutenant, officially took command of the unit on 13th May.

Exactly one month after he took office W/C Bowerman attended a parade at RAF Benson, Oxfordshire, where he and other unit commanders were to be presented with Fahnenbanders (silk streamers, in the colours of the national German flag). The Fahnenbanders were to be attached to the point of the recipient squadron's Standard. The presentation made by Herr Peter Hartmann, the German Ambassador to London, was intended as a token of gratitude from the German people to the squadrons who had preserved peace in their Nation; the RAF 'Peacekeepers'. The end of the ceremony was marked by a flypast of one Puma and two Wessex helicopters, from No.230 and No.60 Squadrons respectively. As the aerial salute flew overhead, the Standard/Color parties of the nine squadrons honored marched off.

The work of training Tornado crews, which had begun with its reformation at Honington, kept the Squadron busy throughout 1994. However, this did not prevent No.XV from undertaking many official engagements, or from finding time to welcome visitors to its northern home.

Although Lossiemouth is situated on the Moray coast, four and a half miles north of Elgin, the distance to the base did not deter a friendly invasion of ex-Squadron members and other invited guests during June 1995. That particular year gave cause for many and various celebrations, the most notable probably being the 50th anniversary of the end of the Second World War. Most of the planning undertaken by No.XV Squadron between the end of 1994 and early 1995, was for a celebration of its own. On 15th

March 1995, No.XV (R) Squadron celebrated the 80th anniversary of its original formation and, to mark the occasion, a series of events were arranged for the weekend of 9th-11th June.

The festivities began on the Friday when W/C Bowerman welcomed the guests arriving at the Squadron crewroom. During the course of the afternoon, many former friendships were rekindled along with the start of some new ones. All eras of the Squadron's history, from the start of the Second World War to the present day, were represented. Officers of all ranks, from Air Marshals down to NCOs' chatted, laughed and joked together.

With everyone at their ease the multitude, which had gathered during the afternoon, repaired to a more congenial location. Both hosts and guests made their way to the Officers' Mess for the start of 'Happy Hour'. The liquid refreshment provided helped to lubricate the tonsils of all those gathered in the bar, thus allowing the friendly chatter to continue unabated. 'Happy Hour' gave way to an 'American Theme' party, when a number of Squadron members and their ladies appeared dressed in various guises of apparel associated with the wild west. To complete the scenario country and western music was provided, whilst the Mess Chef and his staff grappled with the start of a huge barbecue. Soon the aroma of hamburgers, sausages and jacket potatoes wafted on the night air. The party, which lasted well into the night, was only the first stage of the celebrations.

The following day, Saturday, 10th June, RAF Lossiemouth opened its gates for a Families Day. Apart from various ground exhibits and demonstrations, a static aircraft park provided many interesting items. Although the weather tried to intervene, it did not prevent a flying program, with over twenty separate displays, from taking place during the afternoon.

The formal part of the weekend occurred on the Saturday evening, when a sit down buffet was organised for 400 Squadron personnel, ex-members and guests. In order to accommodate such a large number of people, one of the hangers was transformed into a vast dining area. The internal face of the main hanger doors provided a 'canvas' on which a mural depicting the Squadron's 80 years, had been painted. Behind the top table, on which the Squadron silver had been set, was the Squadron Standard. This was guarded by two outward facing Tornado aircraft, one of which was *MacRobert's Reply*.

A number of speeches were made, including one by Air Marshal Sir Michael Simmons, K.C.B., A.F.C., a former C.O., who responded to the toast to the guests. The speeches were in turn followed by the presentation of a number of gifts, some of which alluded to the Squadron's history.

Ensuring the maximum pleasure for their guests, No.XV arranged a bus trip to a few places of interest in the area, including visits to the site of a former airfield and a local museum exhibiting memorabilia appertaining to the airfield.

All too soon the celebrations came to an end, but never in the history of the Squadron had so many members, ex-members and guests gathered in one place to celebrate such an event. Those serving on the Squadron today are tomorrow's ex-members, hopefully they will have the opportunity to reminisce with the serving members at the 100th anniversary of No.XV (R) Squadron; the Squadron which was once known as "Oxford's 'Own'".

Buccaneers at Laarbruch

An unidentified Hawker Siddeley Buccaneer S.2B makes a high speed, low-level, pass across the airfield at Laarbruch. *Author's Collection*

The only indication that an airfield is situated in the forest area of West Germany, close to the German/Dutch border, is the simple board recording the name of Royal Air Force Station Laarbruch. *Author's Collection*

A sequence of photographs showing a Buccaneer in the close confines of a hardened aircraft shelter at Laarbruch. *Author's Collection*

A Buccaneer at Greenham Common

Hawker Siddeley Buccaneer S.2B, XT275, was displayed by No.XV Squadron, in the static park at the International Air Tattoo, at the USAFE Base at Greenham Common, Berkshire, during June 1979. *Author's Collection*

The Return of MacRobert's Reply

Watched by former members of No.XV, Squadron Leader Peter Boggis, D.F.C., RAF (Retired), unveils the MacRobert Family Crest, on Buccaneer, XT287, coded "F", at Mildenhall. *Author's Collection*

A Buccaneer at Mildenhall

A Buccaneer and its cockpit. Members of the Mildenhall Register were able to get a close look at the No.XV Squadron aircraft whilst it was parked on the ramp, at the USAFE base at RAF Mildenhall, Suffolk. The only clue to the identity of the aircraft, which had flown in for a reunion, is the tailcode letter "J". *Author's Collection*

A Hunter at Mildenhall

A Hunter and its cockpit. Hawker Hunter, T.7A, WV318, also seen on the parking ramp at RAF Mildenhall. *Author's Collection*

A Barbecue at Laarbruch

Wing Commander Eddie Cox, C.O., No.XV Squadron (left), chats to Pat Russell, at a barbecue held during a weekend visit to RAF Laarbruch. The latter was a bomb aimer with the Squadron towards the end of the Second War World. Apart from the usual operations, Pat also participated in both *'Manna'* and *'Exodus'* operations a fact of which he was very proud. *Author's Collection*

The First Navigator C.O.

Former air gunner Harry Bysouth relates a humorous anecdote to Valerie Ford-Jones, during a period of refreshment, at Laarbruch. Looking on, and enjoying the joke are Rose Pluck (2nd from left) and Percy Pluck (extreme right). The latter was a member of Lancaster groundcrew. *Author's Collection*

Wing Commander Barry Dove had the distinction of being the first navigator to command No.XV Squadron. *Courtesy of and via Barry Dove*

A Lancaster's Salute

Lancaster bomber, PA474, operated by the Battle of Britain Memorial Flight, flies over *MacRobert's Reply*, as if in salute to the Tornado, prior to landing at Mildenhall. *Authors Collection*.

Mike Rudd's Command

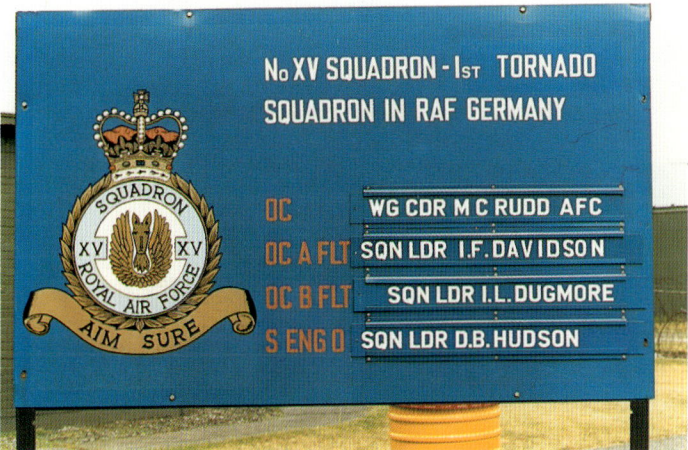

The sign says it all. The signboard outside the No.XV Squadron hangar at Laarbruch, leaves the visitor in no doubt as to the unit's claim to fame. *Author's Collection*

Some of the weaponry and ordnance carried by the Tornado are displayed in a HAS at Laarbruch. *Author's Collection*

A hardened aircraft shelter in No.XV's dispersal area, at Laarbruch. Concealed in the shadows is *MacRobert's Reply*. *Author's Collection*

Wing Commander Mike Rudd, C.O., No.XV Squadron (2nd from left) enjoys a light-hearted, off-duty moment. With him are Group Captain Ken Batchelor, ex-Station Commander, RAF Mildenhall (left), Group Captain McRobbie, Station Commander, RAF Laarbruch (3rd from left) and Don Clarke (extreme right, foreground). *Author's Collection*

Wing Commander John Broadbent, who assumed command of No.XV Squadron from Mike Rudd on 20th January 1989, is seen here with Betty George, M.B.E. (wife of Hugh George) and G/C Ken Batchelor. *Author's Collection*.

Aviation artist Keith Aspinall presents W/C John Broadbent with an original painting of Tornado, ZA549, depicted in its 75th anniversary colors. *Author's Collection*

The real thing. Panavia Tornado, GR.1, ZA549, photographed outside the No.XV Squadron hangar, displayed in its 75th anniversary colour scheme. *Author's Collection*

The tailfin of ZA549 tells the story. Apart from displaying the pertinent dates, the white "XV" Roman numerals are superimposed over a blue painted 75. *Author's Collection*

MacRobert's Reply was on display also wearing a new color scheme. *Author's Collection*

They came from far and wide to join in No.XV's celebrations. A Buccaneer from No.237 Squadron came to join the party. *Author's Collection*

The pilot of McDonnell-Douglas, F.G.R.2, Phantom, XV467, of No.56 Squadron, climbs down from his cockpit. As the navigator's cockpit has already been vacated, it would appear he got to the bar first. *Author's Collection*

A 'Gatecrasher'? Not an uninvited guest from the Soviet Block, but an English Electric (later B.A.C) Lightning fighter painted, for war game purposes, to resemble a Russian aircraft. *Author's Collection*

S/L Peter Boggis, D.F.C. (left), W/C John Broadbent, C.O., No.XV Squadron (center) and Air Marshal Sir Michael Simmons, K.C.B., A.F.C., at the official parade for the anniversary march-past. *Author's Collection*

The Band of the Royal Air Force Germany march into the No.XV Squadron hangar at the start of the official 75th Anniversary Parade. The Band was followed by the Color Party and escort, which comprising of F/L J B MacDonald, Standard Bearer, and Sgt S Roberts, Sgt A McQuiggan and W/O J Whelan as Escort to the Standard. *Author's Collection*

Paying homage to departed merit. During the Remembrance Service at Reichswald Forest War Cemetery in 1990, serving members of No.XV marked the last resting places of former members of the Squadron who were killed during the Second World War, by standing behind the headstones. Less than a year later, F/L Stephen Hicks (left) would be mourned, having paid the supreme sacrifice in action during the Gulf War. *Author's Collection*

Squadron Leader Peter Boggis, D.F.C., RAF (Retired), who flew the original Stirling bomber named *MacRobert's Reply* and S/L Pablo Mason (right), a Tornado GR.1 pilot with the Squadron. *Author's Collection*

Warrant Officer J Whelan stands guard over the Squadron Color, whilst it adorns the Stone of Remembrance in Reichswald Cemetery. *Author's Collection*

The Gulf War Period

F/L Stephen Hicks (left) together with his pilot Rupert Clark, photographed by the artwork of Tornado, ZA471, named 'Emma'. Rupert Clark appears to be practicing a pre-flight ritual. *Courtesy of and via Lyn Hicks*

F/L Stephen Hicks photographed in flight over the Arabian Desert. This is possibly one of the last photographs taken of the No.XV navigator. *Courtesy of and via Lyn Hicks*

Taken from an original painting by artist Keith Aspinall, Tornado, ZA471, named 'Emma' (in the foreground) leads Tornado, ZA491, named 'Nikki' on a low-level sortie across the Saudi Arabian desert during the Gulf War. The former aircraft flew a total of 35 missions, which was a record for No.XV Squadron. The latter machine originally flew with No.20 Squadron. *Courtesy of Keith Aspinall*

Tornado pilot F/L Rupert Clark, who was shot down and captured during the Gulf War, chats with a member of the public in the Friends of International Air Tattoo enclosure, during an air display at RAF Fairford. *Author's Collection*

Flight Lieutenant John Peters (right) and F/L John Nichol became public figures, after they were shot down during the early stages of the Gulf War. *Author's Collection*

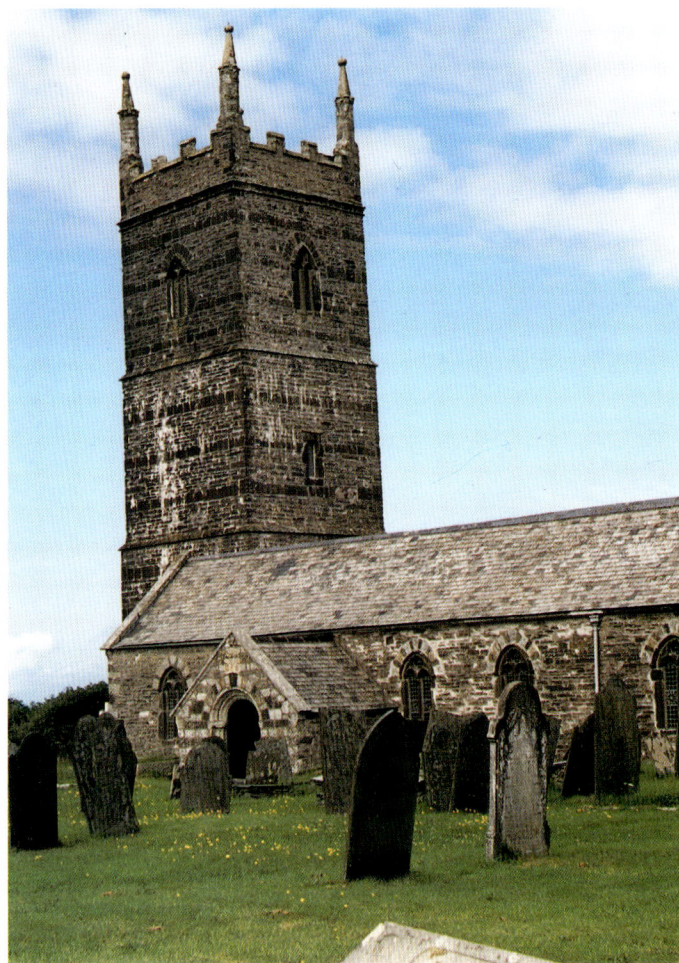

The last resting place of Stephen Michael Hicks, who was killed in action on 14th February 1991. F/L Hicks was flying as navigator to F/L Rupert Clark, when their aircraft was shot down during a Gulf War sortie. Note the No.XV Squadron crest on the headstone. *Authors Collection*

Flight Lieutenant Stephen Hicks is buried in the cemetery at St. Eval Parish Church, near St. Mawgan, Cornwall. His grave is to the left of the photograph, behind the church tower. The site, which overlooks the Atlantic Ocean, was specially chosen by Lyn his widow, as Stephen always wanted to be buried within sight of the sea. *Author's Collection*

A dramatic photograph, taken by aircraft enthusiast Alan Chapman, of Panavia Tornado, ZA410, coded "EX", departing from RAF Fairford. The aircraft, crewed by F/L John Peters and F/L John Nichol, had been on static display in a Gulf War Salute, during an airshow weekend at the USAFE airfield. *Courtesy of Alan and Janet Chapman*

Back to Honington

Panavia Tornado, ZA600, coded "F", *MacRobert's Reply*, undergoes servicing in a hangar at RAF Honington. Note the TWCU emblem on the tailfin, together with the white roman numerals. *Author's Collection*

The airbrakes are extended and the flaps are down in this picture showing details of the rear fuselage of Tornado, ZA600. *Author's Collection*

Two views of Tornado, ZA595, coded "TV", displayed in the hangar at Honington, with some of the ordnance carried by the aircraft. *Author's Collection*

With memories abundant in the No.XV Squadron briefing room, in the form of photographs, maps, models and a Roll of Honor, Air Marshal Sir Michael Simmons, K.C.B., A.F.C., ponders on memories of his own of the time he was unit's C.O. Author's *Collection*

The 80th Anniversary

A sign of welcome is displayed outside hangar K16 at RAF Lossiemouth, where No.XV Squadron held the formal proceeding for the 80th anniversary celebrations. *Author's Collection*

A sign of friendship is displayed in the Officers' Mess, amongst ex-Squadron members from two different generations. From left to right are: John Cox; John Peters; 'Robbi' Roberts; 'Rocky' Knight; Eric Smith and John Nichol. *Photograph taken and supplied to Author's Collection by Elizabeth Cox*

The crewroom in hangar K17, where members of the Squadron can relax and chat, or catch up on the news, surrounded by items appertaining to the unit's past history. Included in the latter, displayed on the end wall of the crewroom, is the original banner used by the MacRobert Family. *Author's Collection*

Amongst the aircraft parked in the static area, for the Families Day, was Hawker Siddeley Nimrod MR2P, XV 255, from the RAF Kinloss Maritime Reconnaissance Wing. *Author's Collection*

Another visitor was Hawker Siddeley Hawk T1A, XX247, from No.1 TWU/234 Squadron, based at Brawdy. *Author's Collection*

De Havilland Tiger Moth, DF155, did not have far to travel as the aircraft, which has the civilian registrations G-ANFV, was based at Lossiemouth. *Author's Collection*

A civilian visitor to the Families Day was Douglas DC3 Dakota, G-ANAF, operated by Air Atlantique. *Author's Collection*

Hangar K17 at Lossiemouth, photographed during the period prior to the guests arriving for the 80th anniversary celebrations. The No.XV Squadron crewroom is to the right of the picture. *Author's Collection*

Primed and ready to go. Panavia Tornado GR1, ZA601, coded "TI", parked on the ramp behind hangar K16, was one of No.XV's aircraft on view at the anniversary 'bash'. *Author's Collection*

The internal faces of the main doors of hangar K16, were decorated with a huge mural depicting the Squadron's eighty years of aviation and engineering. The centerpiece depicts a Hawker Hind aircraft, with a Stirling bomber to the left and a Tornado to the right. *Author's Collection*

Ticket No.230, which gave the bearer admittance to the Squadron's hangar party. *Author's Collection*

Tornado, ZA559, "F", *MacRoberts Reply* also went to the party. The aircraft, which had been provided by No.XV Squadron as the GR1 display aircraft for 1994, was specially painted for the occasion. Still wearing the color scheme a year later, albeit with the addition of the dates '1915-1995' on the tailfin, *MacRoberts Reply* was one of the two aircraft which stood guard over the Squadron Standard in the hangar. *Author's Collection*

Ex-flight engineer Ken James, an amateur artist of some ability, presents a painting of his former aircraft to, W/C Graham Bowerman, as an 80th anniversary gift for the Squadron. *Author's Collection*

As a gesture of thanks for her help towards arranging travel to Lossiemouth for a large group of ex-Squadron members, Graham Bowerman presented Betty George with a personalised print of a No.XV Squadron Tornado. Valerie Ford-Jones admires the print, whilst Hugh George (who was a Blenheim man) tries to look disinterested. *Author's Collection*

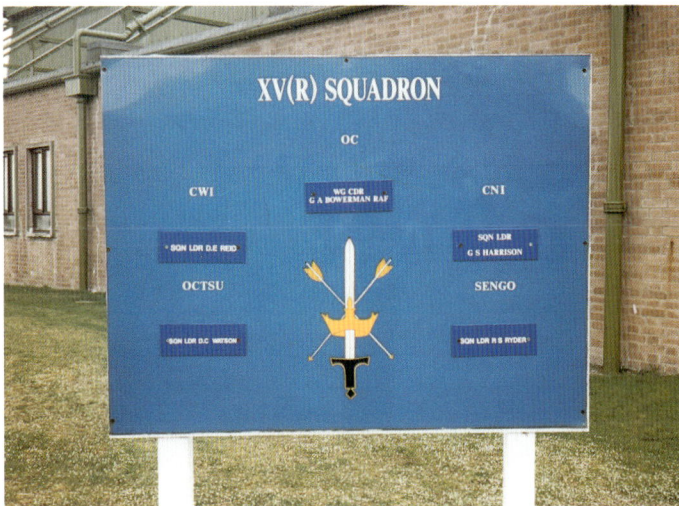

A more formal signboard for No.XV Squadron, displayed outside hangar K17. Apart from bearing the TWCU emblem and the name of Graham Bowerman as C.O., the board also records the names of S/L Reid, the Chief Weapons Instructor; S/L Harrison, the Chief Navigation Instructor; S/L Ryder, the Senior Engineering Officer and S/L Watson, the Officer Commanding the Tornado Servicing Unit. *Author's Collection*

Above and below: Panavia Tornado GR1, ZA601, coded "TI", on the ramp at Lossiemouth during 1996. Also in the photograph are Tornado, ZA563, coded "TC" and Tornado, ZA556, coded "TA". *Author's Collection*

Wing Commander Graham Bowerman welcomes the author to Lossiemouth, at the start of a the start of a private visit during April 1996. *Author's Collection*

The nosewheel undercarriage assembly of Tornado, ZA614, coded "TB" gets a close inspection from a male member of the servicing team. *Author's Collection*

Left: The servicing teams and trades involved in keeping the aircraft in flying condition include a number of women. Tornado, ZA563, coded "TC" benefits from a walk round external inspection by a member of the Womens Royal Air Force. *Author's Collection*

With the canopy raised and the access steps in place, who could resist a look in the cockpit. Valerie Ford-Jones takes advantage of the situation to learn more about the 'office' layout of Tornado GR1, ZA614, from Flight Lieutenant Torben Harris who explains the functions of some of instruments. *Author's Collection*

Heads and tails. A line up of No.XV Squadron aircraft, with Tornado GR1, ZA562, coded "TT", nearest the camera. The aircraft is conspicuous against the other machines, due to the blue/grey color scheme which adorns it. Note also the red painted aircraft tailcodes. *Author's Collection*

A three-quarter front view of Panavia Tornado GR1, ZA562, on the ramp at Lossiemouth. *Author's Collection*

Tailpiece. A close-up of the tailfin of Tornado, ZA562, displaying the Squadron's white Roman numerals and the unorthodox red code letters. *Author's Collection*

Postscript

Shortly after its formation in 1915, whilst still working-up to operational readiness, No.15 Squadron, Royal Flying Corps, was detailed to train crews for service with other squadrons. This situation was brought about by the urgent need for operational aircrew as replacements for those lost in combat, on the Western Front. Eighty Years later, the wheel of historical fortune has turned full circle. Although there is currently no urgent need for replacement aircrew, No.XV (R) Squadron, Royal Air Force, is training the pilots and navigators who will become 'Tomorrow's Peacekeepers'. Following the anniversary celebrations in June 1995, the Squadron returned to its allotted task. During the remaining eighteen months of his tenure as commanding officer, W/C Bowerman oversaw the training of more than 100 Tornado aicrew. The fact that, during that period, over 4,500 flying hours were amassed, over 17,500 rounds of ammunition were fired and over 9,100 bombs were dropped bears witness to the commitment of the Squadron.

During this same period, many official engagements were carried out, including providing aircraft for a number of official flypasts. The latter involving a deployment to Reims, to enable the Squadron to participate in the Bastille Day flypast over Paris.

Other deployments, as part of the on-going NATO exercise program, included visits to Goose Bay, Las Vegas, and Turkey.

On 28th February 1997, Graham Bowerman relinguished command of the Squadron, having been posted to the Joint Warfare Staff at HMS Dryad, near Portsmouth. The city of Portsmouth, on the south coast of England, has been home to the Royal Navy for hundreds of years. It is hardly the place to be seen wearing an RAF uniform! However, the move did have some compensations, Graham Bowerman was awarded an O.B.E. in the New Year's Honors List, and later received promotion to the rank of Group Captain. Wing Commander Graham Dixon, like his predecessor, was no stranger to the Squadron. Apart from having served with No.XV in Germany, Graham Dixon had also seen service as an instructor with the Tactical Weapons Conversion Unit. Under his command the training continued unabated, with nearly seventy crews passing through the course. Rumour has it that amongst that number, were one Air Vice-Marshal and one Air Commodore!

Throughout the year, the Squadron juggled its commitments between training of aircrew, civil duties in and around Lossiemouth, receiving and entertaining visitors and undertaking detachments to various locations. Apart from NATO exercises in Denmark, the latter also included exchange visits to Germany and training in Sardinia.

Towards the end of 1998, the Royal Air Force announced that No.XV Squadron was to provide an aircraft and crew to form the Tornado GR1 Display Team for the 1999 airshow season.

With No.XV again thrust into the public eye on the airshow circuit, the Squadron selected as crew members two officers whose careers had followed similar courses. As youngsters, both had been members of the Air Training Corps, both had gained their Private Pilots Licence prior to joining the RAF, and both had completed their initial officer training at Royal Air Force College Cranwell. Flight Lieutenant Andy Myers, who had seen service as the Unit Test Pilot at RAF Brawdy, was selected to fly the aircraft, whilst F/L Chris Stradling, a Gulf War veteran, was chosen to occupy the rear seat. Apart from serving at Brawdy, on the Pembrokeshire coast in Wales, Andy Myers had also served with No.360 Squadron at Wyton, where he flew Canberra's, operating in an Electronic Warfare Training role. He converted to the Tornado aircraft in 1993, prior to being posted to No.IX Squadron based in Germany.

By comparison, on completion of his navigational and tactical weapons training, Chris Stradling was posted to No.17 (Fighter) Squadron based at RAF Bruggen. In 1991, when Iraqi Forces invaded Kuwait, No.17 was detached to the Middle East, where Chris undertook a number of operational missions. Following a two year ground appointment at Bruggen, Chris Straddling was posted back to No.17 Squadron for a further three year tour of duty. Flight Lieutenant Myers and F/L Stradling, who had both joined No.XV Squadron as instructors during 1996, were suitably qualified for the task ahead of them. Collectively, they had over 4,500 flying hours, of which 3,200 were accumulated on the Tornado GR1. After many meetings and discussions on procedures, timing and maneuvers they began their display flying training in January 1999. Initially, the programme was practiced at an altitude of 5,000', but this was brought down to lower altitudes as the team became more proficient.

During the planning of No.XV's return to the airshow scene nothing was left to chance. Four aircraft, each with the names of the two man crew painted on the canopy rails, were made available in the event of one or more aircraft going u/s. Likewise, as a precaution against the vagaries of the weather, three versions of their proposed display routine were adopted. A full display for perfect weather conditions, a rolling display for reasonable weather and a flat display for the cloudy days. Given that F/L Myers and F/L Stradling were scheduled to participate in over thirty separate displays and three flypasts, in seven different countries, the chances of variable weather conditions were only to be expected. Whilst Andy Myers and Chris Stradling faced up to the challenge of their first season as display aircrew, the Squadron faced a more daunting task; a major expansion in both men and machines. No.XV has, due to the closure of the Tri-National Tornado Training Establishment at the end of March 1999, become the largest Squadron in the Royal Air Force. On 1st April, responsibility for the initial training of both Tornado aircrews and groundcrews, previously undertaken by the TTTE, passed to No.XV. A change of leadership followed on 23rd July when Graham Dixon, having been promoted to the rank of Group Captain, handed over command of the Squadron to W/C Simon Dobb.

As the echos of the airshow displays fade away and 1999 slips into oblivion, who knows how many of the youngsters, who marveled at the flying display executed by F/L Andy Myers and F/Lt Chris Stradling, will be inspired to follow their example and join the RAF. In training those young men and women for tomorrow's Royal Air Force, the Squadron is fully aware of the high level of responsibility placed upon it. In its new role, No.XV (R) Squadron will undoubtedly continue to *"AIM SURE"* into the next century.

The Vargaries of the English Weather

The low cloud and dark April skies over RAF Lossimouth are not conducive for airshow flying practice. *Author's Collection via F/L Chris Stradling*

Photo-Call in the Hangar

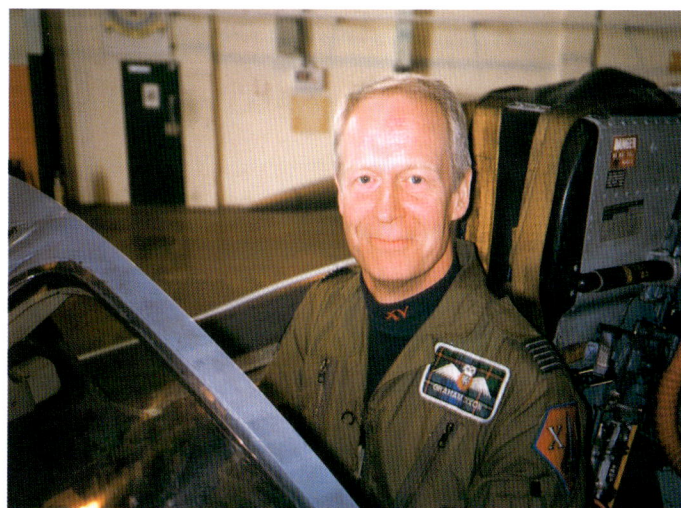

With the weather over the base unsettled, the aircraft undergo maintenance in one of the Squadron's two hangars. *Author's Collection via F/L Chris Stradling*

The Boss. Wing Commander Graham Dixon, who took command of the Squadron in February 1997. *Author's Collection via F/L Chris Stradling*

"I'll Take The Front Seat, You Go in the Back"

Flight Lieutenant Andy Myers will occupy the pilot's seat of the No.XV Squadron Tornado display aircraft for the 1999 airshow season ... *Author's collection via F/L Chris Stradling*

... whilst Flight Lieutenant Chris Stradling will occupy the navigator's seat. *Author's Collection via F/L Chris Stradling*

XV Squadron Remembered

It goes without saying that with a history spanning over 80 years, there are many artifacts or items of memorabilia relating to No.XV still in existence. Some of those items are in the care of the Squadron at its base at RAF Lossiemouth, whilst other items are in the collections of private individuals. The majority of artifacts owned by the latter are keepsakes or mementos of their service with the Squadron, probably during the Second World War or shortly after.

Over recent years a number of ex-No.XV Squadron members have graciously donated their mementos and other various items of memorabilia to the author's collection. The generosity of those donors has ensured that, together with the items secured by the author from other sources, a large part of No.XV Squadron's heritage has been preserved. The photographs below illustrate a small percentage of that collection.

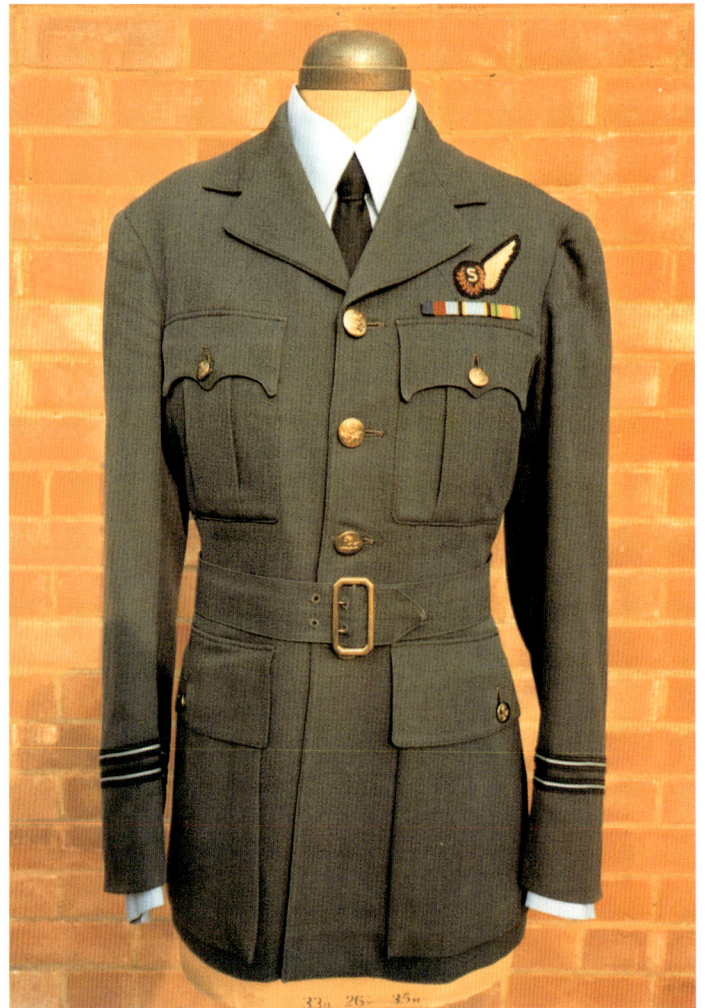

Left: The uniform and cap of an air Vice-Marshal, worn by AVM Joe Cox, C.O. of No.XV Squadron between June and December 1940. *Author's Collection* Right: An RAF tunic worn by F/L George Wright who, as a Sergeant wireless operator, flew with F/S Gil Marsh on Stirling bombers. After Gil Marsh was severely wounded in action, on 23rd August 1943, George Wright transferred to the crew of F/O Stoddart with who he completed his tour of 31 missions. Note the Signaler Brevet above the medal ribbons and the flight lieutenant rank tapes on lower part of the sleeves. *Author's Collection*

Left: The battle-dress blouse worn by Sgt Doug Fry, who was shot down on the night of 30th/31st July 1943. The hole, made by a flying fragment of shrapnel which wounded Doug Fry in the stomach, can be seen to the lower right portion of the jacket. *Author's Collection* Right: A general issue Irvin leather flying jacket, worn by George Wright who flew as a wireless operator on Stirling bombers. The collar, when turned up, forms an integral hood with a yellow painted exterior. *Author's Collection*

Left: The upper part of the Novex 14 flying suit worn by F/L Rupert Clark, who was shot down during the Gulf War. Apart from the brevet displaying the pilot's 'wings' together with his name, a No.XV Squadron 75th anniversary patch is also evident at the top of the sleeve. *Author's Collection* Right: The Novex 14 flying suit worn by John Craig, a navigator on Tornados. Under the suit, which bears an earlier form of Squadron patch, is an aircrew sweatshirt. *Author's Collection*

Left: The remains of the leather flying helmet worn by Harry Jackson who flew with the Squadron, as an air gunner, during 1944. The inside of the helmet was used as a log-book, with all the operations flown by Sgt Jackson being recorded in ink. *Author's Collection*

Left: Two fragments of camouflaged metal, recovered from the crash site of Stirling bomber, BK657, LS-C. The aircraft, which was shot down at 02.15 hours, on 27th April 1943, crashed at Breukelen, Holland. Also recovered from the site was the remains of an incendiary bomb. The items were donated to the Author's collection by Jan Uithol and Henk Rebel of the Crash Research Aviation Society Holland (CRASH). *Author's Collection* Right: Fragments of metal recovered from the crash site of Stirling bomber, BF386, LS-Q, which crashed during an air test on 29th October 1942. The aircraft, which had eleven personnel on board, dived into the ground at Salters Lode, five miles south-west of Downham Market, Suffolk. The items were donated to the Author's collection by John Reid, Stirling Aircraft Research Library. *Author's Collection*

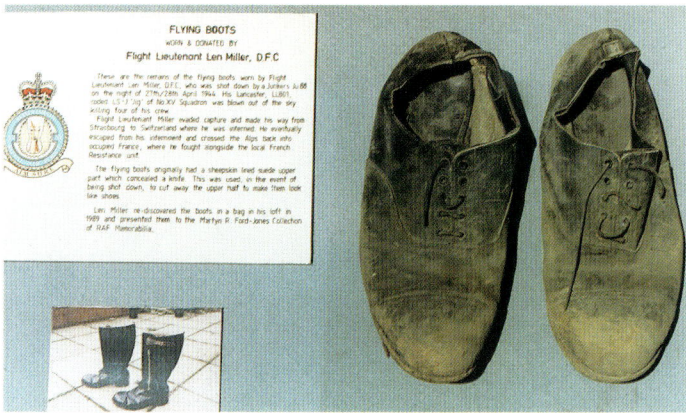

The mud-splattered remnants of a once smart pair of RAF aircrew flying/escape boots. The upper part of the boots, which were of suede with an off-center zip fastener, were designed to be cut away to form a sturdy pair of walking shoes. Those seen in the photograph were worn by F/L Len Miller, who was shot down over France. Having evaded capture, he walked to Switzerland where he was interned, but later escaped back into France. He was eventually repatriated to England, having spent some time in the company of a French resistance unit. Following his return home, Len Miller kept the boots as a memento of his adventures, until finally donating them to the Author's collection. *Author's Collection*

Right: A number of artifacts relating to No.XV Squadron's history include, a brass button stick used and donated by Hugh George, aircrew 'Dog Tags' worn and donated by George Wright and silk RAF inner flying gloves donated by Ken James. The latter were worn under leather flying gauntlets. The samples of 'Window' (Chaff), first used during July 1943, were donated by 'Robbi' Roberts and John Reid. *Author's Collection*

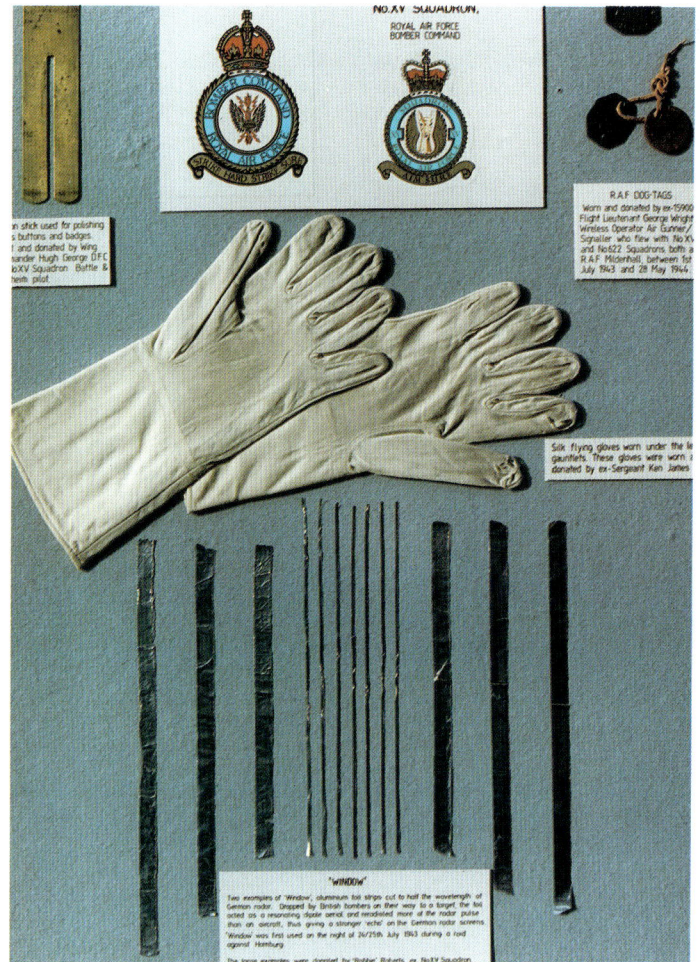

A large piece of metal, measuring 20"x12", recovered from the crash site of Lancaster, HK773, LS-W, the last aircraft of its type to be lost by No.XV Squadron during World War Two. The bomber developed engine trouble minutes after taking-off from RAF Mildenhall, and crashed in the forest at Mundford, Suffolk. The aircraft exploded on impact with the ground. *Author's Collection*

Left: *Modern Memorabilia.* A selection of shoulder patches issued and worn by members of No.XV Squadron aircrew and groundcrew, spans a period of twenty years commencing with the Buccaneer. The pin badges in the center of the picture often worn with pride by possible future generations of No.XV Squadron personnel. The lower section of the picture shows patches worn and donated by F/L Mike Toft. Mike Toft was a navigator who originally flew with No.360 Squadron, but was later posted to No.XV Squadron. He was based at Laarbruch, West Germany, but saw action with the Squadron during the Gulf War period. *Author's Collection*

XV Squadron on Canvas

Over the years the aircraft flown by No.XV Squadron have, like many other squadrons of the Royal Air Force, been 'captured' on canvas. The majority of paintings familiar to the public have, in the main, been painted by artists such as Frank Wootton, P.G.Av.A, AVM Norman Hoad, C.V.O.,C.B.E., A.F.C., Terence Cuneo and Charles Cundall. The latter artist was responsible for a painting, completed in 1942, showing No.XV Squadron's Stirling bombers on an airfield at sunset. However, a number of lesser known artists have also chosen the aircraft of No.XV Squadron, as the subject for their particular canvases. Many of these latter pictures adorn the walls of the Squadron crewroom, the C.O.'s office or the artist's own studio. Equally, many of the paintings are commissioned or purchased by ex-Squadron members, who proudly display them in their homes. The pictures shown below, which provide a chronological record of some of the aircraft taken into combat by No.XV Squadron, fall into one of the categories mentioned above.

A Fairey Battle bomber, coded LS-Y. *Painted by and courtesy of Ken James.*

Bristol Blenheim Mk.IV bomber. *Painted by and courtesy of Ken James*

Vickers Wellington, L7889, LS-T. After service with No.XV Squadron, this aircraft joined No.1504 Flight. *Painted by and of courtesy of Ken James*

A painting by Maurice Gardner, captures the moments leading to the destruction of Short Stirling bomber, EF427, LS-A, which was shot down on the night of 31st July 1943, whilst participating in an attack against Remscheid. The painting was commissioned for use as a dust-cover for the book, "Bomber Squadron – Men who flew with XV". *Author's Collection*

Stirling over the Rhur. *Painted by and courtesy of Ken James.*

An action painting detailing the attack by Stirling bombers of No.XV Squadron, against German warships in Brest harbor, on 18th December 1941. To the right of the picture is Stirling bomber, N6086, LS-F, *MacRobert's Reply. Painted by and courtesy of Group Captain David Luck, RAF(Retired)*

Stirling bomber, BK816, LS-X, was named *Madame X* by pilot Gil Marsh and his crew. The picture, by John Whittock, captures the aircraft at dispersal. *Courtesy of Gil Marsh*

A painting by Alan Dyer shows Stirling bomber, BK611, LS-U, flying low over the mist shrouded Cambridgeshire fens. Ely Cathedral is evident in the background. *Courtesy of John Reid*

'Evening at Dispersal' by Douglas Webber, shows Avro Lancaster, LL752, LS-A, of "A" Flight, No.XV Squadron, with the crew waiting to board the aircraft, prior to commencing an operation. *Photograph from the Author's Collection*

A painting commissioned by Len Miller, captures the last moments of his former aircraft, Lancaster, LL801, LS-J, which was shot down by Oberleutnant Martin Becker, on the night of 27th/28th April 1944. *Painted by and courtesy of Keith Aspinall*

'Calm before the Storm' by Keith Aspinall shows four Lancasters bombers of No.XV Squadron, heading towards enemy territory. *Author's Collection*

A No.XV Squadron Lancaster, coded LS-S, survives the ordeal of another big raid. *Painted by and courtesy of Keith Aspinall*

Two-ship Formation. Tornado, ZA491, coded "N"-Nikki, leads a sortie with Tornado, ZA471, coded "E"-Emma flying as wingman. *Painted by and courtesy of Keith Aspinall*

Panavia Tornado GR1, ZA462, coded "EM", flying as wingman, skims low over the sea on a mission during the Gulf War period. *Painted by and courtesy of Keith Aspinall*

Appendices

APPENDIX I
ROLL OF HONOR
for No.15/XV SQUADRON

WORLD WAR ONE

NAME	RANK	CREW	FORCE	DIED	AGE
Adams, M.C, Matthew	Lt	Pil	RAF	07.08.18	25
Allardice, Alexander	A.M.II	Obs	RFC	13.11.16	19
Barton, Frederick	Sgt	Pil	RFC	16.10.16	?
Bradford, George	2/Lt	Pil	RFC	04.02.17	?
Brereton, Herbert	Lt	Pil	RFC	21.12.16	22
Brooke-Murray, Kenneth	Capt	Obs	RFC	23.09.16	25
Brooking, Walter	2/Lt	Obs	RFC	19.01.16	18
Browne, Harold	Lt	Pil	RAF	03.05.18	18
Burleigh, Robert	Lt	Pil	RFC	29.08.16	23
Carre, Edward	Lt	Obs	RFC	16.10.16	22
Colinsky, Jacob	A.M.II	Obs	RAF	23.05.18	24
Conway, John H. De	Lt	Pil	RFC	15.06.17	?
Cumming, Alfred	2/Lt	?	RFC	07.06.17	25
Cunningham, James	2/Lt	Pil	RFC	14.03.16	21
Dalgleigh, Neil	2/Lt	Obs	RAF	30.10.18	29
Davey, Wilfred	Lt	Obs	RFC	21.11.17	?
Davis, Horace	2/Lt	Obs	RFC	06.02.17	23
Derrick, Leslie	2/Lt	Obs	RAF	03.05.18	?
Desborough, Laurence	Lt	Obs	RFC	30.11.17	22
Focken, Leslie	2/Lt	Pil	RFC	26.10.16	21

This officer served as 2/Lt L. C. Fawkner

Fear, Robert	2/Lt	Obs	RFC	05.03.18	?
Fine, Solomon	Lt	Pil	RAF	18.05.18	19
Fox, Walter	2/Lt	Obs	RAF	22.08.18	22
Fraser, Robert	2/Lt	Obs	RAF	18.05.18	19
Gillespie, Robert	2/Lt	Obs	RAF	07.08.18	?
Guppy, Edgar	A.M.I	Obs	RAF	27.10.18	26
Haarer, Philip	2/Lt	Pil	RFC	22.11.16	20
Hare, Edward	2/Lt	Obs	RFC	24.03.17	29
Harry, Reginald	2/Lt	Obs	RFC	29.08,16	?
Holmes, John	Lt	Pil	RAF	09.11.18	19
Honeyman, Herbert	2/Lt	Obs	RFC	10.12.17	24
Howlett, Percy	A.M.II	Obs	RAF	04.05.18	24
Laird, Andrew	2/Lt	Obs	RFC	2.12.16	22
Lee, Richard T.	A.M.II	Obs		10.12.17	28
Leggat, Matthew	2/Lt	Obs	RFC	26.03.18	22
MacKenzie, Adrian	2/Lt	Obs	RFC	01.04.17	?
Malcomson, Thomas	Capt	Pil	RFC	10.12.17	28
Manders, A.	2/Lt	Pilot	RAF	16.11.18	?

Miller, Walter	Lt		RFC	02.10.16	24
Newton, John	A.M.I	Obs	RFC	14.03.16	19
Pateman, CdeG(F), Henry	2/Lt	Pil	RFC	06.02.17	20
Powell, Cecil	2/Lt	Obs	RFC	15.06.17	22
Prestwich, Joseph	Lt	Pil	RFC	07.02.16	23
Proctor, Julien	Cpl	?	RAF	26.09.18	31
Reading, Vernon	2/Lt	Pil	RFC	26.03.18	22
Richardson, M.M., Edward	2/Lt	Obs	RAF	09.11.18	23
Ritter, MiD, William	Lt	Obs	RFC	02.06.17	23
Rutley, Harold	A.M.II	Obs	RAF	02.11.18	?
Sayer, James	2/Lt	Pil	RFC	03.04.17	19
Sealy, Laurence	A.M.II	Obs	RAF	02.11.18	20
Settle, Reginald	2/Lt	Pil	RFC	23.07.16	24
Shaw, Frederick	Sgt	Obs	RFC	04.02.17	25
Stretton, Sidney	Lt	?	RFC	27.03.17	?
Thierry, Leonard	2/Lt	?	RFC	10.12.17	?
Vinson, Albert	Capt	?	RFC	22.03.18	?
Wadham, Vivian	Capt	Pil	RFC	17.01.16	24
Welsford, Geoffrey	Lt	Pil	RFC	30.03.16	20
Wilson, ?	2/Lt	Pil	RFC	30.03.16	?
Wilson, Ronald	A.M.II	Pil	RAF	08.09.18	34
Wylie, M.M.,MiD, Alan	2/Lt	Obs	RFC	20.11.17	22
Young, George	2/Lt	Pil	RFC	20.11.17	?

INTER-WAR PERIOD

Shennan, Peter	P/O	Pil	RAF	05.04.39	23

WORLD WAR TWO

Adams, Ernest	Sgt	F/E	RAFVR	08.05.44	19
Adkins, Henry	Sgt	B/A	RAFVR	08.12.42	31
Aiken, Ronald	Sgt	?	RAFVR	21.11.42	24
Aitken, Robertson	W/O	A/G	RAFVR	08.06.44	22
Allen, Reginald	P/O	Pil	RAFVR	12.06.43	?
Amies, Alan	F/L	Pil	RAFVR	12.05.44	21
Amos, Orison	F/S	Nav	RNZAF	03.03.43	30
Anderton, James	Sgt	Pil	RAFVR	23.09.43	28
Andrews, Ronald	Sgt	?	RAFVR	13.01.44	21
Appleby, Eric	Sgt	F/E	RAFVR	29.01.44	20
Apps, Percy	F/S	WAG	RAF	19.09.42	21
Archibald, William	P/O	Nav	RAFVR	14.02.43	?
Armer, George	Sgt	Nav	RAFVR	02.03.43	30
Armstrong, James	F/S	WOp	RAAF	08.06.44	22

Name	Rank	Role	Service	Date	Age
Arnott, Kenneth	P/O	Obs	RAFVR	16.07.42	22
Arnott, Patrick	Sgt	B/A	RAFVR	25.05.43	19
Ashcroft, Francis	F/S	Nav	RAFVR	16.07.44	?
Ashdown, Richard	F/S	Pil	RAF	27.02.43	21
Ashill, Denis	S/L	Chap	RAFVR	29.12.42	30
Atkinson, Sam	Sgt	F/E	RAFVR	16.09.42	19
Attenborrow, Eric	Sgt	WOp	RAFVR	16.07.44	?
Austin, Ronald	LAC	A/G	RAF	25.05.40	21
Avent, Douglas	Sgt	Obs	RAF	12.05.40	22
Bagg, Arthur	F/S	Obs	RCAF	09.03.43	22
Baillie, John	Sgt	F/E	RAF	09.05.42	26
Baker, Herbert	Sgt	Nav	RAFVR	12.05.44	?
Ball, John	F/L	Pil	RAF	08.08.44	24
Bamber, Hugh	F/O	Pil	RAF	07.07.40	19
Bance, Eric	Sgt	F/E	RAFVR	08.09.42	30
Bannister, Harry	F/S	Pil	RAFVR	11.09.42	31
Bayner, James	F/S	B/A	RAFVR	05.05.43	33
Barber, George	Sgt	Pil	RAFVR	04.07.43	21
Barkshire, Albert	Sgt	A/G	RAFVR	08.08.44	37
Barnes, Robert	F/S	WAG	RAFVR	23.04.44	22
Barrett, Denis	Sgt	Pil	RAFVR	16.07.42	21
Barrett, William	F/S	WAG	RAF	13.08.41	26
Barrie, Charles	Sgt	A/G	RAFVR	29.10.42	22
Barton-Smith, B.Sc, Hugh	F/S	Pil	RAFVR	28.08.42	26
Bassett, Thomas	F/O	Pil	RAFVR	12.05.40	22
Batchellor, LL.B, Cecil	Sgt	?	RAF	29.06.41	23
Bate, Howard	Sgt	F/E	RAFVR	16.11.44	25
Batham, Raymond	Sgt	A/G	RAFVR	29.01.44	19
Baxter, William	Sgt	Obs	RAF	18.05.40	22
Beard, H. N.	Sgt	?	RAF	04.08.40	?
Beare, Royston	Sgt	Pil	RAFVR	15.06.45	22
Beazley, Harry	P/O	B/A	RCAF	08.06.44	26
Bebbington, Richard	Sgt	Pil	RAFVR	03.07.43	24
Beck, George	Sgt	B/A	RAFVR	31.07.43	20
Bee, Charles	F/S	A/G	RCAF	19.09.42	22
Bell, John	Sgt	A/G	RAFVR	11.06.44	20
Bell, DFC, Walter	F/L	Pil	RAFVR	21.07.44	25
Belton, Robert	Sgt	A/G	RAFVR	04.09.43	24
Benjamin, Thomas	Sgt	A/G	RAFVR	08.05.44	21
Benny, John	Sgt	Pil	RAFVR	28.09.43	?
Benson, Charles	P/O	Pil	RNZAF	20.02.44	29
Bente, John	Sgt	F/E	RAF	18.12.41	27
Bentley, Sidney	Sgt	F/E	RAF	13.11.41	?
Bergin, Joseph	F/S	Nav	RNZAF	21.01.44	33
Bessette, Bertie	Sgt	A/G	RCAF	16.04.43	?
Beswick, Arthur	Sgt	A/G	RAF	29.10.42	23
Beverton, Harry	F/S	A/G	RAFVR	12.09.44	?
Bidgood, Harry	Sgt	WAG	RAFVR	09.03.42	19
Billington, Edward	F/O	B/A	RAFVR	22.06.43	21
Birchall, Albert	P/O	A/G	RCAF	19.09.42	?
Bird, Peter	P/O	Pil	RAFVR	11.05.41	20
Blackburn, John	Sgt	Nav	RAFVR	17.04.43	?
Blackhall, William	LAC	Gdcw	RAF	??.??.43	?
Blackmore, Lester	Sgt	WAG	RAFVR	13.08.42	23
Blanchard, Rex	Sgt	A/G	RAFVR	04.11.43	19
Bland, Arthur	Sgt	A/G	RAFVR	27.03.43	20
Blignaut, Jochemus	Sgt	Pil	RAFVR	08.12.42	?
Bliss, John	Sgt	WOp	RAFVR	19.11.43	20
Bloomer, Peter	Sgt	Obs	RAF	25.05.40	22
Bond, William	Sgt	Pil	RAFVR	30.05.42	26
Booth, Alfred	F/S	B/A	RAFVR	16.11.44	22
Booth, Neville	F/L	Pil	RAFVR	18.05.42	25
Booth, Reginald	Sgt	WAG	RAFVR	02.11.42	?
Borrett, Arnold	Sgt	A/G	RAAF	19.02.43	25
Bottomley, Handley	Sgt	A/G	RAF	24.08.43	?
Bovett, George	F/O	WOp	RAFVR	08.08.44	?
Bowden, Peter	F/O	B/A	RAF	18.11.43	20
Bowen, David	Sgt	F/E	RAFVR	27.09.43	21
Bowen, John	Sgt	F/E	RAFVR	21.04.44	20
Bowen, William	Sgt	F/E	RAFVR	02.03.43	27
Bowers, Henry	Sgt	WOp	RAFVR	15.08.40	23
Bowers, Robert	Sgt	A/G	RAFVR	18.11.43	19
Bowyers, Clifford	S/L	Pil	RAFVR	14.05.43	21
Bradford, Jack	P/O	Pil	RAFVR	31.08.43	23
Bragg, Wilfred	Sgt	F/E	RAF	08.04.43	23
Brandt, Augustus	Sgt	Obs	RAF	07.04.42	?
Brennan, James	Sgt	A/G	RAFVR	21.07.44	23
Brennan, Patrick	F/L	A/G	RAFVR	19.02.43	?
Briggs, James	Sgt	A/G	RAFVR	24.03.44	21
Brock, Derek	Sgt	F/E	RAFVR	31.08.43	20
Brockett, William	Sgt	F/E	RAFVR	27.08.44	30
Brodie, William	Sgt	Pil	RAF	09.05.42	22
Bromley, Thomas	Sgt	A/G	RAF	17.04.43	22
Brook, Philip	F/S	A/G	RAAF	01.09.43	20
Brookes, B.A. Joseph	Sgt	Obs	RAFVR	06.11.42	27
Broomfield, Thomas	Sgt	A/G	RAFVR	21.07.44	21
Brophey, Burton	Sgt	A/G	RCAF	26.03.44	19
Brown, Dennis	Sgt	Nav	RAFVR	31.07.43	21
Brown, Douglas	Sgt	?	RAFVR	29.06.41	21
Brown, Herbert	Sgt	F/E	RAF	05.05.43	?
Brown, James	P/O	?	RAFVR	16.09.42	22
Brown, Maurice	F/S	A/G	RAFVR	21.04.44	20
Brown, Roy	W/O	Pil	RAFVR	16.01.45	25
Brown, Thomas	Sgt	F/E	RAF	19.09.42	24
Broyd, Stanley	Sgt	F/E	RAF	08.08.41	33
Buchanan, James	Sgt	F/E	RAF	02.10.42	21
Bunce, Gordon	F/O	Pil	RAF	18.12.41	?
Burcham, Alfred	P/O	Nav	RAFVR	29.01.44	26
Burgess, James	Sgt	A/G	RNZAF	06.11.42	31
Burke, Wil	S/L	Pil	RAF	08.06.40	31
Burkill, Clifford	F/S	WOp	RAFVR	09.03.43	34
Burrell, George	Sgt	WAG	RAFVR	20.06.42	?
Burton, Frederick	Sgt	A/G	RAF	11.04.43	20
Burtt, Marcus	Sgt	WAG	RAFVR	27.08.42	?
Busby, Peter	Sgt	F/E	RAFVR	12.08.42	18
Bushell, DFM. Jack	P/O	A/G	RAFVR	19.07.41	33
Butcher, Sidney	Sgt	A/G	RAFVR	06.11.42	20
Butler, Robert	P/O	Pil	RAFVR	21.01.44	23
Butterworth, John	Sgt	WAG	RAF	18.05.42	22
Byrne, Phillip	Sgt	F/E	RAF	26.07.42	28
Cairns, John	Sgt	A/G	RAFVR	12.05.44	20
Calder, James	P/O	Pil	RNZAF	18.11.43	25
Campbell, Daniel	P/O	Obs	RNZAF	11.05.41	25
Campbell, Donald	F/S	A/G	RCAF	26.06.43	?
Campbell, DFC. James	F/O	Nav	RAFVR	02.11.44	23
Campbell, Robert	F/O	Pil	RAAF	23.07.41	27
Canday, Charles	Sgt	A/G	RAFVR	07.07.44	37
Christopher	Sgt	A/G	RAFVR	15.03.43	24
Care, Donald	Sgt	B/A	RAFVR	02.03.43	21
Carlyle, William	Sgt	?	RAF	10.03.45	24

Name	Rank	Role	Service	Date	Age
Cairns, John	Sgt	A/G	RAFVR	12.05.44	20
Calder, James	P/O	Pil	RNZAF	18.11.43	25
Campbell, Donald	F/S	A/G	RCAF	26.06.43	?
Campbell, Daniel	P/O	Obs	RNZAF	11.05.41	25
Campbell, DFC. James	F/O	Nav	RAFVR	02.11.44	23
Campbell, Robert	F/O	Pil	RAAF	23.07.41	27
Canday, Charles	Sgt	A/G	RAFVR	06.07.44	37
Cantwell, Christopher	Sgt	A/G	RAFVR	15.06.44	24
Care, Donald	Sgt	B/A	RAFVR	02.03.43	21
Carlyle, William	Sgt	?	RAF	10.03.45	25
Carmichael, Ernest	Sgt	A/G	RAFVR	16.07.44	19
Carrott, James	F/S	Nav	RAFVR	26.02.44	21
Carruthers, Carl	F/S	WOp	RCAF	09.03.42	?
Carson, Lawrence	F/L	B/A	RCAF	19.02.43	29
Carter, John	W/O	A/G	RCAF	07.10.43	?
Caselton, Victor	Sgt	F/E	RAFVR	04.09.43	20
Cash, Noel	Sgt	?	RAFVR	19.05.42	?
Cato, Hugh	Sgt	Pil	RAAF	05.11.44	21
Cavanagh, William	LAC	WOp	RAF	12.05.40	22
Caveney, Thomas	Sgt	A/G	RAFVR	03.03.43	20
Chalker, Charles	Sgt	F/E	RAF	04.07.43	20
Chalmers, Frederick	F/O	B/A	RCAF	20.02.44	?
Chambers, William	Sgt	A/G	RAFVR	30.05.42	21
Champ, Wilfred	Sgt	WOp	RAFVR	12.06.43	21
Chancellor, B.A. Roland	P/O	?	RAFVR	18.12.41	24
Chandler, William	F/S	B/A	RAAF	25.05.44	20
Channer, Richard	Sgt	B/A	RAFVR	27.04.41	24
Chapman, Kenneth	F/S	Pil	RNZAF	06.11.42	24
Chapman, Maurice	F/O	Pil	RAFVR	26.06.43	24
Mentioned in Despatches x 2.					
Chapman, Paul	F/L	Pil	RAF	18.05.40	32
Charbonneau, Ivan	F/S	Pil	RCAF	09.05.42	20
Chatteris, William	F/O	Nav	RAFVR	10.03.45	25
Chave, Owen	F/L	Pil	RAFVR	14.02.43	30
Childs, Jack	F/S	Nav	RAFVR	18.11.43	33
Childs, John	F/L	Pil	RAFVR	28.07.43	21
Christie, Charles	F/S	?	RCAF	13.01.44	24
Church, Cecil	Sgt	Nav	RAFVR	22.03.45	?
Clarke, Ronald	F/O	Pil	RAF	11.06.40	20
Clarke, George	P/O	Pil	RAFVR	27.01.44	?
Clayton, Henry	Sgt	A/G	RAFVR	16.11.44	24
Cleaver, Donald	Cpl	Gdcw	RAFVR	29.10.42	36
Clegg, Herbert	Sgt	Obs	RAFVR	27.08.42	28
Close, James	Sgt	Nav	RAFVR	04.07.43	21
Cobby, Arthur	Sgt	F/E	RAF	31.07.43	21
Cobell, Walter	Sgt	A/G	RAF	13.04.42	?
Cockburn, William	F/S	WAG		23.07.41	?
Colbourn, Cecil	Sgt	Obs		18.05.40	26
Colbourne, Charles	P/O	Pil	RAF	13.10.41	?
Cole, Edward	F/O	B/A	RCAF	23.09.43	?
Cole, James	LAC	Gdcw	RAFVR	30.08.43	28
Coles, Raymond	Sgt	WAG	RAFVR	27.09.43	21
Collins, Norman	Sgt	Nav	RAFVR	24.08.43	?
Condron, John	Sgt	WOp	RAFVR	26.06.43	23
Conn, William	Sgt	WAG	RAFVR	14.08.41	27
Cook, James	Sgt	F/E	RAF	14.02.43	34
Cooley, Peter	Sgt	A/G	RAFVR	22.03.45	20
Coop, Samuel	Sgt	Obs	RAFVR	03.09.442	?
Cooper, Ernest	LAC	WAG	RAF	12.05.40	20
Cooper, Francis	F/S	Pil	RAFVR	07.04.42	?
Cope, George	Sgt	WOp	RAFVR	22.03.45	21
Corbett, John	Sgt	WAG	RAFVR	08.08.41	24
Cordell, Ronald	Sgt	Nav	RAFVR	27.09.43	20
Cornell, Eric	P/O	Pil	RAFVR	24.08.43	21
Cowen, Leon	Sgt	F/E	RAF	16.09.42	?
Cowie, James	F/L	Pil	RCAF	09.02.45	22
Cowlick, Andrew	W/O	Pil	RNZAF	03.06.42	26
Crapp, Francis	Sgt	Nav	RAF	08.12.42	20
Crawford, Bernard	F/O	Pil	RNZAF	19.02.43	?
Creed, William	Sgt	A/G	RAFVR	3.09.42	25
Crighton, Frederick	Sgt	WAG	RAFVR	03.06.42	21
Criswick, Maurice	Sgt	A/G	RAFVR	04.07.43	31
Crone, John	F/O	?	RAAF	17.01.45	21
Cronk, Gavin	W/O	B/A	RCAF	12.05.44	29
Cross, William	F/S	Pil	RCAF	09.03.42	?
Crowe, Robert	Sgt	A/G	RAFVR	25.07.44	20
Crozier, Alexander	F/S	WAG	RNZAF	23.06.43	25
Cully, Wilbert	F/S	A/G	RAFVR	28.04.44	22
Curry, David	Sgt	F/E	RAFVR	18.11.43	25
Curtis, Eric	P/O	Pil	RAFVR	22.06.43	?
Dale, Herbert	W/C	Pil	RAF	11.05.41	33
Dalton, Frank	Sgt	WOp	RAFVR	04.07.43	21
Davidson, Frank	F/S	WAG	RCAF	19.09.42	28
Davie, James	F/S	A/G	RCAF	04.07.43	28
Davies, Kenneth	Sgt	F/E	RAF	18.07.41	22
Davis, Frederick	Sgt	A/G	RAFVR	29.06.43	22
Davis, Joseph	F/S	Pil	RAF	26.02.44	23
Davis, Norman	Sgt	WOp	RAFVR	11.06.44	23
Davis, Roy	Sgt	F/E	RAF	22.06.43	20
Davis, Vincent	F/O	Pil	RAAF	04.12.44	22
Dawson, Leon	Sgt	A/G	RCAF	12.06.43	?
Dawson, Thomas	Sgt	A/G	RAFVR	28.07.43	27
Dawson-Jones, Francis	F/O	Pil	RAF	18.05.40	23
Day, Peter	F/O	Nav	RAFVR	09.02.45	21
Dean, William	Sgt	F/E	RAFVR	26.03.44	26
Dee, William	Sgt	F/E	RAFVR	22.03.45	20
Dench, Francis	P/O	Pil	RAF	13.08.40	22
Devereux, Sidney	Sgt	WAG	RAFVR	29.06.43	22
Devitt, Robert	Sgt	WAG	RAFVR	23.09.43	20
Devlin, Richard	Sgt	F/E	RAF	17.01.45	31
Dickinson, George	F/S	B/A	RAFVR	09.02.45	21
Dickinson, William	Sgt	A/G	RAFVR	04.07.43	33
Dickson, Harry	Sgt	WAG	RCAF	13.08.41	20
Dillicar, John	P/O	Pil	RNZAF	31.07.43	26
Dillingham, Horace	P/O	Nav	RAF	04.07.43	22
Disley, Ronald	P/O	Pil	RAFVR	09.03.42	37
Dobson, Alan	Sgt	?	RAFVR	09.02.45	23
Dobson, Herbert	Sgt	A/G	RAFVR	07.10.43	21
Dobson, William	F/L	Pil	RAAF	11.06.44	30
Dolan, Stephen	Sgt	F/E	RAFVR	05.01.45	?
Dolby, Harold	F/S	WAG	RAFVR	21.04.44	22
Dombrain, Peter	P/O	Pil	RAAF	01.06.44	21
Donaldson, George	Sgt	A/G	RAFVR	27.08.44	21
Doughty, Harry	Sgt	A/G	RAFVR	30.05.42	28
Douglas, Stanley	Sgt	Nav	RAuxAF	28.07.43	29
Douglass, Albert	Sgt	Pil	RAFVR	19.05.42	31
Douglass, Peter	F/O	Pil	RAF	12.05.40	?
Dove, Cyril	P/O	Pil	RAFVR	15.02.41	20
Doyle, France	F/O	Pil	RAFVR	30.05.42	?
Drew, B.A. Robert	F/S	B/A	RAFVR	06.11.42	31

Duncanson, William	Sgt	A/G	RAF	28.08.43	?
Dunk, Thomas	F/O	A/G	RAFVR	08.06.44	32
Dyer, Bruce	F/S	WOp	RAAF	27.08.44	20
Dyer, DFC,MiD. Hedley	P/O	Pil	RAF	16.09.41	26
Earley, Bernard	F/L	Pil	RAFVR	02.11.44	24
B.Sc.(London), DFM, MiD.					
East, David	Sgt	A/G	RAFVR	25.04.42	25
Easthope, Frederick	W/O	Nav	RAFVR	08.06.44	22
Eastman, John	Sgt	Nav	RAFVR	28.04.44	20
Eccles, Joseph	Sgt	Nav	RAFVR	04.07.43	?
Edmund, H. V.	Sgt	?	RAF	19.05.42	?
Edwards, Thomas	A.M.II	Gdcw		19.05.42	?
Edwards, William	P/O	Obs	RAF	23.05.40	?
Eldridge, Gordon	Sgt	A/G	RAFVR	19.09.42	21
Ellis, Alfred	Sgt	A/G	RAFVR	19.02.43	?
Elton, Percy	Sgt	F/E	RAFVR	03.03.43	21
Emans, Albert	P/O	?	RAF	??.??.43	?
Emberson, Thomas	P/O	Pil	RAFVR	04.05.43	?
Entwhisle, Herbert	W/O	WOp	RAFVR	15.02.44	
Ethelston, John	P/O	B/A	RAFVR	19.02.43	?
Evans, David	F/L	F/E	RAFVR	21.07.44	?
Evans, Ivor	Sgt	WAG	RAFVR	13.10.41	?
Evans, Kenneth	Sgt	Obs	RAFVR	09.05.42	?
Evans, Robert	Sgt	WAG	RAFVR	16.09.42	22
Evans, Sydney	Sgt	F/E	RAFVR	11.08.43	33
Eve, Percy	Sgt	WAG	RAF	19.07.41	21
Ewen, Thomas	Sgt	Pil	RAFVR	07.10.43	29
Exelby, MiD. Raymond	F/L	WAG	RAF	18.12.21	21
Fagg, Ernest	LAC	WOp	RAF	18.05.40	23
Fagg, John	Sgt	F/E	RAFVR	08.06.44	23
Faint, Reginald	Sgt	A/G	RAFVR	27.08.44	27
Farrelly, James	Sgt	F/E	RAF	09.03.43	20
Fenley, John	F/O	Nav	RAFVR	20.02.44	22
Ferguson, David	Sgt	F/E	RAF	18.12.41	22
Ferguson, Isaiah	P/O	Pil	RAFVR	16.09.42	?
Fiddes, Henry	P/O	B/A	RAFVR	17.04.43	30
Fisher, Charles	S/L	Pil	RAF	29.10.42	30
Fitzgerald, John	F/S	A/G	RCAF	26.07.42	19
Flaherty, Michael	Sgt	B/A	RAFVR	24.05.43	30
Forrest, William	P/O	A/G	RCAF	21.02.44	29
Forster, Kenneth	Sgt	A/G	RAFVR	16.09.42	24
Fortune, Herbert	Sgt	WAG	RAFVR	17.04.43	25
Fothersgill, Wilfred	F/S	Pil	RAF	21.04.44	23
Fowler, Ronald	Sgt	WAG	RAFVR	19.02.43	30
Fowler, Thomas	Sgt	A/G	RAFVR	12.06.43	21
Frame, William	Sgt	B/A	RAFVR	24.08.43	29
France, Ralph	P/O	Nav	RAFVR	31.08.43	20
Francis, Edward	Sgt	Nav	RAFVR	07.10.143	27
Frankish, Claude	P/O	Pil	RAFVR	12.05.40	25
Franklin, George	W/O	WOp	RNZAF	21.09.44	24
Franklin, James	Sgt	A/G	RAFVR	16.11.44	?
Frazer, John	P/O	Obs	RAFVR	27.08.42	22
Frearson, Frederic	F/O	WAG	RAFVR	02.11.44	22
Freedman, Harry	F/O	A/G	RAFVR	17.01.45	20
Friend, George	Sgt	F/E	RAF	23.07.41	20
Frost, Arthur	Sgt	A/G	RAFVR	18.07.41	28
Fuller, W. B.	F/O	A/G	RAFVR	17.01.45	?
Funnell, John	F/L	Pil	RNZAF	21.01.44	33
Furness, B.A. John	P/O	Nav	RAAF	04.12.43	22
Gard, Gilbert	Sgt	A/G	RAFVR	18.11.43	29
Garfit, Joseph	Sgt	F/E	RAFVR	29.06.43	32
Garrett, Leslie	Sgt	A/G	RAFVR	12.12.44	?
Garvey, Peter	Sgt	Pil	RAF	15.08.40	22
Gearing, Leonard	Sgt	F/E	RAFVR	01.06.44	?
Geraghty, William	Sgt	B/A	RAFVR	26.02.44	22
Gericke, DFM. Phillip	F/S	A/G	RAAF	29.01.44	20
Gerrard, R.	W/O	B/A	RAF	23.04.44	?
Gibbons, Anthony	F/S	A/G	RCAF	31.07.43	?
Gibbs, Sidney	Sgt	A/G	RAFVR	28.09.43	?
Gibson, Raymond	F/O	Nav	RAFVR	28.08.44	?
Gilchrist, Campbell	F/S	A/G	RCAF	06.07.44	22
Gill, Albert	Sgt	WAG	RAFVR	08.06.44	22
Gilson, Edmond	W/O	Obs	RCAF	28.08.42	?
Gladwell, Willis	F/O	Nav	RCAF	06.07.44	?
Gladwin, Lewis	Sgt	Pil	RCAF	14.02.43	21
Glanfield, John	F/S	A/G	RAAF	13.06.44	30
Glenday, Lindsay	Sgt	Nav	RNZAF	16.09.43	33
Goddard, Kenneth	F/O	WAG	RAFVR	25.05.44	24
Gold, Edmund	F/S	WAG	RAFVR	07.10.43	22
Golder, Reginald	F/S	A/G	RAAF	07.04.42	?
Golding, James	Sgt	A/G	RAFVR	RAFVR	24
Golub, Michael	F/L	Pil	RCAF	06.07.44	?
Gomersall, Ernest	Sgt	A/G	RAFVR	28.07.43	19
Goodchild, Cyril	Sgt	Nav	RAFVR	19.02.43	19
Goodridge, Noel	P/O	WOp	RAAF	15.06.44	20
Goodwin, George	Sgt	A/G	RCAF	13.10..41	?
Gorbert, Edward	F/S	WOp	RAFVR	24.08.43	21
Gordon, James	P/O	A/G	RAF	25.05.40	30
Gough, Sydney	F/O	A/G	RAF	13.06.44	32
Gould, Edgar	Sgt	WAG	RAFVR	11.04.42	?
Gould, John	Sgt	WAG	RAFVR	16.04.43	21
Goulding, Clarence	F/S	A/G	RAF	18.12.41	29
Graham, Leslie	Sgt	F/E	RAFVR	03.10.43	22
Grant, Robert	Sgt	A/G	RAFVR	21.06.44	?
Grant, William	Sgt	A/G	RCAF	29.06.41	29
Gray, Angus	Sgt	Obs	RAF	13.08.40	20
Gray, Roderick	F/S	Pil	RNZAF	31.08.43	22
Green, William	Sgt	A/G	RAF	16.09.43	?
Greenbeck, Stanley	Sgt	WAG	RAFVR	19.09.42	25
Greenwood, John	Sgt	A/G	RAFVR	17.04.43	28
Gregory, John	Sgt	A/G	RAFVR	09.02.45	22
Grove, William	F/L	Pil	RAFVR	24.03.44	24
Grundy, Robert	F/S	Pil	RAF	18.08.43	?
Guild, Edward	P/O	Pil	RAFVR	16.09.41	?
Gunn, Kelsall	Sgt	F/E	RAF	28.08.43	?
Gunning, Frank	Sgt	Obs	RAF	12.06.40	19
Gurr, Anthony	P/O	Pil	RAFVR	08.04.43	20
Gustafson, Roy	F/S	B/A	RCAF	07.10.43	21
Guy, John	Sgt	F/E	RAFVR	25.07.44	21
Gwynne, Lloyd	F/O	B/A	RNZAF	23.06.43	30
Hackett, Arthur	Sgt	?	RAFVR	03.01.44	?
Haigh, Ronald	Sgt	F/E	RAFVR	01.09.43	28
Hains, Gordon	P/O	Pil	RAFVR	20.06.42	25
Hales, Ronald	Sgt	A/G	RAFVR	08.06.44	20
Hall, Hubert	Sgt	Pil	RAF	12.05.40	27
Hall, James	Sgt	A/G	RAFVR	08.04.43	?
Hall, John	Sgt	Pil	RAFVR	04.07.43	21
Hall, DFC, MiD. John	S/L	Pil	RAF	18.05.42	24
Hall, Joseph	Sgt	A/G	RAFVR	09.02.45	32
Hammond, John	Sgt	Obs	RAF	12.08.42	?

Name	Rank	Role	Service	Date	Age
Hanberger, Peter	Sgt	A/G	RAFVR	04.05.43	20
Hance, John	F/S	WOp	RAAF	05.11.44	23
Hannah, Wilfred	F/S	Pil	RNZAF	06.11.42	24
Hansford, Albert	F/S	A/G	RCAF	02.10.42	34
Harbidge, William	Sgt	A/G	RAFVR	26.02.44	19
Hare, Thomas	W/O	Pil	RCAF	07.04.42	22
Harriman, Douglas	P/O	Pil	RAF	25.05.40	20
Harris, Victory	F/L	Pil	RCAF	27.02.43	24
Harris, William	F/L	Pil	RNZAF	16.02.44	23
Harrison, Ernest	F/O	B/A	RAFVR	25.07.44	31
Harrison, Jack	Sgt	WOp	RAF	11.04.43	?
Haswell, George	Sgt	F/E	RAF	08.06.44	21
Hathaway, Maurice	F/S	A/G	RAFVR	09.02.45	22
Hawkins, John	F/O	Pil	RNZAF	23.06.43	?
Hawthorn, Kenneth	Sgt	F/E	RAF	28.08.42	22
Haycock, DFC. Dennis	F/L	Pil	RAFVR	17.04.43	22
Haydock, Douglas	Sgt	F/E	RAF	26.02.44	?
Hayes, Eric	Sgt	WAG	RAFVR	03.06.42	24
Hayes, George	Sgt	Obs	RAFVR	19.09.42	?
Hayles, DFM. Raymond	F/S	WOp	RAFVR	21.02.44	21
Hayward, Frederick	F/S	WOp	RAFVR	08.06.44	28
Head, Peter	Sgt	WAG	RAFVR	29.10.42	21
Headley, George	F/S	?	RAF	??.??.43	?
Hearn, Douglas	F/S	WOp	RAFVR	16.11.44	22
Heathcote, George	P/O	Nav	RNZAF	21.04.44	35
Heathcote, Mid. Gilbert	F/L	Pil	RAF	18.12.41	30
Heathcote, Roy	Sgt	WAG	RAFVR	19.09.42	21
Helyar, Roy	Sgt	A/G	RAFVR	13.08.42	22
Henderson, Ian	F/S	A/G	RAF	19.09.42	?
Henderson, ?	F/O	Pil	RAF	24.05.40	?
Henderson, William	Sgt	A/G	RAFVR	05.11.44	20
Hendry, Douglas	Sgt	B/A	RAFVR	27.09.43	?
Henson, George	Sgt	Obs	RCAF	13.08.41	22
Heurtley, John	F/S	WAG	RNZAF	20.06.42	23
Heward, William	Sgt	WAG	RAFVR	28.08.44	?
Higgins, Clarence	P/O	Obs	RAF	11.09.42	28
Higgins, Robert	F/S	Pil	RNZAF	14.08.42	23
Higgison, Alfred	Sgt	Pil	RNZAF	18.07.41	24
Highland, William	F/S	A/G	RAAF	03.10.43	21
Hill, Eric	F/S	Pil	RAAF	04.09.43	25
Hill, Hugh	P/O	A/G	RCAF	16.12.42	?
Hills, Michael	F/O	Nav	RAFVR	11.06.44	?
Hipps, Arnold	P/O	Obs	RAFVR	19.07.41	25
Hoggard, Robert	F/O	Pil	RAAF	02.11.44	21
Hohnen, M.	P/O	Pil	RAF	04.08.40	?
Holborow, Richard	F/S	WOp	RAFVR	21.01.44	20
Holden, Alfred	Sgt	Pil	RAFVR	28.07.43	19
Holdsworth, John	Sgt	Obs	RAFVR	07.07.40	?
Holland, Frank	F/S	B/A	RAFVR	24.03.44	29
Hollingshead, Ronald	Sgt	Obs	RAFVR	14.09.40	20
Hollinrake, K.	Sgt	F/E	RAF	21.06.44	?
Holmes, Arthur Noel	Sgt	Obs	RAF	24.05.40	20
Holmes, Russell	P/O	WAG	RCAF	16.12.42	?
Holt, B.	Sgt	B/A	RAFVR	31.10.44	?
Honeybill, Ernest	Sgt	F/E	RAFVR	18.08.43	28
Hood, Louis	F/S	A/G	RAFVR	08.06.44	27
Hood, Walter	Sgt	Pil	RNZAF	29.10.142	23
Hooker, Allan	F/S	A/G	RAAF	01.09.43	?
Hopkins, Reginald	Sgt	Obs	RAF	18.05.40	29
Hopkins, William	Sgt	F/E	RAF	27.08.42	?
Hopson, David	F/O	Pil	RAFVR	19.02.43	20
Horton, George	F/O	Nav	RAFVR	08.05.44	29
Horton, Leslie	Sgt	A/G	RAF	26.05..21	21
Houghton, Eric	F/S	Pil	RNZAF	25.07.44	22
Hounsome, Harry	W/O	F/E	RAFVR	05.11.44	25
Houston, Walter	F/S	Pil	RCAF	13.01.44	24
Howitt, Ian	Sgt	B/A	RAFVR	13.09.44	?
Howland, Harold	F/S	Pil	RAFVR	02.03.43	26
Howson, Paul	F/O	B/A	RAFVR	19.022.43	?
Hudson, Colin	Sgt	Nav	RAFVR	18.08.43	?
Hughes, Philip	P/O	A/G	RAFVR	14.09.40	22
Hughes, Robert	Sgt	A/G	RAFVR	25.05.44	19
Humm, Harry	Sgt	WAG	RAFVR	14.04.42	22
Humphries, Douglas	Sgt	B/A	RAFVR	04.07.43	22
Hunt, James	LAC	Gdcw	RAFVR	29.12.42	31
Hunt, Leslie	F/L	Pil	RCAF	04.07.43	22
Hunter, Russell	F/O	WOp	RCAF	22.06.43	?
Hunter, Thomas	Sgt	A/G	RAF	13.09.44	?
Hunter, William	Sgt	F/E	RAFVR	02.11.44	21
Hurley, Max	P/O	Pil	RAAF	20.02.44	20
Hurworth, George	Sgt	A/G	RAFVR	15.02.41	?
Hussey, John	F/O	B/A	RAFVR	29.01.44	23
Hutton, George	Sgt	A/G	RAFVR	23.06.43	?
Hutton, Grantley	Sgt	F/E	RAF	16.12.42	?
Hyde, Donald	Sgt	A/G	RAFVR	16.04.43	22
Hynes, William	W/O	?	RCAF	13.01.44	23
Hyrons, Albert	Sgt	F/E	RAFVR	02.03.43	?
Ingle, Ray	Lt.	Pass	SAAF	31.07.43	25

(Lt Ingle was attached to No.XV Squadron from South Africa House).

Name	Rank	Role	Service	Date	Age
Jackson, Arthur	Sgt	Nav	RAFVR	24.03.44	22
Jackson, Bernard	F/O	Nav	RNZAF	31.07.43	23
Jackson, DFM., Francis	Sgt	F/E	RAF	29.12.42	22
Jager, William	F/S	B/A	RAAF	21.01.44	25
James, Clifford	Sgt	F/E	RAF	19.02.43	?
James, Leslie	Sgt	B/A	RAFVR	16.04.43	22
Jamieson, Gerald	F/O	B/A	RCAF	06.07.44	34
Jamieson, Lawrence	F/S	B/A	RNZAF	01.06.44	26
Jarvis, Alfred	Sgt	WAG	RNZAF	06.11.42	22
Jarvis, Arthur	F/L	Pil	RAFVR	21.06.44	29
Jeans, Donald	Sgt	WAG	RAFVR	13.08.42	25
Jeeves, Dennis	Sgt	Pil	RAFVR	19.07.41	?
Jeffrey, George	Sgt	Obs	RCAF	08.08.41	24
Jeffreys, Kenneth	Sgt	WOp	RAF	18.12.41	31
Jenkins, Thomas	Sgt	A/G	RAFVR	22.03.45	36
Jennings, William	Sgt	A/G	RAFVR	05.05.43	?
Johnson, John	Sgt	F/E	RAF	24.03.44	28
Johnson, William	F/S	A/G	RCAF	16.09.42	29
Johnson, William	Sgt	A/G	RAFVR	29..06.43	20
Johnson, William	P/O	Pil	RCAF	24.05.43	?
Johnston, John	Sgt	?	RAFVR	13.01.44	19
Jones, Ernest	F/S	Pil	RAF	12.05.44	24
Jones, Frank	P/O	Obs	RAF	30.07.40	32
Jones, Howell	Sgt	B/A	RAFVR	26.02.43	20
Jones, John	Sgt	A/G	RAFVR	02.03.43	21
Jones, Martin	Sgt	WOp	RAFVR	06.07.44	27
Jones, Maurice	Sgt	Obs	RAF	11.06.40	23
Jones, Philip	F/O	B/A	RAFVR	08.05.44	23
Jones, Thomas	P/O	Pil	RAFVR	08.05.44	21
Jones, Trevor	Sgt	F/E	RAFVR	16.09.41	22
Jordan, Albert	F/S	A/G	RAFVR	21.04.44	?

Jordan, Sidney	Sgt	B/A	RAFVR	27.09.43	?		Lucas, Eric	Sgt	A/G	RNZAF	11.05.41	26
Judd, George	F/O	Pil	RAFVR	31.07.43	?		Ludgate, John	P/O	A/G	RAFVR	28.08.42	21
Keeble, Kenneth	Sgt	WAG	RAFVR	19.02.43	20		Lutwyche, Percy	Sgt	Nav	RAFVR	08.04.43	?
Keen, Jack	P/O	Pil	RAFVR	29.06.43	27		Lyons, Henry	F/O	Nav	RCAF	24.04.43	32
Keen, Richard	F/S	Nav	RAFVR	13.09.44	29		MacAulay, Wilfred	Sgt	A/G	RAFVR	23.06.43	21
Kelley, Alfred	F/S	A/G	RCAF	08.12.42	?		MacDougal, Allan	W/O	Pil	RAAF	13.09.44	26
Kendall, Alfred	Sgt	A/G	RAFVR	16.09.42	32		MacKenzie, Walter	F/S	Obs	RCAF	16.09.41	25
Kennedy, Frederick	F/O	Obs	RCAF	30.05.42	29		MacKlin, Henry	Sgt	F/E	RAFVR	19.02.43	?
Kennedy, Frederick	Sgt	A/G	RAFVR	29.06.43	23		MacLennan, John	W/O	B/A	RCAF	16.07.44	?
Kent, James	Sgt	A/G	RAFVR	16.09.42	20		McAusland, Kenneth	F/S	Pil	RAFVR	13.08.42	27
Kieswetter, Emerson	P/O	Obs	RCAF	16.12.42	24		McCallum, Donald	Sgt	A/G	RAFVR	09.03.42	29
Kimber, John	Sgt	WAG	RAFVR	08.04.43	?		McCallum, Eric	F/S	A/G	RNZAF	24.03.44	20
King, Desmond	Sgt	A/G	RAFVR	28.04.44	20		McCallum, John	P/O	Obs	RAFVR	18.07.41	?
King, Edward	Sgt	A/G	RAFVR	19.09.42	19		McCaughey, Peter	F/O	B/A	RAFVR	11.04.43	22
Kingston, Harry	Sgt	A/G	RAFVR	13.04.42	24		McCauseland, William	F/S	Pil	RAF	12.08.42	?
Kirk, C. W.	Sgt	Nav	RAFVR	08.06.44	?		McCosh, Hargrave	P/O	Obs	RAFVR	06.04.41	?
Kite, Cuthbert	LAC	Gdcw	RAFVR	01.07.41	25		McDonnell, Patrick	LAC	WAG	RAF	12.05.40	23
Knipe, Humphrey	Sgt	WAG	RAFVR	09.05.42	20		McGoven, John	Sgt	A/G	RAFVR	20.06.42	30
Knox, Kenneth	Sgt	A/G	RAFVR	05.11.44	20		McGrane, James	Sgt	WOp	RAFVR	03.03.43	22
Lacy, John	Sgt	F/E	RAF	16.04.43	22		McIntosh, Alexander	Sgt	F/E	RAF	29.10.42	?
Lake, Aston	P/O	Nav	RAFVR	12.06.43	20		McIntosh, Robert	F/S	Nav	RAFVR	24.03.44	?
Lake, George	Sgt	Nav	RAF	17.01.45	31		McKay, Ronald	P/O	Pil	RNZAF	13.04.42	22
Lamb, Frederick	Sgt	A/G	RAFVR	13.10.41	?		McKay, William	F/O	Pil	RAAF	23.04.44	20
Lamb, James	Sgt	B/A	RAF	02.03.43	21		McKie, Kenneth	Sgt	F/E	RAFVR	13.09.44	33
Lambert, Frederick	Sgt	B/A	RAFVR	08.04.43	20		McKillop, Robert	Sgt	B/A	RNZAF	16.12.42	25
Lambie, William	F/O	Nav	RAFVR	05.05..43	28		McLaggan, Alexander	P/O	Obs	RAFVR	26.07.40	20
Land, James	P/O	Pil	RCAF	19.09.42	22		McLaren, James	Sgt	Obs	RAFVR	20.06.42	22
Lander, Robert	Sgt	Nav	RAFVR	04.09.43	22		McMaster, Gordon	F/S	A/G	RAAF	20.02.44	25
Law, Ian	F/O	Nav	RCAF	02.11.44	30		McMillan, Roderick	W/O	Nav	RCAF	13.06.44	?
Lawrence, Hector	S/L	Pil	RAF	18.05.40	26		McNee, John	Sgt	A/G	RAFVR	13.09.44	?
Lawrence, Victor	F/S	WAG	RAF	28.08..42	20		McNulty, peter	Sgt	A/G	RAFVR	04.05.43	21
Lawson, Ralph	Sgt	WOp	RAFVR	04.09.43	20		McQuillan, Francis	F/S	Pil	RAF	22.06.43	21
Lax, Frederick	Sgt	F/E	RAF	03.02.43	21		McRae, D. W.	P/O	B/A	RCAF	21.06.44	?
Leadley, Thomas	P/O	Nav	RAFVR	31.08.43	21		McSparrow, Ernest	Sgt	WAG	RAFVR	30.05.42	?
Leah, Edmund	F/O	Nav	RAFVR	08.08.44	21		McWalter, Thomas	Sgt	B/A	RAFVR	31.03.41	22
Lee, David	P/O	Nav	RAAF	05.11.44	25		Maddock, Frederick	F/S	A/G	RCAF	25.05.44	?
Lee, Samuel	F/O	B/A	RAFVR	12.12.44	32		Maginn, H. H.	Sgt	?	RAF	11.09.42	?
Lee, Stanley	Sgt	B/A	RAFVR	28.04.44	27		Malcolm, John	Sgt	F/E	RAFVR	09.02.45	19
Lemky, Ronald	W/O	B/A	RCAF	13.06.44	23		Malley, Donald	Sgt	A/G	RAFVR	28.08.43	?
Letbe, Thomas	Sgt	Nav	RAFVR	11.04.43	?		Maloney, Terence	Sgt	WOp	RAF	04.07.40	18
Levesque, Fernand	F/S	A/G	RCAF	22.08.42	?		Mansfield, Sidney	Sgt	A/G	RAFVR	11.09.42	20
Lewis, Aubrey	F/S	Pil	RNZAF	11.08.43	30		Margetts, Edward	Sgt	WAG	RAF	27.08.42	20
Lewis, Donald	P/O	Obs	RNZAF	23.07.41	23		Markovitch, Alfred	P/O	F/E	RAFVR	02.11.44	21
Lewis, T. D.	Sgt	?	RAF	19.05.42	?		Marsh, Charles	Sgt	F/E	RAFVR	21.02.44	27
Lewis, Victor	F/S	B/A	RAAF	27.03.44	28		Marsh, Roy	F/L	Pil	RAFVR	12.12.44	21
Lilley, Alfred	Sgt	A/G	RAFVR	08.06.44	19		Marsh, Thomas	P/O	Pil	RAFVR	26.03.44	20
Lilley, Geoffrey	P/O	B/A	RAFVR	02.11.44	23		Marshall, Sydney	F/O	Pil	RAAF	18.07.41	24
Ling, John	Sgt	Obs	RAFVR	29.10.42	22		Martin, James	Sgt	A/G	RAFVR	22.06.43	21
Llewelyn, J. D.	S/L	Pil	RAF	23.05.40	?		Martin, William	Sgt	A/G	RAFVR	02.03.43	?
Lockhart, Henry	Sgt	WAG	RAFVR	16.07.42	23		Masters, James	P/O	Pil	RAF	23.05.40	19
Logan, Kenneth	Sgt	B/A	RAFVR	27.08.44	28		Masters, Robert	F/S	?	RAAF	24.03.45	24
Long, Arthur	F/S	Nav	RAAF	01.06.44	21		Masur, Dennis	P/O	A/G	RCAF	27.08.42	21
Long, Clarence	F/O	Nav	RCAF	19.02.43	?		Mathews, Arthur	F/S	WOp	RAFVR	28.04.44	?
Long, Joseph	W/O	F/E	RAFVR	07.10.43	27		Matlock, Elmer	F/S	A/G	RCAF	08.03.43	21
Long, Sidney	Sgt	B/A	RAFVR	31.07.43	26		Matthews, Glyndwr	Sgt	B/A	RAFVR	29.06.43	19
Longworth, Herbert	Sgt	F/E	RAFVR	24.03.44	27		Matthews, Martin	Sgt	B/A	RAFVR	22.03.45	22
Loudon, Arthur	Sgt	F/E	RAFVR	11.04.43	22		Matthews, William	Sgt	A/G	RAAF	27.02.43	26
Lovell, Robert	Sgt	A/G	RAFVR	17.10.41	23		Matthews, William	Sgt	F/E	RAFVR	06.07.44	22
Lowe, James	P/O	B/A	RAAF	04.12.44	22		Maycock, Ronald	Sgt	Nav	RAFVR	18.05.42	21
Lowrie, Robert	Sgt	F/E	RAF	19.09.42	19		Mayor, Gordon	Sgt	WAG	RAFVR	18.07.41	22

Name	Rank	Role	Unit	Date	Age
Meijer, Adolph	P/O	WAG	RAFVR	03.03.43	36
Dutch Cross of Merit					
Meller, Harry	F/O	Nav	RAFVR	02.03.43	31
Melville, Robert	P/O	Pil	RAAF	16.07.42	25
Meredith, James	F/O	Pil	RNZAF	02.10.42	24
Meredith, Owen	P/O	WOp	RAAF	02.11.44	21
Merrie, John	Sgt	F/E	RAFVR	28.07.43	27
Metaxa, Anthony	Sgt	Pil	RAFVR	29.06.41	?
Middlemas, Neville	Sgt	Obs	RAF	12.05.40	26
Middleton, Kenneth	Sgt	A/G	RAFVR	31.08.43	19
Miles, J.	Sgt	A/G	RAFVR	12.08.42	?
Miller, John	Sgt	A/G	RAFVR	02.11.44	26
Millen, Frank	P/O	Pil	RCAF	16.12.42	21
Mills, Terence	Sgt	A/G	RAFVR	12.08.42	?
Milner, Joseph	W/O	Pil	RCAF	01.09.43	?
Minns, Douglas	F/O	B/A	RAFVR	21.08.44	22
Mitchell, Arthur	Sgt	Pil	RAFVR	23.07.41	23
Mitchell, Desmond	F/O	A/G	RAFVR	23.08.43	20
Mobbs, Richard	Sgt	A/G	RAFVR	13.06.44	20
Moffat, Robert	P/O	Obs	RAFVR	08.06.40	20
Moffat, William	F/O	Pil	RAFVR	03.03.43	21
Mohr-Bell, Harold	P/O	Pil	RAFVR	13.10.41	?
Monteith, John	P/O	Pil	RCAF	19.02.43	20
Moorcroft, Albert	Sgt	A/G	RAFVR	03.06.42	22
Moore, Donald	W/O	WOp	RAFVR	13.09.44	?
Moran, William	F/L	Pil	RAAF	27.08.44	20
Morellly, Max	W/O	Nav	RCAF	16.09.42	21
Moroni, Hubert	Sgt	A/G	RAFVR	20.02.44	21
Morris, Frederick	Sgt	F/E	RAFVR	30.05.42	23
Morris, Frederick	Sgt	F/E	RAFVR	24.08.43	19
Morris, George	W/O	A/G	RAFVR	02.11.44	23
Morris, Hugh	Sgt	WAG	RAFVR	26.05.43	20
Morrison, George	Sgt	A/G	RAFVR	08.08.44	19
Moss, George	Sgt	A/G	RAFVR	02.03.43	19
Moss, Leonard	Sgt	WAG	RNZAF	28.08.42	28
Mounteney, Archibald	Sgt	WAG	RAFVR	09.05.42	?
Mugridge, Herbert	Sgt	A/G	RAFVR	04.05.43	20
Muir, John	F/O	WAG	RAFVR	14.02.43	25
Mumford, Ronald	Sgt	Obs	RAF	16.09.42	?
Munn, Stanley	Sgt	A/G	RAF	04.07.43	35
Munt, Victor	Sgt	WAG	RAFVR	09.05.42	24
Murphy, Patrick	Sgt	A/G	RAF	30.07.40	22
Murray, L. B.	Lt.	Pass	RAF	26..02.42	?
Murray, Robert	P/O	WOp	RAAF	12.12.44	33
Myland, DFC. D. E.	P/O	Pil	RAF	04.09.40	?
Needham, Arthur	Sgt	WAG	RAFVR	16.09.41	21
Needham, Frank	F/O	Pil	RAFVR	08.08.41	29
Newell, Alfred	Sgt	WOp	RAFVR	26.03.44	?
Newlyn, Raymond	P/O	WOp	RAFVR	29.01.44	21
Newman, Jack	P/O	Nav	RAFVR	19.09.42	?
Newman, Raymond	P/O	B/A	RAFVR	26.05.43	25
Newport, Jack	F/S	Pil	RAFVR	23.06.43	20
Newton, Frederick	P/O	Pil	RAAF	22.03.45	24
Niall, Alexander	W/O	Pil	RNZAF	16.09.43	30
Nicholls, James	Sgt	A/G	RAFVR	14.02.43	21
Nicholls, Robinson	Sgt	F/E	RAF	16.07.42	?
Nicholson, George	F/L	Pil	RAF	09.03.42	?
Nicholson, Robert	Sgt	A/G	RAFVR	18.05.42	23
Nicklin, Arthur	F/S	B/A	RAFVR	05.03.45	22
Nixon, Frederick	F/S	B/A	RCAF	12.08.42	26
Nixon, George	Sgt	F/E	RAF	03.06.42	21
Nixon, Thomas	F/S	A/G	RAFVR	15.06.44	19
Noel, John	Capt	Pass	R.A.	13.04.42	35
Norris, Raymond	Sgt	WOp	RAFVR	01.06.44	21
Nuttall, Norman	Sgt	WOp	RAFRV	11.05.41	20
Nystrom, Stanley	F/S	A/G	RAFVR	01.06.44	21
Oakes, Fred	W/O	B/A	RCAF	21.07.44	20
Oakley, Albert	F/O	Pil	RAF	12.05.40	25
Oakley, Eric	Sgt	WAG	RAFVR	07.04.42	27
O'Hare, Leonard	P/O	Pil	RAFVR	16.09..42	?
Oliver, Richard	Sgt	A/G	RAFVR	08.12.42	29
O'Mara, Sidney	Sgt	A/G	RAFVR	29.06.41	?
O'Neill, William	F/S	Pil	RCAF	18.12.41	34
Orchard, Leslie	Sgt	F/E	RAF	19.07.41	21
Orchard, Thomas	Sgt	A/G	RAFVR	26.05.43	18
Ordish, DFC., Charles	F/O	Pil	RAFVR	31.12.42	23
O'Riordan, Dennis	Sgt	F/E	RAFVR	26.05.43	?
Orr, Thomas	Sgt	Obs	RAFVR	16.09.42	28
Osman, Patrick	Sgt	A/G	RAFVR	18.12.41	?
Oswin, Arthur	Sgt	A/G	RAFVR	28.08.42	?
Overend, Norman	F/O	Pil	RNZAF	12.09.44	21
Overend, William	Sgt	B/A	RAFVR	03.09.42	22
Oxenbridge, Edward	Sgt	Pil	RAFVR	19.02.43	21
Palmer, Edward	P/O	A/G	RCAF	03.09.42	?
Palmer, William	F/L	Pil	RAFVR	08.06.44	?
Parker, Geoffrey	F/S	A/G	RAFVR	21.06.44	20
Parkhouse, Evan	Sgt	A/G	RAFVR	05.01.45	?
Paterson, John	Sgt	A/G	RAFVR	24.09.44	19
Patterson, Eric	P/O	Pil	RCAF	28.08..42	?
Pattison, Leslie	P/O	WAG	RAFVR	05.05.43	28
Paul, Frank	Sgt	A/G	RAFVR	08.06.44	23
Pavely, DFM. Robert	P/O	Nav	RAF	26.06.43	22
Pawlyk, A.	F/S	A/G	RCAF	30.06.44	?
Payne, Gerald	Sgt	A/G	RAFVR	28.08.42	19
Payne, N. F.	F/S	?	RAF	19.05.42	?
Peach, Dermott	Sgt	F/E	RAFVR	25.05.44	20
Pearsall, P.	Sgt	F/E	RAFVR	13.06.44	?
Pelham, Maurice	Sgt	F/E	RAFVR	13.06.44	20
Penmen, David	Sgt	WAG	RAFVR	18.12.41	21
Perline, David	Sgt	WAG	RAF	12.06.40	21
Perring, Clive	P/O	Nav	RAFVR	16.04.43	22
Perrin, Edward	Sgt	Obs	RAF	12.05.40	23
Perry, Harry	F/S	B/A	RCAF	04.09.43	24
Peters, John	F/S	A/G	RAF	18.12.41	21
Peters, Thomas	Sgt	WAG	RAFVR	16.09.43	21
Petrie, Peter	Sgt	WAG	RAF	04.09.40	?
Peuleve, D. H.	Sgt	WAG	RAF	12.06.40	21
Phillips, John	F/S	WAG	RCAF	09.03.42	?
Phillips, Sidney	Sgt	A/G	RAFVR	27.02.43	21
Phillips, Simon	F/O	Pil	RNZAF	13.07.44	22
Piche, Kenneth	P/O	Pil	RCAF	16.04.43	21
Pittard, Ronald	Sgt	F/E	RAFVR	26.05.43	22
Pittendrigh, Wilfred	Sgt	WAG	RAFVR	11.09.42	21
Plumb, Stanley	Sgt	F/E	RAFVR	11.05.41	23
Ponting, Roland	Sgt	?	RAF	29.10.42	?
Poole, Frank	Sgt	B/A	RAFVR	11.08.43	?
Porteous, Kenneth	F/S	A/G	RAFVR	21.01.44	28
Portsmouth, Robert	Sgt	A/G	RAFVR	11.08.43	19
Powys-Jones, Hugh	Sgt	Obs	RAF	04.09.40	22
Price, William	Sgt	A/G	RAFVR	04.09.43	?

Priddle, Terrance	F/S	B/A	RAFVR	05.11.44	22	Sharp, Frank	Sgt	WAG	RAFVR	18.05.42	?
Prime, Norman	Sgt	?	RAF	11.09.42	?	Sharp, Reginald	Sgt	?	RAF	19.05.42	?
Pritchard, William	Sgt	WOp	RAFVR	02.03.43	26	Shaw, Wilfred	Sgt	WAG	RAFVR	26.02.43	21
Probert, William	F/S	WAG	RCAF	19.09.42	24	Shea, Victor	F/S	B/A	RCAF	04.07.43	20
Pryke, George	Sgt	F/E	RAF	29.06.43	21	Shearer, Robert	Sgt	A/G	RAF	18.12.41	?
Purry, Ronald	F/L	Pil	RAAF	15.06.44	21	Shearer, William	Sgt	A/G	RAFVR	24.09.44	22
Pybus, Alfred	Sgt	F/E	RAFVR	29.01.44	27	Sherratt, Harold	Sgt	A/G	RAFVR	03.03.43	19
Pye, James	Sgt	F/E	RAF	19.09.42	?	Shewen, John	Sgt	F/E	RAFVR	04.12.44	19
Quinn, George	P/O	Pil	RAFVR	03.09.42	?	Shiells, DFM. James	P/O	Pil	RAFVR	16.04.43	31
Radcliffe, George	W/O	A/G	RAFVR	11.06.44	21	Shoemaker, DFC. Wilbert	P/O	Pil	RCAF	26.07.43	?
Ragless, John	F/O	B/A	RAFVR	15.02.44	27	Shoesmith, Terence	Sgt	A/G	RAFVR	13.08.42	?
Ralph, Basil	F/S	A/G	RAuxAF	15.02.44	33	Shortland, Wilfred	Sgt	Obs	RAF	12.05.40	28
Ralph, Joe	P/O	Pil	RAF	30.09.39	19	Sills, James	P/O	WOp	RAFVR	24.03.44	24
Rampton, Albert	P/O	Obs	RAFVR	09.03.42	24	Simcox, James	P/O	Nav	RAFVR	13.06.44	22
Ramsey, Ian	F/S	WOp	RNZAF	31.07.43	31	Simpson, Greville	Sgt	WAG	RAF	15.02.41	?
Ratcliffe, Edward	F/O	Nav	RAFVR	19.02.43	30	Sinclair, Anthony	Sgt	WAG	RAFVR	01.03.43	20
Rate, Bernard	Sgt	B/A	RAFVR	28.07.43	?	Skelton, Robert	Sgt	WOp	RAFVR	08.12.42	27
Raymond, Lloyd	F/S	A/G	RCAF	23.09.43	?	Skilbeck, Robert	P/O	WOp	RAAF	04.12.44	20
Read, George	F/O	A/G	RAFVR	31.08.43	?	Skillin, Robert	Sgt	F/E	RAFVR	16.09.43	20
Reardon, Glyndwr	Sgt	WAG	RAFVR	11.02.41	?	Sleven, Arthur	Sgt	A/G	RAFVR	21.02.44	20
Recchia, Edward	Sgt	WAG	RCAF	29.10.42	?	Slingsby, Harold	P/O	A/G	RAFVR	03.02.45	38
Rees, Thomas	Sgt	F/E	RAFVR	11.06.44	21	Smith, Allen	Sgt	A/G	RAFVR	26.06.43	?
Reid, Frank	F/S	A/G	RAAF	01.06.44	21	Smith, Augustus	Sgt	B/A	RAFVR	24.03.44	?
Reid, George	Sgt	A/G	RAFVR	07.07.40	28	Smith, DFM. Charles	P/O	Obs	RAF	18.12.41	23
Relph, Henry	Sgt	B/A	RAFVR	12.06.43	30	Smith, Charles	W/O	Obs	RCAF	26.05.43	22
Rennie, Charles	F/S	Nav	RAFVR	27.08.44	22	Smith, Cyril	Sgt	F/E	RAFVR	21.01.44	33
Renshaw, Richard	P/O	Pil	RAFVR	29.06.41	?	Smith, Frank	Sgt	WAG	RAF	11.05.41	21
Reynolds, Thomas	F/L	Pil	RAFVR	25.07.44	21	Smith, George	F/S	A/G	RNZAF	28.08.42	22
Richards, John	Sgt	Obs	RAFVR	19.09.42	?	Smith, Keith	Sgt	Nav	RAFVR	18.07.44	21
Richardson, Jack	Sgt	WOp	RAFVR	19.02.43	?	Smith, Leonard	Sgt	A/G	RAFVR	26.05.43	?
Riordan, Lex	P/O	WOp	RAAF	17.01.45	20	Smith, Paul	Sgt	Pil	RAF	15.02.41	?
Ripley, Jack	P/O	Pil	RCAF	08.03.43	21	Smith, Philip	F/L	Pil	RAFVR	24.09.44	31
Roberts, Edwin	Sgt	Obs	RAF	12.05.40	26	Smith, Ronald	Sgt	Pil	RAFVR	30.06.41	27
Robinson, Charles	P/O	Pil	RAF	26.07.40	21	Smith, Thomas	Sgt	WAG	RAFVR	19.09.42	21
Robinson, John	Sgt	F/E	RAF	28.08.42	23	Snead, F. C.	Sgt	A/G	RCAF	16.09.41	?
Robinson, John	F/S	B/A	RAFVR	11.06.44	29	Soper, Sinclair	F/L	Pil	RCAF	28.04.44	26
Rodway, George	Sgt	WOp	RAFVR	04.05.43	22	Soper, Thomas	Sgt	WOp	RAFVR	30.05.42	?
Rogers, Peter	Sgt	Nav	RAFVR	26.02.43	22	Sparkes, Eric	Sgt	F/E	RAFVR	15.02.44	20
Rolfe, Ronald	Sgt	F/E	RAFVR	15.06.44	21	Spenceley, Frederick	Sgt	F/E	RAF	20.06.42	21
Rolls, Henry	Sgt	A/G	RAF	15.08.40	19	Spencer, Ronald	Sgt	WOp	RAF	11.06.40	19
Rose, David	F/S	A/G	RAF	20.06.42	26	Spice, Maurice	F/S	Nav	RAF	26.03.44	?
Ross, Robert	F/S	WAG	RAFVR	08.08.41	25	Spooner, Daniel	P/O	Nav	RAFVR	05.05.43	31
Rowley, Kenneth	Sgt	WAG	RAFVR	08.08.41	24	Spriggs, Anthony	Sgt	F/E	RAF	18.05.42	22
Rugless, George	F/S	WOp	RAAF	05.01.45	22	Spriggs, Eric	F/S	Nav	RAFVR	21.06.44	22
Russell, CGM. Joseph	F/L	Pil	RCAF	21.02.44	24	Stark, George	Sgt	A/G	RAFVR	02.10.42	27
Russell-Collins, Charles	F/S	Pil	RAF	19.09.42	?	Steer, Reginald	Sgt	WAG	RAFVR	13.04.42	22
Rutherford, George	Sgt	Nav	RAFVR	04.05.43	28	Stephen, Raymond	W/O	F/E	RAFVR	29.07.44	24
Ruthven, James	F/S	Pil	RCAF	18.12.41	24	Stephens, Arthur	Sgt	Nav	RAFVR	22.06.43	25
Ryan, John	F/O	Pil	RCAF	18.05.42	30	Stephens, DFC. John	W/C	Pil	RAFVR	30..08.43	25
Ryan, Thomas	F/S	B/A	RCAF	01.09.43	?	Stephenson, George	Sgt	A/G	RAFVR	07.04.42	27
Salter, Harry	F/L	?	RAF	29.10.42	34	Stevenson, Charles	F/O	Nav	RNZAF	16.11.44	23
Sanders, Frederick	F/L	Pil	RNZAF	16.11.44	22	Stevenson, Ian	F/O	B/A	RNZAF	16.09.43	27
Saunders, Arthur	P/O	Pil	RAFVR	29.06.43	?	Stocks, Thomas	Sgt	A/G	RAFVR	14.02.43	19
Scarisbrick, Leonard	Sgt	WOp	RAFVR	16.09.43	22	Stone, Noel	Sgt	B/A	RNZAF	17.10.42	32
Scott, Horace	Sgt	WOp	RAFVR	13.06.44	21	Storie, Richard	F/S	A/G	RAFVR	30.06.41	19
Scrase, Edgar	Sgt	A/G	RAF	13.08.40	19	Stowell, James	P/O	Pil	RNZAF	05.05.43	25
Seeley, Duncan	W/O	A/G	RCAF	19.02.43	?	Strachan, Robert	Sgt	F/E	RAAF	13.08.41	?
Self, Alfred	Sgt	B/A	RAFVR	14.02.43	21	Strickland, Algernon	Sgt	A/G	RAFVR	09.05.42	22
Sharman, Peter	Sgt	F/E	RAFVR	05.05.43	21	Stringer, George	F/S	A/G	RAFVR	23.09.43	20
Sharman, Peter	Sgt	WAG	RAFVR	28.08.42	20	Stringfellow, John	Sgt	Nav	RAFVR	01.09.43	19

Name	Rank	Role	Force	Date	Age
Stubbs, Thomas	Sgt	A/G	RAFVR	13.06.44	20
Stuckey, Victor	P/O	Nav	RAAF	12.12.44	27
Sturgess, John	P/O	A/G	RCAF	13.08.42	21
Summers, Andrew	Sgt	A/G	RAFVR	21.01.44	24
Surridge, Gordon	Sgt	WAG	RAFVR	28.04.42	25
Sutherland, Ian	F/O	?	RAF	04.08.40	21
Swainston, John	Sgt	F/E	RAFVR	12.12.44	23
Sweatman, Peter	F/S	WOp	RAFVR	21.07.44	20
Sykes, Clifford	Sgt	A/G	RAFVR	16.07.44	?
Symondson, Sidney	Sgt	WAG	RAFVR	19.07.41	?
Takideli, Alexander	P/O	Pil	RAF	12.06.40	20
Talbot, Edward	F/S	A/G	RAFVR	28.08.42	30
Talbot, Kenneth	Sgt	?	RAFVR	13.01.44	21
Tanner, Arthur	Sgt	Pil	RNZAF	17.10.42	22
Tarbin, Dennis	Sgt	F/E	RAFVR	16.07.44	19
Taylor, Guy	Sgt	Pil	RAFVR	13.08.41	?
Taylor, Kenneth	Sgt	WAG	RAFVR	07.04.42	20
Taylor, Stanley	Sgt	A/G	RAFVR	04.12.44	20
Taylor, Thomas	F/S	WAG	RAFVR	17.10.42	?
Temperton, Evan	P/O	A/G	RNZAF	03.02.45	22
Thomas, Douglas	Sgt	A/G	RAFVR	29.06.43	?
Thomas, Hamilton	Sgt	F/E	RAFVR	23.09.43	31
Thompson, Allan	Sgt	A/G	RAFVR	24.03.44	24
Thompson, Carlton	P/O	Pil	RCAF	13.06.44	?
Thompson, Clifford	Sgt	WAG	RAF	23.05.40	19
Thompson, George	Sgt	WAG	RAF	08.06.40	?
Thompson-Horan, Robert	Sgt	F/E	RAF	07.04.42	?
Thompson, Ivan	F/O	Pil	RCAF	26.05.43	?
Thomson, John	Sgt	A/G	RAFVR	24.08.43	19
Thomson, B.A. John	P/O	A/G	RAFVR	28.09.41	40

The inscription on the headstone of Thomson's grave reads, "Killed whilst flying 18 September 1941".

Name	Rank	Role	Force	Date	Age
Thorton, John	P/O	Pil	RAFVR	27.08.42	?
Tickle, Dalton	F/S	Nav	RNZAF	23.06.43	24
Tilley, Arthur	F/S	Pil	RAFVR	02.03.43	20
Tipping, Gerald	F/O	Nav	RAFVR	25.07.44	27
Totty, Norman	F/O	Nav	RAFVR	15.02.44	31
Town, Joseph	Sgt	AG	RAFVR	23.09.43	?
Towse, DFM	P/O	Pil	RAFVR	26.07.43	29
Tree, Raymond	Sgt	WAG	RAFVR	12.08.42	20
Trezise, Eric	Sgt	Pil	RAFVR	11.04.43	22
Tuck, Douglas	Sgt	F/E	RAF	09.03.42	22
Tucker, Eugene	F/S	Obs	RAF	23.05.40	24
Turley, Raymond	Sgt	WAG	RAFVR	03.09.42	22
Turner, Jack	F/O	Nav	RAFVR	25.05.44	22
Turner, John	F/S	Nav	RAAF	08.06.44	23
Turner, Ronald	P/O	B/A	RAFVR	03.03.43	20
Tvrdeich, Ivan	P/O	A/G	RNZAF	24.03.44	?
Tyler, Frank	F/S	A/G	RAFVR	08.05.44	?
Tyler, George	Cpl	Arm	RAFVR	14.04.44	31
Upton, James	W/O	Pil	RAAF	16.07.44	30
Vasil, George	Sgt	WAG	RAFVR	19.09.42	21
Venton, Peter	Sgt	Nav	RAFVR	29.06.43	21
Vincent, James	P/O	Nav	RAF	01.10.42	27
Vivian, John	S/L	Pil	RAF	08.08.41	27
Wade, John	Sgt	A/G	RAFVR	12.12.44	?
Wadman, Lester	F/O	A/G	RCAF	08.03.43	23
Waite, Henry	Sgt	F/E	RAF	24.05.43	18
Wakefield, Harry	Sgt	A/G	RAFVR	11.04.43	21
Wakefield, Kenneth	F/S	Obs	RAFVR	28.08.42	?

Name	Rank	Role	Force	Date	Age
Walker, Bryan	P/O	WAG	RAFVR	16.09.42	?
Wallace, Alexander	Sgt	WAG	RAFVR	16.09.41	22
Wallen, Emanuel	Sgt	WAG	RAFVR	31.08.43	26
Walrond, Arthur	Sgt	WAG	RAFVR	29.06.43	29
Warbey, Terence	Sgt	F/E	RAFVR	26.06.43	?
Ward, Dennis	Sgt	F/E	RAFVR	08.08.44	23
Wareham, Bernard	Sgt	WAG	RAFVR	13.10.41	21
Warner, John	P/O	WAG	RAF	23.07.41	?
Warrell, Ronald	Sgt	WAG	RAFVR	04.07.43	21
Watson, Cecil	F/S	B/A	RAAF	08.06.44	29
Watson, Reginald	F/S	Nav	RAAF	12.05.44	26
Watson, Robert	Sgt	B/A	RAFVR	28.04.44	19
Watson, William	Sgt	WOp	RAF	04.08.40	19
Watts, Bernard	Sgt	F/E	RAFVR	28.04.44	?
Watts, Cyril	LAC	WOp	RAF	18.05.40	21
Waylan, John	Sgt	WAG	RAFVR	16.07.42	31
Weaver, Ronald	Sgt	A/G	RAFVR	19.02.43	?
Webber, Harold	F/S	A/G	RAFVR	01.10.42	22
Webber, Norman	Sgt	Obs	RAFVR	03.06.42	22
Webster, Morven	Sgt	A/G	RAFVR	23.06.43	?
Weir, James	F/O	Nav	RAFVR	15.06.44	?
Weir, John	P/O	A/G	RAF	03.06.42	?
Wellesley, Charles	F/S	A/G	RAAF	19.02.43	23
Wells, Francis	Sgt	WOp	RAFVR	24.03.44	?
Wells, William	Sgt	WOp	RAFVR	31.07.43	?
Werner, Raymond	P/O	Pil	RAF	11.06.40	27
West, Harry	Sgt	A/G	RAFVR	15.02.44	30
Wheeler, George	F/S	WAG	RAFVR	25.07.44	21
Wheeler, Leslie	F/S	Pil	RAFVR	24.03.44	21
Whitcher, Arthur	Sgt	A/G	RAF	23/07/41	22
White, AFM. Charles	F/S	WAG	RAF	18.07.41	23
White, Godfrey	W/O	B/A	RCAF	27.01.44	37
Whiting, Geoffrey	Sgt	A/G	RAFVR	25.07.44	?
Whittaker, Gordon	Sgt	A/G	RAFVR	27.04.43	22
Whittaker, John	Sgt	A/G	RCAF	12.05.44	20
Wigley, Horace	Sgt	F/E	RAFVR	06.11.42	21
Wilkins, Leonard	F/S	B/A	RAFVR	17.01.45	22
Wilbourn, Bertram	Sgt	F/E	RAFVR	19.02.43	?
Williams, David	F/S	Pil	RAFVR	05.01.45	23
Williams, Francis	Sgt	F/E	RAFVR	23.06.43	22
Williams, H. E.	Sgt	?	RAFVR	11.09.42	?
Williams, Ivor	Sgt	A/G	RAFVR	08.04.43	20
Williams, DFC. John	F/O	A/G	RAFVR	19.02.43	?
Wilson, Donald	Sgt	F/E	RAFVR	12.05.44	?
Wilson, Jack	F/S	Pil	RAAF	26.05.43	21
Wilson, Matthew	S/L	Pil	RAF	11.04.42	28
Wilson, Patrick	Sgt	A/G	RAFVR	04.12.44	19
Wilton, Raymond	F/O	B/A	RAFVR	13.06.44	20
Winchurch, John	P/O	F/E	RAFVR	31.08.43	21
Wood, George	Sgt	A/G	RAF	28.04.44	?
Wood, Thomas	Sgt	Obs	RAFVR	17.10.42	22
Woodford, Alan	F/S	WOp	RAAF	20.02.44	20
Woodhams, John	Sgt	WAG	RAFVR	29.06.41	22
Woodhouse, C.	Pte	Pass	Army	20.05.42	
Woodley, Charles	F/O	Pil	RCAF	08.06.44	24
Woodruff, DFC. Dennis	F/L	Pil	RAFVR	29.01.44	28
Woods, Dennis	LAC	Obs	RAF	12.05.40	19
Woodward, Arthur	F/O	B/A	RAF	26.06.43	22
Wooldridge, Wilfred	F/S	WAG	RAF	18.12.41	20
Woollard, Peter	F/S	A/G	RAFVR	02.11.44	20

Wootton, Thomas	P/O	Obs	RAF	13.10.41	?
Worling, John	Sgt	Nav	RAFVR	13.04.42	20
Wratten, Jack	Sgt	F/E	RAFVR	26.02.43	?
Wrenshall, Bernard	F/O	B/A	RCAF	08.08.44	31
Wright, Edward	Sgt	WAG	RAFVR	01.09.43	19
Wright, Sidney	Sgt	F/E	RAFVR	12.06.43	21
Yeomans, Oswald	Sgt	Pil	RAF	14.09.40	26
Young, William	F/O	Nav	RAFVR	05.01.45	?

POST WAR

Collins, D. J.	W/C	Pil	RAF	25.03.71	?
Kelly, Paul	F/L	Nav	RAF	25.03.71	?
Rust, F. A.	F/L	Pil	RAF	05.01.53	?
Ruston, Charles	F/L	Nav	RAF	07.02.80	?
Tait, Ken	S/L	Pil	RAF	07.02.80	?

Four unnamed members of aircrew, killed during a night flight, on Washington bomber, WF553, on 05.01.53

GULF WAR

| Hicks, Stephen | F/L | Nav | RAF | 14.02.91 | 29 |

IN MEMORIAM

A few of the many who paid the supreme sacrifice whilst serving with No.15/XV Squadron.

Second Lieutenant Robert Alexander Fraser, (right) photographed shortly before being killed in action, on 18th May 1918. The officer to the left of the picture is thought to be Lieutenant Solomon Fine, the pilot with whom Robert Fraser flew and with whom he died. *Courtesy of Moira Miche*

Second Lieutenant Edward Earle Richardson, M.M., killed in action on 9th November 1918. *Author's collection*

Left: The last resting place of Squadron Leader The Reverend Denis Edward Guy Ashill, the Squadron's Chaplain at RAF Bourn, who was killed in a flying accident, on 29th December 1942. As with many earth bound RAF personnel, Squadron Leader Ashill took the opportunity to fly on an air test with F/L Ordish. Twenty minutes after the aircraft took-off, Stirling bomber, serial W7585, coded LS-U, crashed on the edge of Bassingbourn aerodrome, Cambridgeshire, following an engine fire. Squadron Leader Ashill was buried in Cambridge City Cemetery. The inscription on the stone cross marking his resting place is written in Latin. His name is given as Guido Ashill. *Author's Collection* Center: Sergeant Dennis Brown, killed on 31st July 1943, whilst flying as navigator on Stirling bomber, serial EF427, coded LS-A, during an attack against Remscheid, Germany. The aircraft, which was shot down by flak, crashed to the north-east of the village of Manheim. *Courtesy of Doug Fry* Right: Sergeant James Edward "Geordie" Carrott, killed on 26th February 1944, whilst flying as navigator on Lancaster, serial ED363, coded LS-C, which crashed at RAF Lakenheath, on return from an attack against Augsburg. *Author's Collection via the late Harry Bysouth*

Pilot Officer John Crone, RAAF (center), and his crew were killed when Lancaster, PB802, LS-F, crashed whilst returning to Mildenhall on 17th January 1945. Apart from the pilot, the crew consisted of, P/O Lex Riordon, RAAF, W/Op; Sgt Richard Devlin, F/E; Sgt George Lake, Nav.; F/Sgt Leonard Wilkins, B/A; F/O Fuller, A/G (age unknown); F/O Harry Freedman, A/G (age 20). *Author's Collection via Bob Collis*

Right: Sergeant Joseph "Tony" Davis, was the pilot of Lancaster bomber, serial ED363, coded LS-C, which crashed at 05.10 hours, on the morning of 26th February 1944, whilst attempting to land at RAF Lakenheath. *Author's Collection via the late Harry Bysouth*

Left: Sergeant Sidney John Devereux, Wireless Operator/Air Gunner, aged 22, was killed on 29th June 1944. His aircraft, Short Stirling, serial EH888, coded LS-Z, was shot down during an attack against Cologne. The bomber crashed into the River Ijssel, at Heeschwijk, Netherlands. *Author's Collection via the Devereux Family* Center: Sergeant Leslie Garrett, was killed on 29th June 1943, whilst flying as flight engineer on Stirling bomber, EH888, LS-Z, during an attack against Cologne. *Courtesy of Gordon Garfit* Right: Sergeant William "Bill" Geraghty, age 22, was the bomb aimer of Lancaster, serial ED363, coded LS-C, which crashed at RAF Lakenheath, on return from an attack against Augsburg, on 26th February, 1944. *Author's Collection via the late Harry Bysouth*

Left: *The Airman who was 'killed' twice.* Sergeant John William Greenwood, was posted to No.XV Squadron in October 1942, where he joined the crew of Sgt Monteith. Flying as rear gunner, his first mission was a mine-laying (Gardening) operation, on 7th December. Between that date and 16th February 1943, he flew a total of seven missions against targets in France, Germany and Italy. However, on the night of 19th February, for some unrecorded reason, Sgt Greenwood's place as rear gunner was taken by Flying Officer P. Brennan. Unfortunately, P/O Monteith and his crew, including F/O Brennan, were all killed when their Stirling bomber, Serial BF411, coded LS-A, failed to return from the attack against Wilhelmshaven. The aircraft, which was shot down by the nightfighter ace Oberleutenant Hans-Joachin Jabs, crashed at Tershhelling, Holland, at 21.45 hours. An entry was made in John Greenwood's Flying Log Book which recorded, 'Night Ops – A/C failed to return'. This entry implying that Sergeant Greenwood had possibly been killed as a result of enemy action. This of course was not the case, as John Greenwood was not flying that night. In fact it was to be a further five weeks before he flew again, even then no further entries were made in his Log Book. Having joined the crew of F/L Haycock in early March, John Greenwood flew his first mission with his 'new' crew on the night of 27/28 March 1943. History repeated itself, as Sergeant Greenwood (again) flew seven operations with this crew, before fate stepped-in. On the night of 16th April, the Squadron was detailed to attack Mannheim. Being one of a force of eighteen aircraft despatched by No.XV, Stirling bomber, serial BK691, coded LS-F, piloted by F/L Haycock, took-off from RAF Mildenhall, at 21.48 hours. BK691 was shot down during the early hours of 17th April, and crashed at Hetzerath, Germany. The crew, with the exception of Sgt Norman Hobden, the flight engineer, were all killed. Sergeant John William Greenwood, the airman who had been 'killed' twice, rests with his crew at Rheinberg War Cemetery, Germany. He was 28 years old when he died. *Courtesy of June Watkins (nee Greenwood)* Center: Sergeant Ernest "Mick" Harbidge, age 19, was rear gunner on Lancaster, ED383, LS-C. He was severely injured in a crash at Lakenheath, on 26th February 1944, and died in Ely Hospital a few hours later. *Author's Collection* Right: Sergeant Doug Haydock, flight engineer, was killed on 26th February 1944, when Lancaster, ED383, LS-C, crashed at Lakenheath. *Author's Collection*

Left: Flight Sergeant Harold "Guy" Howland, age 26, was killed on 2nd March 1943, when the aircraft he was piloting failed to return from an attack against Berlin. The Stirling bomber, EF347, LS-T, was shot down by Oberleutnant Wolfgang Kuthe, of IV./NJG1. *Courtesy of Ad van Zanntvoort* Center: Pilot Officer George Judd, pilot, was killed on the night of 30th/31st July 1943, when his aircraft was shot down by flak, during an attack against Remscheid. *Courtesy of Doug Fry* Right: Sergeant Aston Crawford Lake, age 20, was killed on 12th June 1943, whilst flying as navigator on Lancaster, BF571, LS-U, during an attack against Dusseldorf. *Courtesy of Roz Dengate, Niece of Aston Lake*

Left: Sergeant Syd Long, age 26, was killed on 31st July, whilst flying as bomb aimer on Stirling bomber, EF427, LS-A, during an attack against Remscheid. The aircraft was shot down by flak. *Courtesy of Doug Fry* Center: Flight Lieutenant Roy Marsh, age 21, was killed on 12th December 1944, whilst piloting Lancaster bomber, HK627, LS-F, during an attack against Witten. *Author's Collection* Right: Sergeant Raymond Geoffrey Norris, age 21, was killed whilst flying as wireless operator/air gunner on Lancaster, LM121, LS-C. The aircraft, which was participating in an attack against Trappes, France, was shot down by a nightfighter. *Courtesy of Geoff Reynolds*

Left: Flight Lieutenant Charles Brian Ordish, age 23, died on 31st December 1942, from injuries received in an accident two days earlier. The aircraft he was piloting, Stirling bomber, W7585, LS-U, developed engine problems during an air test. It crashed in flames on the boundary of Bassingbourn airfield, Cambridgeshire. *Courtesy of Elizabeth Park (nee' Ordish)* Center: Pilot Officer Clive Perring, age 23, was killed on 16th April 1943, whilst flying as navigator on Stirling bomber, BF474, LS-H, during an attack against Mannheim. *Courtesy of and via Steve Smith* Right: Pilot Officer Robert Skilbeck, RAAF, age 20, was killed on 4th December 1944, whilst flying as wireless operator on Lancaster bomber, HK626, LS-W. The aircraft, which was participating in a daylight attack against Oberhausen, exploded in the air following a direct hit from enemy anti-aircraft fire. *Courtesy of Ralph Skillbeck*

Flying Officer Norman Overend, pilot, (RNZAF) and Flight Sergeant Harry Beverton, air gunner (standing center and right respectively) were killed on 12th September 1944, during an attack against Frankfurt. *Author's Collection*

Above: Sergeant Dennis Ward, age 23, was killed on 8th August 1944, whilst flying as navigator on Lancaster bomber, NN700, LS-Q, during an attack against Rocquecourt, France. *Courtesy of Frank Chasemore*

Above right: Sergeant William Wells was killed on 31st July 1943, whilst flying as wireless operator on Stirling bomber, EF427, LS-A. The aircraft was shot down by flak during an attack against Remscheid. *Courtesy of Doug Fry*

Right: Sergeant William 'Bill' Watson, age 19, wireless operator, who was killed on 4th August 1940, whilst flying on Bristol Blenheim, R3771. The aircraft, which was carrying a crew of four, dived into the ground at Norbury, four miles from Whitchurch, Shropshire, at 17.05 hours. At the time of the accident, the Blenheim was engaged in Fighter Affiliation training with Hawker Hurricanes from No.1 Squadron, based at RAF Northolt. The three other airmen who died were the pilot, P/O M. Hohnen, Sergeant H. Beard and Flying Officer Welsh. *Author's Collection*

APPENDIX II
List of Personnel Known to be
WOUNDED & INJURED IN ACTION

WORLD WAR ONE

NAME	RANK	DATE	REMARKS
N. A. Arthur	2/Lt	06.11.17	
H. Atkinson	2/Lt	17.08.18	
L. J. Bayley	2/Lt	22.03.18	
W. G. Bennett	Sgt	04.05.17	Wounded in the leg by machine gun fire.
L. A. Blunden	A.M.II	23.04.17	
A. W. Bolitho	2/Lt	12.04.18	
K. A. Brooke-Murray	Capt.	16.09.16	Wounded twice in the left leg during a fight with three enemy aircraft.
L. R. Brown	2/Lt	11.02.17	Shot in the fore-arm whilst carying out an artillery patrol.
E. S. Burns	2/Lt	01.11.18	
W. R. Carmichael	2/Lt	??.10.16	Wounded when the aircraft crashed behind enemy lines.
C. H. Collyns	2/Lt	09.10.18	
C. R. Cook	2/Lt	07.07.16	
T. C. Creaghan	Lt.	20.11.17	Wounded in action, whilst carying out low-level flying.
A. R. Cross	Capt.	08.10.18	Slightly injured during forced-landing, after being hit by shell fire.
J. H. Davies	2/Lt.	10.06.18	
E. G. Donaldson	Lt.	16.04.16	
F. E. Elliot	Lt.	03.05.17	
W. R. Fox	2/Lt	07.06.18	
G. A. Griffin	Lt.	07.06.18	
H. Griffith	Lt.	??.07.18	
L. L. Handford	2/Lt	11.04.18	Wounded in the knee by two bullets, during a fight with five Pfalz scouts, two of which were shot down by 2/Lt. Handford.
R. G. Hart	2/Lt	12.04.18	Wounded in the ankle by a bullet splinter at the start of combat with five British aircraft returned to the fight and shot down two enemy aircraft.
J. R. Hodgkinson	Lt.	07.12.17	Wounded in leg when shot down by eight Albatross fighters of Von Richthofen's 'Flying Circus'.
F. H. Hodgson	2/Lt	06.09.16	Wounded during a bombing raid against Bertincourt airfield.
F. H. Hudson	2/Lt	21.02.16	Severely wounded in the head by a shell splinter, during an army reconnaissance over the Courtrai Valley.
D. K. Johnston	Lt.	08.08.16	
G. K. Kerr	Lt.	28.07.16	Wounded in action.
W. F. Mayoss	2/Lt	20.11.17	Wounded in action by hostile gun fire, whilst flying at low level.
W. F. Mayoss	2/Lt	09.03.18	
H. A. Milnes	Sgt	28.05.17	
G. M. Moore	Capt.	01.02.17	Wounded by small arms fire from enemy trenches.
J. P. Morkham	2/Lt	21.12.16	Wounded in action.
C. F. Muirhead	Lt.	17.08.18	
C. H. Nott	Cpl	19.01.16	Struck in the eye by a fragment of shell from anti-aircraft fire, and knocked insensible.
A. E. Packe	2/Lt	01.07.16	Wounded in the early hours of the Battle of the Somme.

Name	Rank	Date	Details
G. A. Penny	2/Lt	21.04.18	
G. Pilkington	L/Cpl	12.10.16	Wounded in action.
T. Poland	Lt.	05.05.17	
F. H. Reeve	2/Lt	04.03.17	
A. Rice-Oxley	Lt.	07.10.16	Wounded in action.
J. F. Ridgway	Sgt	24.03.17	Wounded in the leg by anti-aircraft fire.
H. S. Robertson	Lt.	20.11.17	Wounded in action, during low-level operation.
F. C. Rutter	A.M.II	04.04.18	Wounded in action.
C. C. Snow	Capt.	05.07.18	
S. Stretton	2/Lt	01.08.16	Wounded by anti-aircraft whilst strafing.
N. P. Tucker	2/Lt	01.07.16	Wounded in the early hours of the Battle of the Somme.
F. F. Wessel	Lt.	20.04.17	Wounded in the foot, whilst on artillery observation patrol.
G. S. Wood	2/Lt	23.08.16	
A. M. Wynne	Capt.	01.04.17	Wounded in the leg, when attacked by the German ace Ltn Werner Voss.
J. T. Wyre	Lt.	18.05.18	
O. A. Yunk	A.M.I	27.05.18	Admitted to hospital, due to being gassed.

WORLD WAR TWO

Name	Rank	Date	Details
J. R. Arrowsmith	Sgt	04.10.43	Sprained ankle sustained when baling out of aircraft.
A. Belson	Sgt	28.07.43	Slightly injured when the aircraft crash-landed and burst into flames at West Row, near RAF Mildenhall.
C. Bender	F/O	25.02.45	Wounded by flak during an attack against Kamen, Germany.
L. J. Bentley	Sgt	03.10.43	Slightly wounded as a result of combat with an enemy nightfighter, during an attack against Kassel.
A. J. Box	Sgt	08.06.40	Wounded in action. Admitted to hospital.
C. Bradwell	Sgt	28.07.43	Slightly injured when the aircraft crash-landed and burst into flames at West Row, near RAF Mildenhall.
K. Burns	F/S	04.10.43	Seriously wounded as a result of combat with an enemy nightfighter, during an attack against Kassel.
R. B. Clarke	F/O	21.05.40	
? ? Coen	Sgt	21.07.41	Wounded in shoulder.
B. Cooper	F/S	05.03.45	Wounded by flak over target during attack against Gelsenkirchen.
E. F. Duncombe	Sgt	31.03.41	Wounded during attack against Bremen.
P. F. Eames	F/O	12.05.40	Flesh wound in knee.
A. E. Eaton	Sgt	04.05.43	Wounded in shoulder by bullet.
D. L. Evans	Sgt	26.07.42	Aircraft shot down.
H. Flowerday	Sgt	04.05.43	Wounded in leg by cannon fire.
D. Fry	Sgt	31.07.43	Wounded in the stomach by flak splinters.
T. Gookey	Sgt	28.07.43	Slightly injured when the aircraft crash-landed and burst into flames at West Row, near RAF Mildenhall.
J. Hall	Sgt	10.02.41	Wounded during attack against Hannover.
S. Hines	Sgt	28.07.43	Slightly injured when the aircraft crash-landed and burst into flames at West Row, near RAF Mildenhall.
R. P. Howland	Sgt	20.04.43	Face burnt and hit in foot by flak.
M. C. Huggett	Sgt	20.04.43	Wounded in thigh by flak.
R. E. Hunter	LAC	21.05.40	Crash-landed in field.
D. Jackson	Sgt	28.07.43	Slightly injured when the aircraft crash-landed and burst into flames at West Row, near RAF Mildenhall.
S. L. Jeffrey	F/S	02.11.44	Seriously injured when aircraft collided with another Lancaster, from the same Squadron, during a

			daylight attack against Homberg.	B. S. Piff	Sgt	11.11.40	Slightly injured in mid-air collision.
L. Jeffries	F/O	27.08.43	Broken ankle and burns, as a result of nightfighter attack.	S. A. Robertson	F/O	30.05.40	Wounded during attack. Landed at Martlesham and admitted to Colchester Hospital.
W. Jessop, DFM	Sgt	15.02.41	Wounded during attack against Sterkrade.	R. W. Rose	Sgt	04.10.43	Lost right arm as a result of combat with an enemy nightfighter, during an attack against Kassel.
T. J. Maloney	Sgt	11.11.40	Slightly injured in mid-air collision.				
L. C. Masfen	Sgt	16.07.42	Aircraft shot down and crashed into the North Sea.	J. E. Russell	P/O	15.02.41	Wounded during attack against Sterkrade.
H. McCosh	P/O	01.04.41	Sustained bullet wounds, which proved fatal on 06.04.41.	J. C. Samson	P/O	28.07.43	Slightly injured when the aircraft crash-landed and burst into flames at West Row, near RAF Mildenhall.
P. McIlroy	Sgt	28.07.43	Slighty injured when the aircraft crash-landed and burst into flames at West Row, near RAF Mildenhall.				
R. S. Millard	W/O	15.03.44		D. C. Smith	P/O	13.05.43	Wounded in arm by flak splinters.
T. Mosedale	Sgt	22.06.43	Shot down by nightfighter.	J. R. Stanford	Sgt	15.05.40	Slight wound. Hospitalized in Belgium.
D. Murphy	Sgt	15.03.44	Fractured spine and severe bruising.				
C. Nabaus	F/O	15.03.44	Broken leg	P. F. Webster	F/L	12.05.40	Superficial bullet wounds to both feet.
E. Nestor	P/O	27.08.43	Burns due to aircraft on fire.	P. F. Webster	F/L	21.05.40	Crash-landed in field.
G. J. Owen	Sgt	03.02.45	Wounded during an attack against Dortmund.	G. E. Williams	Sgt	01.03.43	Aircraft shot down.
J. F. Perring	Sgt	16.12.42	Lost a leg.				

POST WAR

J. Nichol	F/Lt	17.01.91	Aircraft shot down.
J. Peters	F/Lt	17.01.91	Aircraft Shot down.

APPENDIX III
List of Personnel Known to be
INJURED IN ACCIDENTS

WORLD WAR ONE

NAME	RANK	DATE	REMARKS
R. H. Barratt	2/Lt.	30.03.17	
G. Henderson	Capt.	20.04.16	
C. D. Kershaw	2/Lt.	24.08.16	
G. N. Moore	Lt.	20.04.16	
J. P. Morkham	2/Lt.	21.12.17	
E. H. Penberthy	2/Lt.	02.06.17	

WORLD WAR TWO

NAME	RANK	DATE	REMARKS
H. Bysouth	Sgt	26.02.44	Received facial injuries, bruising and concussion following crash of aircraft on return from operations.
K. R. Dougan	Sgt	18.11.42	Engine failure on take-off. Aircraft swung due to cross winds and under-carriage collapsed as pilot tried to correct. Bomb aimer sustained injuries to right ankle and left knee.
?. Galloway	Sgt	25.04.41	Sustained injuries whilst abandoning aircraft due to lack of fuel.
L. A. Hack	P/O	19.04.42	Crashed shortly after take-off from Wyton. Eight others, including a member of groundcrew and a soldier, were killed.
?. Lancashire	Sgt	01.06.42	Sustained injuries to neck and head, following a landing accident.
?. Read	F/L	05.01.53	Sustained leg injuries following a crash in a Boeing B.29A Washington bomber.

APPENDIX IV
List of
PRISONERS OF WAR AND INTERNEES

WORLD WAR ONE

NAME	RANK	STATUS	REMARKS
W. R. Carmichael	2/Lt	Obs	Wounded
P. J. Casey	2/Lt		Repatriated 13.12.18
T. H. Clarke	Lt		
A. E. Fereman	2/Lt		
W. R. Hall	2/Lt		
W. Joyce	Lt	Obs	
J. C. Lees	2/Lt		Repatriated 30.12.18
R. T. Lees	A.M.II		
V. C. Morris	2/Lt		Repatriated 02.01.19
D. K. Paris, M.C.	Lt		
?. ?. Piper	Sgt	Obs	Repatriated 05.12.16
C. B. Wilson	Lt	Pil	

WORLD WAR TWO

The ranks indicated in the following list are those given at time of capture.

NAME	RANK	CREW	PoW	CAMP	FTR
J D Aitken	Sgt	WOp	23623	L6	15.08.41
A J Ames	Sgt	F/E	Interned in Spain		21.11.42
			Repatriated January 1943.		
D B Annesley	Sgt	A/G	39183	357	30.06.41
G Armstrong	Sgt	A/G	39290	357	19.07.41
J R Arrowsmith	Sgt	WOp	?	?	04.10.43
C S Aynsley	Sgt	WOp	9585	357	08.09.41
M H Bailey	F/S	WOp	569	357	28.08.43
K J Baker	F/S	A/B	250696	4B	05.10.43
J A Baldwin	F/S	F/E	83669	L3	26.07.43
K A Banks	Sgt	A/G	2349	L3	31.07.43
A B Bateman	F/O	Pil	39647	L3	26.04.42
K L Bearnes	F/S	A/G	259848	L3	27.04.43
A Beazley-Long Sgt	F/E		3683	357	27.04.44
B Beecroft	Sgt	Nav	10	L3	26.07.41
R F Beisley	Sgt	?	259850	4B	10.02.41
L J Bentley	F/S	B/A	1594	357	21.02.44
S H Bignell	F/O	Pil	?	?	28.01.45
R Booth	Sgt	Obs	13061	357	12.05.40
H F Brodie	F/S	F/E	83693	L3	26.07.43
J Brown	Sgt	F/E	23622	344	15.08.41
J B Bunce	Sgt	Nav	23620	357	15.08.41

NAME	RANK	CREW	PoW	CAMP	FTR
G H Burland	Sgt	A/G	23618	L6	15.08.41
A R Byrne	Sgt	Nav	1292	357	11.08.43
H Campbell	W/O	Wop	?	?	04.02.45
C S Carter	Sgt	WOp	222491	4B	18.08.43
W J Cave	Sgt	B/A	250699	4B	11.08.43
W T Chamberlain F/S	A/G		250711	4B	05.10.43
K W Chapman	F/O	Nav	8046	L3	08.06.44
L F Clarke	P/O	A/G	?	?	07.02.45
G A Clary	Sgt	WOp	Interned in Spain		21.11.42
			Repatriated January 1943.		
A R Cole	F/S	B/A	222454	4B	28.07.43
J E Conran	F/O	Pil	766	L3	26.04.41
D W Cook	F/O	F/E	?	?	07.02.45
B E Cooper	P/O	Nav	1770	L3	26.05.43
H C Cooper	Sgt	F/E	1174	L6	14.05.43
N S Craddock	Sgt	A/G	27480	344	04.02.43
J B Craggs	Sgt	B/A	1175	L6	14.05.43
W R Crich	P/O	Pil	Interned in Spain		21.11.42
			Repatriated January 1943.		
J S Curtis	F/S	Nav	259863	4B	03.10.43
H S Daborn	P/O	Nav	Interned in Spain		21.11.42
			Repatriated January 1943.		
J J Damboise	Sgt	A/G	577	357	23.06.43
D T Darby	Sgt	F/E	?	?	05.01.45
W E Davies	Sgt	A/G	594	L3	12.05.40
J T Day	Sgt	F/E	3	L6	26.07.41
J E Dodd	Sgt	A/G	9584	L3	08.09.41
J D Duckett	F/S	B/A	308	357	29.06.43
E F Duncombe	Sgt	?	E8	L6	31.03.41
H J Dunnett	Sgt	A/G	9583	L6	08.09.41
W A Dyson	Sgt	Nav	1065	L1	27.04.43
P F Eames	F/O	Pil	160	L3	30.07.40
A E Eaton	F/S	B/A	1127	L1	05.05.43
A W Edgley	Sgt	A/G	222506	4B	26.05.43
Evaded capture for six weeks before being betrayed in Paris.					
N R Elford	Sgt	A/G	453	L6	24.05.43
W R Fairweather Sgt	Nav		83710	4B	26.07.43
G T Farrell	F/S	A/G	250723	4	05.10.43
H S Flowerday	Sgt	A/G	1130	L6	05.05.43
A S Forbes	Sgt	Pil	27484	344	04.02.43
G W Forster	Sgt	A/G	2669844	4B	15.03.44
D F Frame	Sgt	F/E	43302	357	20.02.44

Name	Rank	Role	No.	Camp	Date
D R Fry	Sgt	A/G	6479	L6	31.07.43
A T Gamble	P/O	Wop	?	?	07.02.45
W R Garrioch	Sgt	Pil	497	357	10.02.41
E L Gold	F/S	WOp	967	357	27.01.44
T G Grant	F/O	Nav	2614	L3	23.09.43
H N Guymer	Sgt	A/G	2	357	26.07.41
J Hall	Sgt	?	?	?	10.02.41
G G Hammond	F/L	Pil	?	?	07.02.45
T P Hanrahan	Sgt	WOp	1182	357	14.05.43
R Harper	Sgt	Pil	9649	357	08.09.41
E C Haynes	Sgt	A/G	23621	357	15.08.41
J F Heal	Sgt	B/A	Interned in Spain		21.11.42
			Repatriated January 1943		
H G Hedge	Sgt	A/G	503	357	10.02.41
R K Hemseed	Sgt	A/G	27524	344	04.02.43
H F Henry	Sgt	A/G	Interned in Spain		21.11.42
			Repatriated January 1943		
J V Higgins	W/O	A/G	?	?	28.01.45
H W Hipwell	Sgt	F/E	1070	357	21.04.43
S D Hirst	Sgt	WOp	169	357	24.05.43
N A Hobden	Sgt	Nav	42758	357	17.04.43
R D Hooley	Sgt	B/A	9587	357	08.09.41
R P Howland	Sgt	WOp	1139	357	21.04.43
G A Howard	Sgt	Nav	27445	L4	04.02.43
D L Howell	F/O	Nav	?	?	07.02.45
M C Huggett	Sgt	A/G	1250	L6	21.04.43
E Hurley	Sgt	WOp	222453	4B	28.07.43
L J Jefferies	F/O	Pil	3134	L3	28.08.43
J D Jeffrey	Sgt	Pil	97	357	13.08.41
D J Jeffs	Sgt	A/G	24967	344	26.04.42
W Jessop	Sgt	WAG	556	L6	16.02.41
J M Johnson	Sgt	Pil	23624	357	15.08.41
D G Jones	F/S	B/A	?	?	28.01.45
J D Jones	F/S	?	850	L7	??.??.??
W H Jordan	Sgt	?	506	357	10.02.41
F Keable-Buckle	Sgt	F/E	?		?28.01.45
C E Keik	Sgt	A/G	1185	357	14.05.43
T R Kemp	Sgt	A/G	Interned in Spain		21.11.42
			Repatriated January 1943.		
B C Kelly	Sgt	Pil	558	357	31.03.41
R F King	Sgt	F/E	3040	L4	27.01.44
J W Lacey	F/S	Nav	?	?	28.01.45
P J Lamason	S/L	Pil	8056	L3	08.06.44
A H Law	Sgt	Nav	1144	L6	05.05.43
R R Lawson	Sgt	WOp	246	L3	26.04.42
R O Leonard	Sgt	A/G	307	357	24.05.43
D Lewis	Sgt	A/G	83692	4B	26.07.43
R Lockwood	P/O	B/A	1224	L3	21.04.43
W B Louch	Sgt	?	?	?	30.06.41
C S Lyons	F/L	Pil	1225	L3	21.04.43
M Mac Mackay	Sgt	WOp	250935	L3	05.10.43
W U MacIntosh	F/O	A/G	?	?	07.02.45
L B McCarthy	Sgt	?	39180	357	30.06.41
T Malcolm	Sgt	Pil	1081	L3	05.05.43
H C Mallen	W.O	Nav	330	357	29.06.43
H Mallen	Sgt	Nav	330	357	29.06.43
D Mann	P/O	Nav	?	?	05.10,43
J R Marshall	Sgt	Nav	?	?	21.04.43
			Repatriated on the Arundel Castle 06.02.45		
R T Martin	Sgt	B/A	39298	357	25.06.43
S J Maxted	Sgt	WOp	222529	4B	26.04.43
			Evaded capture for six weeks before being betrayed in Paris.		
W M McLeod	Sgt	Pil	1149	L4	05.05.43
I A McPhee	Sgt	A/G	3464	357	27.01.44
G Mead	F/S	B/A	3690	357	27.04.44
R S Millard	W/O	B/A	?	?	15.03.44
N A Mitchell	Sgt	Nav	1071	357	27.01.44
W C Moir	Sgt	F/E	9662	357	08.09.41
B J Moore	Sgt	?	564	357	31.03.41
J D Moore	Sgt	A/G	83663	4B	26.07.43
J M Morris	F/O	Pil	?	13D	04.02.45
T Mosedale	Sgt	WOp	43349	L4	23.06.43
E B Mossman	Sgt	WOp	27483	344	04.02.43
R Mount	Sgt	?	565	357	31.03.41
D Murphy	Sgt	A/G	?	?	15.03.44
			Repatriated with a spine fracture and severe bruising.		
J E Murphy	F/O	B/A	?	?	07.02.45
E A Nestor	P/O	Nav	2628	L3	28.08.43
C Nabaus	F/O	Nav	?	?	15.03.44
			Repatriated with a broken leg.		
W G Oliver	F/O	Nav	1333	L3	14.04.43
G J Owen	Sgt	F/E	?	?	04.02.45
R B Pape	Sgt	Nav	?	8B	08.09.41
			Military Medal, Author of 'Boldness Be My Friend'.		
N J Pawley	Sgt	WOp	83672	4B	26.07.43
			Distinguished Flying Medal.		
F R Pepper	Sgt	Pil	13039	357	12.05.40
H L Phillips	Sgt	B/A	1089	L4	27.04.43
T W Piper	S/L	Pil	3647	L3	19.07.41
L I Probett	F/S	A/G	1075	L4	27.01.44
D Rees	Sgt	A/G	39183	L1	30.06.41
A M Reid	Sgt	Nav	27735	344	09.03.43
L C Richards	Sgt	F/E	1333	L6	31.07.43
J Richmond	Sgt	A/G	222636	4B	31.08.43
D S Roberts	Sgt	F/E	234	357	23.06.43
E Routh	Sgt	F/E	1157	L4	05.05.43
C A Russell	F/S	Wop	?	?	28.01.45
J E Russell	P/O	A/G	527	L3	16.02.41
C F Ryall	Sgt	Pil	241	357	24..05.43
R A Scandrett	Sgt	A/G	222552	4B	18.08..43
J Scott	LAC	A/G	13076	357	12.05.40
E F Seabolt	Sgt	A/G	?	?	26.05.43
J Shaw	Sgt	A/G	1097	357	21.04.43
R A Skinner	Sgt	?	274	357	26.04.42
D C Smith	P/O	Pil	1335	L3	14.05.43
E Smith	Sgt	F/E	250761	L3	05.10.43
F Smith	Sgt	WOp	9	357	26.07.41
I Spagatner	Sgt	A/G	817	L7	13.09.44
J J Sparrow	Sgt	A/G	631	357	11.08..43
			Evaded capture for 38 days before being betrayed.		
W A Spencer	Sgt	F/E	1100	L6	27.04.43
R R Stewart	Sgt	Pil	23617	357	15.08.41
R T Stephen	Sgt	F/E	?	?	26.04.42
F Stevens	Sgt	A/G	1101	L4	05.05.43
J M Taggart	Sgt	B/A	222786	4B	28.08.43
T H Tayler	F/L	Pil	3666	L3	26.07.41
J A Taylor	F/S	B/A	?	?	04.02.45
R R Taylor	Sgt	A/G	222560	4B	18.08.43
V N Taylor	Sgt	A/G	98	L6	13.08.41
E L Thomas	LAC	A/G	5209	L3	18.05.40

N L Thomas	F/S	Pil	250764	4B	05.10.43
F Thompson	F/L	Pil	3675	L3	26.07.41
G B Thomsom	Sgt	Nav	870	L7	13.09.44
T S Thorkilsen	Sgt	WOp	39184	L6	30.06.41
T M Thoroughgood	F/S	A/G	?	?	28.01.45
L C Titterton	Sgt	B/A	8	L6	26.07.41
W R Torrance	Sgt	Pil	27521	L3	04.02.43
R P Wallace-Terry	F/L	Pil	?	?	08.09.41
P H Ward	Sgt	B/A	27470	344	04.02.43
E E Warner	Sgt	A/G	1205	4B	14.05.43
		Distinguished Flying Medal			
R Watson	P/O	Pil	42826	L6	27.04.43
		Repatriated 02.02.45			
A E Waugh	Sgt	A/G	281	357	22.06.43
D Wharam	F/O	Nav	1633	357	21.02.44
W A Wilkie	F/S	A/G	?	?	28.01.45

J S Wilkinson	F/O	Nav	?	?	04.02.45
G E Williams	Sgt	A/G	1053	357	02.03.43
E Willis	Sgt	WOp	1275	L6	05.05.43
A V Wood	F/O	Pil	259924	L3	03.10.43
L C Wood	Sgt	B/A	222543	4B	18.08.43
G G Wright	F/S	B/A	259929	4B	03.10.43
M Wyatt, DFC	S/L	Pil	Interned in Spain		21.11.42
		Repatriated January 1943.			
A H Young	P/O	Nav	217	L3	26.04.42

GULF WAR

R Clark	F/L	Pil			01.91
J Nichol	F/L	Nav			17.01.91
J Peters	F/L	Pil			17.01.91

APPENDIX V
List of ESCAPERS AND EVADERS

WORLD WAR TWO

Name	RANK	CREW	FTR	REMARKS
J.A. Beattie	Sgt	?	23.04.40	
W. Blott	F/L	Pil	15.03.44	Interned in Switzerland.
A.W Edgeley	Sgt	A/G	26.05.43	Evaded for six weeks before being captured in Paris.
T.W Forster	Sgt	A/G	15.03.44	Interned in Switzerland.
L.H George,	F/O	WOp	08.06.44	
G.D Gill	Sgt	WOp	15.03.44	Interned in Switzerland.
R.M Gilleade	F/S	B/A	14.06.44	Evaded.
A.F Kellet	Sgt	B/A	23.06.43	Evaded.
J. Marpole	F/L	F/E	08.06.44	
G.R Mattock	Sgt	F/E	15.03.44	Interned in Switzerland.
S.J Maxted	Sgt	WOp	26.05.43	Evaded for six weeks before being captured in Paris.
R.S Millard	WO2	B/A	15.03.44	Interned in Switzerland.
L.A Miller	F/L	Pil	27.04.44	Evaded.
C.M Mora	Sgt	WOp	27.04.43	Evaded.
D. Murphy	Sgt	A/G	15.03.44	Interned in Switzerland.
G.A Musgrove	F/O	B/A	08.06.44	Evaded.
C. Nabarro	F/O	Nav	15.03.44	Interned in Switzerland.
J.J Sparrow	Sgt	A/G	10.08.43	Evaded capture for 38 days before being betrayed.
J.M Trend	F/S	W/Op	13.06.44	Evaded.
D.I Turner	P/O	Nav	23.06.43	Evaded.

APPENDIX VI
List of COMBAT AIRCRAFT
used by No.15/XV SQUADRON

WORLD WAR ONE

Serial	Mk.	Code	Remarks
F.E.			
5202	2b		Aircraft named "Newfoundland No.4".
6328	2b		Aircraft also used by No.20 Squadron.
6331	2b		
BRISTOL SCOUT			
4669	C		
5314	D		
B.E.2			
2077	2c		
2104	2c		
2120	2c		Wrecked 23.06.16.
2480	2c		Aircraft named "Punjab No.35, Mariana".
2494	2e		Wrecked 03.08.16.
2532	2c		
2548	2e		
2561	2e		Shot down during ground support operations, by Leutnant Werner Voss, on 01.04.17.
2572	2c		Crashed 20.06.16.
2578	2c		Aircraft named "Rhodesia No.3".
2606	2c		
2617	2c		Damaged by enemy fire in aerial combat, during an artillery ranging patrol on 16.09.16.
2618	2c		
2627	2c		
2637	2c		Crashed 16.09.17.
2639	2c		Struck-off-charge 03.04.17.
2694	2c		
2705	2c		Shot down by a Fokker monoplane, during a reconnaisance escort duty over the German lines on 17.01.16.
2715	2c		Armored version.
2727	2c		To No.39 Squadron.
2761	2c		
2766	2c		
2767	2c		
2775	2e		
2788	2e		
2810	2e		
2836	2e		
2840	2e		
2868	2e		
3166	2e		
4017	2c		
4019	2c		Shot down on 20.11.16..
4085	2c		
4107	2c		Shot down by German fighters, at 08.40 hours, on 19.01.16, whilst flying on escort duty during a Reconnaissance patrol.
4116	2c		Shot down on 30.03.16, by Oberleutnant Max Immelmann; 13th victory.
4123	2c		Damaged in air to air combat on 19.01.16.
4126	2c		
4127	2c		
4129	2c		
4153	2c		Shot down by a Fokker Scout, piloted by Leutnant G. Leffers, whilst flying an escort duty on a reconnaissance patrol over the Somme, on 14.03.16.
4187	2c		Shot down over Thiepval Wood by Oberleutnant H. Bethge, during an artillery spotting operation duty, on 29.08.16.
4190	2c		Shot down behind enemy lines, between St. Pierre Divion and Grandcourt, at 09.00 hours, on 02.10.16.
4198	2c		
4201	2c		Armored version.
4203	2c		
4205	2c		Shot down in aerial combat over the Somme on 26.10.16.
4300	2c		
5835	2e		
6255	2d		
6285	2e		
6286	2e		
6758	2e		
7060	2e		Shot down on 30.04.17.
7105	2e		Shot down in flames, by Leutnant Erwin Bohme, of Jasta Boelcke, on 04.02.17.

7144	2e	Shot down in flames, by enemy Halberstadt Scout aircraft, during an artillery observation patrol, on 06.02.17.	
7172	2e	Crashed to earth on 22.11.16.	
7177	2e		
7178	2e		
7225	2e		
7236	2e	Shot down on 03.04.17, by Hauptmann T. Von Osterroht, of Jasta 12; 2nd victory.	
7254	2e	Damaged in air to air combat, whilst engaged in a photographic reconnaisance operation on 24.03.17.	
7633	2d		
A2780	2e		
A2790	2e		
A2836	2e		
A2840	2e		
A2841	2e		
A2866	2e	Crashed at Courcelles-le-Court, France on 28.05.17.	
A2868	2e	Wrecked 20.04.17.	
A2875	2e		
A2886	2e		
A2895	2e		
A3157	2e	Destroyed by German artillery after force-landing during a photographic reconnaisance patrol, on 06.04.17.	
A3161	2e	Crashed 02.04.17.	
A3166	2e	Aircraft named "Saran".	

R.E.8.

A3452			
A3478			
A3484			
A3494			
A3542			
A3572		"Presented by the Government of Johore.No.5".	
A3695			
A3697			
A3698		Struck-off-charge 11.04.18.	
A3702		Struck-off-charge ??.11.17.	
A3703		Aircraft named "Black Watch No.2".	
A3704			
A3730			
A3862		Shot down 24.09.17.	
A4074			
A4268			
A4278			
A4296			
A4310		Shot down during a reconnaissance operation, by Vizefeldwebel Riessinger, of Jasta 12, on 15.06.17.	
A4452		To No.53 Squadron.	
A4453		Aircraft named "Johore No.11".	
A4588			
A4596			
A4618		Crashed 07.09.17.	
A4702			
A4704	9	'B' Flight.	
A4711		Aircraft named "Malaya No.12".	
A4728			

A4749			
B742		Shot down during an artillery ranging patrol, by Baron Manfred Von Richtofen, on 26.03.18. This aircraft was claimed by the 'Red Baron' as his 70th victory.	
B821			
B836	15	Rebuilt aircraft. Struck-off-charge 21.03.18.	
B2257			
B2258			
B2260	12	Aircraft named "Zanzibar No.15".	
B2268			
B2276	13		
B2277			
B2280			
B3412			
B3422			
B5031		Shot down by ground fire, during a low level operation on 21.11.17.	
B5032		Shot down on 10.11.17.	
B5065			
B5068			
B5897			
B5900			
B6504		Shot down by ground fire during a low level operation on 30.11.17.	
B6518		Struck-off-charge 14.04.18.	
B6521			
B6529			
B6532			
B6661	8	Completed 350 operational hours.	
B7716			
B7887		Rebuilt aircraft.	
C2240			
C2361		Shot down in flames by three enemy aircraft, whilst participating in army co-operation duties over Buire, France, on 03.05.18.	
C2488			
C2489			
C2513			
C2515			
C2719			
C2724			
C2796		Shot down by flak on 09.11.18.	
C2881			
C2915			
C2917			
C2967		Shot in combat with enemy aircraft, during a counter-attack patrol on 09.11.18.	
C3043			
C5041			
C5065			
C5068			
D4733			
D4736			
D4740		Shot down during army co-operation duties on 07.08.18.	
D4847		Shot down on 08.11.18.	

D4923			
D4944			
D6732			
E52			
E91			
E107			
E268			
E1117			
F6014			
F6203			
H7018			Rebuilt aircraft

WORLD WAR TWO

FAIREY BATTLE

K9204	I		To No.150 Squadron.
K9212	I		To No.12 Squadron
K9214	I		To No.63 Squadron
K9224	I		To No.15 E.F.T.S.
K9225	I		To West Raynham.
K9226			To No.4 Bombing and Gunnery School.
K9227	I	E	To No.4 Bombing and Gunnery School.
K9228	I	F	To No.4 Bombing and Gunnery School.
K9229	I		To No.98 Squadron.
K9233	I	J	To No.4 Bombing and Gunnery School.
K9282	I		To No.88 Squadron.
K9300	I		To Royal Canadian Air Force as No.2002 on 11 .06.41
K9301	I	L	To No.142 Squadron.
K9302	I		Crashed on approach to Abingdon on 21.04.39.
K9303	I		To No.4 Bombing and Gunnery School.
K9304	I		To Royal Canadian Air Force as No.1707 on 26.09.40.
K9305	I		Crashed in forced-landing at Steppingley, Bedfordshire, on 22.11.38.
K9311	I	O	To Royal Canadian Air Force as No.1753 on 14.10.40.
K9312	I		To Station Flight West Raynham.
K9358	I	V	To Station Flight West Raynham.
K9359	I		To No.4 Bombing and Gunnery School.
K9366	I		To No.142 Squadron.
K9367	I		To No.142 Squadron.
K9368	I		To No.12 Squadron.
K9369	I		To No.150 Squadron.
K9422	I		To No.98 Squadron.
K9457	I		To No.4 Bombing and Gunnery School.
K9479	I		To Abingdon.
L4938	I		To No.150 Squadron.
L4939	I		Crashed on take-off from Abingdon on 20.02.39.
L4940	I		To Royal Canadian Air Force on 01.11.40.
L4980	I		To No.105 Squadron.
L5226	III		To No.142 Squadron.
L5229	III		To Abingdon.
L5230	III		To No.105 Squadron.
L5231	III		To No.142 Squadron.
L5232	III		To No.218 Squadron.
L5233	III		To No.88 Squadron.
L5234	III		To No.103 Squadron.
L5235	III		To No.142 Squadron.
L5236	III		To No.103 Squadron.
L5237	III		To No.218 Squadron.
L5238	III		To No.105 Squadron.
L5239	III		Presumed lost over France ??.05.40.
L5248	I		To No.88 Squadron.
L5474	I		Lost in France ??.05.40
N2024	I		To No.4 Bombing and Gunnery School.
N2025	I		To Royal Aircraft Establishment.
P2177	I	Y	To No.105 Squadron.
P2255	I		To No.226 Squadron.

BRISTOL BLENHEIM

L1196	I		To No.40 Squadron.
L8755	IV		To No.254 Squadron.
L8800	IV		To No.114 Squadron.
L8847	IV		Missing after attack against Maastricht, Holland on 12.05.40.
L8848	IV		To No.218 Squadron.
L8849	IV		Missing after attack against Maastricht, Holland on 12.05.40.
L8850	IV		To Air Fighting Developement Unit.
L8851	IV		Missing after attacking enemy columns on 11.06.40.
L8852	IV	Q	Missing after attacking enemy columns on 18.05.40.
L8853	IV		Crashed near Preux-au-Bois, France on 18.05.40.
L8854	IV		Crashed in forced-landing during snowstorm on 16.01.40.
L8855	IV		To No.264 Squadron.
L8856	IV		Force-landed in Belgium on 15.05.40
L9024	IV		Missing after attacking enemy columns on 11.06.40.
L9030	IV		Missing 20.05.40.
L9204	IV		To No.114 Squadron.
L9208	IV		To No.110 Squadron.
L9217	IV		To No.110 Squadron.
L9323	IV		To No.107 Squadron.
L9403	IV		Missing in Arras area,France on 23.05.40.
L9413	IV		To No.139 Squadron.
L9469	IV		Missing after attacking airfields in Holland and N.W. Germany on 25.07.40.
N3588	IV		To No.21 Squadron.
N3591	IV		To No.40 Squadron.
N3598	IV		To No.57 Squadron.
N3627	IV	L	To No.139 Squadron.
N6151	IV		Missing after attack against Maastricht, Holland on 12.05.40.
N6156	IV		To Royal Aircraft Establishment.
N6166	IV		To No.18 Squadron.
N6177	IV		To No.51 Operational Training Unit.
N6228	IV		To No.107 Squadron.
P6911	IV		Missing after attack against Maastricht, Holland on 12.05.40.
P6912	IV		Missing after attack against Maastricht, Holland on 12.05.40.
P6913	IV		Missing Calais area, France on 25.05.40.
P6914	IV		Missing after attack against Maastricht, Holland on 12.05.40.
P6917	IV	H	Failed to return 18.05.40.
R2786	IV		Failed to return 14.05.40.

R2791	IV		To No.139 Squadron.
R2796	IV		To No.21 Squadron.
R3594	IV		To No.114 Squadron.
R3603	IV		Crash-landed in flames having been attacked by Me.110 over Holland on 18.06.40.
R3604	IV		To No.82 Squadron.
R3614	IV	D	Crashed on landing at RAF Alconbury on 24.05.40.
R3704	IV		To No.139 Squadron.
R3706	IV		Missing 20.05.40.
R3746	IV		Missing over Poix area, France on 08.06.40.
R3747	IV		Missing after attack against Le Bourquet, France on 12.06.40.
R3764	IV		Missing after attack against Diepholz airfield on 30.07.40.
R3766	IV		To No.114 Squadron.
R3767	IV		To No.82 Squadron.
R3768	IV		Missing after attack against Lannion airfield, France on 12.08.40.
R3769	IV		Crashed at Burton Latimer, Northamptonshire on 04.09.40.
R3770	IV		Missing over Foret de Gunes area, France on 18.08.40.
R3771	IV	W	Dived into ground near Whitchurch, Shropshire on 04.08.40.
R3777	IV		To No.45 Squadron.
R3894	IV		To No.8 Squadron.
R3896	IV		Failed to return from a photo-recco over Terneuzen/Ghent, Belgium on 07.07.40.
R3904	IV		To No.82 Squadron.
R3905	IV		To No.139 Squadron.
T1859	IV		To No.17 Operation Training Unit.
T1860	IV		To No.107 Squadron.
T1924	IV		To No.54 Operational Training Unit.
T1956	IV		To No.1 Air Armament School.
T1957	IV		To No.1 Air Armament School.
T1986	IV		To No.1 Air Armament School.
T2002	IV		To No.72 Operational Trainig Unit.
T2127	IV		To No.14 Squadron.
T2138	IV		To No.107 Squadron.
T2226	IV		To No.11 Squadron.
T2227	IV		To No.110 Squadron.
T2231	IV		Lost at sea en route to Takoradi, Africa ??.04.41.
T2276	IV		To No.105 Squadron.

VICKERS WELLINGTON

L4343	I	LS-	To No.311 Squadron.
L7797	Ic	LS-V	To No.218 Squadron.
L7889	Ic	LS-T	To No.1504 Flight.
N2752	Ic	LS-	To No.311 Squadron.
N2843	Ic	LS-G	To No.40 Beam Approach Training Flight.
N2856	Ic	LS-F	To No.99 Squadron.
N2871	IA	LS-	To No.311 Squadron.
N2954	IA	LS-	To No.11 Operational Training Unit.
R1066	Ic	LS-	To No.40 Squadron.
R1163	Ic	LS-	To No.75 Squadron.
R1169	Ic	LS-D	To No.20 Operational Training Unit.
R1218	Ic	LS-H	Abandoned by crew when the aircraft ran out of fuel, whilst returning from an attack against Kiel, on 25.04.41. The aircraft crashed at Sand Hutton, seven miles north east of York.

R1222	Ic	LS-J	To No.115 Squadron.
R1240	Ic	LS-K	To No.40 Squadron.
R1279	Ic	LS-L	To No.9 Squadron.
R1280	Ic	LS-M	To No.115 Squadron.
R1436	Ic	LS-X	To No.218 Squadron.
R1464	Ic	LS-Q	To No.40 Squadron.
R1498	Ic	LS-W	To No.1504 Beam Approach Training Flight.
R1596	Ic	LS-R	To No.218 Squadron.
T2624	Ic	LS-B	To Czechoslovakian Training Unit.
T2702	Ic	LS-H	Shot down by a nightfighter, piloted by Hauptmann Walter Ehle, of II/NJG1, during an attack against Hannover on 10.02.41. The aircraft crashed 15 kms west of Kampen, Netherlands, at 23.35 hours.
T2703	Ic	LS-A	Shot down by a nightfighter, piloted by Feldwebel Karl-Heinz Scherfling, off III/NJG1, during an attack against Bremen on 31.03.41. The aircrfaft crashed near the railway station at Haren-Ems, Germany.
T2715	Ic	LS-E	To No.57 Squadron.
T2806	Ic	LS-N	To No.218 Squadron.
T2847	Ic	LS-R	Struck-off-charge 17.02.41.
T2918	Ic	LS-P	To No.218 Squadron.
T2961	Ic	LS-S	To No.57 Squadron.
W5449	Ic	LS-	To No.218 Squadron.

SHORT STIRLING

Conversion to the Short Stirling bomber began on 11 April 1941.

N3638	I	LS-	To No.149 Squadron
N3642	I	LS-	To No.26 Conversion Flight
N3644	I	LS-	Undercarriage collapsed when taxying at Wyton on 21.05.41
N3646	I	LS-R	To No.214 Squadron
N3654	I	LS-B	Shot down by a nightfighter, piloted by Oberleutnant Prinz Egmont Zur Lippe Weissenfeld, during an attack against Berlin.
N3656	I	LS-H	Crashed at Honington with hydraulics and prop shot away, having returned from an attack against Berlin on 13.08.41
N3658	I	LS-E	Shot down by a nightfighter, piloted by Leutenant Loos, of I/NJG 1 based at Venlo, during an attack against Essen on 08.08.41. The aircraft crashed at Overasselt, Netherlands, at 02.37 hours.
N3659	I	LS-N	Shot down by a nightfighter during an attack against Berlin on 13.08.41. The aircraft crashed at Berxen, Germany.
N3660	I	LS-M	Undercarriage collasped whilst carrying out an overshoot in bad visibility at RAF Warboys on 28.09.41.
N3661	I	LS-Q	Undercarriage collapsed whilst landing at RAF Wyton on return from an attack against Hazebrouch marshalling yards on 11.07.41.
N3665	I	LS-S,B	Failed to return from an attack against Brest harbor on 18.12.41. The aircraft was last seen over the sea with one engine smoking badly having been attacked by an enemy fighter.
N3667	I	LS-T	Overshot during landing at RAF Wyton, on returning from an attack against Nuremberg on 12.10.41, and hit an obstruction.

N3668	I	LS-	Undercarriage collapsed due to landing on one wheel, following a wing dropping on final approach to RAF Alconbury on 08.01.42.
N3669	I	LS-C,H	To No.3637M. Displayed on a bomb site, outside St Paul's Cathedral, as LS-H, during London's 'Wings for Victory' week in March 1943.
N3670	I	LS-V	To No.7 Squadron
N3671	I	LS-H	To No.1651 Conversion Unit.
N3673	I	LS-D	Shot down by flak during an attack against Essen on 09.03.42. The aircraft crashed at Woesteloeve, Apeldoorn, Netherlands.
N3674	I	LS-T	To No.214 Conversion Unit.
N3675	I	LS-S	To No.15 Conversion Flight.
N3676	I	LS-U	To No.1651 Conversion Unit
N3683	I	LS-C	To No.7 Conversion Flight
N3703	I	LS-G	Crashed at Godmanchester after being badly damaged by flak, during an attack against Essen on 11.04.42.
N3704	I	LS-A	To No.75 Squadron.
N3707	I	LS-N	To No.1651 Conversion Unit.
N3728	I	LS-T	Shot down by a nightfighter, piloted by Oberleutnant Reinhold Knacke, of II/NJG2, during an attack against Essen on 03.06.42. The aircraft crashed at Melick-Herkenbosch, Netherlands, at 03.10 hours.
N3756	I	LS-C	Crashed into a pond at Potash Farm, Brettenham, 14 miles from Ipswich, at 03.37 hours, on 12.08.42. The aircraft, having been attacked by two Ju.88 nightfighters during an attack against Mainz, had flown home on three engines. It was attempting to land at RAF Wattisham, when it crashed and burst into flames.
N3757	I	LS-G	Shot down by flak during an attack against Bremen on 30.06.42. The aircraft crashed at Hartward, Esens, Germany.
N3758	I	LS-V	To No.1657 Conversion Unit.
N3759	I	LS-Q	The aircraft crashed into the North Sea during a 'Gardening' operation in the Baltic on 19.09.42.
N6004	I	LS-F	To No.1427 Flight.
N6007	I	LS-	To No.7 Squadron.
N6015	I	LS-A	Shot down by flak during an attack against Hamburg on 30.06.41. The aircraft crashed into the sea off Kiel.
N6016	I	LS-G	Shot down by a nightfighter, during an attack against Hamburg on 30.06.41. The aircraft crashed at Ellerbeck, Germany.
N6018	I	LS-C	Shot down by an enemy fighter during a daylight attack against Lille on 19.07.41. The aircraft crashed at Killem, France.
N6021	I	LS-D	Shot down by a nightfighter, during an attack against Hamburg on 16.09.41. The aircraft crashed at Hemslingen, Germany.
N6024	I	LS-K	To No.1651 Conversion Unit.
N6029	I	LS-K	Ditched in the North Sea due to a shortage of fuel, having been attacked by a Ju.88 during a raid against Berlin on 26.07.41
N6030	I	LS-P	Ditched in the North Sea whilst returning from an aborted attack against Wessel on 18.07.41.
N6038	I	LS-R	Ditched into the sea, 50 miles off Milford Haven, Pembrokeshire, whilst returning from an attack against La Rochelle, France, on 23.07.41. It is thought the aircraft ran short of fuel.
N6040	I	LS-C	Undercarriage collapsed on landing at RAF Wyton on 25.10.41.
N6043	I	LS-G	Crashed north of Ramsey St. Mary, Huntingdonshire, on 14.08.41, when taking-off from RAF Alconbury for an attack against Hannover. During climb out the port inner engine lost power, the undercarriage could not be raised and the aircraft would not maintain height.
N6044	I	LS-O,X,E	To No.15 Conversion Flight.
N6045	I	LS-U	Shot down by flak and crashed during an attack against Berlin on 08.09.41. The aircraft crashed at Hengelo, Netherlands.
N6047	I	LS-P	Shot down by a nightfighter during an attack against Nuremberg on 13.10.41. The aircraft crashed at Mariembourg, Belgium.
N6065	I	LS-G	To No.149 Conversion Flight.
N6067	I	LS-E	Force-Landed at Beck Lodge Farm, Mildenhall, Suffolk, due to a shortage of fuel on 26.02.42.
N6076	I	LS-D	Crash-landed at Newmarket due to undercarriage defect on return from an attack against Dortmund on 15.04.42.
N6086	I	LS-F	To No.101 Conversion Flight. Aircraft purchased by Lady MacRobert in memory of her three sons, two of whom were pilots with the Royal Air Force. The Stirling carried the name 'MacRobert's Reply' together with the MacRobert family crest.
N6088	I	LS-G,Q,X	To No.15 Conversion Flight.
N6092	I	LS-O	To No.214 Squadron.
N6093	I	LS-P,C	Crashed at Wyton with damaged undercarriage, on return from Munster on 23.01.42, having been attacked fifteen times by a nightfighter.
N6094	I	LS-R	Undercarriage collapsed when the aircraft swung to avoid a house, having landed in bad visibility at RAF Wyton on 25.03.42
N6096	I	LS-	To No.26 Conversion Unit.
N6097	I	LS-C	Hit trees whilst taking-off for an attack against Kiel on 15.11.41.
N6098	I	LS-G	Caught fire whilst being refuelled at Lossiemouth, Scotland on 29.01.42.
R9144	I	LS-Q,R	To No.1657 Conversion Unit.
R9151	I	LS-Q,U	Crash-landed at Docking, having been badly damaged by flak during an attack against Osnabruck on 18.08.42.
R9153	I	LS-U	Shot down and crashed at Mesmont, France, during an attack against Nuremberg, on 29.08.42.
R9168	I	LS-T	Shot down by a nightfighter and crashed at Epe, Netherlands, during an attack against Diepholz Airfield on 17.17.42
R9192	I	LS-E	To No.1657 Conversion Unit.
R9193	I	LS-S	To No.1651 Conversion Unit.
R9195	I	LS-P	To No.1657 Conversion Unit.
R9201	I	LS-U	Crashed at St. Andre des Eaux, France, during a mining operation in the Gironde on 07.11.42
R9268	I	LS-R	To No.1665 Conversion Unit.
R9274	I	LS-B	Shot down by a nighfighter, piloted by Hauptmann Dormann, of III/NJG1, during an attack against Hamburg, on 04.02.43. The aircraft crashed at Renkum, Nijmegen, Netherlands, at 20.58 hours.

R9279	I	LS-J	Failed to return from an attack against Cologne on 27.03.43. The crash site is unknown.
R9302	I	LS-F	To No.15 Conversion Flight.
R9303	I	LS-P	To No.214 Squadron.
R9304	I	LS-U	To No.1651 Conversion Unit.
R9308	I	LS-P	Overshot whilst landing at RAF Waterbeach,in bad weather, having been attacked and damaged by a Ju.88, on return from Vegesack on 20.07.42.
R9310	I	LS-	To No.149 Squadron.
R9311	I	LS-	To No.218 Squadron.
R9312	I	LS-C	Crashed at Pont du Cens, Nantes, France, during a mining operation in the Bayonne on 17.10.42.
R9313	I	LS-	To No.218 Squadron.
R9314	I	LS-	To No.149 Squadron.
R9315	I	LS-O	To No.1657 Squadron.
R9318	I	LS-B,J	Crashed near Amsterdam, during an attack against Essen, Germany on 17.09.42.
R9319	I	LS-	To No.214 Squadron.
R3951	I	LS-R	Crashed in the Great Belt between Nyborg and Korsoer, Denmark, during a mining operation in the Baltic on 19.09.42.
R9352	I	LS-T	Crashed at Siemolten, France, during an attack against Emden on 20.06.42.
R9353	I	LS-B	To No.1657 Conversion Unit.
W7426	I	LS-V	To No.26 Conversion Flight.
W7427	I	LS-B	To No.26 Conversion Unit.
W7428	I	LS-F,Z	Failed to return from an attack against Brest on 18.12.41, having been attacked over the sea by a Messerchmitt Bf.109.
W7429	I	LS-J,X	Undercarriage collapsed whilst landing at RAR Warboys, with port engines u/s, on return from an attack against Pilsen on 29.10.41.
W7431	I	LS-A	Crashed at Catsholm Farm, Methwold, having run short of fuel on return from an attack against Bremen on 21.10.41.
W7432	I	LS-L	To No.1651 Conversion Unit.
W7435	I	LS-W	Undercarriage collapsed when the aircraft swung off the runway at RAF Alconbury, during the take-off run for an attack against Madgeburg on 14.08.41.
W7437	I	LS-L	Crashed into the North Sea during an attack against Madgeburg on 15.08.41.
W7439	I	LS-N	To No.1615 Conversion Flight.
W7441	I	LS-	To No.7 Squadron.
W7443	I	LS-W,J	To No.1651 Conversion Unit.
W7447	I	LS-	To Telecommunication Flying Unit.
W7448	I	LS-N,E	Crashed into the North Sea during an attack against Essen on 07.04.42.
W7450	I	LS-A	Undercarriage collapsed following a bounced landing and swing at RAF Warboys on 25.11.41. The aircraft, which was participating in an attack against the Rhur, had returned early with an engine defect.
W7455	I	LS-	To No.214 Squadron.
W7460	I	LS-	To No.149 Squadron.
W7463	I	LS-B	To No.15 Conversion Flight.
W7464	I	LS-H	To No.218 Squadron.
W7504	I	LS-A	Undercarriage collapsed following an overshoot whilst landing at RAF Wyton on 27.07.42, on return from an attack against Hamburg. The aircraft was badly holed by heavy flak during the attack.
W7505	I	LS-V	To No.1651 Conversion Unit.
W7511	I	LS-T	Swung whilst taking-off for an air test at RAF Wyton on 08.04.42. The undercarriage collapsed and the port engine caught fire.
W7513	I	LS-R	To No.149 Squadron.
W7514	I	LS-B	Crashed at Kravlund, Denmark, at 01.10 hours, during an attack against Rostock on 26.04.42.
W7515	I	LS-Q	Crashed into the North Sea during a mine-laying operation in the Frisians on 30.05.42.
W7516	I	LS-S	To No.1651 Conversion Unit.
W7518	I	LS-U,C	Shot down by a nightfighter and W,G crashed at 01.00 hours in the Muye Polder, at Maartensdijk in the Province of Zeeland, Netherlands, during at attack against Berlin on 02.03.42.
W7519	I	LS-O	Shot down by a nightfighter and crashed into the North Sea, during a mine-laying operation on the Wangeroog area, Netherlands on 14.04.42.
W7523	I	LS-C	Crashed into trees along the roadside, 1.5 miles north east of Graveley, Cambridgeshire, due to engine failure during take-off from RAF Wyton on 19.05.42.
W7524	I	LS-D	Shot down by flak and crashed into the North Sea off Esbjerg, Denmark, during an attack against Lubeck on 17.07.42.
W7525	I	LS-E	Stalled on approach, after the port outer engine failed, whilst landing at RAF Bourn on 22.08.42.
W7528	I	LS-G	Crashed at Brodersby, Germany, during an attack against Warnemunde on 09.05.42.
W7531	I	LS-F	Shot down by flak and crashed at Galsklint, Denmark, during a mine-laying operation in the Copenhagen Sound on 18.05.42. This aircraft was the replacement Stirling which carried the name "MacRobert's Reply".
W7536	I	LS-G	Overshot whilst landing in bad weather at RAF Wyton , on 22.05.42, and hit a ditch. The aircraft was then struck by Stirling R9312.
W7561	I	LS-F	To No.1651 Conversion Unit.
W7576	I	LS-G	Shot down by a nightfighter and crashed at Horst, Netherlands on 26.07.42, during an attack against Duisburg.
W7578	I	LS-A	Crashed at Noyes-le-Val, France, on 20.09.42, during an attack against Munich.
W7585	I	LS-T,U	Crashed on the boundary of Bassingbourn airfield, Cambridgeshire, and caught fire following engine failure during an air test on 29.12.42.
W7588	I	LS-J	Landed at RAF Coltishall and caught fire, having been badly shot up during an attack against Hamburg on 29.07.42.
W7611	I	LS-F	Crashed at Hesdin, France, during an attack against Karlsruhe on 03.09.42.
W7624	I	LS-E	Shot down by a nightfighter and crashed at Ambt Delden, Netherlands, at 00.08 hours, during an attack against Kassel on 28.08.42.
W7633	I	LS-P	To No.1657 Conversion Unit.
W7634	I	LS-G	Crashed into the sea near Peenemunde during an attack against Lubeck on 02.10.42.
W7635	I	LS-V	Crashed into the North Sea during a mine-laying operation in the Baltic on 09.12.42.
BF311	I	LS-	To No.1651 Conversion Unit.
BF327	I	LS-D	Shot down by a nightfighter and crashed near

			Beusichem, 6 kms east of Culemborg, Netherlands, at 23.55 hours, during an attack against Kassel on 27.08.42.
BF329	I	LS-A	Shot down by a nightfighter and crashed at Romedenne, Belgium, during an attack against Mainz on 13.08.42.
BF340	I	LS-	To No.7 Squadron.
BF347	I	LS-J	Crashed at West Malling, Kent, having requested priority landing, on return from an attack against Dusseldorf on 11.09.42.
BF350	I	LS-O	To No.1657 Conversion Unit.
BF352	I	LS-U	Undercarriage collapsed when the aircraft swung on landing, having returned early with an engine defect, from a mine-laying operation near the Wangeroog, on 09.09.42.
BF353	I	LS-E	Crashed at Ost Flevoland in the Ijsselmeer, 10 kms east of Harderwijk, Netherlands, during an attack against Essen on 16.09.42.
BF355	I	LS-F	To No.1657 Conversion Unit.
BF356	I	LS-D	Overshot and hit trees whilst landing at RAF Bourn, on return from a mine-laying operation in the Frisians on 17.12.42. The port outer engine was u/s and the port inner cut out on touchdown.
BF376	I	LS-N	To No.90 Squadron.
BF378	I	LS-F,T	Shot down by a nightfighter, piloted by Oberleutnant Hans-Joachim Jabs, of IV/NJG1, during an attack against Wilhelmshaven on 19.02.43. The aircraft crashed in the North Sea, 10 kms north west of Schiermonnikoog, Netherlands, at 21.14 hours.
BF380	I	LS-B	Crashed one mile north of Bourn, having attempted a three-engine landing on return from a mine-laying operation off Bayonne on 18.12.42.
BF384	I	LS-R	Undercarriage collapsed when the aircraft swung during take-off from RAF Bourn, in gusty winds, for an attack against Turin on 18.11.42.
BF386	I	LS-Q	Crashed at Salter's Lode, south west of Downham Market, Suffolk, on 29.10.42. The aircraft, which was flying on an air test, was seen to emerge from cloud and execute a steep turn which deleveped into a dive from which it did not recover.
BF389	I	LS-	To No.149 Squadron.
BF392	I	LS-	To No.149 Squadron.
BF411	I	LS-A	Shot down by a nightfighter, piloted by Oberleutnant Hans-Joachim Jabs, of IV/NJG1, during an attack against Wilhelmshaven, on 20.02.43. The aircraft crashed into the North Sea, 50 kms north of Terschelling, Netherlands.
BF412	I	LS-	To No.75 Squadron.
BF435	I	LS-P	To No.90 Squadron.
BF436	I	LS-E	To No.1651 Conversion Unit.
BF439	I	LS-D	To No.1653 Conversion Unit.
BF448	I	LS-T	Shot down by a Messerschmitt Bf.110 nightfighter, crewed by Oberfeldwebel Schellwat and Unteroffizier Willman, of V/NJG1, during an attack against Cologne on 14.02.43. The aircraft, which was flying on the outbound route at an altitude of 4.500m, exploded in the air at 20.56 hours.
BF457	III	LS-B	Shot down by a nightfighter, piloted by Oberleutnant Hans-Joachim Jabs, of IV/NJG1, during an attack against Wilhelmshaven on 20.02.43. The aircraft crashed in the North Sea, near Ameland, Netherlands.
BF460	III	LS-C,F	Shot down by a nightfighter, piloted by Hauptmann Johannes Hager, of Stab II/NJG1, during an attack against Nuremberg on 11.08.43. The aircraft crashed at Doische, Belgium.
BF465	III	LS-	To No.75 Squadron.
BF469	III	LS-	To No.214 Squadron.
BF470	III	LS-G	Shot down and crashed at Haste, 8 kms north east of Osnabruck, during an attack against Kassel on 04.10.43.
BF474	III	LS-H	Crashed at St. Erme, France, during an attack against Mannheim on 17.04.43.
BF475	III	LS-T	Crashed at St. Genevieve, France, during an attack against Frankfurt on 11.04.43.
BF476	III	LS-D,P	Crash-landed at Kragelund, 10 miles north of Vejle, Denmark, during an attack against Rostock on 21.04.43. The crew set fire to the aircraft.
BF482	III	LS-R	Shot down by flak and crashed at Dortmund, during an attack on the city on 24.05.43.
BF521	III	LS-P	To No.622 Squadron.
BF533	III	LS-H,K	To No.1654 Conversion Unit.
BF534	III	LS-E,L	Destroyed by the force of the explosion when Halifax bomber, JB387, coded LN-D,of No.77 Squadron, was attacked by a Messerschmitt nightfighter, of NJG1, during an attack against Dusseldorf on 26.05.43. The Stirling crashed at Julich, Germany.
BF569	III	LS-C,Y,V	Crashed at Vaux, France, during an attack against Monlucon on 16.09.43.
BF571	III	LS-U	Shot down by a nightfighter and crashed in the Waddenzee, near Ameland, Netherland,at 03.04 hours, during an attack against Dusseldorf on 12.06.43.
BF579	III	LS-V	Shot down by a nightfighter and crashed into the North Sea, during a mine-laying operation in the eastern Frisians, Netherlands on 04.07.43.
BK595	I	LS-A	Crashed at Playa de Oro, Spain, due to engine failure whilst returning from an attack against Turin, Italy on 21.11.42.
BK597	I	LS-	To No.149 Squadron.
BK611	I	LS-U	Shot down by both flak and a nightfighter during an attack against Dusseldorf on 26.05.43. The aircraft crashed near Grubbenvorst, Netherlands.
BK648	III	LS-J	Crashed at Menden, Germany, during an attack against Cologne on 04.07.43.
BK652	III	LS-V,Q	To No.622 Squadron.
BK654	III	LS-W	To No.1661 Conversion Unit.
BK656	III	LS-A	Shot down by a nightfighter, piloted by Hauptmann Wilhelm Dorman, of III/NJG1, during an attack against Mulheim on 23.06.43. The aircraft crashed near the statue of Christiaan De Wet, in the National Park "De Hoge Veluwe", 5 kms south west of Otterlo near Ede, Netherlands, at 02.07 hours.
BK657	III	LS-C	Attacked by a nightfighter in the vacinity of north west of Utrecht, on the outbound leg of an attack

			against Duisberg on 27.04.43. The bomber was shot down, and crashed in flames, near Portengen, Netherlands, at 02.15 hours.
BK658	III	LS-K	Shot down by a nightfighter and crashed at Midwolda, Netherlands, during an attack against Dortmund on 05.05.43.
BK667	III	LS-H	Returned early from an attack against St. Nazaire, France, after the starboard outer and port inner engines failed. The bomber overshot in poor visibility and crash-landed at Clyffe Pypard airfield, Wiltshire on 22.03.43.
BK691	III	LS-F	Crashed at Hetzerath, Germany, during an attack against Mannheim on 17.04.43.
BK694	III	LS-O,C	Shot down by a nightfighter, crewed by Hauptmann Hoffmann and Oberfeldwebel Hofler, of IVNJG5, during an attack against Cologne on 29.06.43. The bomber crashed at Lommel, "Blauwe Kei", 12 kms north-north-east of Bourg Leopold, Province of Limburg,Belgium, at 02.18 hours.
BK695	III	LS-X	To No.75 Squadron.
BK697	III	LS-P	Crashed at Campneuville, France, during an attack against Nuremberg on 09.03.43.
BK698	III	LS-	To No.149 Squadron.
BK699	III	LS-E	Shot down by a nightfighter, during an attack against Gelsenkirchen on 26.06.43. The bomber crashed in the Waddenzee, 8 kms east of the island of Texel, Netherlands, at 02.21 hours.
BK703	III	LS-	To No.149 Squadron.
BK704	III	LS-Z	Shot down by a nightfighter and crashed at Barlo, Germany, during an attack against Bochum on 14.05.43,
BK707	III	LS-C,G	Crashed at Sousain, France, during an attack against Mannheim on 19.11.43.
BK719	III	LS-B	To No.1661 Conversion Unit.
BK764	III	LS-R	Crashed at Uyl, near Wassenburg, Germany, during an attack against Munchen Gladbach on 31.08.43.
BK766	III	LS-T	To No.622 Squadron.
BK774	III	LS-T,K	Crashed into the North Sea off Esbjerg, Denmark, during a mine-laying operation in the Kattegat on 04.09.43.
BK782	III	LS-X	Shot down by a nightfighter and crashed in the Houten/Schalkwijk area, Netherlands, at 01.45 hours, during an attack against Dortmund on 04.09.43.
BK805	III	LS-U	Shot down by a nightfighter, piloted by Hauptmann Wilhelm Dormann, of III/NJG1, during an attack against Essen on 26.07.43. The bomber crashed at Osterwick Kreis Ahaus, Germany at 00.40 hours.
BK816	III	LS-X,Y,E	To No.622 Squadron.
BK818	III	LS-O,R	To No.1661 Conversion Unit.
EE877	III	LS-	To No.149 Squadron.
EE907	III	LS-C	To No.1661 Conversion Unit.
EE908	III	LS-V	Shot down during an attack against Peenemunde on 18.08.43. The aircraft crashed into the sea near Griefswalde, Germany.
EE910	III	LS-	To No.199 Squadron.
EE912	III	LS-U	Crashed at Roskow, Germany, during an attack against Berlin on 01.09.43.
EE913	III	LS-	To No.199 Squadron.
EE940	III	LS-Y	Crashed at Ronnenberg, Germany, during an attack against Hannover on 28.09.43.
EE954	III	LS-J	Crashed after being struck by bombs during an attack against Frankfurt on 05.10.43.
EE974	III	LS-O	To No.90 Squadron.
EF131	III	LS-	Undercarriage collapsed when the aircraft attempted to swing, during an overshoot, whilst on acceptance test at RAF Mildenhall on 19.09.43.
EF133	III	LS-U	To No.218 Squadron.
EF161	III	LS-Y	To No.199 Squadron.
EF177	III	LS-S	To No.1661 Conversion Unit.
EF183	III	LS-D	To No.90 Squadron.
EF186	III	LS-W	To No.1661 Conversion Unit.
EF195	III	LS-	Undercarriage collapsed when the aircraft swung on take-off from RAF Mildenhall, foran air test on 15.10.43.
EF333	I	LS-X	Shot down by a nightfighter during a attack against Hamburg on 04.03.43.
EF339	I	LS-Y	Crash-landed at RAF Coltishall, on return from an attack against Hamburg, following an engine failure in flight on 30.07.43.
EF345	I	LS-M	Shot down by flak during an attack against Dortmund on05.05.43. The aircraft crashed at Anholt, Germany.
EF347	I	LS-T	Shot down by a nightfighter, piloted by Oberleutnant Kuthe, of IV/NJG1, during an attack against Berlin on 02.03.43. The aircraft crashed at Mantgum, Netherlands, at 00.58 hours.
EF348	I	LS-N	Shot down by a nightfighter, piloted by Oberleutnant Autenrieth, of II/NJG1, during an attack against Mulheim onn23.06.43. The aircraft crashed at Kessenich, 7 kms north-north-east of Maaseik, in the Province of Limburg, Belgium, at 01.35.
EF351	I	LS-L	To No.620 Squadron.
EF354	I	LS-Q,C	To No.1665 Conversion Unit.
EF355	I	LS-A	To No.1665 Conversion Unit.
EF359	I	LS-B	Shot down during an attack against Duisburg on 09.04.43. The aircraft crashed at Woltershoff, Germany.
EF391	I	LS-M	To No.622 Squadron.
EF399	I	LS-Y	To No.75 Squadron.
EF411	III	LS-	To No.149 Squadron.
EF412	III	LS-	To No.149 Squadron.
EF427	III	LS-A	Shot down by flak during an attack against Remscheid on 31.07.43. The aircraft crashed near the village of Manheim, to the south west of Cologne.
EF428	III	LS-N	Shot down during an attack against Remscheid on 31.07.43. The aircraft crashed at Kleinenbroich, Germany.
EF437	III	LS-Z	Crashed on approach to RAF Mildenhall, due to both port and starboard engines cutting out, on return from an attack against Hamburg on 28.07.43.
EF453	III	LS-	To No.199 Squadron.
EF459	III	LS-S	To No.90 Squadron.

Serial	Mark	Code	Fate
EF460	III	LS-B,N	To No.622 Squadron.
EF461	III	LS-	To No.622 Squadron.
EF490	III	LS-B	To No.622 Squadron.
EF518	III	LS-B	To No.1661 Squadron.
EH875	III	LS-S	Shot down during an attack against Berlin on 23.08.43. The aircraft crashed in the area of the target.
EH879	III	LS-	To No.149 Squadron.
EH888	III	LS-Z	Shot down by a nightfighter, during an attack against Cologne on 29.06.43. The aircraft crashed into the River Ijssel, at Heeschwijk, 2 kms north east of Monsfoort, Netherlands, at 02.40 hours.
EH890	III	LS-U	Ditched into the sea off Clacton, Essex, whilst returning from an attack against Wuppertal, Germany, on 25.06.43.
EH893	III	LS-J	Shot down by flak and nightfighter, during an attack against Hamburg on 28.07.43. The aircraft crashed at Hamburg-Ochsenwaerder, Germany.
EH897	III	LS-Z	To No.622 Squadron.
EH929	III	LS-F	To No.1661 Conversion Unit.
EH930	III	LS-A	To No.199 Squadron.
EH940	III	LS-U,H	To No.218 Squadron.
EH941	III	LS-V	Shot down during an attack against Mannheim on 24.09.43. The aircraft crashed at Hassloch, Germany.
EH980	III	LS-	To No.1654 Conversion Unit.
EH985	III	LS-O	Shot down by a nightfighter, during an attack against Nuremburg on 28.08.43. The aircraft crashed at Hasselburg, Germany.
EH990	III	LS-K	Crashed into the North Sea, during a mine-laying operation in the Kattegat on 08.10.43.
LJ451	III	LS-K	To No.622 Squadron.
LJ453	III	LS-	To No.75 Squadron.
LJ462	III	LS-	To No.75 Squadron.
LJ464	III	LS-	To No.1654 Conversion Unit.
LK386	III	LS-	To No.149 Squadron.
LK393	III	LS-B	To No.1661 Conversion Unit.
MZ264	III	LS-A	To No.622 Squadron.

AVRO LANCASTER

Conversion to the Avro Lancaster began during December 1943. "C" Flight of No.XV Squadron wore the code DJ.

Serial	Mark	Code	Fate
L7527	I	LS-A	Commenced as Manchester bomber, but completed by Messers A. V. Roe as a Mk.I. Lancaster bomber. Failed to return from an attack against Essen on 27.03.44.
R5490	I	LS-U	To the Signals School on 24.01.48.
R5508	I	LS-C	Struck-off-charge 15.01.47.
R5692	I	LS-S	To No.75 Squadron.
R5739	I	LS-K	Failed to return from an attack against Leipzig on 20.02.44.
R5846	I	LS-	To No.622 Squadron.
R5896	I	LS-N	Struck-off-charge 07.04.44.
R5904	I	LS-G,L	Failed to return from an attack against Homberg on 21.07.44.
R5906	I	LS-D	To No.622 Squadron.
W4174	I	LS-P	To No.75 Squadron.
W4181	I	LS-Q	To No.3 Lancaster Finishing School.
W4272	I	LS-P	To No.622 Squadron.
W4355	I	LS-A	Failed to return from an attack against Stuttgart on 16.03.44.
W4852	I	LS-B	Failed to return from an attack against Madgeburg on 22.01.44.
W4885	I	LS-	To No.5 Lancaster Finishing School.
W4980	I	LS-R,W	To No.3 Lancaster Finishing School.
ED310	I	LS-M,N	To No.75 Squadron.
ED323	I	LS-D	Failed to return from an attack against Berlin on 28.01.44.
ED376	I	LS-F	To No.3 Lancaster Finishing School.
ED383	I	LS-C	The bomber ran off the runway into soft ground and turned over onto its back, having overshot twice, whilst landing at RAF Lakenheath, on return from an attack against Augsburg on 26.02.44.
ED395	I	LS-K,M	To No.3 Lancaster Finishing School.
ED473	I	LS-H,D	Failed to return from an attack against Nantes, France, on 08.05.44.
ED610	I	LS-C	Failed to return from an attack against Berlin on 29.01.44.
ED628	III	LS-O	Failed to return from an attack against Berlin on 16.02.44.
ED727	III	LS-	To No.622 Squadron.
ED808	III	LS-	To No.622 Squadron.
ED826	III	LS-W	Dived in to the Wash, on the east coast of England, following engine failure whilst on a loaded climb and cross-country exercise on 13.01.44.
ED908	III	LS-	Missing after a daylight attack against Foret du Croc, France on 20.07.44.
HK612	I	LS-L	Collided with Lancaster, PB115, also from No.XV Squadron, whilst flying on the outward leg for a daylight attack against Homberg, Germany. The aircraft crashedat Keldonk, 2 kms south-south-west of Erp, Netherlands on 02.11.44.
HK614	I	LS-	To No.622 Squadron.
HK615	I	LS-	To No.622 Squadron.
HK616	I	LS-	To No.622 Squadron.
HK617	I	LS-	To No.622 Squadron.
HK618	I	LS-G	Shot down during an attack against Cologne on 28.01.45. The aircraft failed to reach the target.
HK619	I	LS-Y,O,V	Struck-off-charge 19.10.45.
HK620	I	LS-V,W	Failed to return from an attack against Hohenbudberg, Germany on 09.02.45.
HK622	I	LS-	To No.90 Squadron.
HK625	I	LS-	To No.90 Squadron.
HK626	I	LS-M,W	Hit by flak and exploded in mid-air during an attack against Oberhausen, Germany, on 04.12.44. The aircraft, which was 99% destroyed, crashed onto the railway station at Osterfeld.
HK627	I	LS-Q,F	Shot down by flak during an attack against Witten, Germany, on 12.12.44. The aircraft crashed in flames on Stocking Rly Station, West Witten, Germany.
HK628	I	LS-	Struck-off-charge 31.12.46.
HK647	I	LS-E,L,K	To Central Gunnery School.
HK648	I	LS-Z,H,F	Struck-off-charge.
HK693	I	LS-B	To No.1659 Conversion Unit.
HK695	I	LS-V	To G-H Flight.
HK765	I	DJ-Z	Struck-off-charge 23.07.46

HK772	I	LS-G,A,B	To No.44 Squadron.
HK773	I	LS-W	Starboard engine caught fire immediately after take-off for an attack against Bocholt, Germany, on 22.03.45. The aircraft, which was carrying a full bomb load, crashed and exploded in Brandon Wood, Norfolk.
HK789	I	LS-R	Struck-off-charge 23.09.46.
HK799	I	LS-D	Struck-off-charge 11.11.46.
JB475	III	LS-M	To No.195 Squadron.
LL752	I	LS-A	Failed to return from an attack against Louvain (Leuven), France on 12.05.44.
LL754	I	LS-P	Failed to return from an attack against Cologne on 21.04.44. The aircraft crashed near St Joseph's Church, at Koln-Nippes, Germany.
LL781	I	LS-B,L	Crash-landed at Friston, England, on return from an attack against Massy-Palaiseau on 08.06.44, having been attacked by a nightfighter.
LL801	I	LS-J	Shot down by a nightfighter, during an attack against Friedrichshafen on 28.04.44. The aircraft, which exploded in mid-air, crashed near Schoenau, on the French bank of the River Rhine.
LL805	I	LS-M	Shot down by a nightfighter, during an attack against Friedrichschafen on 28.04.44. The aircraft crashed near Geisingen, Germany.
LL806	I	LS-J	Struck-off-charge 05.12.45.
LL827	I	LS-O,P,Q	Failed to return from an attack against Chalons-Sur-Marne, France on 16.07.44. The aircraft crashed at St Gibrien, 5.5 kms, north-west of Chalons-Sur-Marne.
LL854	I	LS-S,Q	To No.184 Squadron.
LL858	I	LS-N	Crashed whilst taking-off from RAF Mildenhall, Suffolk, for a night cross country exercise on 31.04.44.
LL889	I	LS-L,B	Failed to return from an attack against Le Havre, France on 15.06.44.
LL890	I	LS-T	Failed to return from an attack against Wizernes, France on 07.07.44.
LL923	I	LS-O	Failed to return from a daylight attack against Ludwigshafen on 05.01.45.
LL945	I	LS-M	Failed to return from an attack against Massy-Palaiseau, France on 08.06.44.
LM109	I	LS-E	Shot down by flak during an attack against Calais on 24.09.44.
LM110	I	LS-G	Missing after an attack against Frankfurt on 13.09.44.
LM113	I	LS-K	Struck-off-charge 22.05.47.
LM121	I	LS-C	Missing after an attack against Trappes, France on 01.06.44.
LM142	I	LS-A	Missing after an attack against Stuttgart on 25.07.44.
LM156	I	LS-R	Missing after an attack against Gelsenkirchen on 13.06.44.
LM160	I	LS-D	Struck-off-charge 17.10.47.
LM167	I	LS-R	To No.622 Squadron.
LM233	I	LS-	Struck-off-charge.
LM238	I	LS-T	To No.44 Squadron.
LM240	I	LS-R	To No.149 Squadron.
LM441	I	LS-T	Shot down by flak, and crashed 3 kms north of Bonn, during an attack against Berlin on 25.05.44.

LM456	I	LS-C	Missing after an attack against Stuttgart on 21.02.44.
LM465	I	LS-U	Missing after an attack against Gelsenkirchen on 13.06.44.
LM468	I	LS-F	Crashed near Sainte Gemme-Moronval, France, during an attack against Dreux on 11.06.44.
LM473	I	LS-P	To No.3 Lancaster Finishing School.
LM490	I	LS-L	Shot down by nighfighters, in the region of Tetlow, during an attack against Berlin on 25.03.44.
LM533	III	LS-Y	Missing after an attack against Lisieux on 07.06.44.
LM534	III	LS-A	Crashed near Bonnelles, France, during an attack against Massy-Palaiseau on 08.06.44.
LM575	III	LS-H	Crashed at Plasier, near Pontchartrain, France, during an attack against Massy-Palaiseau on 08.06.44.
LM576	III	LS-D	Dived into the ground, near West Row, Mildenhall, Suffolk, on 21.08.44.
ME434	I	LS-D	Missing after an attack against Wanne-Eickel, Germany on 07.02.44.
ME455	I	LS-O	To No.1659 Conversion Unit.
ME695	I	LS-R	To No.1653 Conversion Unit.
ME844	I	LS-C,W	To No.44 Squadron.
ME847	I	LS-	To No.103 Squadron.
ME848	I	LS-N	To No.103 Squadron.
ME849	I	LS-F,L,C	To No.44 Squadron.
ME850	I	LS-D,O	Damaged by American anti-aircraft fire and struck-off-charge 01.01.45. Damaged beyond repair.
ND345	III	LS-	To No.156 Squadron.
ND763	III	LS-W	Hit by flak and crash-landed at Woodbridge, Suffolk on 23.04.44.
ND955	III	LS-W	Crashed at 01.15 hours, near the crossroads at Kerkweg-Heitveldsestraat, at Somerson, Netherlands, during an attack against Aachen on 25.05.44.
ND958	III	LS-H	To No.3 Lancaster Finishing School.
NF916	I	LS-Z	Missing after an attack against Solingen, Germany on 05.11.44.
NF952	I	LS-Q	Missing after an attack against Kiel on 27.08.44.
NF953	I	LS-A	To No.149 Squadron.
NF957	I	LS-X	To No.1659 Conversion Unit.
NF958	I	LS-M	Shot down and crashed south of Wieblingen railway station, 3 kms north-west of Heidelburg, at 23.00 hours on 13.09.44.
NG168	I	LS-	Struck-off-charge 15.05.47.
NG338	I	LS-M	Struck-off-charge 15.05.47.
NG339	I	DJ-G	To No.44 Squadron.
NG340	I	DJ-U	To No.44 Squadron.
NG357	I	LS-K,G	Struck-off-charge 18.02.46.
NG358	I	DJ-H,U	Struck-off-charge 19.10.45.
NG364	I	LS-P	Struck-off-charge 19.10.45.
NG365	I	LS-N	To No.138 Squadron.
NG443	I	LS-	Struck-off-charge 03.09.47.
NG444	I	LS-Y	Struck-off-charge 22.05.47.
NG445	I	DJ-O,E	To No.44 Squadron.
NG489	I	DJ-M	To No.44 Squadron.
NG494	I	DJ-P,B	To No.44 Squadron.

NN700	I	LS-Q	Missing after an attack against Rocquecourt, France on 08.08.44. The aircraft possibly crashed into the English Channel.
NN704	I	LS-	To No.218 Squadron.
NN709	I	LS-	Struck-off-charge 15.05.47.
NX559	I	LS-V	Struck-off-charge 12.03.48.
NX561	I	LS-L	Struck-off-charge 15.05.47.
NX687	VII	LS-A	To Empire Flying School.
PA170	I	LS-N	Crashed during an attack against Oberhausen, Germany on 04.12.44.
PA235	I	LS-E	Struck-off-charge 20.11.46.
PA445	I	LS-	To No.1653 Conversion Unit.
PB112	III	LS-K	To No.195 Squadron.
PB115	III	LS-W	Collided with HK612, also from No.XV Squadron, during a daylight attack against Homberg, Germany on 02.11.44. The aircraft crashed between Veghel and Eindhoven, Netherlands, at approximately 14.00 hours.
PB137	III	LS-U	Shot down during an attack against Heinsberg, Germany on 16.11.44.
PB139	III	LS-B	To No.195 Squadron.
PB259	III	LS-Q	To No.218 Squadron.
PB674	I	LS-	To No.218 Squadron.
PB802	I	LS-F	Dived into the ground at Harling, Norfolk on 17.01.45.
PD119	I	LS-	To Royal Aircraft Establishment. (B.I. Special. Designed to carry a 22,00lbs Grand Slam bomb).
PD121	I	LS-	Struck-off-charge 19.05.47. (B.I. Special. Designed to carry a 22,000lbs Grand Slam Bomb).
PD122	I	LS-	Struck-off-charge 25.03.48. (B.I. Special. Designed to carry a 22,000lbs Grand Slam Bomb).
PD125	I	LS	Struck-off-charge 25.03.48. (B.I. Special. Designed to carry a 22,000lbs Grand Slam Bomb).
PD126	I	LS-	Struck-off-charge 25.03.48. (B.I. Special. Designed to carry a 22,000lbs Grand Slam Bomb).
PD127	I	LS-S	Struck-off-charge 16.10.47. (B.I. Special. Designed to carry a 22,000lbs Grand Slam bomb).
PD128	I	LS-R	To No.44 Squadron. (B.I. Special. Designed to carry a 22,000lbs Grand Slam Bomb).
PD131	I	LS-V	Struck-off-charge 19.05.47 (B.I. Special. Designed to carry a 22,000lbs Grand Slam Bomb).
PD225	I	LS-	Sold for scrap 07.05.47.
PD234	I	LS-E	To No.138 Squadron.
PD238	I	LS-	To No.617 Squadron.
PD285	I	LS-Z,Y	To No.622 Squadron.
PD371	I	LS-	To No.617 Squadron.
PD404	I	LS-H	To No.44 Squadron.
PD419	I	LS-P,V	Missing after an attack against Dortmund, Germany on 04.02.45.
PP664	I	LS-U,A	Struck-off-charge on10.11.46.
PP672	I	LS-N	Swung on take-off from Juvincourt, and under-carriage collapsed, during an Exodus Operation (Repatriation of prisoners of war) on 13.05.45.

RA543	I	LS-A	Struck-off-charge 20.01.47.
RF140	I	LS-	To No.44 Squadron.
RF184	I	LS-	To No.90 Squadron.
RF185	I	LS-	To No.90 Squadron.

POST WAR AIRCRAFT

AVRO LINCOLN

Serial	Mark	Code	Remarks
RE341	II	AA	
RF370	II	LS-A	
RF392	II	LS-C	
RF395	II	LS-E	
RF449	II	LS-	
RF503	II	LS-F	
RF512	II	LS-E,H	
RF514	II	LS-B	
RF532	II	LS-D	

BOEING B-29A WASHINGTON

Serial	Mark	Code	Remarks
WF497	I	LS-A	From No.149 Squadron.
WF499	I	LS-B	From No.149 Squadron.
WF504	I	LS-C	Delivered on 15.01.51.
WF505	I	LS-D	Delivered on 18.01.51.
WF506	I	LS-E	Delivered on 19.01.51.
WF507	I	?	Delivered on 14.02.53
WF552	I	?	From No.57 Squadron.
WF553	I	?	From No.57 Squadron. Crashed in Lincolnshire on 05.01.53.

ENGLISH ELECTRIC CANBERRA

Serial	Mark	Code	Remarks
WD951	BII	?	From No.61 Squadron
WD961	BII	?	From No.61 Squadron
WD964	BII	?	From No.10 Squadron
WD980	BII	?	From No.50 Squadron
WF916	BII	?	From No.44 Squadron
WH724	BII	?	Converted to T.II
WH725	BII	?	To No.50 Squadron
WH731	BII	?	To No.50 Squadron
WH850	T4	?	To Station Flight RAF Honington
WH872	BII	?	From No.10 Squadron
WH907	BII	?	To No.61 Squadron
WJ575	BII	?	To No.57 Squadron
WJ647	BII	?	To No.61 Squadron
WJ717	BII	?	To No.61 Squadron
WJ724	BII	?	To No.61 Squadron
WJ972	BII	?	To No.6 Squadron
WJ974	BII	?	To No.57 Squadron
WJ976	BII	?	To No.44 Squadron
WJ977	BII	?	To No.57 Squadron
WJ985	BII	?	To No.104 Squadron
WK107	BII	?	From No.12 Squadron
WK122	BII	?	From No.61 Squadron
WK132	BII	?	From No.61 Squadron
XA536	BII	?	From No.50 Squadron

HANDLEY PAGE VICTOR

Serial	Mark	Code	Remarks
XA925	B1A		
XA935	B1		
XA938	B1	?	To Mod(PE) RAE Farnborough
XA939	B1		
XA940	B1		
XA941	B1	?	Collected from Handley Page on 07.10.58 by W/C David Green, O.C. No.XV Squadron
XH588	B1	?	Collected from Handley Page on 31.10.58 by W/C David Green, O.C. No.XV Squadron
XH589	B1	?	Collected from Handley Page by S/L McGillivray, 'B' Flight Commander
XH590	B1	?	Collected from Handley Page on 01.12.58 by W/C David Green, O.C. No.XV Squadron
XH591	B1	?	Collected from Handley Page on 19.01.59 by W/C David Green, O.C. No.XV Squadron
XH592	B1	?	Collected from Handley Page on 01.01.59 by W/C David Green, O.C. No.XV Squadron
XH593	B1	?	Collected from Handley Page on ??.02.59 by W/C David Green, O.C. No.XV Squadron
XH594	B1	?	Collected from Handley Page on 27.02.59 by W/C David Green, O.C. No.XV Squadron
XH587	B1A		
XH591	B1A		
XH613	B1A	??	Crashed on approach to RAF Cottesmore, on 14.06.62, due to electrical malfunction.
XH616	B1A		
XH618	B1A		
XH620	B1A		
XH648	B1A	??	To the Imperial War Museum.
XH651	B1A		

HAWKER HUNTER

Serial	Mark	Code	Remarks
WV318	T7A	??	RAF Laarbruch Station Flight.

HAWKER SIDDELEY BUCCANEER

No.XV Squadron reformed at RAF Honington, on 1st October 1970

Serial	Mark	Code	Remarks
XN977	S2B	??	Engine exploded and caught fire causing irreparable damage, during training sortie on 08.03.82.
XT275	S2B	A	Displayed by No.XV Squadron in the static park, at the International Air Tattoo, at Greenham Common, on 22.06.79.
XT279	S2B	C	Delivered to No.XV 26.10.65
XT287	S2B	F	Carried the name 'MacRobert's Reply' together with the MacRobert family crest. Named by S/L P.J. Boggis at official ceremony at RAF Mildenhall, on ??.05.80.
XV332	S2B	?	Delivered to No.XV 16.01.67
XV345	S2A	??	Broke up in low-level flight near Nellis Air Force Base, whilst participating in a 'Red Flag' exercise 07.02.80. S/L Ken Tait and F/L Charles Rushton when killed when structural failure of the main spar occurred during a maneuver at an altitude of approximately 100'.
XV349	S2B	??	From No.12 Squadron

Serial	Mark	Code	Remarks
XW525	S2B	??	This aircraft was the first new build aircraft to be received by the Royal Air Force.
XW526	S2B	??	This aircraft was the first of its type to received by the Royal Air Force.
XW527	S2B	??	This aircraft was displayed in the static park at the Paris Air Show in May 1971.
XW528	S2B	??	First delivered 04.08.70
XW530	S2B	??	First delivered to No.12 Squadron.
XW531	S2B	??	First delivered 03.12.70
XW532	S2B	??	Crashed on 25.03.71, during low level flight, shortly after taking-off from RAF Laarbruch, West Germany. The pilot, W/C D Collins and the navigator, F/L P Kelly, were both killed.
XW533	S2B	??	First delivered to No.237 OCU
XW534	S2B	??	This aircraft was used for a demonstration flight by Air Chief Marshal Sir John Barraclough, KCB, CBE, DFC, AFC, on 10.09.73. The pilot was F/L Moses.
XW535	S2B	??	First delivered 14.05.71
XW536	S2B	??	Crashed into North Sea, of Danish coast, having been abandoned by crew following a mid-air collision with Buccaneer XW528.
XW537	S2B	E	First delivered 27.07.71
XW540	S2B	??	Aircraft used by A&AEE, Boscombe Down, for cold weather trials at Goose Bay, Canada, prior to delivery to No.XV Squadron.
XW541	S2B	??	First delivered 10.12.71
XW542	S2B	??	First delivered 11.01.72
XW543	S2B	??	First delivered 22.05.72
XW544	S2B	??	First delivered 26.06.72
XW545	S2B	??	First delivered 15.08.72
XW546	S2B	??	First delivered 01.10.72
XW547	S2B	??	First delivered 01.11.73
XW548	S2B	??	First delivered 03.01.73
XW549	S2B	??	First delivered 14.02.73
XW550	S2B	??	First delivered 04.04.73
XX887	S2B	??	First delivered 29.08.74
XX888	S2B	D	First delivered 21.11.74
XX890	S2B	??	Crashed near Laarbruch, Germany, following loss of control during final approach.
XX891	S2B	??	First delivered 17.07.75
XX893	S2B	H	First delivered 20.10.75
XX894	S2B	??	From No.12 Squadron
XZ432	S2B	??	The last aircraft of its type to be delivered to No.XV Squadron.

The last Buccaneer sortie, flown by No.XV Squadron occurred during June 1983.

PANAVIA TORNADO

Serial	Mark	Tailcode	Remarks
ZA392	GR1	EK	Flew as 'K' with No.27 Squadron in Gulf War. Crashed near Shaibah, Iraq, on 17.01.91, following an attack on an enemy airfield. Two casualties.
ZA396	GR1	GE	Flew as 'F' with No.20 Squadron in Gulf War Shot down on 20.01.91.
ZA399	GR1	GA	Flew as 'G' with No.20 Squadron in Gulf War.
ZA409	GR1T	EW	Delivered to No.XV Squadron at Laarbruch 1983.
ZA410	GR1	EX	To No.12 Squadron
ZA446	GR1T	EF	To Boscombe Down
ZA447	GR1	EA	To No.12 Squadron

ZA448	GR1	EB	Crashed at Nellis Air Force Base, on 29.03.88, during Green Flag exercise. Crew escaped unhurt.	ZA587	GR1	TD	From No.20 Squadron
ZA450	GR1	EC	To No.12 Squadron	ZA588	GR1	TM	From TWCU/45 Squadron
ZA451	GR1	??	Aircraft abandoned near Jever, Germany, on 06.02.84, after being struck by lightning.	ZA589	GR1	TE	From No.31 Squadron
				ZA592	GR1	TC	From No.27 Squadron
ZA453	GR1	EG	Flew in Gulf War	ZA594	GR1T	TU	From TWCU/45 Squadron
ZA454	GR1	EH	Crew ejected safely after engine fire, Goose Bay, Canada, 30.04.90	ZA595	GR1	TV	From TWCU/45Squadron
				ZA597	GR1	TA	From BAe Warton
ZA455	GR1	EJ	Flew as 'J' in Gulf War	ZA598	GR1	TN	From No.617 Squadron
ZA456	GR1	GB	Flew as 'M' with No.20 Squadron in Gulf War.	ZA600	GR1	TH	From No.17 Squadron
ZA459	GR1	EL	Flew as 'L' in Gulf War.	ZA601	GR1	TI	From No.31 Squadron
ZA461	GR1	DK	To No.12 Squadron	ZA602	GR1	TX	From No.9 Squadron
ZA462	GR1	EM	To No.17 Squadron	ZA604	GR1	TY	From TWCU/45 Squadron
ZA463	GR1	GL	Flew as 'Q' with No.20 Squadron in Gulf War.	ZA607	GR1	TJ	From No.17 Squadron
ZA468	GR1	??	Crew ejected after take-off from Laarbruch, on 20.07.89.	ZA608	GR1	TK	From No.617 Squadron
				ZA609	GR1	??	From No.617 Squadron
				ZA611	GR1	TG	From No.27 Squadron
ZA469	GR1	GD	Flew as 'I' with No.20 Squadron in Gulf War.	ZA612	GR1	TZ	From No.17 Squadron
ZA470	GR1	FL	Flew with No.16 Squadron in Gulf War.	ZA613	GR1	TL	From No.617 Squadron
ZA471	GR1	ER	Flew as 'E' in Gulf War. Aircraft named "Emma".	ZA614	GR1	TB	From No.617 Squadron
ZA472	GR1	EE	To No.17 Squadron	ZD713	GR1	TW	From TWCU/45 Squadron
ZA475	GR1	FC	Flew as 'P' with No.16 Squadron in Gulf War.	ZD717	GR1	CD	Flew as 'C' with No.17 Squadron. Shot down by SA2 missle on 14.02.91, whilst engaged on attack against Al Taqaddum.
ZA477	GR1	FA	Flew in Gulf War				
ZA491	GR1	GC	Flew as 'N' with No.20 Squadron in Gulf War. Aircraft named "Nikki".				
				ZD745	GR1	BM	Flew with No.14 Squadron in Gulf War.
ZA541	GR1	TO	From TWCU/45 Squadron	ZD790	GR1	DL	Flew as 'D' with No.14 Squadron in Gulf War.
ZA544	GR1	TP	From No.14 Squadron	ZD791	GR1	BG	Flew as 'B' with No.14 Squadron in Gulf War. Shot down by flak on 17.01.91. F/L John Peters, the pilot, and F/L John Nichol, the navigator, both ejected safely but were made prisoners of war.
ZA548	GR1	TQ	From TWCU/45 Squadron				
ZA549	GR1	TR	From No.27 Squadron				
ZA552	GR1	TS	From No.2 Squadron				
ZA556	GR1	TA	From TWCU/45 Squadron				
ZA559	GR1	F	From No.617 Squadron	ZD809	GR1	BA	Flew as 'A' with No.14 Squadron in Gulf War.
ZA562	GR1	TT	From No.27 Squadron	ZD890	GR1	AE	Flew as 'O' with No.9 Squadron in Gulf War.
ZA563	GR1	TC	From No.27 Squadron	ZD892	GR1	BJ	Flew as 'H' with No.14 Squadron in Gulf War. Aircraft named "Helen".

APPENDIX VII
List of GERMAN AIRCREW
Known to have shot down
No.15/XV SQUADRON AIRCRAFT

WORLD WAR ONE

NAME	RANK	UNIT	REMARKS
Bethge, H.	Oblt		
Boelcke, Oswald	Hptm	Jasta 2	
Bohme, Erwin	Ltn	Jasta Bolcke	
Immelmann, Max	Ltn		
Kirmaier, Stefan	Oblt	Jasta 2	
Leffers, Gustav	Ltn		
Osteroht, T. Von	Hptm	Jasta 2	
Richthofen, Manfred von	Rittm	JG.1	
Riessinger, ?	Vfbl	Jasta 12	

WORLD WAR TWO

NAME	RANK	UNIT	REMARKS
Augenstein, Hans-Heinz	Hptm	IV./NJG1	
Autenrieth, Hans	Oblt	II./NJG1	
Bauer, Victor	Oblt	III./NJG1	
Becker, Martin 'Tino'	Hptm	II./NJG6	
Dorman, Wilhelm	Hptm	III./NJG1	
Ehle, Walter	Hptm	II./NJG1	
Geislinger, Walter	Offz	VI./NJG1	
Gildner, Paul	Obfw	II./NJG2	
Hager, Johannes	Hptm	StII./NJG1	
Hoffmann,	Hptm	IV./NJG5	
Jabs, Hans Joachim	Oblt	IV./NJG1	Shot down three No.XV Squadron Stirling bombers, within 45 minutes, on the night of 19.02.43
Jaeckel, Ernst	Fw	2./JG26	First German pilot credited with shooting down a four-engined bomber.
Kalinowski,	Fw	VI./NJG1	
Knacke, Reinhold	Oblt	II./NJG2	
Kuthe, Wolfgang	Oblt	IV./NJG1	
Lau, Fritz	Oblt	III./NJG1	
Loos, Kurt	Lt	I./NJG1	
Luschner,	Uffz	III./NJG3	
Pfeiffer, Karl	Uffz	IV./NJG1	
Schellwat,	Obfw	V./NJG1	
Scherfling, Karl-Heinz	Fw	III./NJG1	
Sothe, Fritz	Hptm	/NJG4	
Szameitat, Paul	Oblt	II./NJG3	
Weissenfeld, Prinz Egmont Zur Lippe	Oblt	IV./NJG1	

WORLD WAR ONE

Left: Hauptmann Oswald Boelcke, who shot down Lt. Edward Carr and Sgt Frederick Barton, on 28th October 1916. *Courtesy of and via Norman Franks* Center: Leutnant Max Immelman shot down a B.E.2c, piloted by Lieutenant Geoffrey Welsford, on 30th March 1916. Although the British pilot died in the attack his observer, Lieutenant Wayland Joyce, survived. *Courtesy of and via Norman Franks* Right: Leutnant Stephan Kirmaier. *Courtesy of and via Norman Franks*

Left: Hauptmann Von Osteroht shot down B.E.2e, 7236, on 3rd April 1917. The pilot 2/Lt Sayer was killed, but 2/Lt Morris survived to be taken prisoner of war. *Courtesy of and via Norman Franks* Center: Manfred Freiherr von Richthofen claimed R.E.8, B742, as his 70th victim on 26th March 1918. The No.15 Squadron aircraft was crew by 2/Lt Vernon Reading and 2/Lt Matthew Leggett, both of whom were killed. *Courtesy of and via Norman Franks* Right: Lieutenant Werner Voss, who killed Lt Adrian Mackenzie with a bullet through the heart, on 1st April 1917. The aircraft, piloted by Captain Wynne who survived a crash landing, was the German ace's 23rd victory. *Courtesy of and via Norman Franks*

WORLD WAR TWO

Oberleutnant Reinhold Knacke in hunting mood, photographed in the cockpit of his Me.110. *Courtesy of and via Ad Van Zantvoort*

Right: Oblt Reinhold Knacke, of II./NJG2, in more relaxed mood. *Courtesy of and via Ad Van Zantvoort*

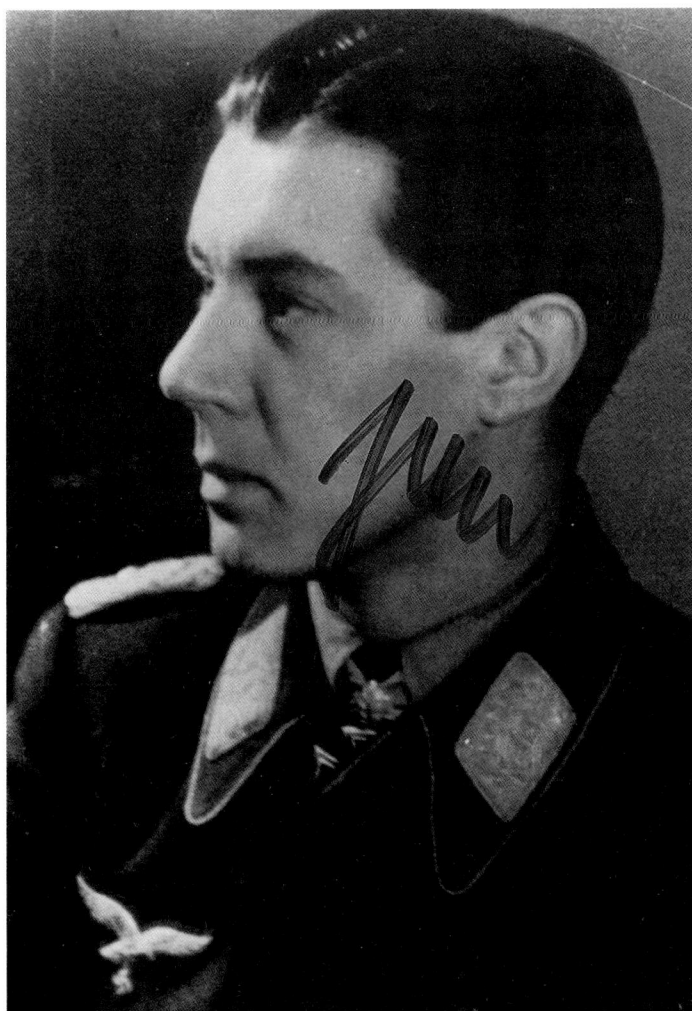

Oblt Hans Joachim Jabs, of IV./NJG1, who shot down three No.XV Squadron aircraft, within 45 minutes, on the night of 19th February 1943. *Courtesy of Hans Joachim Jabs*

A gathering of Luftwaffe Eagles at Leeuwarden during 1942. From left to right (foreground) are: Oberleutnant Prinz Zur Lippe-Weissenfel; Hauptmann Helmet Lent and Oberleutnant Ludwig Becker. *Courtesy of and via David Williams*

APPENDIX VIII
List of known
AIRCRAFT/PERSONNEL LOSS STATISTICS
for No.15/XV SQUADRON

WORLD WAR ONE AIRCRAFT LOSSES

Aircraft Type	S/Down	Crashed	Wrecked	Damaged
B.E.2	13	5	2	3
				1 F-L
R.E.8.	9	1	-	-
Total	22	6	2	4

WORLD WAR TWO AIRCRAFT LOSSES

Aircraft Type	Individual Sorties	Squadron Missions	Aircraft Losses
Blenheim	543	97	19 Missing
			3 FTR
			5 Crashed
			1 Force-Landed
Wellington	173	38	2 Shot Down
			1 Crashed
Stirling	2231	353	3 FTR
			54 Shot Down
			43 Crashed
			7 Force-Landed
			4 Ditched
			22 Damaged
Lancaster	2840	226	13 Missing
			17 FTR
			10 Shot Down
			13 Crashed
			2 Force-Landed
			2 Damaged
Overall Total	5787	714	32 Missing
			23 FTR
			66 Shot Down
			62 Crashed
			10 Force-Landed
			4 Ditched
			24 Damaged

GULF WAR AIRCRAFT LOSSES

Aircraft Type	
Tornado	2 Shot Down
Total	2 S/Down

PERSONNEL STATISTICS

WORLD WAR ONE PERSONNEL LOSSES

Table of the 62 known losses recorded in the Roll of Honor.

RANK/CREW CATEGORY LOSSES

	Pilot	Obs	Unk	Total
Officers	24	21	4	49
NCOs	1	1	1	3
Other Ranks	-	10	-	10
Total	25	32	5	62

RANK CATEGORY LOSSES

	RFC	RAF	Total
Officers	38	11	49
NCOs	2	1	3
Other Ranks	4	6	10
Total	44	18	62

AIRCREW CATEGORY LOSSES

	RFC	RAF	Total
Pilot	19	5	24
Observers	21	12	33
Unknown	4	1	5
Total	44	18	62

WORLD WAR TWO

CREW/NATIONALITY LOSSES

Table of 964 known losses recorded in the Roll of Honor.

	RAF	RCAF	RNZAF	RAAF	SAAF	RAuxAF	ARMY	TOTAL
Pilots	122	26	24	17	-	-	-	189
F/Eng	114	-	-	1	-	-	-	115
Nav	63	7	7	6	-	1	-	84
B/A	55	16	5	5	-	-	-	81
A/G	188	30	6	11	-	1	-	236
WAG	90	7	5	1	-	-	-	103
W/Op	47	1	1	9	-	-	-	58
Obs	49	8	2	-	-	-	-	59
Gdcw	6	-	-	-	-	-	-	06
Unknown	26	2	-	2	-	-	-	30
Pass	-	-	-	-	1	-	1	02
Chaplain	1	-	-	-	-	-	-	01
Total	761	97	50	52	1	2	1	964

Table of 1036 total known losses recorded in the Roll of Honor.

RANK CATEGORY LOSSES (All Nationalities)

	W.W.I	W.W.II	POST 1945	TOTAL
Officers	49	257	6	312
Warrant/Offs	-	30	-	30
NCOs	3	663	-	666
O/Ranks	10	14	-	24
Unknown	-	-	4	4
Total	62	964	10	1036

APPENDIX IX
List of
MISSIONS FLOWN BY LANCASTER, LL806, "J" - JIG

DATE	PILOT	TARGET			
01.05.44	Sparks, M. P/O	Chambly	09.08.44	Moran, W. F/O	Lille
08.05.44	Sparks, M. P/O	Cap Griz Nez	11.08.44	Cato, H. F/O	Lens
10.05.44	Sparks, M. P/O	Courtrai	12.08.44	MacDougall, A. W/O	Falaise
11.05.44	Sparks, M. P/O	Louvaine	14.08.44	MacDougall, A. W/O	St Questin
19.05.44	Sparks, M. P/O	Le Mans	15.08.44	Jennings, R. F/O	St Trond
21.05.44	Sparks, M. P/O	Duisburg	16.08.44	Leslie, W. F/L	Stettin
22.05.44	Sparks, M. P/O	Dortmund	18.08.44	Leslie, W. F/L	Bremen
24.05.44	Fisher, S. F/L	Aachen	26.08.44	Leslie, W. F/L	Kiel
27.05.44	Ferguson, W. F/S	Boulogne	29.08.44	Leslie, W. F/L	Stettin
28.05.44	Fisher, S. F/L	Angers	31.08.44	Leslie, W. F/L	Pont Remy
30.05.44	Sparks, M. P/O	Boulogne	03.09.44	MacDougall, A. W/O	Enidhoven
02.06.44	Ferguson, W. F/S	Wissant	05.09.44	Leslie, W. F/L	Le Havre
03.06.44	Ball, J. F/L	Calais	06.09.44	Jennings, R. F/O	Le Havre
05.06.44	Ferguson, W. F/S	Quistreaham	08.09.44	Jennings, R. F/O	Le Havre
06.06.44	Ball, J. F/L	Lisieux	17.09.44	Jennings, R. F/O	Boulogne
07.06.44	Sparks, M. F/L	Massey-Palaiseau	20.09.44	Jennings, R. F/O	Calais
08.06.44	Sparks, M. F/L	Fougeres	23.09.44	Jennings, R. F/O	Neuss
10.06.44	Payne, B. F/L	Dreux	24.09.44	Cato, H. F/L	Calais
12.06.44	Sparks, M. F/L	Gelsenkirchen	27.09.44	Marsh, R. F/O	Calais
17.06.44	Sparks, M. F/L	Montdidier	28.09.44	Jennings, R. F/O	Calais
21.06.44	Sparks, M. F/L	Domleger	05.10.44	Percy, P. F/L	Saarbrucken
30.06.44	Sparks, M. F/L	Villers Bocage	06.10.44	Percy, P. F/L	Dortmund
31.06.44	Leslie, W. F/O	Trappes	07.10.44	Percy, P. F/L	Kleve
02.07.44	Sparks, M. F/L	Beauvoir	14.10.44	Cato, H. F/L	Duisburg (A.M.)
05.07.44	Stokes, G. F/O	Wizernes	14.10.44	Cato, H. F/L	Duisburg (P.M.)
07.07.44	Payne, B. F/L	Vaires (Paris)	15.10.44	Hopper-Cuthbert, F/O	Wilhelmshaven
09.07.44	Sparks, M. F/L	Linzeux	18.10.44	Cato, H. F/L	Bonn
10.07.44	Sparks, M. F/L	Nucourt	19.10.44	Kelly, D. F/O	Stuttgart
15.07.44	Bell, W. DFC. F/L	Chalons-Sur-Marne	22.10.44	Jones, B. F/O	Neuss
18.07.44	Bell, W. DFC. F/L	Caen	23.10.44	Hastings, R. F/S	Essen
18.07.44	Bell, W. DFC. F/L	Aulnoye	25.10.44	Percy, P. F/L	Essen
23.07.44	Sparks, M. F/L	Montcandon	28.10.44	Marsh, R. F/O	Flushing
23.07.44	Mason, W. F/S	Kiel	15.11.44	Buchanan, I. F/L	Dortmund
24.07.44	Johnston, M. F/L	Stuttgart	16.11.44	Hopper-Cuthbert, F/O	Heinsberg
25.07.44	Stewart, S. F/O	Stuttgart	20.11.44	Hopper-Cuthbert, F/O	Homberg
28.07.44	Kelly, H. F/O	Stuttgart	21.11.44	Hopper-Cuthbert, F/O	Homberg
30.07.44	Payne, B. F/L	Amaye-Sur-Seulles	23.11.44	Hopper-Cuthbert, F/O	Gelsenkirchen
01.08.44	Leslie, W. F/L	Coulon Villiers	26.11.44	Hopper-Cuthbert, F/O	Fulda
05.08.44	Leslie, W. F/L	Bassens (Bordeaux)	27.11.44	Hopper-Cuthbert, F/O	Cologne
07.08.44	Leslie, W. F/L	Rocque-Court	29.11.44	Noble, C. F/O	Neuss
08.08.44	Marshall, L. F/S	Foret De Lucheaux	30.11.44	Noble, C. F/O	Bottrop
			02.12.44	Marriot, L. F/O	Dortmund

04.12.44	Hopper-Cuthbert, F/O	Oberhausen		01.03.45	Meikle, A. W/O	Kamen
05.12.44	Slaughter, J. P/O	Schwammenauel Dam		02.03.45	Ayres, C. F/O	Cologne
06.12.44	Giles, B. F/S	Meresberg		05.03.45	Baxendale, L. F/S	Gelsenkirchen
08.12.44	Giles, B. F/S	Duisburg		07.03.45	Wright, A. F/S	Dessau
11.12.44	Clayton, N. F/O	Osterfeld		09.03.45	MacDonald, A. F/S	Detteln
12.12.44	Clayton, N. F/O	Witten		10.03.45	Wright, A. F/S	Gelsenkirchen
16.12.44	Hopper-Cuthbert, F/O	Siegen		11.03.45	Wright, A. F/S	Essen
19.12.44	Hopper-Cuthbert, F/O	Trier		12.03.45	Tenger, V. W/O	Dortmund
21.12.44	Hopper-Cuthbert, F/O	Trier		14.03.45	Meikle, A. W/O	Datteln
23.12.44	Hopper-Cuthbert, F/O	Trier		22.03.45	Sievers, W. F/S	Bocholt
24.12.44	Bignell, S. F/O	Bonn (Hangelar)		27.03.45	Tenger, V. F/O	Altenbogge
28.12.44	Burns, N. F/O	Cologne		04.04.45	Woodman, W. W/O	Merseberg
29.12.44	Hopper-Cuthbert, F/O	Coblenz		09.04.45	Sievers, W. F/S	Kiel
31.12.44	Hopper-Cuthbert, F/O	Vohwinkel		13.04.45	Sievers, W. F/S	Kiel
02.01.45	Hopper-Cuthbert, F/O	Nurnberg		14.04.45	Sievers, W. F/S	Potsdam
03.01.45	Bignell, S. F/O	Dortmund		18.04.45	Sievers, W. F/O	Heligoland
05.01.45	Hopper-Cuthbert, F/O	Ludwigshaven		22.04.45	Sievers, W. F/O	Bremen
06.01.45	Bignell, S. F/O	Neuss		30.04.45	Blaxendale, L. F/O	Rotterdam 'Manna' Operation
07.01.45	Gray, L. F/O	Munich		02.05.45	Sievers, W. F/O	The Hague 'Manna' Operation
11.01.45	Hopper-Cuthbert, F/O	Krefeld		07.05.45	Woodman, W. F/O	The Hague 'Manna' Operation
15.01.45	Hopper-Cuthbert, F/O	Enkerschwick		17.05.45	Sievers, W. F/O	Juvincourt - 'Exodus'
16.01.45	Hagues, C. F/L	Wanne-Eickel		24.05.45	Sievers, W. F/O	Juvincourt - 'Exodus' Operation
22.01.45	McHardy, I. F/S	Sterkrade		??.05.45	?	Juvincourt - 'Exodus' Operation
28.01.45	Gray, L. F/O	Cologne		01.06.45	Hall, C. F/O	'Baedeker' Sortie
01.02.45	Hunt, D. F/O	Munchen Gladbach		05.06.45	Strickland, F. F/O	'Baedeker' Sortie
03.02.45	Tenger, V. F/S	Dortmund		15.07.45	Woodcock, A. F/O	'Baedeker' Sortie
09.02.45	Bruce, J. SGT	Hohenbudberg		03.08.45	Woodcraft, R. F/L	'Dodge' Operation
14.02.45	Burns, N. F/O	Chemnitz		21.09.45	Wright, R. F/O	'Baedeker' Sortie
22.02.45	Burns, N. F/O	Buer		07.10.45	Umwin, F. F/O	'Dodge' Operation
23.02.45	Meikle, A. F/S	Gelsenkirchen				
25.02.45	Meikle, A. F/S	Kamel				
26.02.45	Hunt, D. F/O	Dortmund				
27.02.45	Meikle, A. F/S	Gelsenkirchen				
28.02.45	Ayres, C. F/O	Gelsenkirchen				

Avro (Armstrong Whitworth built) Lancaster Mk.I., LL806, LS-J, was struck-off-charge on 5th December 1945, having completed 134 operational missions, 3 'Manna' operations, 3 'Exodus' missions and at least two 'Dodge' sorties.

APPENDIX X
List of BATTLE HONORS
for No.15/XV SQUADRON

The Honors marked * are those displayed on the Squadron Standard.

WORLD WAR ONE

Arras	1915
Western Front*	1915-1918
Battle of the Somme*	1916
Cambrai*	1917
Somme	1918
Hindenburg Line*	

WORLD WAR TWO

France and the Low Counties	1939-1940
Muese Bridges*	1940
Dunkirk	1940
Invasion Ports	1940
Fortress Europe	1941-1944
Ruhr*	1941-1945
Berlin*	1941-1945
Biscay Ports	1940-1945
Normandy*	1944
France and Germany	1944-1945

POST WAR

Gulf War	1991

APPENDIX XI
List of HONORS AND AWARDS
for No.15/XV SQUADRON

Dates quoted for the award of a Distinguished Flying Cross in this list are given as recorded in the *London Gazette*. However, in some cases, these dates differ from those given in the Squadron's records.

FIRST WORLD WAR

NAME	RANK	AWARD	DATE
J. C. Alexander	F/S	MSM	03.06.18
W. C. Barker	Lt	M.C. and Bar	10.01.17
R. H. Barratt	Lt	MiD	
K. R. Binning	Capt	M.C.	20.10.16
N. Booth	Cpl	M.M.	
W. Buckingham	2/Lt	M.C.	26.05.17
J. C. Burchett	Lt	M.C.	22.06.18
H. A. Chippendale	Lt	M.C.	22.06.18
H. A. Coysh	Lt	DFC	03.12.18
? Cox	Sgt	CdeG(B)	
J. A. Craig	Capt	DFC	03.12.18
A. R. Cross	Capt	DFC	03.12.18
W. R. Cox	Lt	M.C.	25.05.17
C. S. Goodfellow	2/Lt	M.C.	18.07.17
G. A. Griffen	Lt	DFC	03.12.18
L. F. Handford	2/Lt	M.C.	
R. G. Hart	2/Lt	M.C.	26.07.18
A. M. Hill	Capt	DFC	03.12.18
G. L. Hobbs	Lt	M.C.	22.06.18
F. N. Hudson	2/Lt	M.C.	30.03.16
L. Jenkins,	Capt	M.C.	03.06.16
		Bar to M.C.	??.09.16
? Kirdik	Cpl	M.M.	
F. Knot	Cpl	DCM	
F. G. Manville	Lt	M.C.	10.01.17
?. ?. Moore	F/S	MiD	01.01.17
G. M. Moore	Capt	M.C.	10.01.17
M. Moore	1/AM	CdeG(B)	
? Moore	F/S	M.M.	
T. D. Mountford	F/St	DCM	04.06.17
W. G. Pender	Capt	M.C.	10.01.17
H. B. Pett	Capt	M.C.	22.06.18
T. G. Poland	F/O	M.C.	18.06.17
? Potts	F/S	CdeG(B)	
J. P. Ridgway	Sgt	DCM	11.05.17

E Smalley	S/M	MSM	03.06.18
H. V. Stammers	Maj	DFC, CdeG(B)	08.02.19
B. E. Sutton	Capt	M.C.	16.09.18
? Wilkins	F/S	MiD	
M. Wilks	1/AM	CdeG(B)	

SECOND WORLD WAR

P. Alderson	Sgt	DFM	22.11.40
A. A. Aleandri	F/O	DFC	13.04.45
G. Allen	W/O	DFC	17.10.44
S. A. Anderson	W/O	DFC	25.05.45
J. Angus	F/S	DFM	17.02.44
A. W. Armstrong	F/S	DFM	12.11.44
M. J. Bagan	W/O	DFC	17.04.45
C. H. Baigent	A/F/L	DFC	06.11.42
J. Baker	Sgt	DFM	14.10.43
R. Baker	W/O	MiD	
R. P. Baker, MiD	A/F/L	DFC	13.10.44
W. C. Baker	Sgt	DFM	22.11.40
C. W. Ball	F/O	DFC	27.03.45
R. R. Banks	W/O	DFC	17.04.45
A. E. Barford	F/S	DFM	15.02.45
L. R. Barr	A/F/L	DFC	26.05.42
L. R. Barr,	F/L	Bar to DFC	07.08.42
J. Barrass	P/O	DFC	13.08.43
A. E. Barrett	F/S	DFM	23.03.45
J. F. Barron	F/S	DFM	26.05.42
E. Bartholomew	P/O	DFC	25.09.45
R. I. Bell	P/O	DFC	12.12.44
W. J. Bell	A/F/L	DFC	14.07.44
C. A. Bender	F/O	DFC	21.09.45
N. A. Bennitt	A/F/O	DFC	07.08.42
L. G. Bentley	Sgt	DFM	12.11.43
S. A Bicknell	F/S	DFM	13.10.44
D. R. Bishop	Sgt	DFM	08.12.44
D. A. Boards	Sgt	DFM	19.08.43
P. J. Boggis	F/O	DFC	09.01.42
W. J. Bolduc	P/O	DFC	30.06.44
D. T. Bone, DFM	A/F/L	DFC	17.07.45
A. J. Box	Sgt	DFM, MiD	13.09.40
D. P. Boyle	Sgt	BEM	17.03.41

E. M. Brigg	Sgt	DFM	10.02.42		G. W. Gabel	P/O	DFC	15.10.43
O. V. Brooks	P/O	DFC	19.05.44		G. A. Gallop	F/L	DFC	08.12.44
J. H. Brown	W/O	DFC	17.04.45		J. C. Galley	Sgt	DFM	22.08.41
V. W. Brown	F/S	DFM	16.07.45		J. W. Gaylor	Sgt	DFM	03.05.43
I. S. Buchanan	A/F/L	DFC	13.04.45		L. H. George	F/O	DFC	19.05.44
B. C. Bull, DFM	P/O	DFC	25.05.45		W. H. George	A/F/L	DFC, MiD	22.10.40
N. W. Burns	A/F/O	DFC	25.09.45		P. K. Gerricke	F/S	DFM	12.02.44
J. R. Burrett	P/O	DFC	20.02.45		J. A. Gibson, DFC	A/S/L	DSO	16.03.45
J. R. Bushell	Sgt	DFM	06.06.41		R. S. Gilmour	P/O	DFC	22.11.40
H. T. Bysouth	F/S	DFM	21.05.45		E. F. Glover	P/O	DFC	17.10.44
V. F. Cage	A/F/L	DFC	22.05.45		G. L. Goodall	F/S	DFM	11.01.43
C. C. Calder	W/C	OSC*	??.11.45		R. Gourley	F/O	DFC	16.11.45
		DSO, DFC. * National Order of the Southern Cross.			E. Grimshaw	P/O	DFC	17.10.44
R. W. Cameron	A/S/L	DFC	12.04.45		C. N. Hagues	A/S/L	DFC	21.09.45
P. J. Camp	Sgt	DFM	22.10.40		A. N. Halkett	F/S	DFM	02.11.42
G. A. Cantrell	F/O	DFC	13.10.44		W. H. Hall	A/F/L	DFC	23.03.45
D. L. Capel	A/F/O	DFC	20.02.45		F. F. Hamilton	F/S	DFM	12.12.42
P. D. Carden	A/S/L	DFC	08.12.44		R. H. Hardy	F/S	DFM	09.01.42
M. A. Chapman	P/O	MiDx2	??.??.??		D. H. Haycock	A/F/L	DFC	01.06.45
P. R. Child	F/S	DFM	13.10.44		A. A. Haydon	F/L	DFC	17.07.45
D. E. Clark	F/L	DFC	12.01.43		R. P. Hayles	F/S	DFM	12.02.44
F. Clark	Sgt	DFM	19.04.43		R. A. Hearne	F/S	DFM	12.11.44
G. M. Claydon	A/F/L	DFC	14.07.44		E Henry	W/O	DFC	16.11.45
A. G. Cochrane	A/F/O	DFC	22.08.41		E. S. Henzel	F/O	DFC	17.10.44
R. J. Cook	Sgt	DFM	10.02.42		J. V. Hislop	F/O	DFC	16.02.45
E. G. Cook	F/L	DFC	06.11.45		C. Houlgrave	F/S	DFM	16.07.4
J. Cope, DFM	A/F/L	DFC	20.04.43		H. M. Hunt	F/O	DFC	20.02.45
P. H. Cope	P/O	DFC	18.02.44		R. E. Hunter	LAC	DFM	09.07.40
F. Cork	F/L	DFC	21.09.45		W. A. Irwin	P/O	DFC	20.04.43
J. E. Cowell	A/F/O	DFC	13.10.44		F. C. Jackson	Sgt	DFM	19.04.43
J. Cox	W/C	DFC	24.12.40		R. D. Jennings	A/F/O	DFC	25.05.45
W. R. Cox	A/F/L	DFC	16.01.45		W. Jessop	Sgt	DFM	22.11.40
A. R. Craddock	A/F/L	DFC	14.05.43		M. J. Johnston	A/F/L	DFC	17.10.44
D. Craven	Sgt	DFM	15.06.43		W. P. Jolly, DFM	A/F/L	DFC	16.11.45
L. W. Crowe	F/O	DFC	20.07.45		B. Jones	F/O	DSO	??.12.44
P. Curtis	Sgt	DFM	28.12.42		E. T. Jones	F/O	DFC	08.12.44
J. R. Cunningham	Sgt	DFM	28.10.42		H. A. Jones	F/O	DFC	20.07.45
D. Dancy	W/O	DFC	06.11.45		H. R. Jones	Cpl	DFM	23.07.40
R. F. Davis	Sgt	DFM	19.07.40		R. T. Jones	F/S	BEM	24.09.41
F. H. Dengate	A/F/L	DFC	13.10.44		W. H. Kendall	P/O	DFC	27.03.45
F. Denton, DFM	W/O	DFC	20.02.45		W. J. Klufas	A/F/L	DFC	15.10.43
P. Devine	Sgt	DFM	18.05.43		P. J. Lamason	A/S/L	Bar to DFC	27.06.44
J. J. de Willimoff	A/F/L	DFC	11.04.44				DFC, MiD.	
W. W. Diggins	Sgt	DFM	02.10.42		R. Lambert, DFM	W/O	DFC	26.05.42
J. T. Dollison	A/F/L	DFC	22.05.45		K. A. Lawrence	F/L	DFC	18.01.44
J. E. Dorie	F/S	DFM	13.08.43		D. J. Lay	A/W/C	DSO	03.12.42
K. T. Dorsett	Sgt	DFM	16.04.45		K. N. Lewis	A/F/L	DFC	21.09.45
K. Dunlop	F/O	DFC	25.05.45		J. A. Lithgow	Lt	DFC	16.02.43
N. Edgecombe	P/O	DFC	16.11.45		D. Lowe	W/O	DFC	17.04.45
W. E. Egri	F/S	DFM	12.12.42		R. F. Lown	A/F/L	DFC	16.11.43
L. B. Elias	F/O	DFC	13.10.44		J. C. MacDonald	A/G/C	Bar to DFC	16.03.43
J. Ell	F/O	DFC	14.11.44				DFC, AFC.	
A. W. England	F/O	DFC	17.07.45		N. MacFarlane	A/W/C	DSO	17.07.45
R. F. Escreet	Sgt	DFM	15.06.43		W. D. MacMonagle	F/S	DFM	12.01.43
J. C. Fabian, DFC	A/F/L	Bar to DFC	19.05.44		J. N. Mahler	A/S/L	DFC	22.11.40
W. M. Ferguson	A/F/O	DFC	19.01.45		D. M. Mansel-Pleydel	F/O	DFC	19.05.44
P. H. Firth	LAC	MiD			J. Marpole	F/L	DFC	22.05.45
S. Fisher	F/O	DFC	13.10.44		L. H. Marriott	A/F/L	DFC	22.05.45
G. Foster	F/O	DFC	16.11.45		L. W. Marshall	F/O	DFC	13.04.45
J. W. French	A/F/L	DFC	22.05.45		W. E. McAlpine	P/O	DFC	29.12.42

E. H. McCaffrey	A/F/L	DFC	06.11.42		J. V. Russell	F/S	CGM	25.10.43
D. R. McFadden	P/O	DFC	27.03.45		R. A. Sargeant	P/O	DFC	17.10.44
E. W. McLachlan	P/O	DFC	20.02.45		A. Scholfield	P/O	DFC	20.07.45
T. J. McQuaid	F/O	DFC	25.09.45		D. R. Scott	A/F/L	DFC	31.12.43
H. Meades	Sgt	DFM	02.10.42		B. D. Sellick, DFC	W/C	Bar to DFC	26.05.42
R. R. Megginson	Sgt	DFM	22.10.40		A. C. Sellwood	A/F/O	DFC	25.05.45
R. R. Megginson, DFM	A/S/L	DFC	15.10.43		J. R. Shiells	F/S	DFM	01.06.45
S. W. Menual	A/S/L	DFC, MiD	22.08.41		W. A. Shoemaker	P/O	DFC	07.08.42
D. W. Mepham	Sgt	DFM	13.07.43		G. L. Sim	P/O	DFC	17.04.45
D. Midgley	A/F/L	DFC	31.12.43		J. Slaughter	A/F/L	DFC	21.09.45
H. T. Miles	F/O	DFC	06.11.42		J. D. Sleeman	A/F/L	DFC	13.04.45
L. A. Miller	P/O	DFC	22.02.44		J. L. Smale	Sgt	DFM	18.08.43
H. W. Milton	P/O	DFC	17.10.44		W. Sneddon	F/S	DFM	16.10.44
F. E. Mitchell	Sgt	DFM	11.01.43		D. D. Soderquist	F/S	DFM	12.12.42
E. J. Moore	A/F/L	DFC	14.07.44		E. G. Spannier	W/O	DFC	17.10.44
W. M. Morris	A/S/L	DFC	18.07.41		M. J. Sparks	A/F/L	DFC	13.10.44
A. W. Muirhead	F/O	DFC	13.10.44		N. Spratt	W/O	DFC	06.11.45
R. Munns	F/O	DFC	09.02.43		J. D. Stephens	A/W/C	DFC	??.07.43
G. A. Musgrove	F/O	DFC	12.12.44		W. J. Stephens	Sgt	DFM	30.07.40
P. A. Nettleton	A/F/L	DFC	16.01.45		S. W. Stewart	A/F/L	DFC	16.02.45
D. W. Newall	F/O	DFC	10.02.43		G. W. Stokes	A/F/L	DFC	24.10.44
C. B. Noble	F/O	DFC	25.09.45		R. A. Stone	Sgt	DFM	09.07.40
R. T. Noonan	F/S	DFM	11.06.43		G. A. Stubbings	P/O	DFC	25.05.45
A. R. Oakeshott	A/F/L	DFC	30.07.40		J. Sutcliffe	Sgt	DFM	30.07.40
H. J. O'Conner	F/S	DFM	03.05.43		I. C. Swales, DFM	F/O	DFC	13.03.42
C. A. O'Donnell	Sgt	DFM, MiD	22.10.40		W. H. Swent	P/O	DFC	19.01.45
P. B. Ogilvie, DSO	W/C	DFC	09.01.42		A. K. Sykes	A/F/L	DFC	13.08.43
E. C. Orchard	W/O	DFC	17.11.44		O. O. Sylvestre	F/S	DFM	20.02.45
C. B. Ordish	A/F/L	DFC	09.02.43		G. B. Tait	P/O	DFC	15.06.43
R. V. Ostler	A/F/L	DFC	20.04.45		P. A. Tanton	Sgt	DFM	19.08.41
T. R. Palmer	F/O	DFC	17.04.45		A. J. Taylor	Sgt	DFM	30.07.40
R. Pape	Sgt	M.M	??.??.??		G. Thomas	F/S	DFM	26.03.45
W. C. Parke	A/F/L	DFC	08.12.44		N. L. Thomas	F/S	DFM	12.11.43
D. A. Parkin	A/F/L	DFC	23.06.42		C. E. Thompson	P/O	DFC	19.07.40
R. F. Paveley	A/Sgt	DFM	22.11.40		J. A. Thornhill	P/O	DFC	25.09.45
N. J. Pawley	Sgt	DFM	17.07.43		H. Tilson	F/O	DFC	01.01.43
B. G. Payne	A/S/L	DFC	13.04.45		W. Towse	Sgt	DFM	17.07.43
S. A. Pearce	Sgt	DFM	19.04.43		V. F. Trehearne	Sgt	DFM	30.07.40
M. S. Pepper	Sgt	DFM	10.02.42		L. H. Trent	A/F/L	DFC	09.07.40
W. K. Perry	A/F/L	DFC	16.02.45		W. D. Turner	F/O	DFC	27.03.45
K. W. Peters	Sgt	DFM	15.06.43		C. A. Vernieux	F/O	DFC	09.01.42
E. J. Phillips	F/O	DFC	15.10.43		G. Ware	P/O	DFC	14.05.43
R. H. Phillips	F/O	DFC	29.12.42		E. F. Warren	F/S	DFM, MiD	12.01.43
D. W. Pidsley	Maj	DFC	10.11.42		W. D. Watkins	A/W/C	DSO	05.11.44
T. W. Pierce	F/O	DFC	12.01.43				DFC, DFM.	
K. M. Pincott	F/S	DFM	19.05.44		D. J. Watson	Sgt	DFM	19.08.43
C. J. du Preez	Sgt	DFM	15.05.42		F. J. Watson	P/O	DFC	25.05.45
W. T. Poole	P/O	DFC	19.01.45		R. Waugh	F/O	DFC	20.08.43
G. A. Potter	Sgt	DFM	13.07.43		P. F. Webster	A/S/L	DFC	09.07.40
E. G. Powell	Sgt	DFM	11.01.43		J. Wellings	F/S	DFM	16.11.44
W. F. Prewer	W/O	DFC	25.05.45		F. Wenton, DFM	W/O	DFC	19.02.45
A. J. Putt	A/F/L	DFC	23.09.41		L. Whiskie	F/S	DFM	25.09.45
D. Rainton	F/S	DFM	12.11.44		H. J. Whittingham	A.F.L.	DFC	21.09.45
C. Raymond	A/F/L	DFC	06.06.41		H. C. Wilkie	A/F/L	DFC	27.07.43
S. C. Readhead	Sgt	DFM	13.09.40		J. Williams	F/O	DFC	24.04.45
V. C. Reid	W/O	DFC	17.11.44		W. J. Woodhouse	P/O	DFC	17.10.44
I. W. Renner	P/O	DFC	14.05.43		D. C. Woodruff	A/F/L	DFC	15.02.44
W. H. Roach	Sgt	DFM	18.08.43		F. Wright, DFC	F/L	Bar to DFC	13.04.45
P. B. Rosenhain	F/O	DFC	25.09.45		H. H. Wright	Sgt	DFM	28.12.42
H. T. Rudall	F/O	CdeG(F)	17.04.47		J. R. York	F/O	DFC	13.10.44
					M. M. Yudelman	Lt	DFC	16.02.43

POST WAR AWARDS

M.P. Bowker	?	Queen's Commendation ??.??.75		A.T. Hudson	W/C	OBE	??.??.??
J. Broadbent	W/C	DSO	??.06.91	J. McClure	?	BEM	08.06.73
G. Buckley	S/L	DFC	29.06.91	N. Risdale	S/L	DFC	29.06.91
M. Carlile	?	BEM	??.??.??	M.G. Simmons	W/C	AFC	01.01.76
J. Craig	W/C	MBE	??.05.96	C.C. Tavner	F/L	Queen's Commendation 26.03.74	
R. Harden	F/L	Queen's Commendation		T. Torrens	W/C	MBE	??.06.91
				H.B. Williamson	S/L	AFC	01.01.74

APPENDIX XII
List of BRAVERY AWARDS
made to FORMER No.XV SQUADRON MEMBERS

NAME	RANK	AWARD	XV SQUADRON
John A. Liddell	Lt	Victoria Cross	1915
William Barker	2nd/Lt	Victoria Cross	1916
Thomas Grey	Sgt	Victoria Cross	1933/36
Hughie Edwards	A/Comm	Victoria Cross	1936
Leonard Trent	G/Cpt	Victoria Cross	1939
Jack Bailey	Sgt	C.G.M.	1943

APPENDIX XIII
List of COMMANDING OFFICERS
of No.15/XV SQUADRON

RANK	NAME	TENURE
Maj	Philip B Joubert de la Ferte	25.04.15 - ??.08.15
Capt	L. Daws (Temporary Command)	??.08.15 - 13.09.15
Maj	Edgar Ludlow-Hewitt	13.09.15 - 30.10.15
Maj	H. le M Brock, DSO	07.11.15 - 17.12.16
Maj	George I Carmichael, DSO	19.12.16 - 16.02.17
Maj	H. S. Walker	19.02.17 - 14.01.18
Maj	H. V. Stammers, DFC	14.01.18 - 27.11.18
Maj	C. C. Durston	27.11.18 - 31.12.19
S/L	Pat C. Sherren, MC	20.03.24 - 07.11.27
S/L	Charles E. H. James, MC	07.11.27 - 02.09.29
W/C	J. K. Wells, AFC	02.09.29 - 03.03.30
S/L	G. H. Martingell, AFC	03.03.30 - 23.03.33
S/L	Ernest S. Goodwin, AFC	23.03.33 - 08.03.33
S/L	Robert M. Foster, DFC, AFC	08.05.33 - 31.05.34
S/L	Thomas W. Elmhirst, AFC	01.06.34 - 12.08.35
S/L	Charles H. Cahill, DFC, AFC	12.08.35 - 01.01.36
S/L	Frank G. Robinson	01.01.36 - 05.02.36
S/L	Cyril D. Adams	05.02.36 - 02.05.38
S/L	J. G. Llewelyn	02.05.38 - 21.05.39
	(S/L Llewelyn became deputy commanding officer on appointment of J. L. Wingate, who was promoted to the rank of W/C).	
W/C	J. L. Wingate	21.03.39 - 21.12.39
S/L	Ralph W. Lywood	21.12.39 - 02.06.40
	(Promoted to Wing Commander 1 January 1940).	
W/C	Joe Cox, DFC	02.06.40 - 14.12.40
W/C	Herbert R. Dale	14.12.40 - 10.05.41
W/C	Patrick B. Ogilvie, DSO, DFC	16.05.41 - 07.01.42
W/C	John C. MacDonald, DFC, AFC	07.01.42 - 06.06.42
W/C	Douglas J. H. Lay, DSO, DFC	06.06.42 - 07.12.42
W/C	Stewart W. Menaul, DFC, AFC	07.12.42 - ??.04.43
W/C	John D. Stephens, DFC	07.05.43 - 30.08.43
W/C	A. J. Elliot	03.09.43 - 15.04.44
W/C	W. D. Watkins, DFC, DFM	15.04.44 - 16.11.44
W/C	Nigel G. Macfarlane, DSO	21.11.44 - 12.03.46
W/C	D. T. Witt, DSO, DFC, DFM	12.03.46 - 19.08.46
W/C	G. B. Bell, OBE	19.08.46 - 11.04.47
S/L	L. F. Kneil, DFC	11.04.47 - 24.08.48
S/L	J. H. Blount, DFC	24.08.48 - 15.12.49
S/L	J. R. Denny, MBE, DFC	15.12.49 - 17.05.53
F/L	J. Vnoucek (Temporary Command)	17.04.53 -05.06.53
S/L	J. M. Ayshford, DFC	05.06.53 - 28.03.55
S/L	A. R. Scott, DFC	28.03.55 - 01.04.57
W/C	David A. Green, DSO, OBE, DFC	01.09.58 - 01.04.60
W/C	J. G. Matthews, AFC	01.04.60 - 01.12.61
W/C	N. G. Marshall	01.12.62 - 01.10.64
W/C	D. J. Collins	01.10.64 - 31.10.64
W/C	D. J. Collins	01.10.70 - 25.03.71
W/C	Roy Watson	06.04.71 - 09.04.73
W/C	Michael G. Simmons	09.04.73 - 02.01.76
W/C	Peter D. Oulton	02.01.76 - 04.07.78
W/C	T. Nattrass, AFC	04.07.78 - 06.07.81
W/C	Eddie Cox	06.07.81 - 01.09.83
W/C	Barry Dove, AFC	01.09.83 - 01.08.86
W/C	Mike C. Rudd, AFC	01.08.86 - 20.01.89
W/C	John A. Broadbent, DSO	20.01.89 - 23.08.91
W/C	Andrew D. White	23.08.91 - 01.04.92
W/C	Alan T. Hudson, OBE	01.04.92 - 13.05.94
W/C	Graham A. Bowerman, OBE	13.05.94 - 28.02.97
W/C	Graham P. Dixon	28.02.97 - 23.07.97
W/C	Simon Dobb	23.07.99 -

APPENDIX XIV
List of
FORMER OFFICERS OF No.XV SQUADRON
who achieved AIR RANK

NAME	RANK
The Lord Elworthy	Marshal of the R.A.F
GCB, CBE, DSO, MVO, DFC, AFC, MA.	
Sir Philip B Joubert de la Ferte	Air Chief Marshal
KCB, CMG, DSO.	
Sir Edgar R Ludlow-Hewitt	Air Chief Marshal
GCB, GBE, CMG, DSO, MC.	
Sir John Whitworth-Jones	Air Chief Marshal
GBE, KCB.	
Sir Robert M Foster	Air Chief Marshal
KCB, CBE, DFC, AFC.	
Sir Raymond G Hart	Air Marshal
KBE, CB, MC.	
Sir Charles E N Guest, KBE, CB.	Air Marshal
Sir Leslie Dalton-Morris, KBE, CB.	Air Marshal
Sir Thomas W Elmhirst, KBE, CB, AFC.	Air Marshal
T A Langford-Sainsbury	Air Vice-Marshal
CB, OBE, DFC, AFC.	
Ernest S Goodwin, CB, CBE, AFC	Air Vice-Marshal

NAME	RANK
Joseph Cox, CB, OBE, DFC	Air Vice-Marshal
H le M Brock, CB, DSO	Air Commodore
G K Vasse, CBE	Air Commodore
Cyril D Adams, CB, OBE	Air Commodore
Sir Hughie I Edwards	Air Commodore
VC, KCMG, CB, DSO, OBE, DFC.	
John C MacDonald, CB, CBE, DFC, AFC.	Air Commodore
J H L Blount, DFC	Air Commodore
David A Green, DSO, OBE, DFC.	Air Commodore
John G Matthews, CBE, AFC.	Air Commodore
Sir David Cousins, KCB, AFC, BA.	Air Chief Marshal
Robert P O'Brien, CB, OBE, FRAes.	Air Vice-Marshal
Peter C Norriss, CB, AFC.	Air Marshal
R V Morris, AFC.	Air Commodore
Sir Michael Simmons, KCB, AFC.	Air Marshal
Peter D Oulton	Air Commodore
Trevor Nattrass, AFC.	Air Commodore
Michael Rudd, AFC.	Air Commodore
Andrew D White	Air Commodore

APPENDIX XV
List of SQUADRON LOCATIONS

WORLD WAR ONE

LOCATION	COUNTRY	DATE
South Farnborough	England	March 1915
Hounslow	England	April 1915
Swingate Down	England	May 1915
Lydd (Detachment)	England	June/July 1915
St Omer	France	December 1915
Droglandt	France	May 1916
Vert Galand	France	March 1916
Marieux	France	March 1916
Lealvillers	France	October 1916
Courcelles-le-Comte	France	June 1917
La Gorque	France	July 1917
Savy	France	August 1917
Longavesnes	France	August 1917
Lechelle	France	October 1917
Dapaume (ALG)	France	November 1917
Lechelle	France	November 1917
Lavieville	France	March 1918
La Houssoye	France	March 1918
Fienvillers	France	March 1918
Vert Galand	France	April 1918
Senlis	France	September 1918
Quatre Vents Ferme	France	October 1918
Selvigny	France	October 1918
Vignacourt	France	December 1918
Fowlmere	England	February 1919

Squadron disbanded December 1919; Reformed 1924

INTER WAR YEARS

Martlesham Heath	England	March 1924
Abingdon	England	June 1934

WORLD WAR TWO

Betheneville	France	September 1939
Conde-Vaux	France	September 1939
Wyton	England	December 1939
Alconbury	England	April 1940
Wyton	England	May 1940
Bourn	England	August 1942
Mildenhall	England	April 1943

POST WAR YEARS

Wyton	England	August 1946
Marham	England	November 1950
Coningsby	England	January1951
Cottesmore	England	May 1953
Honington	England	February 1955

Squadron disbanded April 1957; Reformed September 1958

Cottesmore	England	September 1958

Squadron disbanded October 1964; Reformed October 1970

Honington	England	October 1970
Laarbruch	West Germany	January 1971

Squadron disbanded July 1983; Reformed September 1983

Laarbruch	West Germany	September 1983
Muharraq	Bahrain	December 1990
Laarbruch	West Germany	March 1991

Squadron disbanded 1992; Reformed 1992 as the Tornado XV (R) Squadron/Tactical Weapons Conversion Unit.

Honington	England	April 1992
Lossiemouth	Scotland	November 1993

APPENDIX XVI
List of EQUIVALENT RANKS
of RAF, LUFTWAFFE AND USAAF

RAF	LUFTWAFFE	USAAF
	Reichsmarshall	
Marshal of the RAF	Generalfeldmarschall	General (5 Star)
Air Chief Marshal	Generaloberst	General (4 Star)
Air Marshal	General der Flieger	Lieutenant General
Air Vice Marshal	Generalleutnant	Major General
Air Commodore	Generalmajor	Brigadier General
Group Captain	Oberst	Colonel
Wing Commander	Oberstleutnant	Lieutenant Colonel
Squadron Leader	Major	Major
Flight Lieutenant	Hauptmann	Captain
Flying Officer	Oberleutnant	First Lieutenant
Pilot Officer	Leutnant	Lieutenant
Warrant Officer	Stabsfeldwebel	Warrant Officer
Flight Sergeant	Oberfeldwebel	Master Sergeant
	Unterfeldwebel	
Corporal	Unteroffizier	Staff Sergeant
	Hauptgefreiter	Sergeant
Leading Aircraftman	Obergefreiter	Corporal
Aircraftman 1st Class	Gefreiter	Private 1st Class
Aircraftman 2nd Class	Flieger	Private

APPENDIX XVII
List of ABBREVIATIONS

WORLD WAR ONE - RANKS

ABBREVIATION   RANK

Maj	Major
Capt	Captain
Lt	Lieutenant
2nd/Lt	2nd/Lieutenant
L/Cpl	Lance Corporal
A.M.I	Air Mechanic First Class
A.M.II	Air Mechanic Second Class

WORLD WAR TWO and POST WAR

MRAF	Marshal of the Royal Air Force.
ACM	Air Chief Marshal
AM	Air Marshal
AVM	Air Vice-Marshal
A/Cdre	Air Commodore
G/C	Group Captain
W/C	Wing Commander
S/L	Squadron Leader
F/L	Flight Lieutenant
F/O	Flying Officer
P/O	Pilot Officer
W/O	Warrant Officer
F/S	Flight Sergeant
Sgt	Sergeant
Cpl	Corporal
LAC	Leading Aircraftman
A/	Acting Rank

IMPERIAL GERMAN AIR SERVICE/LUFTWAFFE

Obslt	Obersleutnant
Oblt	Oberleutnant
Hptm	Hauptmann
Ltn	Leutnant
Fw	Feldwebel

CREW STATUS

Pil	Pilot
Nav	Navigator
F/E	Flight Engineer
B/A	Bomb Aimer
WOp	Wireless Operator
WAG	Wireless Operator/Air Gunner
A/G	Air Gunner

Obs	Observer
Arm	Armorer
Gdcw	Groundcrew
Pass	Passenger

AIR FORCES

RAF	Royal Air Force
RAFVR	Royal Air Force Volunteer Reserve
RAuxAF	Royal Auxiliary Air Force
RCAF	Royal Canadian Air Force
RAAF	Royal Australian Air Force
RNZAF	Royal New Zealand Air Force
SAAF	South African Air Force

HONORS, ORDERS, AWARDS AND DECORATIONS

VC	Victoria Cross
KGCB	Knight Grand Cross of the Order of the Bath
KCB	Knight Commander of the Order of the Bath
CB	Companion of the Order of the Bath
KCMG	Knight Commander of the Order of St Michael and St George
CMG	Companion of the Order of St Michael and St George
MVO	Member of the 4th or 5th Class of the Royal Victorian Order
GBE	Knight Grand Cross of the Order of the British Empire
KBE	Knight Commander of the Order of the Bath
DBE	Dame Commander of the Order of the British Empire
CBE	Commander of the Order of the British Empire
OBE	Officer of the Order of the British Empire
MBE	Member of the Order or the British Empire
BEM	British Empire Medal
DSO	Distinguished Service Order
DSC	Distinguished Service Cross
MC	Military Cross
DFC	Distinguished Flying Cross
AFC	Air Force Cross
DCM	Distinguished Conduct Medal
CGM	Conpicuous Gallantry Medal
MM	Military Medal
DFM	Distinguished Flying Medal
AFM	Air Force Medal
MSM	Military Service Medal
Bar	Second Award of a Decoration
MiD	Mentioned in Despatches
CdeG(B)	Croix de Guerre (Belgian)
CdeG(F)	Croix de Guerre (France)

Bibliography

Carter, Nick and Carol. *The Distinguished Flying Cross and How It Was Won 1918-1995*. London, England; Savannah Publications. 1998.

Chorley, W.R. *Bomber Command Losses*, Vols I-VI. Leicester, England; Midland Counties Publications.

Clarke, M.B.E., Don. *The Mildenhall Register Newsletters*. Bedford, England, 1980 - 1999.

Commonwealth War Grave Registers. London, England; Commonwealth War Graves Commission.

Ford-Jones, Martyn R. *Bomber Squadron; Men who flew with XV*. London, England; Kimber, William & Co Ltd. 1987

Ford-Jones, Martyn R. *No.XV Squadron Roll of Honor 1915-1990*, Volumes I-VI. Unpublished. 1995.

Gomersall, Bryce B. *The Stirling File*. Kent, England; Air- Britain Publications, 1979, (revised) 1987.

Halley, James J. *The Lancaster File*. Kent, England; Air-Britain Publications, 1985.

H.M.S.O. *Officers Died In The Great War*. Suffolk, England; J. B. Hayward & Son, Enlarged Edition 1988.

H.M.S.O. *Prisoners of War Naval & Air Forces of Britain and the Empire 1939-1945*. Suffolk, England; J. B. Hayward & Son. 1990.

Hobson, Chris. *Airmen Died In The Great War 1914-1918*. Suffolk, England; J. B. Hayward & Son. 1995.

Jones, Terry. F/L. *Aim Sure: 75 Years of XV (Bomber) Squadron*. Weeze, West Germany. 1990.

Mason, Pablo. S/L. *Pablo's War*. London, England. Bloomsbury Publishing Ltd. 1992.

Mason, Tim. *British Flight Testing*, Marltesham Heath 1920 -1939. Great Britain. 1993.

Middlebrook, Martin and Everit, Chris; *The Bomber Command War Diaries*, England; Viking. 1985.

No.3 Group, Bomber Command, Royal Air Force, *Roll of Honor 1939-1945*.

No.XV Squadron Association Newsletters (Various Editors).

No.XV Squadron, Royal Air Force, Unrestricted Papers and Files, Lossiemouth, Scotland.

Peters, John and Nichol, John. *Tornado Down*. London, England. Penquin Group. 1992.

Roberson, Norman J. *The History of No15/XV Squadron*. Weeze, West Germany; 1975.

RAF Aircraft Accident Cards. London, England; Air Historical Branch, Ministry of Defence.

Royal Air Force. Air 27, Form 540, *No.XV Squadron*. London, England; Public Record Office.

RAF Bomber Command Loss Cards. London, England; Royal Air Force Museum.

Tavender, I.T. *The Distinguished Flying Medal, A Record of Courage, 1918-1982*. Great Britain; J. B. Haywood and Son. 1990.

Vann, Ray & Bowyer, Chas. *XV Squadron RFC/RAF 1915-19*. England; Cross and Cockade.

Index

United States of America

ROYAL FLYING CORPS/ROYAL AIR FORCE UNITS.

Royal Air Force Training Units.